Phytoremediation of Environmental Pollutants

Phytoremediation of Environmental Pollutants

Edited by
Ram Chandra
N. K. Dubey
Vineet Kumar

CRC Press
Taylor & Francis Group
Boca Raton London New York

CRC Press is an imprint of the
Taylor & Francis Group, an **informa** business

CRC Press
Taylor & Francis Group
6000 Broken Sound Parkway NW, Suite 300
Boca Raton, FL 33487-2742

First issued in paperback 2020

ISBN 13: 978-0-367-57253-2 (pbk)
ISBN 13: 978-1-138-06260-3 (hbk)

Library of Congress Cataloging-in-Publication Data

Names: Chandra, Ram (Biotechnology professor), editor. | Dubey, N. K., editor. | Kumar, Vineet (Vineet Kumar Rudra), editor.
Title: Phytoremediation of environmental pollutants / editors, Ram Chandra, N.K. Dubey and Vineet Kumar.
Description: Boca Raton : Taylor & Francis, CRC Press, 2018. | Includes bibliographical references and index.
Identifiers: LCCN 2017029046| ISBN 9781138062603 (hardback : alk. paper) | ISBN 9781315161549 (ebook)
Subjects: LCSH: Phytoremediation.
Classification: LCC TD192.75 .P47835 2018 | DDC 628.4--dc23
LC record available at https://lccn.loc.gov/2017029046

Visit the Taylor & Francis Web site at
http://www.taylorandfrancis.com

and the CRC Press Web site at
http://www.crcpress.com

Contents

Preface

The discharge of complex environmental pollutants from various industrial sectors has created a challenge for scientists and environmentalists concerning the sustainable development of mankind. Currently, the management of pollutants present in soil, water, air, and food is a serious problem worldwide due to its direct effect on human health. Consequently, the problems of global warming, climate change, change in biodiversity, and extinction of numerous important flora and fauna have been observed by scientific communities the world over.

Therefore, international researchers have reoriented their minds for environmental waste management and bioremediation of toxic compounds present in the environment. However, microbial technologies' current cleanup techniques are not offering real solutions for removal of complex toxic pollutants and safe recycling of natural resources. There is a growing need among remediation professionals for effective, affordable, nonpolluting alternatives to energy-intensive engineering processes for sustainable development. Therefore, phytoremediation has recently arisen as an important technique to augment the bioremediation process because it is an emerging green technology that uses a broad range of plants to remediate soil, sediment, surface water, and groundwater environmental contaminants with toxic metals, organics, pesticides, and radionuclides.

Consequently, convergences of phytoremediation and microbial bioremediation strategies have been suggested as a more successful approach for remediation of contaminants, particularly organic compounds. Microbe-assisted phytoremediation with naturally occurring microbes deliberately stimulated via seed inoculation has been investigated in *in situ* and *ex situ* conditions globally for the real solution of problems for sustainable development. Further, to investigate the mechanism of plant–microbe interaction for detoxification of complex pollutants, a variety of contaminant-degrading enzymes (i.e., peroxidases, dioxygenases, laccases, P450 monooxygenases, phosphotases, dehalogenases, and nitroreductases) has been reported in plants, fungi, and endophytic and rhizospheric bacteria. But, this knowledge is fragmentary, specific, and limited to only a few researchers. Therefore, the application of suitable technology for sustainable development is still urgently warranted.

Hence, the aim of this book is to provide and disseminate detailed, up-to-date knowledge regarding the various aspects of phytoremediation and plant–microbe interaction to researchers, students, and industry professionals to inspire collective responsibility to develop innovative technology for ecological restoration and environmental health. Therefore, this book has compiled timely knowledge of various process and molecular mechanisms for industrial waste detoxification during phytoremediation in wetland plants. In addition, the book has highlighted phytoextraction speciation potential and adaptation of hyperaccumulator wetland plants for the detoxification of heavy metals from organic waste for ecorestoration. Further, the book has illustrated details regarding siderophore formation and nutrient availability during plant–microbe interaction in rhizospheric zones of plants growing in polluted environments.

The book has also provided the rare knowledge for *in situ* and *ex situ* phytoextraction potential of common weeds growing on heavy metal–polluted sites for ecorestoration and recycling of natural resources. The book has placed special emphasis on the role of endophytic bacteria for phytoremediation of environmental pollutants during plant–microbe interaction in the presence of heavy metals. Lastly, the book has imbibed the special knowledge on constructed wetland treatment systems as a promising technology for treatment and recycling of hazardous wastewater, which is a plant–microbe-based technology required in the current scenario for treatment of complex wastewater generated from textile and agrobased industries. The book has also covered adequate knowledge on phytocapping, vermicomposting, and life cycle assessment of environmental pollutants as a new technique and tools for monitoring of environmental pollutants. Thus, this book will be a unique compendium of up-to-date knowledge on phytoremediation of environmental pollutants contributed by various experts in this area from universities, government laboratories, and other academic global institutions.

Editors

Ram Chandra is currently professor and dean, Department of Environmental Microbiology, School for Environmental Sciences, Babasaheb Bhimrao Ambedkar (A Central) University, Lucknow, Uttar Pradesh, India. He obtained his MSc from Banaras Hindu University, India, in 1987. Subsequently, his PhD was awarded in 1994. He has led work for bacterial degradation of lignin from pulp paper mill waste and molasses melanoidins from distillery waste. He has published more than 100 original research articles in peer-reviewed journals of high impact. He has also published six books and 22 book chapters on biodegradation and bioremediation of industrial pollutants. He has been named a fellow of the Academy of Environmental Biology (FAEB), fellow of the Association of Microbiologists of India (FAMI), and fellow of the Biotech Research Society, India (FBRSI).

N. K. Dubey is currently working as a professor in the Center of Advanced Studies in Botany, Department of Botany, Banaras Hindu University, Varanasi, Uttar Pradesh, India. He obtained his MSc and PhD degrees from Gorakhpur University, Uttar Pradesh, India. He has produced more than 227 publications, four patents, and six books. He has 33 years of teaching experience and is the recipient of several awards, such as the National Academy of Science, India, SR memorial gold medal for best teacher of the faculty of science, BHU, and the young scientist award of the Indian Science Congress Association. He has also worked in several administrative positions at the university. He has visited Japan, France, South Korea, Belgium, Malaysia, the United Kingdom, and Italy. Currently, he is on the editorial boards of several national and international publications. He has guided several MSc and PhD students in their studies.

Vineet Kumar is currently a PhD scholar in the Department of Environmental Microbiology, School for Environmental Sciences, Babasaheb Bhimrao Ambedkar (A Central) University, Lucknow, Uttar Pradesh, India. He received his MSc and MPhil from Chaudhary Charan Singh University, Meerut, Uttar Pradesh, India, in 2010 and 2011. His research is focused on understanding the mechanism of rhizospheric bacterial communities in the degradation and detoxification of organic and inorganic pollutants of complex industrial wastewater. Key components of his research include the use of rhizospheric bacteria to address metal uptake by wetland plants and assessment of bioaugmentation/integration technologies for the phytoremediation of wastewater pollutants from the environment. He is a life member of several scientific societies, including the Association of Microbiologists of India, the Biotech Research Society, India, and the Indian Science Congress Association. He has authored or coauthored of five research papers, twelve book chapters, and five scientific articles.

Contributors

Munees Ahemad
Department of Agricultural Microbiology
Faculty of Agricultural Sciences
Aligarh Muslim University
Aligarh, Uttar Pradesh, India

Surabhi Awasthi
Plant Ecology and Environmental Science
    Division
CSIR-National Botanical Research Institute
Lucknow, Uttar Pradesh, India

Prosenjit Chakraborty
Molecular Plant Pathology Laboratory
Department of Botany
University of North Bengal
West Bengal, India

Ram Chandra
Department of Environmental
    Microbiology
School for Environmental Sciences
Babasaheb Bhimrao Ambedkar (A Central)
    University
Lucknow, Uttar Pradesh, India

Reshu Chauhan
Plant Ecology and Environmental Science
    Division
CSIR-National Botanical Research Institute
Lucknow, Uttar Pradesh, India

Shibu Das
Molecular Plant Pathology Laboratory
Department of Botany
University of North Bengal
West Bengal, India

N. K. Dubey
Department of Botany
Institute of Science
Banaras Hindu University
Varanasi, Uttar Pradesh, India

Anupa Fonia
Laboratory of Algal Biotechnology
Department of Botany
School of Life Sciences
Hemwati Nandan Bahuguna Garhwal
    University
Srinagar, Uttarakhand, India

V. K. Garg
Department of Environmental Science and
    Engineering
Guru Jambheshwar University of Science
    and Technology
Hisar, India
and
Center for Environmental Science and
    Technology
Central University of Punjab
Bathinda, Punjab, India

Sanjay P. Govindwar
Department of Biochemistry
Shivaji University
Vidyanagar, Kolhapur, Maharashtra, India

Fengxiang X. Han
Department of Chemistry and
    Biochemistry
Jackson State University
Jackson, Mississippi

Jawed Iqbal
Department of Microbiology and
    Immunology
H. M. Bligh Cancer Research Laboratories
Rosalind Franklin University of Medicine
    and Science
Chicago Medical School
Chicago, Illinois

Rahul V. Khandare
Department of Biotechnology
Shivaji University
Vidyanagar, Maharashtra, India

Abhishek Khapre
Solid and Hazardous Waste Management
  Division
CSIR-National Environmental Engineering
  Research Institute
Nagpur, Maharashtra, India

Ravindra Nath Kharwar
Department of Botany
Banaras Hindu University
Varanasi, Uttar Pradesh, India

Dhananjay Kumar
Department of Environmental Sciences
Babasaheb Bhimrao Ambedkar (A Central)
  University
Lucknow, Uttar Pradesh, India

Dhananjay Kumar
Laboratory of Algal Biotechnology
Department of Botany
School of Life Sciences
Hemwati Nandan Bahuguna Garhwal
  University
Srinagar, Uttarakhand, India

Narendra Kumar
Department of Environmental Sciences
Babasaheb Bhimrao Ambedkar (A Central)
  University
Lucknow, Uttar Pradesh, India

Sanjeev Kumar
Department of Environmental Sciences
Babasaheb Bhimrao Ambedkar (A Central)
  University
Lucknow, Uttar Pradesh, India

Sunil Kumar
Solid and Hazardous Waste Management
  Division
CSIR-National Environmental Engineering
  Research Institute
Nagpur, Maharashtra, India

Vineet Kumar
Department of Environmental
  Microbiology
School for Environmental Sciences
Babasaheb Bhimrao Ambedkar (A Central)
  University
Lucknow, Uttar Pradesh, India

Vineet Kumar Mishra
Molecular Microbiology and Systematics
  Laboratory
Department of Biotechnology
Mizoram University
Aizawl, Mizoram, India

Bhanu Pandey
Department of Botany
Banaras Hindu University
Varanasi, Uttar Pradesh, India

Ajit Kumar Passari
Molecular Microbiology and Systematics
  Laboratory
Department of Biotechnology
Mizoram University
Aizawl, Mizoram, India

Niraj R. Rane
Department of Biotechnology
Shivaji University
Vidyanagar, Maharashtra, India

Aniruddha Saha
Molecular Plant Pathology Laboratory
Department of Botany
University of North Bengal
West Bengal, India

Dipanwita Saha
Department of Biotechnology
University of North Bengal
Siliguri, West Bengal, India

Kavita Sharma
Department of Environmental Science and
    Engineering
Guru Jambheshwar University of Science
    and Technology
Hisar, Haryana, India

Pooja Sharma
Department of Environmental
    Microbiology
School for Environmental Sciences
Babasaheb Bhimrao Ambedkar (A Central)
    University
Lucknow, Uttar Pradesh, India

Rajesh Kumar Sharma
Department of Botany
Banaras Hindu University
Varanasi, Uttar Pradesh, India

Bhim Pratap Singh
Molecular Microbiology and Systematics
    Laboratory
Department of Biotechnology
Mizoram University
Aizawl, Mizoram, India

Kshitij Singh
Department of Environmental
    Microbiology
School for Environmental Sciences
Babasaheb Bhimrao Ambedkar (A Central)
    University
Lucknow, Uttar Pradesh, India

Manvi Singh
Department of Environmental Sciences
CSIR-National Botanical Research Institute
Lucknow, Uttar Pradesh, India

Pradyumna Kumar Singh
Plant Ecology and Environmental Science
    Division
CSIR-National Botanical Research Institute
Lucknow, Uttar Pradesh, India

Preeti Singh
Laboratory of Algal Biotechnology
Department of Botany
School of Life Sciences
Hemwati Nandan Bahuguna Garhwal
    University
Srinagar, Garhwal, Uttarakhand, India

Vijetna Singh
Department of Biotechnology
Indira Gandhi National Tribal University
Amarkantak, Madhya Pradesh, India

B. B. Maruthi Sridhar
Department of Environmental and
    Interdisciplinary Sciences
Texas Southern University
Houston, Texas

Pankaj Kumar Srivastava
Department of Environmental Sciences
CSIR-National Botanical Research Institute
Lucknow, Uttar Pradesh, India

Sudhakar Srivastava
Institute of Environment and Sustainable
    Development
Banaras Hindu University
Varanasi, Uttar Pradesh, India

Yi Su
School of Science and Computer
    Engineering
University of Houston-Clear Lake
Houston, Texas

Supriya Tiwari
Department of Botany
Institute of Science
Banaras Hindu University
Varanasi, Uttar Pradesh, India

Bhumi Nath Tripathi
Department of Biotechnology
Indira Gandhi National Tribal University
Amarkantak, Madhya Pradesh, India

Preeti Tripathi
Plant Ecology and Environmental Science
    Division
CSIR-National Botanical Research Institute
Lucknow, Uttar Pradesh, India

Rudra Deo Tripathi
Plant Ecology and Environmental Science
    Division
CSIR-National Botanical Research Institute
Lucknow, Uttar Pradesh, India

Sonam Tripathi
Department of Environmental
    Microbiology
School for Environmental Sciences
Babasaheb Bhimrao Ambedkar (A Central)
    University
Lucknow, Uttar Pradesh, India

Shivani Uniyal
Department of Botany
Banaras Hindu University
Varanasi, Uttar Pradesh, India

Anuprita D. Watharkar
Department of Biochemistry
Shivaji University
Vidyanagar, Kolhapur, Maharashtra, India

Sangeeta Yadav
Department of Environmental
    Microbiology
School for Environmental Sciences
Babasaheb Bhimrao Ambedkar (A Central)
    University
Lucknow, Uttar Pradesh, India

Zothanpuia
Molecular Microbiology and Systematics
    Laboratory
Department of Biotechnology
Mizoram University
Aizawl, Mizoram, India

1

Phytoremediation: A Green Sustainable Technology for Industrial Waste Management

Ram Chandra and Vineet Kumar

CONTENTS

1.1 Introduction

Global industrialization, urbanization, and population in the last two decades have resulted in the generation of huge quantities of toxic waste. This hazardous waste includes a variety of organic and inorganic compounds which pose serious threats to ecosystems. Organic contaminants include different compounds such as petroleum hydrocarbons (e.g., benzopyrene), chlorinated solvents (i.e., polychlorinated benzenes, also known as PCBs), linear halogenated hydrocarbons (e.g., trichloroethylene), volatile organic carbons, and explosives such as trinitrotoluene. Inorganic compounds include nitrates, phosphates, metals and metalloids, such as mercury (Hg), arsenic (As), lead (Pb), cadmium (Cd), chromium (Cr), copper (Cu), nickel (Ni), selenium (Se), silver (Ag), and zinc (Zn), and nonradioactive or radioactive nuclides, like uranium (U), strontium (Sr), and cesium (Cs). Despite requirements for pollution control measures, these wastes are generally dumped on land or discharged into water bodies (rivers, canals, lakes, etc.) without adequate treatment, and thus they become a large source of environmental pollution and health hazards. It is, therefore, urgent to adequately remove these pollutants from contaminated sites. Contamination of soil and water poses major environmental concerns in the present scenario. A wide range of methods based on not only physical and chemical but also biological means have been available for the remediation of soil and water for decades, but environmental preservation requires development of more sustainable approaches that promise thorough, economical, and environmentally friendly ways, compared to conventional methods (Salt et al. 1995).

In recent decades, phytoremediation has provided a cost-effective, long-lasting, and aesthetic solution for remediation of hazardous pollutants from contaminated sites (Kramer 2005, Suresh and Ravishankar 2004). The idea that plants can be used for environmental remediation is very old and cannot be traced to any particular source (Raskin et al. 1997). About 300 years ago, plants were proposed for use in the treatment of wastewater. The term phytoremediation ("phyto" meaning plants and the Latin suffix "remedium" meaning to clear or restore) refers to a diverse collection of plant-based green technologies that use either naturally occurring or genetically engineered plants to remove, transfer, stabilize, and/or degrade contaminants located in soil, sediment, and water (Cunningham et al. 1997, Newman and Reynolds 2004). Phytoremediation has also been called green remediation, botano-remediation, agroremediation, and vegetative remediation (Chaney et al. 1997). It is socially accepted by surrounding communities and regulatory agencies as a potentially effective and beautiful technology (Newman and Reynolds 2005). A number of green plants, including herbs, shrubs, and trees (both terrestrial and aquatic) have been reported to be endowed with magnificent abilities for restoration and reclamation of contaminated environments (Yoon et al. 2006, Gupta and Sinha 2007a, Qixing et al. 2011). Plants naturally provide roots, stems, and leaves as habitats for a wide array of microorganisms which can break down contaminants, enhancing the treatment process. Plants, through several natural biophysical and biochemical processes, such as adsorption, transport and translocation, hyperaccumulation or transformation, and mineralization, can remediate pollutants. Recently, Environment Canada released a database, PHYTOREM (PHYTOREmediation of Metals), which contains a worldwide inventory of 775 terrestrial and aquatic plants with capabilities for accumulating or hyperaccumulating one or several of the 19 key metallic elements (Padmavathiamma and Li 2007). The U.S. Environmental Protection Agency (EPA) also maintains a website for researchers and the general public with information on the growing field of phytoremediation (http://www.clu-in.org).

The purpose of this chapter is to provide a concise discussion of the processes associated with the use of phytoremediation as a cleanup and decontamination technique for remediation of hazardous industrial waste-contaminated sites. In this chapter, we describe various processes of phytoremediation, including phytoextraction, phytofiltration, phytostabilization, phytovolatilization, phytodegradation, rhizodegradation, and phytodesalination, with special emphasis on remediation of organic and inorganic pollutants. We also discuss the potential and challenges of phytoremediation strategies for the removal of organic and inorganic pollutants from contaminated sites, and we provide some information on costs. Furthermore, the use of transgenic plants in phytoremediation progress is highlighted.

1.2 Categories of Phytoremediation, Their Principles, and Their Applications

Phytoremediation is a broad expression comprising different strategies used by plants to decontaminate soil, sludge, sediment, and wastewater. Depending on the contaminants, the site conditions, the level of cleanup required, and the type of plants, phytoremediation technology can be divided into different categories, namely, phytoextraction, phytofiltration, phytostabilization, phytovolatilization, and phytodegradation, with each category having a different mechanism of action for remediating organic and inorganic pollutants from contaminated soil, sludge, sediment, groundwater, surface water, and wastewater (Raskin et al. 1997, Alkorta et al. 2004). The different mechanisms of phytoremediation are illustrated in Figure 1.1

1. Phytoextraction: The use of plants that absorb pollutants and accumulate them in organs removed from fields, together with crops, in order to purify soil from heavy metal and organic substance contamination

2. Phytofiltration: Use of plant roots, seedlings, and excised shoots, which also take up stored contaminants from an aqueous growth matrix

3. Phytostabilization: Plant-mediated immobilization or binding of contaminants into the soil matrix, thereby reducing their bioavailability

4. Phytovolatilization: Contaminant uptake by plants and volatilization into the atmosphere

5. Phytodegradation: The use of plants and associated microorganisms to degrade organic pollutants

6. Rhizodegradation: Enhancement of degradation of organic pollutants in the plant root zone by microorganisms

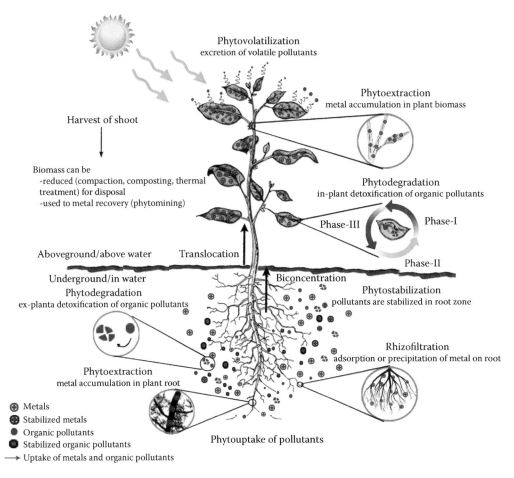

FIGURE 1.1
Schematic representation of various phytoremediation strategies.

1.2.1 Phytoextraction

Phytoextraction, also called phytoaccumulation, phytoabsorption, or phytosequestration, refers to the use of plants to absorb, translocate, and store toxic contaminants from soil, sediments, and/or sludge in the root and shoot tissues (Salt et al. 1998, Garbisu and Alkorta 2001). Phytoextraction is the most recognized and applied phytoremediation technique for the removal of toxic metals from contaminated environments (Figure 1.2). The idea of using plants to extract metals from contaminated soil was introduced and developed by Utsunamyia (1980) and Chaney (1983). Phytoextraction requires long-term maintenance and routine harvesting of the plants, as well as safe disposal of polluted plant materials. The cost involved in phytoextraction, when compared with the those of conventional soil remediation techniques, is 10-fold lower. This means that phytoextraction is a cost-effective technique (Salt et al. 1995). Phytoextraction does not work for some metals, including Pb, but it can be effective for As, Cd, and Ni removal. Phytoextraction occurs in the root zone of the plant. The root zone may typically be relatively shallow, with the bulk of the root at a shallower rather than a deeper depth. This can be a limitation of phytoextraction. Numerous factors, including pH of wastewater and sediment, mobilization and uptake from soil, compartmentalization and sequestration within

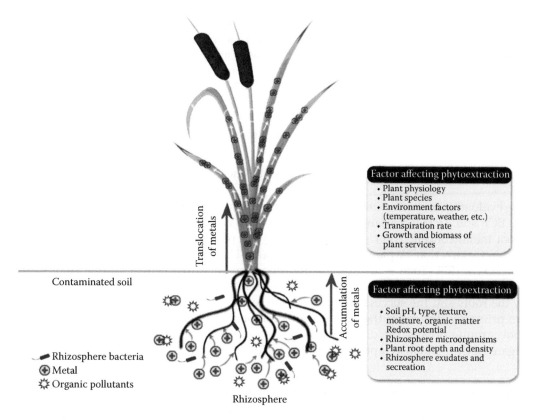

FIGURE 1.2
Illustration of phytoextraction and factors that affect its efficacy.

the root, efficiency of xylem loading and transport (transfer factors), distribution between metal sinks in the aerial parts, sequestration and storage in leaf cells, and plant growth and transpiration rates can also affect the remediation process of a contaminated site. Generally, the ideal plants to be used in phytoextraction should have the following characteristics

1. Tolerates high levels of metal concentration
2. Fast growth rate and high biomass production
3. Accumulates high level of metals in harvestable parts
4. Widely distributed and with a deep root system
5. Resistance to disease and pests and is unattractive to animals
6. Easy cultivation, harvesting, and processing
7. Low cultivation requirements
8. Repulsive to herbivores, to avoid food chain contamination

1.2.1.1 Bioconcentration and Translocation Factors

Two important factors that make a plant an efficient phyoextractor are its biomass production and its bioconcentration efficiency (McGrath and Zhao 2003). The bioconcentration efficiency of a plant is its ability to accumulate metals from contaminated soils and can be estimated using the bioconcentration factor (BCF), which is defined as the ratio of the metal concentration in the root of the plant to that in the soil/sludge (Yoon et al. 2006, Gupta and Sinha 2007b). A plant's ability to translocate metals from the root to the shoot is measured using the translocation factor (TF), which is defined as the metal concentration in the shoots versus that in the roots (Deng et al. 2004, Yoon et al. 2006). The BCF and TF are calculated by the following equations.

$$\text{BCF} = (\text{metal concentration in plant root})/(\text{metal concentration in soil or sludge})$$

$$\text{TF} = (\text{metal concentration in plant shoot})/(\text{metal concentration in plant root})$$

The evaluation and selection of plants for phytoremediation purposes entirely depend on the BCF and TF (MacFarlane et al. 2007). The BCF is a more important measure than the shoot metal concentration when considering the potential of a given candidate species for phytoextraction. A plant with both BCF and TF values of >1 has potential for use in phytoextraction (Yoon et al. 2006). In addition, a plant with a BCF of >1 and TF of <1 has potential for phytostabilization (Fitz and Wenzel 2002). However, the heavy metal uptake by the roots and the successive translocation to the aboveground plant parts and organs are primarily driven by the plant's transpirational pool. Together with translocation to aboveground organs, the distribution and accumulation of heavy metals in the different organs of the plant also represent key factors for successful phytoremediation.

1.2.1.2 Bioaccumulation Factor and Phytoextraction Rate

The processes of phytoextraction generally require translocation of heavy metals to the easily harvestable green plant parts. Enrichment occurs when a contaminant taken up by a plant is not degraded rapidly, i.e., it accumulates in the plant. The enrichment factor, also known as the bioaccumulation factor (BAF) or biological accumulation coefficient (BAC), is calculated based on the ratio of the plant shoot concentration versus the soil concentration (Li et al. 2007, Cui et al. 2007):

$$\text{BAF} = (\text{metal concentration in plant shoot})/(\text{metal concentration in soil})$$

Accumulation and translocation of metals depend on various biological or physico-chemical factors and change according to environmental conditions, such as temperature, pH, water ion content, salinity conditions, availability of heavy metals, and so on. To make a crude evaluation of the general phytoextraction efficiency of plants, the depth of the rooting zone, the density of soil, the biomass production, and mortality of the aboveground biomass component that is harvested are taken to into account to calculate the metal phytoextraction rate (PR) of the growing plant. Zhao et al. (2003) suggested that phytoextraction efficiency can be estimated by calculating the percentage of metals or nutrients in the harvestable biomass versus the concentration of metals in the soil mass where the species were planted, as shown in the following equation:

$$\text{PR}(\%) = \left[(C_{\text{plant}} \times M_{\text{plant}})/(C_{\text{soil/sludge}} \times M_{\text{rooted zone}}) \right] \times 100$$

where M_{plant} is the mass of the harvestable aboveground plant biomass produced in one harvest, C_{plant} is the metal concentration in the harvested component of the plant biomass,

$M_{\text{rooted zone}}$ is the mass of the soil volume rooted by the species under study, and C_{soil} is the metal concentration in the soil volume.

1.2.1.3 Metal Extraction Amount and Phytoremediation Time

The metal extraction amount and phytoremediation time also can be used to evaluate plant phytoextraction efficiency; these measures of efficiency are calculated as follows (Zhang et al. 2010):

$$\text{Metal extracted } (\mu g/\text{plant}) = (\text{metal concentration in plant tissue}) \times (\text{plant biomass})$$

$$\text{Phytoremediation time (years)} = (\text{metal concentration} \times \text{soil mass})/$$
$$(\text{metal concentration in plant shoot} \times \text{plant shoot biomass} \times 4)$$

The total metal distribution rate for a plant part is determined to identify roots or aerial parts that are accumulators; the rate is calculated using the following equation:

$$\text{Rate} = \left([M]_{ij} \times DW_i \times 100\right) \bigg/ \left(\sum I[M]_i \times DW_i\right)$$

where $[M]_{ij}$ represents the j heavy metal content in plant part i (i.e., roots, stems, leaves, and shoots) and DW_i represents the weight of part i (Santos-Jallath et al. 2012).

1.2.1.4 Types of Phytoextraction

According to Salt et al. (1998), there are two basic strategies of phytoextraction:

1. Chelate-assisted phytoextraction
2. Continuous phytoextraction

1.2.1.4.1 Chelate-Assisted Phytoextraction

Chelate-assisted phytoextraction is also called induced phytoextraction. The phytoextraction mechanism has its own limitations, e.g., low mobility and low bioavailability of some heavy metals (especially Pb) in polluted environments. An increase of heavy metal mobility can be achieved by adding synthetic chelating agents which are capable of solubilizing and complexing with heavy metals in a soil solution as well as promoting heavy metal translocation from roots to the harvestable parts of the plant (Evangelou et al. 2007). Chelating agents have been used as soil extractants and to maintain the solubility of trace elements in hydroponic solutions. This strategy of phytoextraction is based on the fact that the application of metal chelates to a soil significantly enhances metal accumulation in plants. Enhancing metal accumulation in existing high-yielding crop plants without diminishing their yield is the most feasible strategy in the development of phytoremediation. To improve the metal accumulation capacities and uptake speed of nonhyperaccumulating plants, the addition of chelating agents has been proposed. Several studies have reported that the application of chelating agents, such as ethylene diamine triacetic acid (EDTA), *N*-(2-hydroxyethyl)-ethylene diaminetriacetic acid (HEDTA), diethylene–tetramine-pentaacetate acid (DTPA), ethylenediamine di-*o*-hydroxy phenylacetic acid (EDDS), ethylene glycol-*O,O'*-bis-(2-amino-ethyl) *N,N,N',N'*-tetraacetic acid (EGTA), and citric acid, can enhance the effectiveness of phytoextraction by mobilizing metals and increasing metal accumulation in plants (Chen and Cutright 2001, Turgut et al. 2004, Nowack et al. 2006).

1.2.1.4.2 Continuous Phytoextraction

The process of phytoextraction via use of hyperaccumulator plants is called continuous phytoextraction. Continuous phytoextraction generally depends on the natural ability of plants to accumulate, translocate, and resist high amounts of metals over the complete growth cycle (Garbiscu and Alkorta 2001). Continuous phytoextraction is also environmentally friendly, as it leaves the site suitable for cultivation of other plants. In urbanized areas, continuous phytoextraction may be used in two types of sites. One type comprises degraded soils in postindustrial areas, while the other, which is a highly promising future application of phytoextraction, is connected with soils in the vicinity of transportation routes and in urban green areas. The phytoextraction potentials of ornamental plant species that are most frequently planted in urban locations are under investigation in many research centers worldwide. Such species include *Tagetes erecta* L.

1.2.2 Phytofiltration

Phytofiltration is a cost-competitive technology that is primarily used to remediate or adsorb pollutants, mainly metals, from water and aqueous waste streams (Dushenkov et al. 1995). Phytofiltration may be achieved via rhizofiltration, blastofiltration, or caulofiltration.

1.2.2.1 Rhizofiltration

Rhizofiltration is a root zone technology in which terrestrial and aquatic plants interact with pollutants present in wastewater (Figure 1.3). It reduces the mobility of contaminants and prevents their migration to the groundwater, thus reducing bioavailability for entry into the food chain. Rhizofiltration can be used for metals that are retained only within the roots (U.S. EPA 2000a). It is particularly effective and economically compelling when low concentrations of contaminants and a large volume of water are involved. An ideal plant for rhizofiltration should comprise rapidly growing roots with the ability to remove toxic metals from solution over an extended period of time. The advantages of rhizofiltration include it ability to be used as in *in situ* or *ex situ* applications, and species other than hyperaccumulators can also be used. The production of hydroponically grown transplants and the maintenance of successful hydroponic systems in the field requires the expertise of qualified personnel, and the facilities and specialized equipment required can increase overhead costs.

Rhizofiltration works via several physical and biochemical processes, such as adsorption, precipitation, rhizodegradation, and bioaccumulation. Rhizofiltration of metals involves the absorption into or adsorption or precipitation onto plant roots of metals present in a complex industrial wastewater. In addition to surface absorption, other slower mechanisms underlying rhizofiltration may also occur: these include biological processes (intracellular uptake, deposition in vacuoles, and translocation to the shoot) or precipitation of the metal from the wastewater by the action of plant exudates. Based upon the nature of pollutants present in wastewater, rhizofiltration processes may occur with phytoextraction, phytostabilization, or phytovolatilization processes. The efficiency of the mechanism

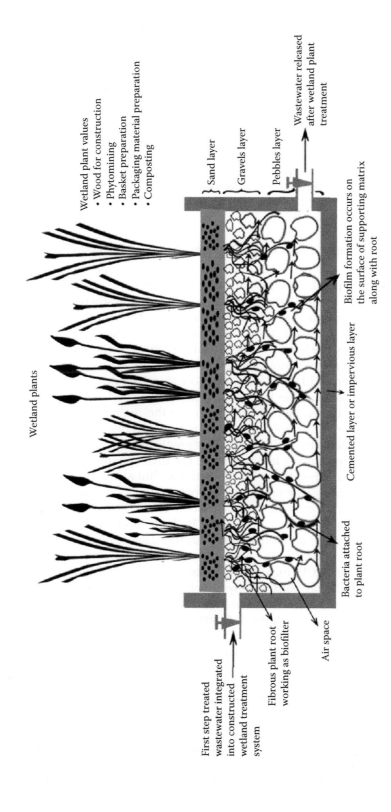

FIGURE 1.3
Simplified schematic of a subsurface horizontal flow CW used for wastewater treatment.

of rhizofiltration lies in the efficiency of roots to synthesize certain chemicals which cause heavy metals to accumulate in the plant body and then precipitate on the plant root surface. Several aquatic species have been identified and tested for their phytoremediation capacities in removing heavy metals from polluted waters. These species include sharp dock (*Polygonum amphibium* L.), duck weed (*Lemna minor* L.), water hyacinth (*Eichhornia crassipes*), water lettuce (*P. stratiotes*), water dropwort [*Oenathe javanica* (BL) DC], calamus (*Lepironia articulate*), pennywort (*Hydrocotyle umbellate* L.) (Zaranyika and Ndapwadza 1995, Zayed et al. 1998). However, these plants have limited potential for rhizofiltration, because they are not efficient in removing metals due to their small, slow-growing roots. Sunflower (*Helianthus annus* L.) and Indian mustard (*Brassica juncea* Czern.) are the most promising terrestrial candidates for removing metals from contaminated water. The roots of Indian mustard have been found to be effective in removal of Cd, Cr, Cu, Ni, Pb, and Zn, and sunflower can remove Pb, U, ^{137}Cs, and ^{90}Sr from hydroponic solutions (Dushenkov et al. 1995, Prasad and Freitas 2003).

1.2.2.1.1 Rhizofiltration and Constructed Wetlands for Phytoremediation of Pollutants

A constructed wetland (CW) is an engineered wetland, in a more controlled environment, that has been applied for treatment of wastewaters for more than five decades. Phytoremediation with wetland plants is an eco-friendly, aesthetically pleasing, cost-effective, solar-driven, reliable, and robust wastewater treatment technology that is useful for cleaning up nutrient overloads, toxic metals, and organic pollutants present in wastewaters, municipal solid waste leachates, industrial wastewaters, agricultural, aquacultural, and dairy farm effluent, storm runoff, sludges, pharmaceutical waste, and mine drainage, as well as polluted river and lake waters with low to moderate levels of contamination (Vymazal 2014). In addition, these treatment systems are good at removing pathogenic microorganisms from wastewater. Macrophytes are the main biological component of wetland ecosystems; they contribute directly to pollution reduction through uptake and assimilation and indirectly by facilitating the growth of important pollutant-reducing microorganisms through complex interactions in the rhizosphere (Guan et al. 2015). They not only assimilate pollutants directly into their tissues but also act as catalysts for purification reactions by increasing the microbial diversity in the root zone through the release of oxygen and exudates and promotion of a variety of chemical and biochemical reactions that enhance purification (Stottmeister et al. 2003). Wetland plants have fibrous root systems with large contact areas and the ability to accumulate large metal concentrations in their organs compared to plants in the surrounding water. However, the principal pollutant removal mechanisms in CW include biological processes, such as microbial metabolic activity and plant uptake, as well as physico-chemical processes, including sedimentation, adsorption, filtration, volatilization, and precipitation of the waste–sediment, root–sediment, and plant–water interfaces (Kadlec 2003). Although filtration is considered an important process in these removal mechanisms, additional interactions occur among media, plants, and water. The complex microbial community associated with filter material roots, created by interactions with wastewater, is mainly responsible for the degradation efficiency of pollutants and ecosystem stability (Stottmeister et al. 2003, Desta et al. 2014). In addition, wetland plants have high remediation potentials for macronutrients due to their generally fast growth and high biomass production. The common plants in wetlands are the common reed (*Phragmites* spp.), cattails (*Typha* spp.), rush (*Juncus* spp.), and bulrush (*Scirpus* spp.). These plants can withstand extreme environmental conditions, including the presence of toxic contaminants such as heavy metals. The capacity of constructed wetlands for accumulation of heavy metals in shoots and roots of plants and the CW bed have been demonstrated in many studies. These systems provide habitats for wildlife and as a consequence, increased biodiversity; thus, they can be implemented to restore degraded areas. The advantages of wetland plants include their ability to be used as *in situ* or *ex situ* applications and can include plant species other than hyperaccumulators. However, one of the limitations for CW is the requirement of a large land area to achieve proper pollutant removal, due to the low hydraulic loading rate. It is necessary then to find the optimal CW design characteristics in order to maximize pollutant removal efficiencies and to keep the treatment area to a minimum. The main parameters that affect the removal efficiency of CW are the hydraulic residence time and temperature. A well-designed CW should be able to maintain the wetland's hydraulic properties, namely, the hydraulic loading rate (HLR) and the hydraulic retentention time (HRT), as these factors affect the treatment performance of a CW. The HRT is determined by the mean surface area of the wetland system (*A*), the flow depth (*y*), and the porosity of the substrate (*p*), i.e., the space available for water to flow through the media, roots, and other solids in a CW system (Ghosh and Gopal 2010). A CW requires a large land area to achieve proper pollutant removal, due to low HLRs:

$$\text{HRT} = V/Q = (A \times y \times p)/Q$$

where Q (in liters per day) is the design flow rate, an assumed constant A (in square meters) is the mean surface area of the system, V (in cubic meters) is the system volume, and y (in meters) is the flow depth.

In these systems, the wastewater is fed into the system at an inlet and it flows slowly through the porous media under the surface of the bed in a more or less horizontal path until it reaches the outlet zone, where it is collected before leaving a level control arrangement at the outlet (Figure 1.3). During its passage through the system, the wastewater comes into contact with a network of aerobic, anoxic, and anaerobic zones. There are different CW types, depending on the flow type, and these can be divided into the subsurface flow (which includes vertical and horizontal subsurface flow CWs, depending on the direction of the flow) and the surface flow (which includes the free water table on top of the soil).

1.2.2.2 Caulofiltration

The term caulofiltration (with the Latin root *caulis*, for shoot) was first proposed by Mesjasz-Przybyłowicz et al. in 2004, when excised plant shoots were used for removal of heavy metals from contaminated wastewater (Figure 1.4a). In the study, the metals were absorbed or adsorbed, and thus their movement in underground water was minimized. Mesjasz-Przybyłowicz et al. (2004) used a nickel-hyperaccumulating plant, *Berkheya coddii*, for phytoextraction and phytofiltration of the heavy metals from the wastewater; they immersed the cut ends of shoots (excised shoots) in concentrated solutions of Cd, Ni, Pb, or Zn and observed that large amounts of these metals accumulated in the leaves.

1.2.2.3 Blastofiltration

Blastofiltration is based on a concept similar to that of rhizofiltartion, but with blastofiltration plant seedlings are used for the removal of heavy metals from aqueous solutions. Blastofiltration may turn out to be an alternative plant-based wastewater treatment technology that takes advantage of the fast increase in the surface area-to-volume ratio

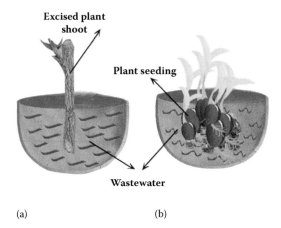

FIGURE 1.4
Removal of heavy metals in contaminated wastewater through caulofiltration (a), and blastofiltration (b).

after germination and the ability of the seedlings employed to absorb large quantities of heavy metals, thus making this method suitable for wastewater treatment (Figure 1.4b). The mechanisms involved in biosorption processes include chemisorption, complexation, ion exchange, microprecipitation, hydroxide condensation onto a biosurface, and surface adsorption. Lin et al. (2002) researched six kinds of crop seedlings for the capacity to remove Pb from water via the blastofiltration method. Under the selected experimental conditions, they found that a concentration of 100 mg/L Pb was reduced to below 5 mg/L Pb in 72 h. In the dry roots of sunflower, pea, or castor oil plants, Pb accumulated to 91.6, 40.7, and 52.8 mg/g, respectively, while corn and Chinese sorghum also exhibited an ability for accumulation of Pb in the roots. Several traditional crop seedlings with a "super-accumulation tendency" have been suggested for better blastofiltration prospects in the phytoremediation of Pb pollution in water. It was recently reported that young seedlings of Indian mustard may be as effective as roots in removing heavy metals from water when grown in aquaculture in aerated water.

1.2.3 Phytostabilization

Phytostabilization, also called *in situ* or in place inactivation or phytimmobilization, is another strategy by which plant species can reduce the mobility and biovailability of pollutants in the environment, via either immobilization or prevention of migration (U.S. EPA 2000a,b). In this process, certain plant species immobilize contaminants in the soil through adsorption and accumulation by the roots, adsorption on roots, or precipitation within the root zone and physical stabilization of soil, reducing the risk to human health and environment (Barcelo and Poschenrieder 2003). The phytostabilization technique is useful for immobilizing many heavy metals, for example, Pb, As, Cd, Cr, Cu, and Zn. In addition, phytostabilization can be used to restore vegetation cover in areas with limited vegetation that has resulted due to high levels of pollution. The microorganisms (bacteria and mycorrhiza) living in the rhizosphere of these plants also have an important role in these processes: not only can they actively contribute to change the trace element speciation, but also they can assist the plants in overcoming phytotoxicity, thus assisting in the revegetation process (Mastretta et al. 2006). The process is applied in situations where there are potential human health impacts and exposure to substances of concern can be reduced to acceptable levels by contaminants. Phytostabilization requires the growth of a healthy and strong layer of plants before human activity can resume on the treated land. However, there may be potential for phytostabilization of organic pollutants, since some organic pollutants and metabolic by-products of these pollutants can be attached to or incorporated into plant components, such as lignins. This form of phytostabilization has been called phytolignification (Cunningham et al. 1995). One difference between the two methods, however, is that phytostabilization of metals is generally intended to occur in the soil, whereas phytostabilization of organic pollutants through phytolignification occurs aboveground. Unlike phytoextraction or hyperaccumulation of metals into shoots or root tissues, phytostabilization primarily focuses on sequestration of the metals within the rhizosphere but not in plant tissues (Figure 1.1). Characteristics of plants appropriate for phytostabilization at a particular site include the following: tolerance to high levels of the contaminant(s) of concern; high production of root biomass that will be able to immobilize the contaminants through uptake, precipitation, or reduction; retention of applicable contaminants in roots, as opposed to transfer to shoots, to avoid a need for special handling and disposal of shoots. However, the amendments used for stabilizing trace elements in contaminated soils commonly include liming agents, phosphates (H_3PO_4, triple calcium phosphate, hydroxyapatite, phosphate rock), trace element (Fe/Mn) oxyhydroxides, organic materials (e.g., biosolids, sludge, composts), natural and synthetic zeolites, cyclonic and fly ash, and steel shots (Vangronsveld et al. 2009). Yadav et al. (2009) reported that the use of organic amendments, such as dairy sludge increased plant growth and reduced bioavailability of As, Cr, and Zn in soil, whereas biofertilizers reduced the uptake of As, Cr, and Zn by plants. *Jatropha curcas* grows in Cr-contaminated soils and accumulates Cr in roots followed by shoots. Accumulation of Cr induces oxidative stress in *J. curcas*, and the plant is able to tolerate this stress through hyperactivity of an antioxidant defense system (Yadav et al. 2010). The examples of some commonly reported heavy metal hyperacccumulator plants are provided in Table 1.1.

TABLE 1.1

Some Plants Capable of Phytostabilization of Heavy Metals

Plant Species	Metal(s)	Reference
Isocoma veneta (Kunth) Greene	Cd, Cu, Mn, Pb, Zn	Gonzalez & Gonzalez-Chavez 2006
Teloxys graveolens	Cd, Cu, Mn, Pb, Zn	Gonzalez & Gonzalez-Chavez 2006
Euphorbia sp.	Cd, Cu, Mn, Pb, Zn	Gonzalez & Gonzalez-Chavez 2006
Dalea bicolor Humb. & Bonpl. ex Willd.	Cd, Cu, Mn, Pb, Zn	Gonzalez & Gonzalez-Chavez 2006
Lygeum spartum L.	Cu, Pb, Zn	Conesa et al. 2006
Piptatherum miliaceum L. Coss.	Cu, Pb, Zn	Conesa et al. 2006
Atriplex lentiformis (Torr.) S. Wats.	As, Cu, Mn, Pb, Zn	Mendez et al. 2007
Atriplex canescens (Pursh) Nutt.	As, Hg, Mn, Pb	Rosario et al. 2007
Bidens humilis	Ag, As, Cd, Cu, Pb, Zn	Bech et al. 2002
Baccharis neglecta Britt.	As	Flores-Tavizon et al. 2003
Schinus molle L.	Cd, Cu, Mn, Pb, Zn	Gonzalez & Gonzalez-Chavez 2006

1.2.4 Phytovolatilization

The release of volatile contaminants to the atmosphere via plant transpiration is known as phytovolatilization (Figure 1.5). Phytovolatilization is a specialized form of phytotransformation that can be used only for those contaminants that are highly volatile in nature (Terry et al. 2000). It is a complex process driven by biophysical mechanisms of mass transport through vegetation. Contaminants like Hg, As, and Se, once they are taken up by plant roots, can be converted into nontoxic forms and volatilized into the atmosphere from the roots, shoots, or leaves (U.S. EPA 2000b). During the process of Se phytovolatilization, plants metabolize various inorganic or organic species of Se (e.g., selenate, selenite, and Se-Met [Met]) into a gaseous form (Berken et al. 2002). Se volatilization in the form of methyl selenate was proposed as major mechanism of Se removal from soil. Indian mustard (*Brassica juncea*) has a high rate of Se accumulation and volatilization, and also a fast growth rate, making it a promising species for Se remediation (Pilon-Smits et al. 1999). Recently, S-adenosyl-L-Met:L-Met S-methyltransferase, an enzyme involved in the

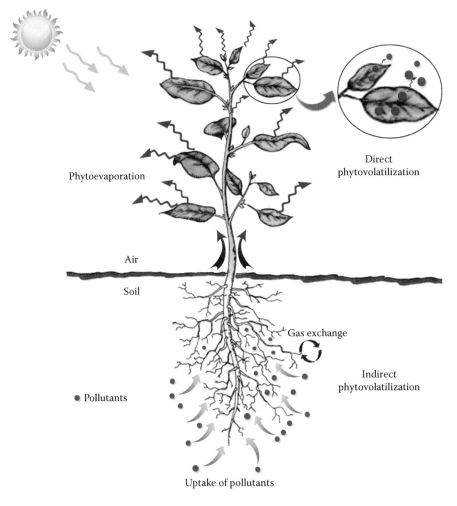

Phytoevaporation

Direct phytovolatilization

Air

Soil

Gas exchange

Indirect phytovolatilization

• Pollutants

Uptake of pollutants

FIGURE 1.5
Various processes in the phytovolatilization method.

methylation of Se-Met to Se-methyl-Met, was identified as the catalyst for an important rate-limiting step in the Se phytovolatilization process (Tagmount et al. 2002). In addition to metals, phytovolatilization this has been established for organic contaminants also. For example, the poplar tree (*Liriodendron* sp.) is a phytovolatizer and was shown to volatilize up to 90% of trichloroethane absorbed from contaminated soil (McGrath and Zhao 2003). Although transfer of contaminants to the atmosphere may not achieve the goal of complete remediation, phytovolatilization may be desirable in that prolonged soil exposure and the risk of groundwater contamination are reduced. The main advantage of phytovolatilization is that it can completely remove the pollutant from the site, without the need for plant harvesting and disposal, as with the other methods. This is the most controversial of phytoremediation technologies, because Hg and Se are toxic and there is doubt whether the volatilization of these elements into the atmosphere is desirable or safe.

Plants can volatilize both organic and inorganic contaminants provided that the inorganic contaminants do not form methyl or hydride derivatives. Chemical characteristics such as Henry's law volatility constant (K_H) and the vapor pressure determine whether organic contaminants will volatilize. Henry's law volatility constant is the measure of a contaminant's ability to move in water. Pollutants with K_H values of $>10^{-4}$ tend to move in the air space between soil particles, and those with K_H values of $<10^{-6}$ move predominantly in water (Reger et al. 2009). Hence, pollutants with K_H values between 10^{-4} and 10^{-6} can move in both water and air and diffuse passively into the atmosphere in less toxic forms through transpiration streams (i.e., opening of stomata). Plants that show high levels of efflux of a pollutant towards the atmosphere through the transpiration process are applicable for the phytovolatilization mechanism. Moreover, there are two pathways for phytovolatilization in plants. In the first mechanism, pore water with dissolved contaminants is taken up by roots via transpiration. The chemical is then translocated to aboveground tissues via xylem sap flow, where it diffuses to the atmosphere through stems or vaporizes from leaves. Transpiration-driven transport has been shown to drive trichloroethylene volatilization from trees in summer. The transpiration stream concentration factor (TSCF) of a compound, defined as the ratio of the compound concentration in the stem xylem to the concentration in the soil solution in contact with roots (Cherian and Oliveira 2005). The TSCF is traditionally modeled as a function of the octanol–water partition coefficient (K_{ow}), with uptake maximized for compounds with a log K_{ow} of ~2.5. Compounds with greater hydrophobicity are assumed to sorb strongly to plant lipid tissues, while more hydrophilic compounds are thought to be unable to pass the lipid membranes of the root. In the second volatilization pathway, dissolved volatile compounds in the pore water diffuse across the root epidermis into gas-filled root cortical tissues (aerenchyma). Aerenchyma are porous root tissues that are found only in wetland plants and are adapted to enhance oxygen (O_2) transport from the atmosphere to roots but which are also effective conduits for gas-phase transport in the opposite direction.

According to Limmer and Buken (2016), phytovolatilization can exist in two different forms: direct and indirect phytovolatilization.

1.2.4.1 Direct Phytovolatilization

In direct phytovolatilization, plants take up volatile contaminants from polluted soil or water and release them into the atmosphere through the stem or trunk and leaves via transpiration. In direct phytovolatilization, plants also create physical changes within the subsurface that may lead to a reduction in contaminants in the soil and water. The physical processes influencing direct volatilization rates can include groundwater table fluctuations, transpiration rates, and preferential pathways created by the tree roots within the subsurface. These processes may enhance the rate at which contaminants are directly volatilized through the soil, with significant remediation implications (Limmer and Burken 2016). This mechanism is believed to be dependent on many environmental factors (pressure, rainfall, temperature), soil characteristics (permeability, moisture, porosity), and the presence of trees. Direct phytovolatilization requires the plant to take up, translocate, and volatilize the compound, and so volatilization of compounds produced or transformed by the plant are not considered directly phytovolatilized. The increased flow rate of groundwater to a plant's roots may provide more opportunity for mass transfer due to greater exposure of the contaminated water to the gas phase, giving a larger mass of contaminants a chance to volatilize out of the water and into the gas-phase pore space. The rate of direct phytovolatilization is calculated by the following equation (Limmer and Burken 2016):

$$\text{Direct phytovolatilization (\%)} = (\text{mass phytovolatilized})/$$
$$(\text{mass phytovolatilized} + \text{mass in harvestable part of plant})$$

1.2.4.2 Indirect Phytovolatilization

Indirect phytovolatilization is volatile contaminant flux from the subsurface as a result of plant root activities. Plants move vast quantities of water (globally, ~62,000 km^3/year) and concurrently explore large volumes of soil (Jasechko et al. 2013). These processes cause profound changes in subsurface chemical fate and transport. The activities of plant roots can increase the flux of volatile contaminants from the subsurface through the following mechanisms (Limmer and Burken 2016):

1. Lowering the water table
2. Advection with gas fluxes caused by diel water table fluctuations
3. Increased soil permeability
4. Chemical transport via hydraulic redistribution
5. Advection with water towards the surface

1.2.5 Phytodegradation

Phytodegradation, which is also known as phytotransformation, is a process in which contaminants taken up by plants are broken down through metabolic processes within the plant or through the effects of enzymes produced by the plants on contaminants in close proximity to the plant (Vishnoi and Srivastava 2008). Plants are able to produce enzymes that catalyze and accelerate degradation of organic pollutants. Hence, organic pollutants are broken down into simpler molecular forms through what are termed *ex planta* metabolic processes and are incorporated into plant tissues to aid plant growth (McGrath and Zhao 2003). From this point of view, green plants can be regarded as a "green liver" for the biosphere. Remediation of a site by phytotransformation is dependent on direct uptake of contaminants from the media and accumulation in the vegetation. Direct uptake of chemicals into plant tissue via the root system is dependent on uptake efficiency, transpiration rate, and concentration of the chemical in the soil and water. Moreover, uptake efficiency depends on chemical speciation, physico-chemical properties, and plant characteristics, whereas transpiration rate depends on plant type, leaf area, nutrients, soil moisture, temperature, wind conditions, and relative humidity. Certain enzymes produced by plants are able to break down and convert chlorinated solvents (e.g., trichloroethylene), ammunition wastes, and herbicides. This technology also can be used to remove contaminants from petrochemical sites and storage areas, fuel spills, landfill leachates, and agricultural chemicals. Successful implementation of this technology requires that the transformed compounds that accumulate within the plant be nontoxic or significantly less toxic than the parent compound(s).

1.2.6 Rhizodegradation

Rhizodegradation (also called rhizoremediation, phytostimulation, plant-assisted biore-mediation, microbe-assisted phytoremediation, plant-aided *in situ* biodegradation, and enhanced rhizosphere biodegradation) is a process in which plant-supplied substrates stimulate the growth of microbial communities in the rhizosphere to break down organic pollutants in the soil (Vishnoi and Srivastava 2008). A plant's rhizosphere provides a unique environment for microorganisms that are capable of breaking down hazardous pollutants into nontoxic and harmless products through their metabolic activity. This is a slower process than phytodegradation. In rhizodegradation, the process of biodegradation results from plant-released nutrients in the form of root exudates. The microbial population and activity in the rhizosphere can increase due to the presence of these exudates, and they can result in increased degradation of organic pollutants in the soil. Rhizospheric microorganisms in turn promote plant growth via nitrogen fixation, nutrient (i.e., phosphorus) mobilization, production of plant growth regulators, decreased plant stress hormone levels, protection against plant pathogens, and degradation of pollutants before they negatively impact the plant. A schematic for phytostabilization is provided in Figure 1.6. As a consequence of these

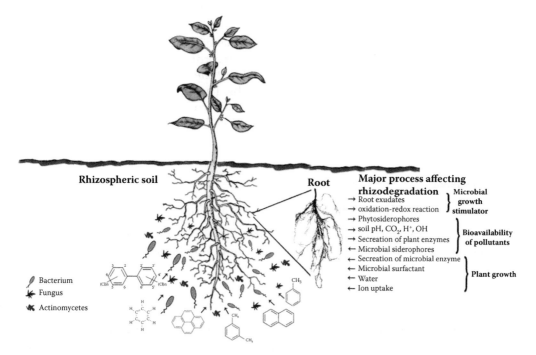

FIGURE 1.6
Schematic mechanism of rhizoremediation of organic pollutants in polluted soil.

mutual interactions, also known as rhizosphere effects, there are elevations in the number, diversity, and metabolic activities of microbes able to degrade contaminants or support plant growth in close vicinity of the roots compared to these parameters in bulk soil. In many cases, rhizosphere microbes are the main contributors to the contaminant degradation process. Even enhanced rhizoremediation might be considerably slower than *ex situ* treatments due to environmental restrictors at field sites, such as competition by weed species which are better adapted to the site, limited plant growth in heavily and unevenly contaminated soil, presence of plant pathogens, and other biotic and abiotic stressors. Furthermore, rhizoremediation is effective only in the rooting zone of plants and is unsuitable for deeper subsurface layers. Some toxic contaminant metabolites can also bioaccumulate in plants, making strict regulations for disposal of the plant material necessary. Appealing features of rhizodegradation include the destruction of contaminats *in situ* the potential complete mineralization of organic contaminants; in addition, thetranslocation of the compound(s) to the plant or atmosphere is less likely than with other phytoremediation technologies.

1.2.7 Phytodesalination

Phytodesalination refers to the use of halophytic plants for removal of salts from saline soils so that the soils can support normal plant growth. Halophytic plants have been suggested to be naturally better adapted to cope with heavy metals compared to glycophytic plants. They are adapted to live in a saline environment, such as seawater, a saltwater marsh, or a salt desert, by means of mechanisms that include osmotic adjustment and ion compartmentalization in their vacuoles. While the high tolerance of halophytes to salinity is recognized as the common characteristic of these species, there is no universally agreed-upon definition for these plants. Some authors base their definition on the ability of such plants to complete the life cycle in water with a rather high salinity level of about 11.7 g NaCl/L (200 mM NaCl) (Flowers and Colmer 2008). A recent database review presented a broader list of halophytes by including species that can tolerate a minimum salinity of 4.7 g NaCl/L (80 mM NaCl) (Santos et al. 2015). According to published estimates, two halophytes, *Suaeda maritime* and *Sesuvium portulacastrum*, can remove 504 and 474 kg of NaCl from 1 ha of saline soil in a period of 4 months. Therefore, *S. maritime* and *S. portulacastrum* have been sucessfuly used to accumulate NaCl from highly saline soils and enable use of the soils for crop production after a few repeated cultivation and harvest cycles.

1.3 Categories of Plants That Grow on Metal-Contaminated Sites

Plants have been categorized according to three basic survival strategies for growth on heavy metal-contaminated sites: metal excluders, indicators, and accumulators (Figure 1.7) (Baker 1981, Ghosh and Singh 2005).

1.3.1 Metal Excluders

Metal excluders are plants which effectively limit the levels of heavy metal translocation within them and maintain relatively low levels in their shoots over a wide range of soil and sludge contaminant levels, thus protecting the leaf tissues, particularly the metabolically active photosynthetic cells, from heavy metal damage. However, they can still contain

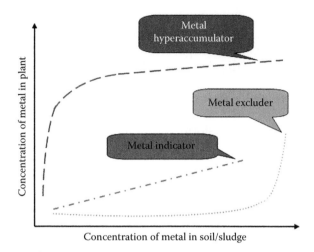

FIGURE 1.7
Three response strategies of plant to increasing metal concentrations in soil and sludge. *Source:* Ghosh M., Singh S.P., *Applied Ecology and Environmental Research* 3(1):1–18, 2005.

large amounts of metals in their roots. In excluder plants, metal concentrations in aerial parts are kept low (the ratio of metal concentration in the plant to that in the soil is $\ll 1$) and constant over a wide range of soil contaminant concentrations.

1.3.2 Metal Indicators

Metal indicators are plants that accumulate heavy metals in their aboveground tissues, and the metal levels in the tissues of these plants generally reflect metal levels in the soil or sludge. However, under continued uptake of heavy metals, these plant species die. Indicator plants are able to provide indirect or direct information on the impact of pollutants in the environment. These plants are ecologicaly important, since they are pollution indicators and also, like accumulators, they absorb pollutants. In indicator plants, the uptake and transport of metals are regulated in such a way that the ratio of metal concentration in the plant to that in the soil is near 1.

1.3.3 Accumulators

Metal accumulators are plant species that concentrate metals in their aboveground tissues to levels far exceeding those present in the soil or in the nonaccumulating plant species growing nearby. If these plants continue to take up metals, they will eventually die. In accumulator plants, the concentration ratio of the metal in the plant to that in the soil is >1.

1.4 Mechanisms for Heavy Metal Tolerance, Accumulation, and Detoxification in Plants

The term heavy metal is used in describing transition metals with an atomic mass over 20 and specific gravity above 5. Heavy metals are the main group of inorganic contaminants in the environment, and a considerably large areas of land are contaminated with them due to the use of sewage and industrial sludge, municipal compost, pesticides, and fertilizers, and also emissions from municipal waste incineration, car exhaust, residues from metalliferous mines, and smelting industries. Each source of contamination has its own damaging effect on plants, animals, and, ultimately, human health, but those sources that add heavy metals to soils and water are of serious concern due to the persistence of these metals in the environment and their carcinogenicity to human beings. They cannot be destroyed biologically but are transformed from one oxidation state or organic complex to another.

1.4.1 Tolerance in Plants

Tolerance to heavy metals in plants can be defined as the ability of a plant to survive in a soil that is toxic to other plants, and this is manifested by an interaction between the plant's genotype and its environment (Macnair et al. 2000). In plants, physical barriers are the first line of defense against metals. Little is known about the role of the plant cell wall and its binding properties in relation to metal tolerance. However, reports on this topic have been somewhat controversial. Mehes-Smith et al. (2013) reported that a significant amount of metal accumulation occurred between the cell wall and the cell membrane. Divalent and trivalent metal cations are able to bind to plants due to the presence of functional groups, such as –COOH, –OH, and –SH in plant cell walls. The most important component of the plant cell wall is the pectin, which consists of carboxyl groups. Under metal stress, divalent and trivalent heavy metal ions bind with the carboxyl group of pectin (Mehes-Smith et al. 2013). The binding of lead to JIM5-P (within the cell wall) restricts the movement of metals to the plasma membrane and acts as a physical barrier in *Funaria hygrometrica* protonemata (Krzesłowska et al. 2009). Krzesłowska et al. (2010) reported that the lead that bound to JIM5-P within the cell was either taken up or remobilized by endocytosis along with the pectin epitope. Elevated levels of Fe, Cu, Zn, and Pb have been observed in cell walls of *Minuartiaverna* sp. *Hercynica* growing on mine dumps (Neumann et al. 1997, Solanki and Dhankhar 2011). The plant plasma membrane functions as the first line of defense for heavy metal contamination, as it is the first cell structure that is exposed to heavy metals. The plasma membrane restricts the uptake and accumulation of metals by inhibiting their entry into the cytoplasm (Mehes-Smith et al. 2013). The restriction of metals to the plasma membrane can be achieved by changing the cell wall binding capacity to metal ions, by modifying the ion channels present on the membrane, by modifying the efflux pumps, or via the root exudates. In addition, biosynthesis of diverse cellular biomolecules is a way to tolerate or neutralize metal toxicity. This includes the induction of a myriad of low-molecular-weight protein metallochaperones or chelators, such as nicotianamine, putrescine, spermine, mugineic acids, organic acids, glutathione, phytochelatins, and metallothioneins, or cellular exudates, such as flavonoid and phenolic compounds, protons, heat shock proteins, specific amino acids, such as proline and histidine, hormones, and other compounds, such as salicylic acid, jasmonic acid, and ethylene. When the above-mentioned strategies are not available to restrain metal poisoning, equilibrium of cellular redox systems in plants is upset, leading to increased induction of ROS. To mitigate the harmful effects of free radicals, plant cells have developed an antioxidant defense mechanism which is composed of enzymatic antioxidants, like superoxide dismutase, catalase, ascorbate peroxidase, guaiacol peroxidase, and glutathione reductase, and nonenzymatic antioxidants, like ascorbate, glutathione, carotenoids, alkaloids, tocopherols, proline, and phenolic compounds (flavonoids, tannins, and lignin), that are all scavengers of free radicals.

1.4.2 Accumulation and Detoxification in Plants

All plants have the ability to accumulate heavy metals, including Mg, Fe, Mn, Zn, Cu, Mo, and Ni, which are essential for plany growth and development. Certain plants, however, have the ability to accumulate Cd, Cr, Pb, Co, Ag, Se, and Hg, which have no known biological function in plants. However, a class of rare plants, called hyperaccumulators, combines extremely high tolerance to and foliar accumulation of trace heavy metals and other elements. Hyperaccumulators have recently gained considerable interest because of their potential use in phytoremediation. A hyperaccumulator can accumulate exceptional concentrations of trace metals in their aerial parts without showing visible signs of toxicity. Plants have developed flexible strategies to cope with fluctuations in their environment in order to minimize the adverse effects of metal deficiency or toxicity. Plant responses can include a significant alteration in gene expression, particularly of membrane transporters that are responsible for the uptake, efflux, translocation, and sequestration of essential and nonessential mineral nutrients. Metal hyperaccumulator plants possess several unique characteristics, such as the ability to take up and translocate exceedingly large amounts of metals to their shoots and hypertolerate the toxic metals (Freeman and Salt 2007). The vacuole is the final destination for practically all toxic substances to which plants can be exposed. Another protein family whose members can mediate transport of metals into the vacuole or other subcellular compartments is the Zn-regulated transporter, Fe-regulated transporter-like proteins, natural resistance-associated macrophage proteins, and the cation diffusion facilitator family, also called the MTP family (Memon and Schröder 2009). Chelation and sequestration processes result in removal of toxic ions from sensitive sites and in accumulation. It is assumed that most of the hyperaccumulated metals are bound to ligands, such as histidine, nicotianamine, citrate, malate, glutathione, phytochelatins, metallothioneins, or proline; these ligands play a major role in metal detoxification and sequestration in plants.

1.5 Mechanisms of Organic Pollutant Detoxification in Plants

1.5.1 Sorption, Uptake, and Translocation of Organic Compounds to Roots

Phytoremediation of organic contaminants may occur directly through uptake, translocation into plant shoots, and metabolism (i.e., phytodegradation), volatilization (i.e., phytovolatilization) or indirectly through plant–microbe–contaminant interactions within plant root zones (rhizospheres). Plant uptake is the first crucial step in plant metabolism of organic compounds. As organic compounds are usually xenobiotics of human origin in plants, there are no transporters for their uptake and the usual mechanism of uptake is by simple diffusion (passive uptake) through cell walls and membranes. When organic compounds come into contact with roots, they may become sorbed to the root surface. The hydrophilic and hydrophobic natures of the organic compounds also determine their potential uptake. Hemicellulose in the cell wall and the lipid bilayer of plant membranes can bind hydrophobic organic pollutants effectively. In terms of constant plant and environmental features, the hydrophobicity of a compound, expressed as its octanol–water partition coefficient (K_{ow}), is the determining factor for root entry and translocation. In practice, an optimum range of hydrophobicity exists ($1 \leq \log K_{ow} \leq 3.0$) outside of which uptake and concomitant translocation of organic molecules are severely restricted. Organic contaminants with a log K_{ow} of <1.0 are considered to be very water soluble, and plant roots do not generally accumulate them at a rate that surpasses passive influx into the transpiration stream (Dietz and Schnoor 2001). Contaminants with a log K_{ow} of >3.5 show high sorption to roots but slow or no translocation to stems and leaves. However, plants readily take up organic contaminants with a log K_{ow} between 0.5 and 3.5, as well as weak electrolytes and easily translocate these compounds to the shoot. Hydrophobic chemicals with log K_{ow} of >3.5 are candidates for phytostabilization or rhizosphere remediation by virtue of their long residence time in the root zone (Dietz and Schnoor 2001). Uptake of organic chemicals through the roots depends on a plant's uptake efficiency and transpiration rate, and also the concentration of the chemical in the soil water, according to the following equation (Burken and Schnoor, 1997):

$$U = \text{TSCF} \times T \times C$$

Where U is the rate of organic compound uptake by the plant (in milligrams per day), the TSCF is the efficiency of organic compound uptake, T is the transpiration rate (in liters per day), and C is the soil water concentration of chemicals (in milligrams per liter). The TSCF is defined as the ratio of the compound concentration in the xylem fluid relative to that in the external solution, and it is a measure of uptake into the plant shoots (Dietz and Schnoor 2001, Cherian and Oliveira 2005). The TSCF is a dimensionless ratio of the concentration in the xylem sap versus the bulk concentration in the root zone solution:

$$\text{TSCF} = (\text{contaminant concentration in xylem sap})$$
$$/(\text{concentration in root zone solution})$$

Because the concentration of a contaminant in the xylem sap is difficult to measure directly for intact plants, the TSCF is often determined based on the measured root concentration; the shoot and tissue concentration is normalized to the amount of water transpired during exposure to the chemical (i.e., the concentration in xylem sap is equal to the milligrams of the compound in the shoots, per liter of water transpired); this is based on an assumption that no phytovolatilization of the chemical occurs. Almost all experimental TSCF values have been obtained under laboratory conditions, where the root zone concentrations can be more easily measured and controlled. Accurate TSCF values are extremely difficult to determine in the field, because of the inherent variabilities associated with root zone exposure.

1.5.2 Enzymatic Transformation of Organic Compounds

Certain enzymes produced by plants are able to metabolize or transform complex and recalcitrant organic compounds into less toxic metabolites. This process is known as phytotransformation. The assimilation of organic compounds in plants is essential in the contact between plant cell enzymes and organic contaminants. The "Green Liver Model" has been used to describe phytotransformation, as plants behave analogously to the human liver when dealing with these xenobiotic compounds (foreign compounds and pollutants)

(Burken 2004). After uptake of the xenobiotics, plant enzymes increase the polarity of the xenobiotics by adding functional groups such as hydroxyl groups (–OH). However, once an organic compound is taken up and translocated, it undergoes one or more phase of transformation (Dietz and Schnoor 2001, Burken 2004):

- Phase I: Chemical conversion via oxidation, reduction, or hydrolysis. Phase I metabolism entails mechanisms similar to the ways that the human liver increases the polarity of drugs and foreign compounds. Whereas in the human liver enzymes such as cytochrome P450 isoforms are responsible for the initial reactions, in plants enzymes such as nitroreductases are responsible for the initial reactions.
- Phase II: Conjugation with glutathione, sugar, or amino acids. In phase II, plant biomolecules such as glucose and amino acids are added to the polarized xenobiotic to further increase the polarity (a process known as conjugation). Phase II reactions serve to increase the polarity and reduce the toxicity of the compounds, although many exceptions to the rule are known. The increased polarity also allows for easy transport of the xenobiotic along aqueous channels in the plant.
- Phase III: Sequestration or compartmentation. In the final stage of phytotransformation (phase III metabolism), sequestration of the xenobiotic occurs within the plant vacuoles or via its binding to the plant cell wall and lignin via ATP binding, ABC transporters, and multidrug resistance proteins to reduce the toxicity of the foreign compound. The xenobiotic can polymerize in a lignin-like manner and develop a complex structure that is sequestered in the plant. This ensures that the xenobiotic is safely stored and will not affect the functioning of the plant. In Figure 1.8, the Green Liver model is illustrated in the example of the process of transformation of explosive contaminants in plants once such compounds have been taken up from the soil.

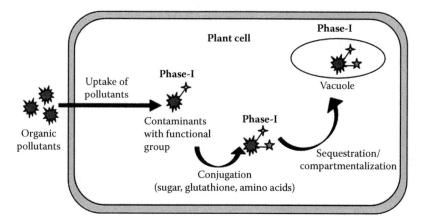

FIGURE 1.8
The Green Liver model for the metabolism of organic pollutants in plant cells.

1.6 Phytoremediation of Complex Industrial Wastes

Complex industrial waste generated from different processing industries, such as distilleries, pulp and paper mills, textile factories, tanneries, and petroleum processing plants, is a main source of surface soil and water pollution in the environment. This complex waste is characterized by its high biological oxygen demand (BOD), chemical oxygen demand (COD), total dissolve solids (TDS), and phenol, sulfate, and heavy metal levels, along with several other complex organic pollutants. The safe disposal of complex industrial waste into the environment is challenging due to high concentrations of heavy metals along with other complex recalcitrant organic pollutants. The degree to which native plants able to take up an element is the most reliable method for prediction of the availability of a contaminated metal to plants and the application of phytoextraction methods to remove heavy metals from contaminated sites. Here, we describe the phytoremediation of various complex industrial wastes.

1.6.1 Distillery Wastes

Distillery industries are sources of heavy metals in the environment due to the discharge of their huge amount of complex wastes (Chandra and Kumar 2016). During the course of alcohol production, around 12–15 L of effluent is generated per liter of alcohol production. This untreated effluent is known as spent wash (Figure 1.9a). The spent wash after anaerobic digestion is known as postmethanated distillery effluent (PMDE). PMDE remains in a dark brown complex waste with a high levels of BOD, COD, TDS, phenol, sulfate, and heavy metals, along with several other complex organic pollutants (Figure 1.9b). In addition, distilleries produce large quantities of postmethanated distillery sludge (PMDS) per

(a)   (b)   (c)

(d)   (e)   (f)

FIGURE 1.9
Types of distillery wastes generated and their environmental impacts. (a) Spent wash; (b) spent wash after anaerobic treatment, known as PMDE; (c) PMDE collection tank; (d) collection tanks after settlement of sludge shown in panel c; (e and f) terrestrial and aquatic pollution due to industrial waste.

day during anaerobic digestion of spent wash, which is characterized by high levels of organic matter (OM), TDS, phenol, sulfate, and heavy metals, along with melanoidins. In addition, distilleries in India produce ~1500 tons of PMDS/day during anaerobic digestion of spent wash (Figure 1.9c) which is characterized by high levels of OM, TDS, phenol, sulfate, and heavy metals, e.g., Cd, Cu, Cr, Mn, Fe, Pb, Ni, and Zn, along with melanoidins (Chandra and Kumar 2017). These wastes, when released into water bodies, damage aquatic ecosystems by reducing the penetration power of sunlight and ultimately reduce the photosynthetic activities and dissolved oxygen content, whereas in agricultural soils it causes inhibition of seed germination and depletion of vegetation by reducing the soil alkalinity and manganese availability (Figure 1.9d to f). As a result of difficulties in treating this complex waste effluent and sludge, people are advocating for application of wastewater recycling and phytoremediation techniques.

Several potential wetland plants that grow on distillery waste-contaminated sites under natural conditions have indicated the phytoremediation potential of use of such plants at contaminated sites as natural hyperaccumulators of heavy metals from complex organic wastes (Chandra and Kumar 2017). Chandra and Kumar (2017) evaluated the phytoextraction potential of 15 potential native plants growing on stabilized distillery sludge (Figure 1.10). These 15 native plants were, namely, *Dhatura stramonium* (dhatura), *Achyranthes* sp. (chaff flower), *Kalanchoe pinnata* (life plant), *Trichosanthes dioica* (parval), *Parthenium hysterophorous* (Congress grass), *Cannabis sativa* (Bhang), *Amaranthus spinosus* L. (amarant), *Croton bonplandianum* (Ban tulsi), *Solanum nigrum* (kali makoy), *Ricinus communis* (castor oil plant), *Sacchrum munja* (munja), *Basella alba* (pui), *Setaria viridis* (green foxtail), *Chenopodium album* (goosefoots), and *Blumea lacera* (janglimuli).

The heavy metal accumulation from distillery sludge and the distribution of these metals in roots, shoots, and leaves follow varied patterns (Figure 1.11). *B. lacera, P. hysterophorous, S. viridis, C. album, C. sativa, B. alba, T. dioica, A. spinosus* L., *Achyranthes* sp., *D. stramonium,*

FIGURE 1.10
Luxuriant growth of native plants on a stabilized distillery sludge dumping site. (a) Sludge dumping site near the industry; (b) *Parthenium hysterophorous*: (c) *Ricinus communis*; (d) *Cannabis sativa*; (e) *Solanum nigrum*; (f) *Chenopodium album*.

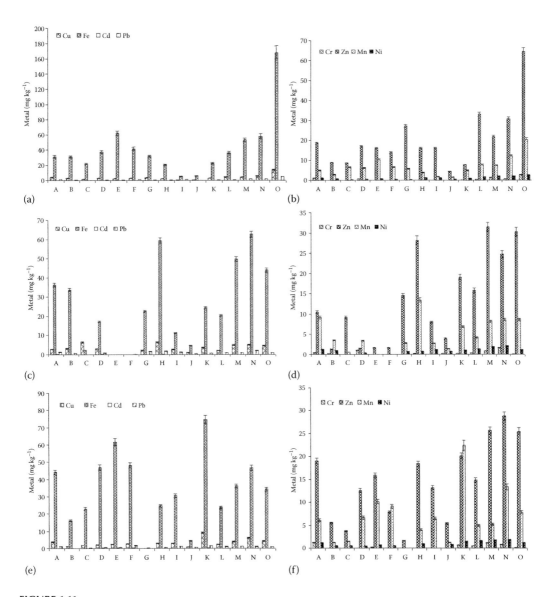

FIGURE 1.11
Accumulation and distribution of heavy metals in different part of growing native plants. (a and b) Root; (c and d) shoot; (e and f) leaf. *Source:* Chandra, R., Kumar, V., *Environmental Science and Pollution Research* 24(3):2605–2619, 2017.

S. munja, and *C. bonplandianum* have been noted as root accumulators for Fe, Zn, and Mn, while *S. munja*, *P. hysterophorous*, *C. sativa*, *C. album*, *T. dioica*, *D. stramonium*, *B. lacera*, *B. alba*, *K. pinnata*, and *Achyranthes* sp. are shoot accumulators of Fe. In addition, *A. spinosus* L. was found to be a shoot accumulator for Zn and Mn. Similarly, all plants were determined to be leaf accumulators of Fe, Zn, and Mn, except for *A. spinosus* L. and *R. communis*. Further, the BCF of all tested plants was <1, while the TF was >1.

Further, observations via transmission electron microscopic (TEM) of roots of *P. hysterophorous* and *C. sativa* showed formation of multinucleolus and multivacuoles and deposition of metal granules in cellular components of roots as a plant adaptation mechanism

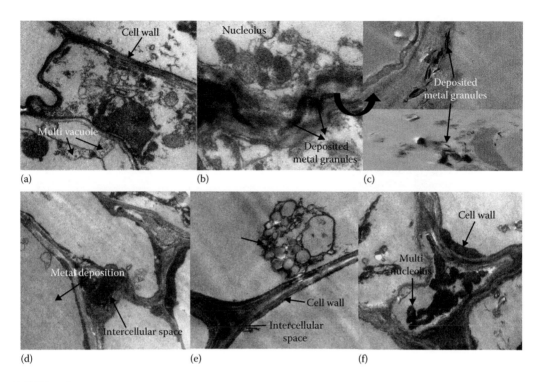

FIGURE 1.12
TEM images of native plant roots after phytoextraction of heavy metals. (a–c) *Parthenium hysterophorous*; (d–f) *Cannabis sativa. Source:* Chandra, R., Kumar, V., *Environmental Science and Pollution Research* 24(3):2605–2619, 2016.

for phytoextraction of a heavy metal-polluted site. Deformities of the cell wall, creation of intercellular spaces, and deposition of metal granules at the surface of the cell wall were also observed as specific features of these plant; however, the plants were found to grow very healthy throughout the year even after variations in weather, as shown by the TEM observations (Figure 1.12d to f). The cellular compartment of roots showed a characteristic feature for formation of vacuoles in a chain. This may be an excretory property of accumulated metals through vacuoles, which appeared very clearly upon TEM observation (Figure 1.12a and e). The observed metal granules were abundant in the cell cytoplasm and also were depositied in cell walls. The metal granule and multinucleolus formation in nuclei, along with projections in the structure of the nuclear membrane caused by metal granules, were observed clearly in the TEM images (Figure 1.12).

1.6.2 Pulp and Paper Mill Wastes

Pulp and paper mills are considered one of the most polluting industries worldwide. This is one of the most important industries in India, not only for economic purposes but also for social purposes. It ranks as the sixth largest polluting source, the world's second largest consumer of chlorine, and the greatest source of toxic organochlorine discharge in wastewater (Figure 1.13a and b). In addition, some highly toxic pollutants, like dibenzo-*p*-dioxins and dibenzofurans, also remain in pulp paper mill effluents; these compouds are generated unintentionally during the processing and paper manufacturing process. All of these compounds get mixed together and form a complex, colored

FIGURE 1.13
Environmental impact of pulp and paper mill industrial waste discharges after secondary treatment. (a) Air;
(b) aquatic; (c) terrestrial pollution; (d) soil pollution; (e) aquatic pollution; (f) health hazards through food chain.

effluent with high COD, BOD, and TDS levels. Hence, discharged effluents after secondary treatment contribute to deteriorating water quality due to the increasing BOD, COD, TDS, and decreased dissolved oxygen. The wastewater from pulp and paper mills has generally low biodegradability due to the presence of recalcitrant compounds. Pulp paper mill effluent and sludge are both important sources of toxic recalcitrant organic pollutants and heavy metal pollution in the environment (Figure 1.13c to f) (Chandra and Singh 2012). Therefore, several recent studies have reported phytoremediation and phytoextraction of heavy metals from contaminated sites by other plants (Lone et al. 2008, Mani and Kumar 2014). The potential for native plants in the uptake of an element is the most reliable method for prediction of availability of a contaminated metal to plants and their application for phytoextraction of heavy metals from contaminated sites (Chandra et al. 2017).

Mishra et al. (2012) assessed the phytoremediation potentials for remediation of heavy metal pollution by six aquatic plants, including *Eichhornia crassipes*, *Hydrilla verticillata*, *Jussiaea repens*, *Lemna minor*, *Pista stratiotes*, and *Trapa natans*, when the plants were grown in paper mill effluent. They found that all the plants caused decreased levels of Cu and Hg in the effluent. Among the above six macrophytes, *L. minor* and *E. crassipes* showed high tolerance to Cu and Hg with increased hyperaccumulation. Mazumdar and Das (2015) investigated the phytoextraction potential of 25 wetland plants species, namely, *Alternanthera sessilis*, *Fimbristylis dichotoma*, *Sonchus arvensis*, *Parthenium hysterophorus*, *Eclipta prostrate*, *Diplazium esculentum*, *Eragrostis atrovirens*, *Ludwigia hyssopifolia*, *Nicotiana plumbaginifolia*, *Chenopodium album*, *Solanum americanum*, *Cynodon dactylon*, *Ricinus communis*, *Leucas lavandulifolia*, *Colocasia esculenta*, *Amaranthus cruentus*, *Centella asiatica*, *Amaranthus spinosus*, *Ipomoea aquatic*, *Eichhornia crassipes*, *Mikania micrantha*, *Phaseolus vulgaris*, *Fimbristylis bisumbellata*, *Eleusine indica*, and *Cyanthillium cinereum*, growing abundantly on pulp paper mill waste-contaminated site. They found that such plants were able to grow and tolerate high concentrations of Pb, Zn, Mg, and Fe from the pulp paper waste contamination, which indicated that wetland plants at the site were able to ameliorate the metal contamination. Chandra et al. (2017) investigated the heavy metal phytoextraction potential of native wetland plants growing on organic pollutant-rich pulp paper sludge. They selected 12 representative native plants based on plant population numbers growing on the pulp paper sludge and evaluated the plants for their phytoextraction potential of heavy metal removal (Figure 1.14).

The results for analysis of heavy metal accumulation in the native plants growing on the sediment of the pulp paper mill effluent drainage system showed variable patterns of accumulation and distribution in different parts of plants (Figure 1.15). The metal analysis of different parts of plants showed variable mobilization patterns of different metals from plant to plant. Overall, the total metal accumulation pattern observed was in the following order: Fe > Zn > Mn > Cu > Cr > Ni > Pb > Cd. Thus, it was observed that all the growing plants had heavy metal phytoextraction efficiencies in the organic pollutant-rich environment. It was also noted that several evaluated plants showed BCFs greater than 1 for accumulated metals, e.g., Pb in *Triticum aestivum* (1.285), *Commelina benghalensis* (5.142), *Cannabis sativa* (3.466), *Phragmitis cummunis* (7.180), and *Ricinus cummunis* (66.80). Furthermore, TF values greater than 1 for all these plants showed strong evidence for the phytoextraction and *in situ* remediation potentials of these plants. All the tested heavy metals accumulated in significant amounts in their different parts (Figure 1.16).

(a) (b) (c)

(d) (e) (f)

FIGURE 1.14
Heavy metal hyperaccumulator wetland plants in use for *in situ* phytoremediation of a pulp paper mill waste-polluted site. (a) *Argemone mexicana*; (b) *Brassica campestris*; (c) *Phalaris arundinacea*; (d) *Alternanthera philoxeroides*; (e) *Commelina benghalensis*; (f) *Parthenium hysterophorus*.

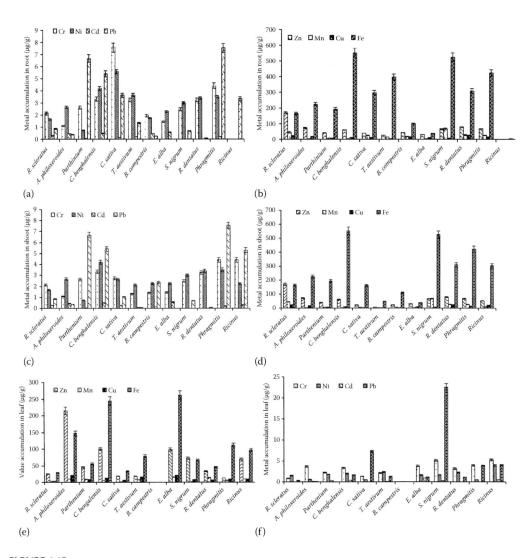

FIGURE 1.15
Heavy metal accumulation patterns in various parts of native wetland plants grown on pulp and paper mill effluent sludge. (a and b) Root; (c and d) shoot; (e and f) leaves. *Source:* Chandra, R. et al., *Ecological Engineering* 98:134–145, 2017.

1.6.3 Tannery Wastes

Manufacture of leather, leather goods, leather boards, and fur produces numerous by-products, solid wastes, and a huge volume of wastewater containing different loads of pollutants; in addition, these industries produce toxic emissions into the air. According to conservative estimates, more than 600,000 tons/year of solid waste, including trimmings of finished leather, shaving dust, hair, fleshing, and trimming of raw hides and skins, are generated worldwide by the leather industry, and approximately 40–50% of the hides are lost to shavings and trimmings. In India, the tannery industry plays a prominent role, as it contributes 15% of the total production capacity of the world at present, with more than 3000 tanneries with an annual processing capacity of 0.7 million tons of hides and

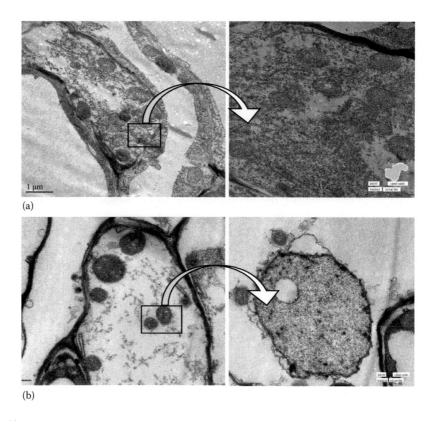

(a)

(b)

FIGURE 1.16
TEM images of roots of hyperaccumulator wetland plants, showing heavy metal granules in cellular components. (a) *Brassica campestris*; (b) *Phlaris arundinacea*. Panels on the right are higher-magnification images of the boxed portions of images on the left.

skins. The uncontrolled release of tannery wastewater and sludge is considered a source of potentially toxic elements, and its disposal is problematic due to the presence of several heavy metals (i.e., Fe, Cr, Cd, Pb, Zn, Mn, and Cu). Metals in tannery waste occur in complex forms and vary widely in their bioavailability to plants. Several reports have emphasized the accumulation of metals in plants growing on tannery waste-contaminated soil (Barman et al. 2000, Sinha et al. 2006). Singh and Sinha (2005) evaluated the accumulation potential and tolerance of the plant *Brassica juncea* (L.) Czern (cv. Rohini) when grown on various amendments of tannery sludge. They found that these plants were effective accumulators for Cr, Fe, Zn, and Mn. The plants of the Brassicaceae family are well known for many metal-accumulating species, and they also have high economic values. *Brassica juncea* and *Brassica campestris* were commonly grown on a tannery waste-amended and -irrigated area of Jajmau, Kanpur, India, due to their tolerance for and bioaccumulation of metals and their high-yield properties (Sinha et al. 2006). The authors also reported that these plants accumulate ample amounts of metals such as Cr and Fe in their root and shoot parts. Further, Sinha et al. (2006) studied the phytoremediation potential of 25 different vegetables and crop plants, namely, 25 different vegetables and crop plants, namely, amaranth (*Amaranthus blitum* L.), bitter gourd (*Momordica charantia* L.), black mustard (*Brassica nigra* L.), bottle gourd (*Lagernaria siceraria* Molina Standley), cauliflower (*Brassica oleracia* L.), chilis (*Capsicum annuum* L.), coriander (*Coriandrum sativum* L.), cucumber (*Cucumis sativus* L.), eggplant (*Solanum melongena* L.), fenugreek (*Trigonella foenum-graecum* L.), garlic (*Allium sativum* L.), hemp (*Cannabis sativa* L.), jack tree (*Artocarpus heterophyllus* L.), kidney bean (*Phaseolus vulgaris* L.), maize (*Zea mays* L.), okra (*Abelmoschus esculentus* L.), pig weed (*Chenopodium album* L.), potato (*Solanum tuberosum* L.), spinach (*Spinacia oleracia* L.), Sudan grass (*Sorghum tuberosum* L.), sugarcane (*Saccharum officinarum* L.), turmeric (*Curcuma domestica* Val.), vegetable sponge (*Luffa aegyptiaca* Mill.), wheat (*Triticum aestivum* L. emend. Thell.), and yellow mustard (*Brassica campestris* L.) growing on a tannery waste-contaminated site. They reported that these plants accumulated high amounts of Fe, Cr, Zn, Mn, and Cu in leafy vegetables compared to nonleafy vegetable crops. Recently, Gupta and Sinha (2007a) reported the phytoremediation potential of four plants, i.e. *Sida acuta*, *Ricinus communis*, *Calotropis procera*, and *Cassia fistula*, that were growing at a tannery sludge disposal site. Overall, the plants *S. acuta* and *C. fistula* were found suitable for the decontamination of most of the metals from the tannery waste-contaminated sites. Furthermore, Gupta and Sinha (2007b) also reported high TF values for all the tested metals (Fe, Mn, Zn, Cr, Cu, Pb, Ni, Cd) in the plants of *Chenopodium album* grown on soil amended with 10% tannery sludge. They found that most of the tested metals accumulated in harvestable parts of this plant.

1.6.4 Textile Wastes

The main environmental concern in the textile industry is the amount of water discharged and the chemical load it carries. The textile industry uses high volumes of water throughout its operations, from the washing of fibers to bleaching, dying, and washing of finished products. On average, approximately 200 L of water are required to produce l kg of textiles. The textile dyes and dye intermediates with high aromaticity and low biodegradability have emerged as major environmental pollutants, and nearly 10–15% of the dyes are lost in the dying process and are released in wastewater, which is an important source of environmental contamination (Instituto Tecnologico Agrario de Castilla y Leon, 2000). Dye-synthesizing wastewater and textile wastewater are two types of poorly treated wastewaters that contain organic dyes. In textile industries, about 10–15% of the dye gets lost in the effluent during the dying process. The large volumes of wastewater generated also contain a wide variety of chemicals, used throughout processing. The effluents from the textile sector are characterized by high BOD, COD, TDS, TSS, and color, which may distort the water quality, add odor, and significantly, hinder economic activities, making their proper treatment a great concern (Khandare and Govindwar 2015). These can cause damage if not properly treated before being discharged into the environment. A few studies have reported removal of textile dyes by using various plants, for instance, *Typha angustifolia*, *Phragmites australis*, *Tagetes patula*, *Blumea malcolmii*, *Typhonium flagelliforme*, *Glandularia pulchella*, *Portulaca grandiflora*, *Aster amellus*, *Zinnia angustifolia*, *Petunia grandiflora*, *Ipomoea hederifolia*, *Typha domingensis*, *Thymus vulgaris*, *Rosmarinus officinalis*, *Alternanthera philoxeroides*, *Hydrilla verticillata*, *Nasturtium officinale*, *Rheum rhabarbarum*, and *Brassica juncea* (Khandare and Govindwar 2015, Tahir et al. 2016).

1.6.5 Agrochemicals

The use of agrochemicals, such as chemical fertilizers and pesticides, is integral in the current agriculture production system around the globe. Pollution by agrochemicals is one of the most significant threats to the integrity of the world's surface waters. The agrochemicals of main ecological concern are heavy metal-based fertilizers, fungicides, and pesticides, because they are toxic and persistent in the environment and hence they can eventually bioaccumulate to higher levels that could affect human beings and other living organisms. The applied pesticide can be transported from the sprayed area to nontarget areas away from the crop, which thus affects not only the pest species but potentially nontarget endangered species also. The study of pesticide phytoremediation has recently yielded some interesting results. Li et al. (2002) looked at the uptake of trifluralin and lindane by ryegrass; trifluralin appeared to be metabolized by the plant, whereas lindane underwent minimal metabolism and instead formed bound residues. A number of factors have been shown to affect the phytoremediation of pesticides. For example, White and colleagues (2003) showed that organic acids increased the uptake of 2,2-bis(*p*-chlorophenyl-1,1-dichlorethylene) and Knuteson et al. (2002) observed that younger plants (2 weeks old) exhibited greater uptake of simazine [2-chloro-4,6-bis(ethylamino)-1,3,5-triazine] than plants that were just 2 weeks older. Although it has long been known that the ability to take up and translocate heavy metals is not only species specific, but also ecotype specific, White et al. (2002, 2003) reported that the uptake of 2,2-bis(*p*-chlorophenyl-1,1-dichlorethylene) was also subspecies specific, with certain cultivars showing significantly higher uptake.

1.6.6 Petroleum Waste

The high rate of growth of petroleum product processing has resulted in the generation of enormous amounts of waste that pose a serious threat to environmental quality around the globe. Petroleum hydrocarbons (PHCs) comprise a diverse mixture of hydrocarbons, typically alkanes (linear or branched), cycloalkanes, aromatic hydrocarbons, and more complex chemicals that occur at petrochemical sites and storage areas, waste disposal pits, refineries, and oil spill sites (Qixing et al. 2011). The accumulation of PHCs now seriously affects the safety of ecosystems and human health. Thus, the remediation of PHC-contaminated soils has attracted worldwide attention. Various plants have been identified for their potential to facilitate the phytoremediation of sites contaminated with petroleum hydrocarbons (Ndimele 2010). Ndimele (2008) also reported that water hyacinths (*Eichhornia crassipes*) significantly accumulate PHCs. Prairie grasses are thought to make superior vehicles for phytoremediation because they have extensive, fibrous root systems. A variety of common weed species used in the cleanup of PHC compounds, including tall fescue (*Festuca arundinacea*), ryegrass (*Lolium perenne*), alfalfa (*Brassica campestris*), smooth meadowgrass (*poapretensis*), crabgrass [*Digitaria sanguinalis* (L.) Scop.], bermuda grass [*Cynodon dactylon* (L.) Pers], and switchgrass (*Panicum virgatum* L.) (Qixing et al. 2011). Yateem et al. (2000) investigated the degradation of total petroleum hydrocarbons (TPH) in the rhizosphere and nonrhizosphere soils of three domestic plants, namely, alfalfa (*Medicago sativa*), broad bean (*Vicia faba*), and ryegrass (*Lolium perenne*). Although the three domestic plants exhibited normal growth in the presence of 1% TPH, the degradation was more profound in the case of leguminous plants. They found that the soil cultivated with broad bean and alfalfa was 36.6 and 35.8% respectively, compared with 24% degradation in the case of ryegrass. Whittig et al. (2003) found that poplar cuttings grown in a PAH-amended sand nutrient solution had similar shoot biomass, growth, and leaf water content as controls but that transpiration, nutrient solution uptake, and root mass were reduced.

1.7 Advantages, Limitations, and Disadvantages of Phytoremediation

Phytoremediation has made tremendous gains in market acceptance in recent years. Its primary advantage is that it is approximately 10 times less expensive than conventional strategies (Chappell 1998). In addition, the other advantages of phytoremediation, in comparison with classical remediation methods, can be summarized as follows:

1. It is an aesthetically pleasing, solar energy-driven cleanup technology.
2. Plants can be easily grown without much effort and also can be monitored easily.
3. It may be used in much larger-scale cleanup operations than is possible with other methods.
4. There is minimal environmental disruption, and *in situ* treatment preserves topsoil.
5. Plants act as soil stabilizers, minimizing the amount of contaminated dust that could leave the site and enter surrounding neighborhoods.
6. It is most useful at sites with shallow, low levels of contamination.
7. It avoids excavation and heavy traffic.
8. It has potential versatility to treat a broad range of hazardous environmental pollutants.
9. It is easier to monitor the site.
10. Unlike bioremediation with microbes, phytoremediation is easily visible; the condition of the plants can be visually monitored, and samples of plant tissue can be tested for the presence of the pollutant over time.
11. Other advantages of phytoremediation over engineering or bioremediation methods include the possibility of a useful product, such as wood, pulp, or bioenergy, that could help finance the cleanup.
12. Plants also supply nutrients for rhizospheric bacteria that may also aid in remediation of the pollutants.
13. Phytoremediation also provides wildlife habitat.
14. It has a high probability of public acceptance.

Although phytoremediation is cost-effective, environmentally friendly technology for ecological restoration of industrial waste-contaminated sites, this technology has several disadvantages and limitations that could create hinder implementation of the strategy:

1. It is generally slower than most other treatment technologies and is climate dependent.
2. Plants that absorb toxic heavy metals or persistent chemicals may pose risks to wildlife and contaminate the food chain.
3. The phytoremediation site must be large enough to grow plants and utilize agricultural machinery for planting and harvesting.
4. Environmental conditions also determine the efficiency of phytoremediation, as the survival and growth of plants are adversely affected by extreme environmental conditions, toxicity, and the general conditions of the soil in contaminated lands.

5. Formation of vegetation may be limited by extreme environmental toxicity.

6. Contaminants collected in leaves can be released again to the environment during litter fall.

7. It is a time-consuming process, and it may take at least several growing seasons to clean up a site.

8. The contaminating material should be present within the root zone to be accessible to the root, i.e., if metals are tightly bound to the organic portions of the soil, some of the contamination may not be available to plants.

9. The solubility of some contaminants may be increased, resulting in greater environmental damage or pollutant migration.

10 Plants with low biomass yields and reduced root systems do not support efficient phytoremediation and most likely do not prevent the leaching of contaminants into aquatic systems.

11. Access to the site must be controlled, as the plants may be harmful to livestock and the general public.

Despite some limitations, present day phytoremediation technologies are used worldwide for removing pollutants from contaminated soils and waters, and various research laboratories are engaged to overcome the limitations.

1.8 Challenges and Future Perspectives of Phytoremediation in Environmental Waste Management

In the past decade, phytoremediation has gained wide acceptance as an effective green technology as well as a rigorous field of research. Most of the experiments have taken place under laboratory conditions, where plants grow in hydroponic settings and are treated with heavy metals. While these results are promising, scientists are ready to admit that solution culture is quite different from that of soil. In real soil, metals are tied up in insoluble forms and they are less available, and that is the biggest problem. Although successfully applied in several demonstration projects, large-scale field applications of phytoremediation for organic pollutants are limited by several restrictions: (i) the levels of contaminants tolerated by the plant, (ii) the often limited bioavailability of the contaminants, and (iii) in certain cases, unacceptable levels of evapotranspiration of volatile organic contaminants into the atmosphere.

A possible solution to conquer these constraints is the use of genetically manipulated plants specifically tailored for phytoremediation purposes. Industrial and academic researchers have come together to explore the multifaceted nature of the fascinating biological phenomenon of metal hyperaccumulation, making the field of phytoremediation truly interdisciplinary and collaborative. The field continues to benefit from this approach, as research teams at all organization levels arew involved in studying the remediation of pollutants, from the level of molecules to ecosystems. Moreover, research in phytoremediation is truly interdisciplinary in nature and requires background knowledge in soil chemistry, plant biology, ecology, and soil microbiology, as well as environmental engineering. In view of the current trends of integration of scientific knowledge worldwide, it is hoped that many challenging questions about commercial applications of phytoremediation will be answered in the future. Additional studies should be conducted to better understand interactions among the main rhizosphere factors, which include metals, soil, microbes, and plant roots. Technological progress in spectroscopic and chromatographic techniques should be explored to improve understanding of the fate of metal ions in plant tissues, which in turn will improve our understanding of the processes involved in metal hyperaccumulation and plant tolerance. Since the absorption of metals by plant roots is limited by low solubility, more research is needed for cost-effective and environmentally safe chemicals with chelating properties that will enhance the bioavailability of metals. Plant species should be identified that are capable of rotation in order to maintain effective rates of extraction. Therefore, long-term *in situ* field trials are required for the wide-angle use of phytostabilization a regular effective technology of soil reclamation.

1.9 Conclusions

Industrial waste management is now one of the main issues that must be addressed to ensure a sustainable environment. It is clear that the greatest application of phytoremediation will be in developing countries, where this technology can provide a low-cost means of controlling widespread environmental contamination and restoration of industrial waste-contaminated sites. Plants have built-in enzymatic machineries capable of degrading complex structures and can be used for cleaning up contaminated sites. It is an ecologically sound and sustainable reclamation strategy for bringing polluted sites back into productive use, but it is still in an experimental stage. Therefore, much attention and scientific scrutiny are needed. However, the lack of technical knowhow, weak implementation of environmental policies, and limited financial resources have given rise to serious challenges.

Acknowledgments

Financial assistance from the Department of Science and Technology, Science and Engineering Research Board, New Delhi (file SB/SO/BB-0042/2013) to Ram Chandra and Rajiv Gandhi is acknowledged, and a national fellowship from the University Grant Commission, New Delhi, to Vineet Kumar, Ph.D. Scholar, is highly appreciated.

References

Alkorta, I., Hernandez-Allica, J., Becerril, J., Amezaga, I., Albizu, I., Garbiscu, C. 2004. Recent finding on the phytoremediation of soil contaminated with environmentally toxic heavy metals and metalloids such as zinc, cadmium, lead, and arsenic. *Reviews in Environmental Science and Biotechnology* 3:71–90.

Baker, A.J.M. 1981. Accumulator and excluder-strategies in the response of plant to heavy metals. *Journal of Plant Nutrition* 3:643–654.

Barcelo, J., Poschenrieder, C. 2003. Phytoremediation: Principles and perspectives. *Contributions to Science* 2(3):333–344.

Barman, S.C., Sahu, R.K., Bhargava, S.K., Chaterjee, C. 2000. Distribution of heavy metals in wheat, mustard and weed grown in field irrigated with industrial effluents. *Bulletin of Environmental Contamination and Toxicology* 64:489–496.

Bech, J., Poschenrieder, C., Barcelo, J., Lansac, A. 2002. Plants from mine spoils in the South American area as potential sources of germplasm for phyotremediation technologies. *Acta Biotechnology* 1–2:5–11.

Berken, A., Mulholland, M.M., LeDuc, D.L., Terry, N. 2002. Genetic engineering of plants to enhance selenium phytoremediation. *Critical Reviews in Plant Science* 21:567–582.

Burken, J.G. 2004. Uptake and metabolism of organic compounds: Green-liver model, pp. 59–84. *Phytoremediation: Transformation and Control of Contaminants*, S.C. McCutcheon, J.L. Schnoor (eds.). Hoboken, NJ: John Wiley & Sons, Inc.

Burken, J.G., Schnoor, J.L. 1997. Uptake and metabolism of atrazine by poplar trees. *Environmental Science and Technology* 31:1399–1406.

Chandra, R., Singh, R. 2012. Decolourisation and detoxification of rayon grade pulp paper mill effluent by mixed bacterial culture isolated from pulp paper mill effluent polluted site. *Biochemical Engineering Journal* 61:49–58.

Chandra, R., Saxena, G., Kumar, V. 2015a. Phytoremediation of environmental pollutants: An eco-sustainable green technology to environmental management, pp. 1–29. *Advances in Biodegradation and Bioremediation of Industrial Waste*, R. Chandra (ed.). Boca Raton, FL: CRC Press.

Chandra, R., Yadav, S., Kumar, V. 2015b. Two step treatment by bacteria and rhizofiltration for bioremediation of complex industrial wastewater: A novel approach for safe disposal, pp. 483–519. *Plant-Microbe Interactions*, K. Ramasamy, K. Kumar (eds.). New Delhi: New India Publishing Agency.

Chandra, R., Kumar, V. 2017. Phytoextraction of heavy metals by potential native plants and their microscopic observation of root growing on stabilised distillery sludge as a prospective tool for in situ phytoremediation of industrial waste. *Environmental Science and Pollution Research* 24(3):2605–2619.

Chandra, R., Yadav, S., Yadav, S. 2017. Phytoextraction potential of heavy metals by native wetland plants growing on chlorolignin containing sludge of pulp and paper industry. *Ecological Engineering* 98:134–145.

Chaney, R.L. 1983. Plant uptake of inorganic waste, pp. 50–76. *Land Treatment of Hazardous Waste*, J.E. Parr, P.B. Marsh, J.M. Kla (eds.). Park Ridge, IL: Noyes Data Corp.

Chaney, R.L., Malik, M., Li, Y.M., Brown, S.L., Brewer, E.P., Angle, J.S., Baker, A.J.M. 1997. Phytoremediation of soil metals. *Current Opinion in Biotechnology* 8:279–284.

Chappell, J. 1998. *Phytoremediation of TCE in Groundwater Using Populus*. National Network of Environmental Management Studies Fellowship Program. Washington, DC: U.S. EPA Technology Innovation Office.

Chen, H., Cutright, T. 2001. EDTA and HEDTA effects on Cd, Cr, and Ni uptake by *Helianthus annuus*. *Chemosphere* 45:21–28.

Cherian, S., Oliveria, M.M. 2005. Transgenic plants in phytoremediation: Recent advances and new possibilities. *Environmental Science and Technology* 39(24):9377–9390.

Conesa, H.M., Faz, A., Arnaldos, R. 2006. Heavy metal accumulation and tolerance in plants from mine tailings of the semiarid Cartagena-La Union Mining District (Se Spain). *Science of the Total Environment* 366:1–11.

Cui, S., Zhou, Q., Chao, L. 2007. Potential hyperaccumulation of Pb, Zn, Cu and Cd in endurant plants distributed in an old smeltery, northeast China. *Environmental Geology* 51:1043–1048.

Cunningham, S.D., Berti, W.R., Huang, J.W. 1995. Phytoremediation of contaminated soils. *Trends in Biotechnology* 13:393–397.

Cunningham, S.D., Shann, J.R., Crowley, D.E., Anderson, T.A. 1997. Phytoremediation of contaminated water and soil, pp. 2–19. *Phytoremediation of Soil and Water Contaminants*, ACS Symposium Series 664, E.L. Kruger, T.A. Anderson, J.R. Coats (eds.). Washington, DC: American Chemical Society.

Deng, H., Ye, Z.H., Wong, M.H. 2004. Accumulation of lead zinc, copper and cadmium by 12 wetland plant species thriving in metal contaminated site in China. *Environmental Pollution* 132:29–40.

Desta, A.F., Assefa, F., Leta, S., Stomeo, F., Wamalwa, M., Njahira, M., Appolinaire, D. 2014. Microbial community structure and diversity in an integrated system of anaerobic-aerobic reactors and a constructed wetland for the treatment of tannery wastewater in Modjo, Ethiopia. *PLoS One* 9(12):e115576.

Dietz, A.C., Schnoor, J.L. 2001. Advances in phytoremediation. *Environmental Health Perspectives* 109(1):163–168.

Dushenkov, V., Kumar, P.B.A.N., Motto, H., Raskin, I. 1995. Rhizofiltration: The use of plants to remove heavy metals from aqueous streams. *Environmental Science and Technology* 29:1239–1245.

Evangelou, M.W.H., Ebel, M., Schaeffer, A. 2007. Chelate assisted phytoextraction of heavy metals from soil. Effect, mechanism, toxicity, and fate of chelating agents. *Chemosphere* 68:989–1003.

Fitz, W.J., Wenzel, W.W. 2002. Arsenic transformation in the soil-rhizosphere-plant system, fundamentals and potential application of phytoremediation. *Journal of Biotechology* 99:259–278.

Flores-Tavizon, E., Alarcon-Herrera, M.T., Gonzalez-Elizondo, S., Olguin, E.J. 2003. Arsenic tolerating plants from mine sites and hot springs in the semi-arid region of Chihuahua, Mexico. *Acta Biotechnology* 23:113–119.

Flowers, T.J., Colmer, T.D. 2008. Salinity tolerance in halophytes. *New Phytologist* 179:945–963.

Freeman, J.L., Salt, D.E. 2007. The metal tolerance profile of *Thlaspi goesingense* is mimicked in *Arabdopsis thaliana* heterologously expressing serine acetyl-transferase. *BMC Plant Biology* 7(63):1–10.

Garbiscu, C., Alkorta, I. 2001. Phytoextraction: A cost effective plant based technology for the removal of metals from the environment. *Bioresource Technology* 77:229–236.

Gardea-Torresdey, J.L., de la Rosa, G., Peralta-Videa, J.R. 2004. Use of phytofiltration technologies in the removal of heavy metals: A review. *Pure and Applied Chemistry* 76(4):801–813.

Ghosh, D., Gopal, B. 2010. Effect of hydraulic retention time on the treatment of secondary effluent in a subsurface flow constructed wetland. *Ecological Engineering* 36(8):1044–1051.

Ghosh, M., Singh, S.P. 2005. A review on phytoremediation of heavy metals and utilization of its byproducts. *Applied Ecology and Environmental Research* 3(1), 1–18.

Gonzalez, R.C., Gonzalez-Chavez, M.C.A. 2006. Metal accumulation in wild plants surrounding mining wastes: Soil and sediment remediation (SSR). *Environmental Pollution* 144:84–92.

Guan, W., Yin, M., He, T., Xie, S. 2015. Influence of substrate type on microbial community structure in vertical-flow constructed wetland treating polluted river water. *Environmental Science and Pollution Research International* 22:16202–16209.

Gupta, A.K., Sinha, S. 2007a. Phytoextraction capacity of the *Chenopodium album* L. grown on soil amended with tannery sludge. *Bioresource Technology* 98:442–446.

Gupta, A.K., Sinha, S. 2007b. Phytoextraction capacity of the plants growing on tannery sludge dumping sites. *Bioresource Technology* 98:1788–1794.

Harms, H., Langebartels, C. 1986. Standarized plant cell suspension test system for an ecotoxicologic evaluation of the metabolic fate of zenobiotics. *Plant Science* 45:157–165.

Instituto Tecnologico Agrario de Castilla y Leon. 2000. Sustainable management of arsenic contaminated water and soil in rural areas of Latin America. ARSLAND, project no. 015114.

Jasechko, S., Sharp, Z.D., Gibson, J.J., Birks, S.J., Yi, Y., Fawcett, P.J. 2013. Terrestrial water fluxes 564 dominated by transpiration. *Nature* 496(7445):347–350.

Khandare, R.V., Govindwar, S.P. 2015. Phytoremediation of textile dyes and effluents: Current scenario and future prospects. *Biotechnology Advances* 33:1697–1714.

Knuteson, S.L., Whitwell, T., Klaine, S.J. 2002. Influence of plant age and size on simazine toxicity and uptake. *Journal of Environmental Quality* 31:2096–2103.

Kramer, U., Cotter-Howells, J.D., Charnock, J.M., Baker, A.J.M., Andrew, C., Smith, J. 1996. Free histidine as a metal chelator in plants that accumulate nickel. *Nature* 379:635–638.

Kramer, U. 2005. Phytoremediation: Novel approaches to cleaning up polluted soils. *Current Opinion in Biotechnology* 16:133–141.

Krzeslowska, M., Lenartowska, M., Mellerowicz, E.J., Samardakiewicz, S., Wozny, A. 2009 Pectinous cell wall thickenings formation: A response of moss protonemata cells to lead. *Environmental and Experimental Botany* 65(1):119–131.

Krzesłowska, M., Lenartowska, M., Samardakiewicz, S., Bilski, H., Wozny, A. 2010. Lead deposited in the cell wall of *Funaria hygrometrica* protonemata is not stable: A remobilization can occur. *Environmental Pollution* 158(1):325–338.

Li, H., Sheng, G., Sheng, W., Xu, O. 2002. Uptake of trifluralin and lindane from water by ryegrass. *Chemosphere* 48:335–341.

Li, M.S., Luo, Y.P., Su, Z.Y. 2007. Heavy metal concentration in soil and plant accumulation in a restored management mineland in Guangxi, South China. *Environmental Pollution* 147:168–175.

Limmer, M.A., Burken, J.G. 2016. Phytovolatilization of organic contaminants. *Environmental Science and Technology* 50:6632–6643.

Lin, Q.U.R., Sen, L.D., Qian, D.R., Yao, J.M. 2002. Phytoremediation for heavy metal pollution in water: The blastofiltration of Pb from water. *Journal of Agro-Environment Science* 21(6):499–501.

Lone, M.I., He, Z.L., Stoffella, P.J., Yang, X.E. 2008. Phytoremediation of heavy metal-polluted soils and water: Progresses and perspectives. *Journal of Zhejiang University Science B* 9(3):210–220.

MacFarlane, G.R., Koller, C.E., Blomberg, S.P. 2007. Accumulation and partitioning of heavy metals in mangroves: A synthesis of field-based studies. *Chemosphere* 69(9):1454–1464.

Macnair, M.R., Tilstone, G.H., Smith, S.E. 2000. The genetics of metal tolerance and accumulation in higher plants, pp. 235–250. *Phytoremediation of Contaminated Soil and Water*, Terry, N., Banuelos, G. (eds.). Boca Raton, FL: CRC Press.

Mani, D., Kumar, C. 2014. Biotechnological advances in bioremediation of heavy metals contaminated ecosystems: An overview with special reference to phytoremediation. *International Journal of Environmental Science and Technology* 11(3):843–872.

Mastretta, C., Barac, T., Vangronsveld, J., Newman, L., Taghavi, S., van der Lelie, D. 2006. Endophytic bacteria and their potential application to improve the phytoremediation of contaminated environments. *Biotechnology and Genetic Engineering Reviews* 23:175–207.

Mazumdar, K., Das, S. 2015. Phytoremediation of Pb, Zn, Fe, and Mg with 25 wetland plant species from a paper mill contaminated site in North East India. *Environmental Science and Pollution Research* 22:701–710.

McGrath, S.P., Zhao, F.J. 2003. Phytoextraction of metals and metalloids from contaminated soils. *Current Opinion in Biotechnology* 14:277–282.

Mehes-Smith, M., Nkongolo, K.K., Cholewa, E. 2013. Coping mechanisms of plants to metal contaminated soil. *Environmental Change and Sustainability*, Silvern, S., Young, S. (eds.). InTechOpen. doi:10.5772/55124.

Memon, A.R., Schröder, P. 2009. Implications of metal accumulation mechanisms to phytoremediation. *Environmental Science and Pollution Research* 16:162–175.

Mendez, M.O., Glenn, E.P., Maier, R.M. 2007. Phytostabilization potential of quailbush for mine tailings: Growth, metal accumulation, and microbial community changes. *Journal of Environmental Quality* 36:245–253.

Mesjasz-Przybyłowicz, J. et al. 2004. Uptake of cadmium, lead, nickel and zinc from soil and water solutions by the nickel hyperaccumulator *Berkheya coddii*. *Acta Biologica Cracoviensia Botanica* 46:75–85.

Mishra, S., Mohanty, M., Pradhan, C., Patra, H.K., Das, R., Sahoo, S. 2013. Physico-chemical assessment of paper mill effluent and its heavy metal remediation using aquatic macrophytes— A case study at JK paper mill, Rayagada, India. *Environmental Monitoring and Assessment* 185(5):4347–4359.

Newman, L.A., Reynolds, C.M. 2004. Phytodegradation of organic compounds. *Current Opinion in Biotechnology* 15:225–230.

Newman, L.A., Reynolds, C.M. 2005. Bacteria and phytoremediation: New uses for endophytic bacteria in plants. *Trends in Biotechnology* 23(1):6–8.

Ndimele, P.E. 2008. Evaluation of phyto-remediative properties of water hyacinth (*Eichhornia crassipes* [Mart.] Solms) and biostimulants in restoration of oil-polluted wetland in the Niger Delta. Ph.D. thesis. University of Ibadan, Nigeria.

Ndimele, P.E. 2010. A review on the phytoremediation of petroleum hydrocarbon. *Pakistan Journal of Biological Sciences* 13:715–722.

Neumann, K., Droge-Laser, W., Kohne, S., Broer, I. 1997. Heat treatment results in a loss of transgene-encoded activities in several tobacco lines. *Plant Physiology* 115:939–947.

Nowack, B., Schulin, R., Robinson, B. 2006. Critical assessment of chelant-enhanced metal phytoextraction. *Environmental Science and Technology* 40:5225–523.

Oti Wilberforce, J.O., Nwabue, F.I. 2013. Heavy metals effect due to contamination of vegetables from Enyigba lead mine in Ebonyi State, Nigeria. *Environment and Pollution* doi.org/10.5539/ep.v2n1p19.

Padmavathiamma, P.K., Li, L.Y. 2007. Phytoremediation technology: Hyper-accumulation metals in plants. *Water Air and Soil Pollution* 184:105–126.

Pilon-Smits, E.A.H., Hwang, S., Lytle, M., Zhu, Y., Tai, J.C., Bravo, R.C., Chen, Y., Leustek, T., Terry, N. 1999. Overexpression of ATP sulfurylase in *Brassica juncea* leads to increased selenate uptake, reduction and tolerance. *Plant Physiology* 119:123–132.

Prasad, M.N.V., Freitas, H.M.D.O. 2003. Metal hyperaccumulation in plants: Biodiversity prospecting for phytoremediation technology. *Electronic Journal of Biotechnology* 6(3). doi:10.2225/vol6-issue3-fulltext-6.

Qixing, Z., Zhang, C., Zhineng, Z., Weitao, L. 2011. Ecological remediation of hydrocarbon contaminated soils with weed plant. *Journal of Resources and Ecology* 2(2):97–105.

Raskin, I.I., Smith, R.D., Salt, D.E. 1997. Phytoremediation of metals: Using plants to remove pollutants from the environment. *Current Opinion in Biotechnology* 8(2):221–226.

Reger, D.L., Goode, S.R., Ball, D.W. (eds.). 2009. *Chemistry: Principles and Practice*, 3rd ed. Belmont, CA: Brooks/Cole Publishing.

Rosario, K., Iverson, S.L., Henderson, D.A., Chartrand, S., McKeon, C., Glenn, E.P. et al. 2007. Bacterial community changes during plant establishment at the San Pedro River Mine tailings site. *Journal of Environmental Quality* 36:1249–1259.

Salt, D.E., Prince, R.C., Pickering, I.J. 1995. Mechanisms of cadmium mobility and accumulation in Indian mustard. *Plant Physiology* 109:1427–1433.

Salt, D.E., Smith, R.D., Raskin, I. 1998. Phytoremediation. *Plant and Molecular Biology* 49:643–668.

Santos, J., Al-Azzawi, M., Aronson, J., Flowers, T.J. 2015. eHALOPH, a database of salt-tolerant plants: Helping put halophytes to work. *Plant Cell Physiology* 57(1):e.10.doi:10.1093/pcp/pcv155

Santos-Jallath, J., Castro-Rodrıguez, A., Huezo-Casillas, J. et al. 2012. Arsenic and heavy metals in native plants at tailings impoundments in Queretaro, Mexico. *Physics and Chemistry of the Earth* 37:10–17.

Schmidt, U. 2003. Enhancing phytoextraction: The effect of chemical soil manipulation on mobility, plant accumulation, and leaching of heavy metals. *Journal of Environmental Quality* 32:1939–1954.

Singh, S., Sinha, S. 2005. Accumulation of metals and its effects in *Brassica juncea* (L.) Czern. (cv. Rohini) grown on various amendments of tannery waste. *Ecotoxicology and Environmental Safety* 62(1):118–127.

Sinha, S., Gupta, A.K., Bhatt, K., Pandey, K., Rai, U.N., Singh, K.P. 2006. Distribution of metals in the edible plants grown at Jajmau, Kanpur (India) receiving treated tannery wastewater: Relation with physico-chemical properties of the soil. *Environmental Monitoring and Assessment* 115:1–22.

Solanki, R., Dhankhar, R. 2011. Biochemical changes and adaptive strategies of plants under heavy metal stress. *Biologia* 66:195–204.

Stottmeister, U., Wiessner, A., Kuschk, P., Kappelmeyer, M.K., Bederski, R.A., Müller, H., Moormann, H. 2003. Effects of plants and microorganisms in constructed wetlands for wastewater treatment. *Biotechnology Advances* 22:93-117.

Suresh, B., Ravishankar, G.A. 2004. Phytoremediation: A novel and promising approach for environmental clean-up. *Critical Reviews in Biotechnology* 24(2–3):97–124.

Tagmount, A., Berken, A., Terry, N. 2002. An essential role of S-adenosyl-L-methionine:L-methionine-S-methyltransferase in selenium volatilization by plants. Methylation of selenomethionine to selenium-methyl-L-selenium methionine, the precursor of volatile selenium. *Plant Physiology* 130:847–856.

Tahir, U., Yasmin, A., Khan, U.H. 2016. Phytoremediation: Potential flora for synthetic dyestuff metabolism. *Journal of King Saud University Science* 28:119–130.

Terry, N., Zayed, A.M., De Souza, M.P., Tarun, A.S. 2000. Selenium in higher plants. *Annual Review of Plant Physiology and Plant Molecular Biology* 51:401–432.

Turgut, C., Pepe, M.K., Cutright, T. 2004. The effect of EDTA and citric acid on phytoremediation of Cd, Cr, and Ni from soil using *Helianthus annuus*. *Environmental Pollution* 131(1):147–154.

U.S. EPA. 2000a. *Introduction to Phytoremediation*. EPA report 600/R-99/107. U.S. EPA Office of Research and Development, Cincinnati, OH.

U.S. EPA. 2000b. *Phytoremediation of Contaminated Soil and Ground Water at Hazardous Waste Sites*. Ground Water Forum Issue Paper EPA 540/S-01/500. U.S. EPA Office of Research and Development, Cincinnati, OH.

Utsunamyia, T. 1980. Japanese patent application 55-72959.

Vangronsveld, J., Herzig, R., Weyens, N., Boulet, J., Adriaensen, K., Ruttens, A., Thewys, T., Vassilev A. et al. 2009. Phytoremediation of contaminated soils and groundwater: Lessons from the field. *Environmental Science Pollution Research International* 16(7):765–794.

Vishnoi, S.R., Srivastava, P.N. 2008. Phytoremediation: Green for environmental clean, pp. 1016–1021. *Proceedings of Taal2007: The 12th World Lake Conference*. Shiga, Japan: International Lake Environment Committee Foundation.

Vymazal, J. 2014. Constructed wetlands for treatment of industrial wastewaters: A review. *Ecological Engineering* 73:724–751.

Wenzel, W.W., Bunkowski, M., Puschenreiter, M., Horak, O. 2003. Rhizosphere characteristics of indigenously growing nickel hyperaccumulator and excluder plants on serpentine soil. *Environmental Pollution* 123:131–138.

White, J.C., Wang, X., Gent, M.P., Iannucci-Berger, W., Eitzer, B.D., Schultes, N.P., Arienzo, M., Mattina, M.I. 2003. Subspecies-level variation in the phytoextraction of weathered *p,p'*-DDE by *Cucurbita pepo*. *Environmental Science and Technology* 37:4368–4373.

Wilberforce, J.O.O., Nwabue, F.I. 2013. Heavy metal effect due to contamination of vegetable from Enyigbda Lead Mine in Ebonyi State, Nigeria. *Environmental Pollution* 2:19–26.

Wittig, R., Ballach, H.J., Kuhn, A. 2003. Exposure of the roots of *Populus nigra* L., cv. Loenen to PAHs and its effect on growth and water balance. *Environmental Science and Pollution Research International* 10:235–244.

Yadav, S.K., Juwarkar, A.A., Kumar, G.P., Thawale, P.R., Singh, S.K., Chakrabarti, T. 2009. Bioaccumulation and phyto-translocation of arsenic, chromium and zinc by Jatropha curcas L.: Impact of dairy sludge and biofertilizer. *Bioresource Technology* 100:4616–4622.

Yadav, S.K., Dhotea, M., Kumarb, P., Sharma, J., Chakrabartia, T., Juwarkar, A.A. 2010. Differential antioxidative enzyme responses of *Jatropha curcas* L. to chromium stress. *Journal of Hazardous Materials* 180:609–615.

Yateem, A., Balba, M.T., El-Nawawy, A.S., Al-Awadhi, N. 2000. Plants-associated microflora and the remediation of oil contaminated soil. *International Journal of Phytoremediation* 2:183–191.

Yoon, J., Cao, X., Zhou, Q., Ma, L.Q. 2006. Accumulation of Pb, Cu, and Zn in native plants growing on a contaminated Florida site. *Science of the Total Environment* 368(2):456–464.

Zaranyika, M.F., Ndapwadza, T. 1995. Uptake of Ni, Zn, Fe, Co, Cr, Pb, Cu and Cd by water hyacinth (*Eichhornia crassipes*) in Mukuvisi and Manyame Rivers, Zimbabwe. *Journal of Environmental Science Health A* 30(1):157–169.

Zayed, A., Gowthaman, S., Terry, N. 1998. Phytoremediation of trace elements by wetland plants. 1.Duck weed. *Journal of Environmental Quality* 27(3):715–721.

Zhang, X., Xia, H., Li, Z., Zhuang, P., Gao, B. 2010. Potential of four forage grasses in remediation of Cd and Zn contaminated soils. *Bioresource Technology* 101:2063–2066.

Zhao, F.J., Lombi, E., MacGrath, S.P. 2003. Assessing the potential for zinc and cadmium phytoremediation with the hyperaccumulator *Thlaspi caerulescens*. *Plant and Soil* 249:37–43.

2

Hyperaccumulator versus Nonhyperaccumulator Plants for Environmental Waste Management

Ram Chandra, Vineet Kumar, and Kshitij Singh

CONTENTS

2.1 Introduction

Environmental problems posed by industrial and municipal wastes are well documented. Worldwide, every day, enormous quantities of waste are generated; at present, these wastes are in need of attention. Environmental waste management promotes the proper management and utilization of industrial waste, delivering in-depth, state-of-the-art information on the physico-chemical properties, chemical composition, and environmental risks associated with industrial wastes discharged from various industries, including those in the distillery, pulp and paper, tanning, distilling, textile, petroleum hydrocarbon, and agrochemical sectors. Waste management is generally undertaken to reduce the effects of contamination on human health and the environment and possibly to recover valuable resources from the wastes. Waste management disposes of the products and substances that have been used in a safe and efficient manner. The methods for the management of waste may differ for developed and developing countries. Several traditional methods, such as landfill, incineration and combustion, plasma gasification, pyrolysis, fluidized bed combustion, and composting, are used for the management of wastes to protect the environment. Each type of technology handles a specific composition and quantity of solid waste. It seems to be difficult to propose suitable waste management technologies without determining the quantity and composition of the generated waste. Moreover, these methods are energy-intensive and financially expensive and cause secondary air or groundwater pollution and can destroy waste-degrading microbial communities. Thus, the feasibilities of these treatment methods are defeated. There is an urgent need to reduce the current levels of waste generation and increase material and energy recovery, which are considered essential steps toward an environmentally friendly waste management system.

Phytoremediation is widely viewed as an ecologically responsible alternative to the environmentally destructive physical remediation methods currently in practice, given that phytoremediation is based on the use of green plants to extract, sequester, and detoxify pollutants (Chandra et al. 2015). Phytoextraction is the most commonly used technique of phytoremediation; it involves the utilization of hyperaccumulator and nonhyperaccumulator plants for accumulation, absorption, and degradation of pollutants from the soil, sediment, or water (Chandra and Kumar 2016, Chandra et al. 2017). Hyperaccumulator plants utilize different metabolic processes for the mobilization and uptake of metal ions from soils, based on the efficiency of metal translocation to the plant shoots via the symplast and apoplast (xylem), sequestration of metals within cells and tissues, and transformation of accumulated metals into metabolically less harmful forms (Kumar et al. 1995). At the end of the growth period, the plant biomass is harvested, dried, or incinerated, and the contaminant-enriched material is deposited in a special dump or processed in a smelter. The energy gained from burning the biomass can support the profitability of the technology if the resultant fumes can be cleaned appropriately. Hyperaccumulators make up a subset of the category known as metallophytes, i.e., plants that grow on metal-enriched soils. These species, however, produce little biomass and are slow-growing plants, which makes it unfeasible to use these species in phytoremediation. Therefore, high-biomass crops and/or trees are considered alternatives for hyperaccumulator plants to remediate heavy metal-contaminated sites. This chapter will focus on the role of hyperaccumulator and nonhyperaccumulator plants for removal of heavy metals from contaminated environments.

2.2 Hyperaccumulator versus Nonhyperaccumulator Plants

2.2.1 Hyperaccumulator Plants and Their Characteristics

The fundamental question in phytoremediation is what plant species should be used for phytoremediation? The answer appears to be those species that are able to accumulate

higher amounts of heavy metals in the aboveground parts; these plants have been designated hyperaccumulators. The term hyperaccumulator was coined by Brooks et al. (1977)
to describe plants that contain greater than 0.1% nickel (Ni) in their dried leaves. Since
then, threshold values have also been established for other metals, such as zinc (Zn), lead
(Pb), cadmium (Cd), copper (Cu), chromium (Cr), iron (Fe), manganese (Mn), etc. (Brooks
1998). In general, the term hyperaccumulator defines plants that accumulate >1,000 mg/kg
of Cu, Co, Cr, Ni, or Pb, >10,000 mg/kg of Mn or Zn, or >100 mg/kg of Cd in their aerial
parts (Baker and Brooks 1989). Ni hyperaccumulates in the greatest number of taxa (more
than 75%), while a low number of hyperaccumulators (only five species to date) has been
found for Cd (Rascio and Navari-Izzo 2011). Preferably, the plants for phytoextraction of
heavy metals from any polluted site should have the following characteristics: (i) tolerates high levels of the metal, (ii) accumulates reasonably high levels of the metal in their
aboveground tissues, (iii) rapid growth rate, (iv) produces reasonably high biomass in the
field, and (v) has a profuse root system. Different plant species exhibit different degrees
of metal translocation from root to shoot. Thus far, the majority of the plant species classified as hyperaccumulators fulfill the criteria described above. Moreover, the potential
for phytoextraction may be estimated based on both a bioconcentration factor (BCF) and a
translocation factor (TF). The BCF is defined as the ratio of the metal concentration in the
root of the plant to that in the soil or sludge (Yoon et al. 2006), while the TF is defined as the
metal concentration in the shoots versus that in the roots (Yoon et al. 2006). The BCF and
TF of a plant are calculated by the following equations:

$$BCF = \text{(metal concentration in plant root)}$$
$$/\text{(metal concentration in soil or sludge)}$$

$$TF = \text{(metal concentration in plant shoot)}$$
$$/\text{(metal concentration in plant root)}$$

A plant species with BCF and TF values both greater than 1 is suitable for phytoextraction (Yoon et al. 2006). Several authors have reported that a BCF of >1 or TF of >1 indicates a special ability of a plant to extract and transport metals from the substrate to the
plant parts. Such plant species can be considered hyperaccumulators and may be applied
for phytoextraction of metals (Yoon et al. 2006). Hyperaccumulation of trace elements is
found throughout the whole plant kingdom in temperate as well as tropical climates, but
it is typically restricted to endemic species growing on mineralized soils and related rock
types. The most important types of metalliferous soils hosting hyperaccumulator plants
are (i) serpentine soils (enriched in Ni, Cr, Co), (ii) "calamine" soils (enriched in Zn, Cd, Pb),
(iii) Se-rich soils, and (iv) Cu- and Co-containing soils (Reeves and Baker 2000). In terms of
soil habitats, the hyperaccumulators can be classified into accumulators of (i) Ni, (ii) Zn, Cd,
and Pb, (iii) Se, (iv) Co and Cu, and (v) other elements (e.g., Al, As, Cr, Mn, Ti) (Reeves and
Baker 2000). Over 450 plant species, including trees, vegetable crops, grasses, and weeds,
comprising 101 families (including *Asteraceae, Brassiciaceae, Caryophyyllaceae, Cyperaceae,
Cunnoniaceae, Fabiaceae, Flacourtiaceae, Lamiaceae, Poaceae, Violaceae,* and *Euphobiacea*), have
been identified as hyperaccumulators; these plants can accumulate high concentrations
of two or more metals into their aboveground biomass (Baker and Brooks 1989, Reeves

and Baker 2000, Verbruggen et al. 2009). The *Brassicaceae* family is relatively rich in them, and about 25% of discovered hyperaccumulator plants belong to the family *Brassicaceae*, in particular the genera *Alyssum* and *Thlaspi* (Rascio and Navari-Izzo 2011). *Alyssum* spp. can accumulate over 1% Ni in the shoots, while *Thlaspi* spp. can accumulate more than 3% Zn, 0.5% Pb, and 0.1% Cd in the shoots (Brooks 1998). Reeves and Baker (2000) reported some examples of plants which have the ability to accumulate large amounts of heavy metals and hence can be used in remediation studies. Some of these plants included *Haumaniastrum robertii* (Co hyperaccumulator), *Aeollanthus subacaulis* (Cu hyperaccumulator), *Maytenus bureaviana* (Mn hyperaccumulator), *Minuartia verna* and *Agrostis tenuis* (Pb hyperaccumulators), *Dichapetalum gelonioides*, *Thlaspi tatrense*, *Psycotria vanhermanni*, and *Streptanthus polygaloides* (Ni hyperaccumulators), and *Lecythis ollaria* (Se hyperaccumulator). *Pteris vittata* is an example of a hyperaccumulator that can be used for the remediation of soil polluted with As (Chen et al. 2002).

Some plants can accumulate only a specific metal, while others can accumulate multiple metals. A summary of some of the recent reports on accumulation of heavy metals is provided in Table 2.1. Metal hyperaccumulation is an eco-physiological adaptation to metalliferous soils. Ecological studies of metal hyperaccumulation have been designed to provide insights into how and why plants evolved with metal hyperaccumulation capabilities; these studies have determined the adaptive value of hyperaccumulation and examined how the extraordinary metal concentrations in hyperaccumulators impact the plant species relationships in the habitats in which they have evolved. Five different explanations have been proposed for why hyperaccumulators may have evolved (Boyd 2007). The hypotheses put forward include the following: (i) plants may hyperaccumulate trace elements because storing large quantities of metals may be a means of metal tolerance and disposal, (ii) hyperaccumulating plants may use metals for elemental allelopathy against nearby competitors, (iii) hyperaccumulation of metals may serve as a source of osmotic resistance to drought, (iv) hyperaccumulation may be a mechanism of defense against herbivores and pathogens, and (v) metal accumulation may be accidental. The hyperaccumulator plants show no symptoms of phytotoxicity. Although it is a distinct feature, hyperaccumulation also relies on hypertolerance, an essential property that allows plants to avoid heavy metal poisoning, to which hyperaccumulator plants are as sensitive as nonhyperaccumulators.

2.2.1.1 Native Plants as Natural Hyperaccumulators

The presence of plant communities at a particular site depends on the potential for seedlings to survive and reproduce. Establishment of plant species in a contaminated soil indicates that they are adapted to the contaminant and, as such, can remove or retain it, reducing its toxicity in the soil. However, heavy metal bioavailabilities beyond the permissible limits pose a crucial problem in agriculture and environmental studies. Native plants involved in the uptake of elements can provide the most reliable method for the prediction of availability of metals to plants and the application of the plants for phytoextraction of heavy metals from contaminated sites (Chandra and Kumar 2016). It is important to use the native plants of contaminated sites for phytoremediation, because these plants are naturally adapted in terms of survival, growth, and reproduction under the environmental stresses, compared to plants introduced from another environment. There has been a continuing interest in identifying native plants that are tolerant of heavy metals. However, only a few studies have evaluated the phytoremediation potentials of native plants under *in situ* conditions (Shu et al. 2002, McGrath and Zhao 2003). Identification of native plant

TABLE 2.1

Hyperaccumulator Plants and the Metals They Accumulate

Plant	Family	Common Name	Metal(s) That Accumulates	Reference
Alternanthera philoxeroides	Amaranthaceae	Aligator weed	Cr, Zn, Mn, Ni, Cu, Fe	Chandra et al. 2017
Alyssum sp.	Brassicaceae		Ni	Broadhurst et al. 2004
Alyssum bertolonii	Brassicaceae		Ni	Li et al. 2003
Alyssum caricum	Brassicaceae		Ni	Li et al. 2003
Alyssum lesbiacum	Brassicaceae		Ni	Ingle et al. 2005
Alyssum murale	Brassicaceae	Yellow tuft	Ni	Abou-Shanab et al. 2006
Alyssum serpyllifolium	Brassicaceae		Ni	Ma et al. 2011
Amaranthus viridis L.	Amaranthaceae	Slender amaranth	Pb	Malik et al. 2010
Arabidopsis halleri	Brassicaceae		Zn and Cd	Sarret et al. 2002
Arabidopsis halleri	Brassicaceae		Cd and Zn	Kubota and Takenaka 2003
Austromyrtus bidwillii	Myrtaceae		Mn	Bidwell et al. 2002
Azolla pinnata	Azollaceae	Water velvet	Cd	Rai 2008
Cannabis sativa	Cannabaceae		Cd	Girdhar et al. 2014
Chenopodium album L.	Amaranthaceae	Lamb's quarters	Zn	Malik et al. 2010
Commelina benghalensis	Commelinaceae	Spinder wort	Cr, Zn,Mn, Ni, Cu, Fe, Cd, Pb	Chandra et al. 2017
Corrigiola telephifolia	Caryophyllaceae		As	Garcia-Salgado et al. 2012
Croton bonplandianum	Euphorbiaceae	Ban tulsi	Cr, Zn, Mn, Cu, Fe, Cd, Pb	Chandra et al. 2016
Drosera rotundifolia	Droseraceae	Common sundew	Pb	Fontem Lum et al. 2015
Eleusine indica	Poaceae	Goose grass	Pb	Fontem Lum et al. 2015
Euphorbia cheiradenia	Euphorbiaceae		Pb	Chehregani and Malayeri 2007
Haumaniastrum robertii	Lamiaceae		Co	Brooks 1998
Ipomea alpine	Convolvulaceae	Sweet Potato	Cu	Baker and Walker 1989
Noccaea caerulescens	Brassicaceae	Alpine pennycress	Cd, Zn	Karimzadeh et al. 2012
Parthenium hysterophorus L.	Asteraceae	Congress grass	Cu	Malik et al. 2010
Phragmites cummunis	Poaceae	Common reed	Cr, Zn, Mn, Ni, Cu, Fe, Cd, Pb	Chandra et al. 2017
Pistia stratiotes	Araceae	Water cabbage	Ag, Cd, Cr, Cu, Hg, Ni, Pb, Zn	Odjegba and Fasidi 2004
Pteris cretica	Pteridaceae	Ribbon fern	As	Srivastava et al. 2006
Pteris vittata	Pteridaceae	Ladder brake	As	Ma et al. 2001
Pteris vittata	Pteridaceae	Chinese brake fern	As	Lampis et al. 2015
Ranunculus sceleratus	Ranunculaceae	Cursed buttercup	Zn, Mn, Cu, Fe, Cd, Pb	Chandra et al. 2017

(Continued)

TABLE 2.1 (CONTINUED)

Hyperaccumulator Plants and the Metals They Accumulate

Plant	Family	Common Name	Metal(s) That Accumulates	Reference
Rorippa globosa	Brassicaceae		Cd	Girdhar et al. 2014
Rumex dentatus	Polygonaceae	Toothed dock	Cr, Zn, Mn, Ni, Cu, Fe, Pb	Chandra et al. 2017
Sacchrum munja	Poaceae	Munja	Zn, Mn, Ni, Cu, Fe, Pb	Chandra et al. 2016
Salix viminalis	Salicaceae	Willow	Hg	Wang et al. 2005
Salsola kali			Cr	Gardea-Torresdey et al. 2005
Sedum alfredii	Crassulaceae		Cd, Zn	Zhang et al. 2012
Sedum plumbizincicola	Crassulaceae		Cd, Pb, Zn	Liu et al. 2015
Solanum nigrum	Solanaceae	Black nightshade	Cd	Girdhar et al. 2014
Thlaspi caerulescens	Brassicaceae	Alpine pennygrass	Cd, Ni, Pb, Zn	Baker and Walker 1990
Thlaspi goesingense	Brassicaceae		Ni, Zn	Prasad and Freitas 2003
Thlaspi rotondifolium	Brassicaceae	Pennygrass	Pb	Reeves and Brooks 1983
Turnera subnuda	Turneraceae		Ni	Reeves and Baker 2000
Ricinus communis	Euphorbiaceae	Castor oil plant	Cu, Ni, Zn	Giordani et al. 2005
Ricinus communis L.	Euphorbiaceae	Castor bean	Ni	Adhikari and Kumar 2012

species used for phytoremediation has revealed the potentials of *Brachiaria decumbens* (Australian native grass) (Gaskin et al. 2008) and *Vetiveria zizanioides* L. (Venezuelan native grass) (Brandt et al. 2006). Yoon et al. (2006) evaluated the phytoextraction potential for removal of Pb, Cu, and Zn of 36 by native plants growing on a contaminated site. Among the plants, none of the plant species was identified as a metal hyperaccumulator. *Phyla nodiflora* was the most efficient in accumulating Cu and Zn in its shoots (TFs of 12 and 6.3, respectively), while *Gentiana pennelliana* was the most effective in taking up Pb, Cu, and Zn (BCFs of 11, 22, and 2.6) from a contaminated site and was considered the most promising species for phytoextraction of heavy metal-contaminated sites.

Nouri et al. (2011) collected 12 native plant species grown on a Pb- and Zn-contaminated mine area and analyzed the plants for accumulation of Pb, Zn, Mn, and Fe. They found that among the analyzed plants, none of them was suitable for phytoextraction of Pb, Zn, Mn, or Fe or for phytostabilization of Fe, but *Scrophularia scoparia* was the most suitable plant for phytostabilization of Pb (BCF, 1.43; TF, 0.09); *Centaurea virgata*, *Echinophora platyloba*, and *Scariola orientalis* showed potentials for phytostabilization of Zn (BCFs of 1.73, 1.07, and 1.67 and TFs of 0.24, 0.89, and 0.40, respectively), and *Centaurea virgata* and *Cirsium congestum* were the most efficient in phytostabilization of Mn (BCFs of 1.27 and 1.07; TFs of 0.13 and 0.85).

Lorestani et al. (2011) evaluated the phytoextraction and phytostabilization potentials of plants growing on heavy metal-contaminated soil of a copper mine. They found that none of the collected plant species was suitable for phytoextraction of Cu, Zn, Fe, or Mn, but among the plants, *Euphorbia macroclada* was the most efficient in phytostabilization of Cu and Fe, while *Ziziphora clinopodioides*, *Cousinia* sp., and *Chenopodium botrys* were the most suitable for phytostabilization of Zn, and *Chondrila juncea* and *Stipa barbata* had potentials for phytostabilization of Mn. However, *Euphorbia macroclada* and *Verbascum speciosum* were found to be hyperaccumulators for Fe.

Galfati et al. (2011) evaluated the phytoremediation potentials of 30 native plants growing on industrial waste-contaminated sites from six localities. The results showed that none of the plants was suitable for phytoextraction, because no hyperaccumulator was identified. However, the plants that were most effective in accumulation of metals in leaves were *Malva aegyptiaca* for Cd (TF = 30.7), *Frankenia thymifolia* for Zn (TF = 8.55), *Peganum harmala* for Cu (TF = 29.14), and *Citrulus* sp. for Sr (TF = 10.42). *Anthemis stiparum* was most suitable for phytostabilization of sites contaminated with Cd (BCF = 23.51). The study showed that native plant species growing on contaminated sites may have potentials for phytoremediation.

Recently, Chandra and Kumar (2017) collected 15 native plants growing on stabilized distillery sludge and analyzed them for their heavy metal accumulation potentials in their different parts (Figure 2.1). They found that *Blumea lacera*, *Parthenium hysterophorous*, *Setaria*

FIGURE 2.1
Native hyperaccumulator plants growing in a heavy metal-rich industrial polluted site: (a) *Ricinus communis;* (b) *Chenopodium album;* (c) *Parthenium hysterophorous;* (d) *Rumex dentatus;* (e) *Blumea lacera;* (f) *Amaranthus spinosus* L.; (g) *Sacchrum munja;* (h) *Cannabis sativa;* (i) *Solanum nigrum.*

viridis, Chenopodium album, Cannabis sativa, Basella alba, Tricosanthes dioica, Amaranthus spinosus L., *Achyranthes* sp., *Dhatura stramonium, Sacchrum munja*, and *Croton bonplandianum* were root accumulators for Fe, Zn, and Mn, while *S. munja, P. hysterophorous, C. sativa, C. album, T. dioica, D. stramonium, B. lacera, B. alba, Kalanchoe pinnata*, and *Achyranthes* sp. were shoot accumulators of Fe. In addition, *A. spinosus* L. was found to be a shoot accumulator of Zn and Mn. Similarly, all plants were found to be leaf accumulators of Fe, Zn, and Mn, except for *A. spinosus* L. and *Ricinus communis*. This study recommended that these native plants be used as tools for *in situ* phytoremediation and eco-restoration of industrial waste-contaminated sites. Moreover, some potential hyperaccumulator native plants used as foods or medicines raise health concerns, as such plants may pose health hazards through direct food chain contamination.

2.2.1.2 Weed Plants as Natural Hyperaccumulators

A weed is any plant that requires some form of action to reduce its effect on the economy, the environment, human health, and amenities. Weeds are also known as invasive plants. They are often excellent at surviving and reproducing in disturbed environments and are commonly the first species to colonize and dominate under these conditions. A weed plant can be an exotic species or a native species that colonizes and persists in an ecosystem in which it did not previously exist. Native plants can also become weeds when characteristics within their natural habitat change and enable them to better compete with other species and increase their population size and/or density. In comparison with other native plants, weed species show better tolerance to various adverse environments. They can endure various stress conditions, including heavy metal stress. Therefore, it is a better option to identify the species of weed plants at a site for phytoremediation purposes. The weed species are usually quick growers and have a higher biomass under unfavorable environmental conditions. In most contaminated sites, hardy, tolerant weed species exist, and phytoremediation through these and other nonedible species can restrict the contaminant(s) from being introduced into the food chain. The hyperaccumulating nature of various wild weed species, e.g., *Parthenium hysterophorus* L., *Cannabis sativa, Solanum nigrum, Rorippa globosa, Euphorbia hirta, Amaranthus hybridus*, and *Xanthium strumarium*, could be utilized for heavy metal accumulation from industrial waste-contaminated sites. For instance, Fontem Lum et al. (2015) evaluated the phytoremediation potential of 12 weed species growing on heavy metal-contaminated soil. All plant species, including *Cleome rutidosperma* DC, *Eleusine indica* Gaertn., *Commelina benghalensis* L., *Synedrella nodifolia* Gaertn., *Kyllinga erecta* Schumach., *Asystasia gangetica* (Linn.) T. Anders, *Dissotis rotundifolia* (Sm.) Triana., *Axonopus compressus* (Sw.) P. Beauv., *Paspalum orbiculare* Forst., *Panicum maximum* Jacq., *Cyperus rotundus* Linn., and *Eragrostis tenella* (Linn.) P. Beauv. Ex Roem & Schult, were able to take up and translocate at least one heavy metal from the roots to the shoots. Based on high TF values, *D. rotundifolia* and *K. erecta* were suggested for phytoextraction of Pb, and *P. orbiculare* was suggested for Fe phytoextraction. *Dissotis rotundifolia* was the most efficient species in translocating Al, Pb, Co, Ni, and Sc from plot C in this study. *Eleusine indica* was the most efficient in translocation of Cu, Y, Mn, and Cr in plot C, while *C. benghalensis* was the most efficient for Cu, Ni, Mn, Fe, Y, Sc, and V in translocation in plot A. None of the weeds was identified as a hyperaccumulator, but based on the biological accumulation coefficient (BACs) for *E. indica* and *D. rotundifolia*, these species have potential for hyperaccumulation of Pb.

Recently, Singh et al. (2016) assessed oxidative biomarker responses, antioxidant potentials, and metal accumulation tendencies of weed plants (*Xanthium strumarium, Spilaanthes paniculata, Elipsta alba, Ageratum conyzoides, Peppromia pellucida, Solanum nigrum, Euphorbia hirta,* and *Amaranthus hybridus*) collected from control and metal-contaminated sites. The metal contamination was found to be higher in the soil and plant parts collected from the contaminated site. Based upon metal accumulation tendencies, the weed plants showed a TF of >1, reflecting their potential for use in metal remediation. The metal pollution index values also showed higher tendencies for metal accumulation in the weed plants, particularly in *Solanum nigrum, Euphorbia hirta, Amaranthus hybridus,* and *Xanthium strumarium,* and they also flourished at contaminated sites, with higher production of antioxidants. However, Cd and Ni were higher in *A. hybridus* and *S. nigrum* than in the other species, whereas concentrations of Cu, Pb, and Zn were found to be higher in *E. hirta* and *X. strumarium.* These weed plants can play an important role by reducing metal levels in contaminated areas, as they can grow very well at a contaminated site with their higher antioxidant potentials and metal accumulation tendencies. Along with this, *P. pellucida* can be used for medical purposes, as it has a lower tendency of metal accumulation as well as higher activities of all the antioxidants. Moreover, weeds have a controversial nature. To the agriculturist, they are plants that need to be controlled, in an economical and practical way, in order to produce food, feed, and fiber for humans and animals. Therefore, the negative impacts of weeds indirectly affect all living beings.

2.2.1.3 Wetland Plants

A wetland is a land area that is saturated with water, either permanently or seasonally, such that it takes on the characteristics of a distinct ecosystem. Wetlands are the most important ecosystems on Earth. However, wetlands, acting as natural filters, have the inherent capacity to trap and/or efficiently modify a broad spectrum of contaminants (Mander and Mitsch 2009). Wetland habitats, with their high water levels and increased salt concentrations, are too harsh for many plants. The plants growing in wetlands, often called wetland plants, macrophytes, or hydrophytic plants, are adapted to growing in highly polluted environments. These include aquatic vascular plants and aquatic mosses. One such component of wetlands is the presence of metal-accumulating plant species that can accumulate wide varieties of metals, thus serving as cost-effective alternatives for ameliorating pollution loads (Rai 2008). Such plants can extract and/or stabilize metals in a contaminated site, preventing further spread and transfer into food chains. Moreover, wetland plants are an integral part of constructed wetland (CW) treatment systems. CW systems are a low-cost natural alternative to technical methods of wastewater treatment. They are emerging as a very useful technology for the treatment of a variety of wastewater types. As a CW is a bioremediation system, its metal removal efficiency depends on phytoaccumulation, sedimentation, microbial accumulation, and the nature of the plant species. Wetland plants play important roles in a CW based on their capabilities of degrading and removing nutrients and other organic and inorganic pollutants from contaminated sites. There have been reports of the ability of wetland plants to remove or stabilize pollutants from wastewater and soils in natural and CW systems. An important part of treatment in a CW is attributable to the presence and activities of plants and microorganisms. This is not a new concept, since CWs, reed beds, and floating plant systems have been commonly used for treatment of different wastes for many years. Deng et al. (2004) investigated the concentrations of Pb, Zn, Cu, and Cd that were accumulated by 12 emergent rooted wetland plant species under field conditions from heavily metal-polluted wetland sites in China. They found that these species were not affected by excessive metal contents and possessed various metal resistance or tolerance capabilities. In another study, 10 common regional wetland plant species from a wetland site in Kolkata, India, were analyzed for suitability of remediation of contaminated soil and water with elements like Ca, Cr, Cu, Pb, Zn, Mn, and Fe (Chatterjee et al. 2011). There is evidence that wetland plants, such as *Typha latifolia*, *Phragmites australis*, and *Cyperus esculentus*, grown under such conditions can accumulate heavy metals in their tissues (Ye et al. 2001, Deng et al. 2004, Chandra and Yadav 2011, Yadav and Chandra 2011). Figure 2.2 shows the growth of wetland plants at a contaminated site. Mazumdar and Das (2015) reported phytoremediation of Pb, Zn, Fe, and Mg by 25 wetland plant species, including hyacinth (*Eichhornia crassipes*), water spinach (*Ipomoea aquatica*), water meal (*Wolffia arrhiza*), water chestnut (*Trapa bispinosa*), water lettuce (*Pistia stratiotes*), common arum (*Colocasia esculenta*), common sedge (*Cyperus rotundus*), bulrush (*Scirpus* sp.), arrowhead weed (*Sagittaria montevidensis*), and Bermuda grass (*Cynodon dactylon*), from a pulp and paper mill-contaminated site, and these species were identified as accumulators for their efficiencies in metal accumulation and metal remediation.

Recently, Chandra et al. (2017) evaluated the phytoextraction potential for removal of heavy metals by native wetland plants growing on chlorolignin-containing sludge from pulp and paper mills (Figure 2.3). They observed that *Triticum aestivum* was a root accumulator for all metals except Cu. Similarly, *Brassica campestris* accumulated the maximum amount of all metals in its roots, except for Ni, Fe, and Pb. *Eclipta alba* accumulated the majority of the metals in its shoots and leaves, and *Solanum nigrum* and *Rumex dentatus* accumulated metals in both their roots and shoots. *Rananculus scleratus* accumulated all the tested metals in its aerial parts, except for Cd and Pb. *Cammelina benghalensis* accumulated metals in its roots, except for Cd and Pb. Similarly, *Phragmites cummunis* and *Ricinus cummunis* accumulated the majority of metals in their shoots and leaves. It was also noted that several evaluated plants showed BCF values greater than 1 for accumulated metals, e.g., Pb in *T. aestivum* (BCF, 1.285), *C. benghalensis* (5.142), *Cannabis sativa* (3.466), *P. cummunis* (7.180), and *R. cummunis* (66.80). Thus, all the growing plants were found to have good heavy metal phytoextraction efficiencies in organic pollutant-rich environments. Thus, these plants could be used for *in situ* monitoring and phytoremediation of heavy metals from chlorolignin-containing organic pollutant-rich contaminated sites. The ability of these growing native plants to tolerate and accumulate heavy metals could be exploited for phytoextraction and bioremediation of metal mixed

(a)   (b)   (c)

FIGURE 2.2
Some common wetland plants growing on industrial waste-contaminated sites: (a) *Phragmites australis*; (b) *Typha angustifolia*; (c) *Cyperus esculentus*.

FIGURE 2.3
Heavy metal hyperaccumulator wetland plants showing *in situ* phytoremediation of a pulp paper mill waste-contaminated site: (a) *Commelina benghalensis*; (b) *Ricinus communis*; (c) *Parthenium hysterophorus*; (d) *Alternanthera philoxeroides*; (e) *Brassica campestris*; (f) *Triticum aestivum*.

cholorolignin-contaminated sites. These plants should be prohibited for use as food or fodder due to potential health hazards.

2.2.2 Nonaccumulator Plants

Metal hyperaccumulator plants are potentially useful in soil cleanup, as they can take up significant amounts of metals from contaminated soils, sediments, and water, but their slow growth and low annual biomass production tends to limit their phytoextraction abilities. A possible alternative is use of nonaccumulator plants with either high biomass plants (crops) or fast-growing trees that can be easily cultivated using established practices (Ghosh and Singh 2005, Solhi et al. 2005).

2.2.2.1 Use of Crop Plants in Phytoextraction

The ideal plant species to remediate heavy metal-contaminated soil would be a high biomass producing crop that can both tolerate and accumulate the high metal content present at the site. There are several crops that have good phytoremediation potentials; fast-growing leafy vegetables show high metal uptake and greater accumulation rates (Saglam

2013). Some researchers have identified certain high-biomass crop plants, such as maize (*Zea mays*), tobacco (*Nicotiana tabacum*), Indian mustard (*Brassica juncea*), oat (*Avena sativa*), barley (*Hordeum vulgare*), pea (*Pisum sativa*), oats (*Avena sativa* L.), cabbage (*Brassica rapa* L. subsp. *chinensis*), and sunflower (*Helianthus annuus*), for their ability to take up, translocate, and accumulate heavy metals from contaminated sites (Cui et al. 2004, Turgut et al. 2004, Szabó and Fodor 2006). Ali and Al-Qahtani (2012) and Chang et al. (2014) also reported higher accumulation of heavy metals in leafy vegetables. Nazemi et al. (2012) determined the levels of Pb, As, Cr, Cd, and Ze in vegetables, including leek (*Allium ampeloprasum*), coriander (*Cooriandrum sativum*), parsley (*Petroselinum crispum*), cress (*Lepidium sativum*), basil (*Ocimum basilicum*), and radish leaf (*Beta vulgaris*). The heavy metal concentrations in all vegetable samples were above the permissible limits. Further, the accumulation of As, Cd, Cr, Pb, and Zn in edible vegetables, such as fluted pumpkin (*Telfaria occidentalis*), water leaf (*Talinum triangulare*), pigweed (*Amaranthus hybridus*), bitter leaf (*Vernonia amygdalina*), and garden egg leaf (*Solanum nigrum*), was observed by Oti Wilberforce and Nwabue (2013). Their study showed that vegetables grown in the region were health hazards with regard to human consumption.

As a plant's response to heavy metals may vary depending on the soil type (based on differences in metal bioavailability in different soil types), any subsequent use of a plant species must be confirmed in an experimental system modeled after real world conditions to test the plant's performance in an actual decontamination situation. However, the cropping of contaminated soil with vegetable crops for phytoremediation purposes is unsuitable, because the high contents of heavy metals in these crops have revealed environmental health hazards through food chain contamination. Several heavy metals aggravate human and animal health due to their long-term persistence and accumulation from contaminated environmental sites; the metals are known to cause neural toxicity, renal disorders, asthma, and carcinogenic effects in humans and animals. Thus, several studies have recommended that crop plants can be used for minimization of heavy metals from contaminated soil but they should be prohibited for use as food and fodder, due to the health hazards of various heavy metals that may be recovered from specific parts of the plants through phytomining (Table 2.2).

2.2.2.2 Use of Trees in Phytoextraction

The potential use of trees as suitable vegetation cover for heavy metal-contaminated land has received increasing attention over the past 10 years. Trees potentially are the lowest-cost, sustainable, and ecologically friendly plant type for use in phytoremediation of heavy metal-contaminated sites, especially when it is not economical to use other treatments or if there is no time pressure on reuse of the land (Watson 2002). A number of tree species can grow on land of marginal quality. Establishment of trees on sites with low fertility and poor soil structure will keep costs low for plant establishment. In addition, trees have the most massive root systems, which penetrate the soil for several meters, farther than most herbaceous plants (Stomp et al. 1993). Studies of tree establishment on contaminated land have considered a number of different genera, e.g., *Salix* (willow), *Betula* (birch), *Populus* (poplar), *Alnus* (alder), and *Acer* (Sycamore), and also species such as *Mimosa caesalpiniaefolia*, *Erythrina speciosa*, and *Schizolobium parahyba* (Brazilian leguminous trees). Among them, *Salix* spp. and *Populus* spp. also have great potential in development of phytoextraction methods. *Salix* trees are fast-growing trees and good candidates

TABLE 2.2

Nonhyperaccumulator Crop Plants and the Metals They Accumulate

Plant Species	Family	Common Name	Metal	Reference(s)
Allium ampeloprasum subsp. Persicum	Amaryllidaceae	Wild leek	Cr, Cu, Zn	Taghipour and Mosaferi 2013
Allium cepa	Amaryllidaceae	Onion	Cr, Cu, Zn	Taghipour and Mosaferi 2013
Avena sativa	Poaceae	Oat	Cd	Tanhuanpää et al. 2007
Brassica carinata A. Braun			Pb, Zn, Cd	Hernandez-Allica et al. 2008
Brassica chinensis L.	Brassicaceae	Bok choy	Cd	Chang et al. 2014
Brassica juncea	Brassicaceae	Indian mustard	Ni, Cd	Salt et al. 1995, Giordani et al. 2005
Brassica napus	Brassicaceae	Canola	Cd, Pb	Selvam and Wong 2008
Brassica oleracea var. acephala L.			Pb, Zn, Cd	Hernandez-Allica et al. 2008
Brassica rapa var. rapa L.			Pb, Zn, Cd	Hernandez-Allica et al. 2008
Canavalia ensiformis L.	Fabaceae	Jack bean	Cd	de Andrade et al. 2005
Cynara cardunculus L.			Pb, Zn, Cd	Hernandez-Allica et al. 2008
Festuca arundinacea L.			Pb, Zn, Cd	Hernandez-Allica et al. 2008
Festuca ovina L.			Pb, Zn, Cd	Hernandez-Allica et al. 2008
Festuca rubra L.			Pb, Zn, Cd	Hernandez-Allica et al. 2008
Glycine max	Fabaceae	Soybean	Cu	Khan and Lee 2013
Helianthus annuus	Asteraceae	Sunflower	Cd,	de Andrade et al. 2008
Hordeum vulgare	Poaceae	Barely	Ni,	Giordani et al. 2005
Hordeum vulgare	Poaceae	Barley	Cd, Pb	Belimov et al. 2004
Lactuca sativa	Asteraceae	Iceberg lettuce	Pb, Cd, As, Zn	Cobb et al. 2000
Lactuca sativa L. var. romana Hort		Lettuce	Cd	Chang et al. 2014
Lolium multiflorum L.			Pb, Zn,	Hernandez-Allica et al. 2008
Luffa cylindrica	Cucurbitaceae	Sponge gourd	Ni	Rajkumar et al. 2013
Lycopersicon esculentum	Solanaceae	Tomato	Cr, Cu, Zn	Taghipour and Mosaferi 2013
Melastoma malabathricum			Ni	Syam et al. 2016
Nicotiana tabacum	Solanaceae	Tobacco	Cd	Mench et al. 1989
Phaseolus sp.	Fabaceae	Bean	Pb, Cd, As, Zn	Cobb et al. 2000
Phaseolus vulgaris	Fabaceae	Green bean	Cd, Cu, Ni, Pb Zn	Meers et al. 2007
Pisum sativum	Fabaceae	Pea		Sharma et al. 2010

(Continued)

TABLE 2.2 (CONTINUED)

Nonhyperaccumulator Crop Plants and the Metals They Accumulate

Plant Species	Family	Common Name	Metal	Reference(s)
Poa pratense L.			Pb, Zn, Cd	Hernandez-Allica et al. 2008
Raphanus sativus	Brassicaceae	Radish	Pb, Cd, As, Zn	Cobb et al. 2000
Rumex induratus	Polygonaceae		Hg	Luis et al. 2007
Sarcotheca celebica			Ni	Syam et al. 2016
Sinapis alba	Brassicaceae	White mustard	Cd, Cu, Zn	Płociniczak et al. 2013
Sinapis alba	Brassicaceae	White mustard	Cd, Cu, Zn	Płociniczak et al. 2013
Solanum lycopersicum	Solanaceae	Tomato	Ni	Giordani et al. 2005
Sorghum bicolor (L.) Moench			Pb, Zn, Cd	Hernandez-Allica et al. 2008
Spinacea oleracea	Amaranthaceae,	Spinach	Ni	Giordani et al. 2005
Stizolobium aterrimum L.	Fabaceae	Velvet bean		de Souza et al. 2011
Tritium aestivum			Pb, Zn, Cd	Hernandez-Allica et al. 2008
Trichosanthes dioica	Cucurbitaceae	Parval	Cr, Cu, Pb, Cd	Chandra et al. 2016
Vicia faba L.			Pb, Cd	Hernandez-Allica et al. 2008
Vicia sativa L.			Pb, Cd	Hernandez-Allica et al. 2008
Vigna radiata	Fabaceae	Mung bean	Cd, Ni, Zn	Rani et al. 2013
Zea mays	Poaceae	Corn	Cd Pb	Cortez and Ching 2014

for phytoremediation applications due to their extensive root systems and high rates of water uptake and transpiration (which result in efficient transport of compounds from roots to shoots), rapid growth, and large biomass production. A characteristic of *Salix* spp. which makes them very suitable trees for use in phytoremediation is that they can be frequently harvested by coppicing, yielding as much as 10–15 dry tons/ha/year (Riddell-Black 1993). However, *Populus* trees can be grown in a wide range of climatic conditions and are being used with increasing frequency in short-rotation forestry systems for pulp and paper production.

The majority of *Salix* species grow in lowland wetland habitats and have evolved a number of varieties and hybrids This raises the possibility of using plantations of transgenic poplars across several multiyear cycles to remove heavy metals from contaminated soils (Liphadzi et al. 2003). In addition, a dense tree cover would also prevent erosion and spread of contaminated soil by wind. Nonhyperaccumulator plant species can be used in phytoremediation, despite these species not accumulating high concentrations of metals; their biomass production overcomes by several orders of magnitude the phytoremediation capacities of typical hyperaccumulator plant species (Table 2.3). Therefore, total metal extraction can be higher in nonaccumulator than hyperaccumulator plant species because of the relationship between metal accumulation in aerial parts and the growth potential of these species (Table 2.4).

TABLE 2.3

Examples of Nonaccumulator Crop Plants and the Metals They Accumulate

Plant Species	Family	Common Name	Metal	Reference(s)
Acacia mangium	Fabaceae	Black wattle	Cd, Pb	Ang et al. 2010
Acacia saligna	Fabaceae	Wreath wattle	Cd, Pb	Amira and Abdul 2015
Acer pseudoplatanus L.	Sapindaceae		Cd, Cu, Pb, Zn	Mertens et al. 2004
Alnus glutinosa L. *Gaertn.*	Betulaceae	Black alder	Cd, Cu, Pb, Zn	Mertens et al. 2004
Alnus incana	Betulaceae	Grey alder	Cu, Zn, Cd	Rosselli et al. 2003
Betula pendula	Betulaceae	Silver birch	Cu, Zn, Cd	Rosselli et al. 2003
Cedrus libani	Pinaceae	Lebanon cedar	Pb	Ozen and Yaman 2016
Elaeocarpus decipiens	Elaeocarpaceae	Japanese Blueberry	Zn, Mn, Ni, Co	Dalun et al. 2009
Eucalyptus rostrata	Myrtaceae	Red gum	Cd, Pb	Amira and Abdul 2015
Fraxinus excelsior	Oleaceae	European ash	Cu, Zn, Cd	Rosselli et al. 2003
Fraxinus excelsior L.	Oleaceae	European ash	Cd, Cu, Pb, Zn	Mertens et al. 2004
Hopea odorata	Dipterocarpaceae		Cd, Pb	Ang et al. 2010
Intsia palembanica	Fabaceae		Cd, Pb	Ang et al. 2010
Koelreuteria paniculata Laxm	Sapindaceae	Panicled goldenrain	Zn, Mn, Ni, Co	Dalun et al. 2009
Nerium oleander L.	Apocynaceae	Oleander	Pb, Cd, Zn Cu	Aksoy and Oztiirk 1997, Kaya et al. 2010
Populus nigra	Salicaceae	Black poplar	Pb	Ozen and Yaman 2016
Populus sp.	Salicaceae	Poplar	Cd, Cr, Zn, Ni	Algreen et al. 2014
Robinia pseudoacacio	Fabaceae	Black locust	Cd, Cu, Pb, Zn	Mertens et al. 2004
Salix viminalis	Salicaceae	Basket willow	Cu, Zn, Cd	Rosselli et al. 2003
Salix sp.	Salicaceae	Willows	Cd, Cr, Zn, Ni	Algreen et al. 2014
Sorbus mougeotii	Rosaceae		Cu, Zn, Cd	Rosselli et al. 2003
Swietenia macrophylla	Meliaceae	Mahogany	Cd, Pb	Ang et al. 2010
Thuja sp.	Cupressaceae	Thujas or cedars	Pb	Ozen and Yaman 2016

TABLE 2.4

Differences between Hyperaccumulator and Nonhyperaccumulator Plants

Hyperaccumulator Plants	Nonhyperaccumulator Plants
• They take up exceedingly large amounts of one or more metals without suffering any phytotoxic effects.	• They are take up small amounts of metals from contaminated site.
• They are slow growing and produce low annual biomass amounts.	• They are high annual biomass plants (crops) or fast-growing trees.
• Such plants should have BCF and TF values of >1.	• Plants with BCF and TF values of <1.
• Hyperaccumulators detoxify and sequestrate large amounts of metals in aboveground plant parts.	• They retain most heavy metals in their root cells when taken up from contaminated soil, detoxifying the metals via chelation in the cytoplasm or storing metals in vacuoles.
• Total metal accumulation can be lower in hyperaccumulator plants due to their low biomass production.	• The total metal accumulation can be higher in nonaccumulators because of the relationship between metal accumulation in the aerial parts and the growth potential of these species.
• There is no need for chelating agents.	• The phytoextraction potential of these plants can be improved by the using natural or synthetic chelators.
• They have osmotic resistance to drought.	• They are less tolerant to drought.
• Hyperaccumulator plants mainly include weeds, wetland plants, and some native plants	• Nonaccumulator plants are mainly crop plants and trees.
• They have low growth potential in polluted environments.	• They have high growth potential in polluted environments.
• These plants have very efficient xylem loading of metal ions, leading to sequestration to the shoot.	• In nonaccumulating populations, more of the metal is found in the roots.
• Weak O and N ligands for storage of metal ions in these plants include small organic acids (malate and citrate) and amino acids (histidine), while the storage ligand for transport is mainly nicotianamine.	• Strong storage ligands in nonhyperaccumulator plants are phytochelatins, glutathione, and nicotianamine.
• Most metals accumulate in the shoots, and also in vacuoles of large epidermal storage cells in leaves.	• A low proportion of metals accumulate in shoots, including cytoplasm and chloroplast of mesophylls.
• Hyperaccumulators naturally use metal accumulation as a defense against herbivores and pathogens.	
• Hyperaccumulation is mediated by up to 200 times-higher expression levels of various metal transporter genes.	• These plants have low expression levels of various metal transporter genes.

2.3 Improving Phytoextraction Potentials of Nonhyperaccumulator Plant Species

Chelators can enhance phytoextraction of heavy metals from contaminated soil, and this ability can balance the characteristics of the hyperaccumulating plants. Chelating agents are mainly categorized into two groups: synthetic and natural.

2.3.1 Synthetic Chelating Agents

Some synthetic chelating agents, such as ethylene diamine tetraacetic acid (EDTA), diethylene triamine pentaacetic acid (DTPA), and ethylene glycol tetraacetic acid (EGTA), have been used successfully for induced phytoextraction (Blaylock et al. 1997, Evangelou et al. 2007). A number of different synthetic aminopolycarboxylic acids (APCAs) have been tested, including hydroxyl-EDTA, DTPA, *trans*-1,2-cyclohexylene dinitrilotetraacetic acid (CDTA), EGTA, ethylenediamine-*N,N'*-bis(*o*-hydroxyphenyl)acetic acid, *N*-(2-hydroxyethyl) iminodiacetic acid, and *N,N*-di(2-hydroxybenzyl) ethylenediamine *N,N*-diacetic acid.

Among all the chelating agents, EDTA has been the most commonly used agent for phytoextraction, because of its high efficiency in extracting many metals. It has been used in agriculture as an additive in micronutrient fertilizers since the 1950s. In EDTA-facilitated phytoremediation, the amount of heavy metals taken up by plants is minor compared to the amount mobilized from the soil and the large quantities that leach out of the root zone. Grcman et al. (2001, 2003) found that large quantities of Pb, Zn, and Cd in EDTA-treated soil columns in which cabbage plants were growing had leached into drainage water. Moreover, EDTA and EDTA–heavy metal complexes can be toxic to plants and soil microorganisms and can also persist in the environment due to their low level of biodegradability (Bucheli-Witschel and Egli 2001, Grcman et al. 2003). The *in situ* application of EDTA may pose potential risks of water pollution through the uncontrolled solubilization and migration of metals. To minimize the use of EDTA and the associated potential risk of migration of solubilized metals into groundwater, further research on this topic is still needed. Although EDTA has been shown in several publications to be effective in enhancing phytoextraction, EDTA and EDTA–heavy metal complexes are toxic to soil microorganisms (Grcman et al. 2001) and to plants as they severely decrease shoot biomass. However, the application of such synthetic chelating agents introduces an environmental risk due to their high mobility in the soil, which can result in the transport of the contaminant to surrounding uncontaminated areas; this movement can potentially become a problem due to the high solubility and persistence in the soil of the chelate–heavy metal complexes. Another important consideration for the application of synthetic chelates is that they may cause excessive heavy metal uptake by plants, thus introducing the risk for increased toxicity for the plants.

2.3.2 Natural Chelating Agents

In contrast to the use of high-cost synthetic chelating agents, an interesting alternative based on the same concept of induced phytoextraction is achieved through the use of natural chelating agents, such as EDDS, NTA (nitrilotriacetic acid), and low-molecular-weight organic acids (LMWOA), such as citric acid, oxalic acid, malic acid, and acetic acid. Recent attention has also been focused on the use of biodegradable chelating agents, such as EDDS (ethylene diamine disuccinate). EDDS is a biodegradable isomer of EDTA and is a naturally occurring substance in soil, where it is easily decomposed into less detrimental by-products. It has the potential to be a substitute for EDTA in chelate-assisted phytoremediation. EDDS can readily solubilize metals from soils and has been demonstrated to form strong complexes with metals and to be less toxic to soil microorganisms. EDDS and citric acid, both of which are biodegradable complexing agents, might be promising chelating agents for use as phytoextraction enhancers, because chelants such as EDTA and DTPA might create potential groundwater contamination problems. EDTA, DTPA, and citric acid lead to shoot metal concentrations of Cu and Zn higher than those observed in control plants. In addition, citric acid was able to induce removal of Cu and Zn from soil without increasing the leaching risk. NTA is a biodegradable chelating agent which has been in use for the past 50 years, primarily in detergents. In comparison to synthetic APCAs, NTA was more effective than N-(2-hydroxyethyl)iminodiacetic acid, hydroxyl-EDTA, EDTA, EGTA, CDTA, and DTPA in the extraction of As and Zn from soil (Chiu et al. 2005). It is known that exudation of organic compounds by roots may influence the solubility of essential and toxic ions indirectly and directly. LMWOA present an advantage because of their high rate of biodegradation in soil, which means they do not cause the negative effects potentially caused by the application of synthetic chelating compounds. The use of natural compounds such as LMWOA, which are easily biodegradable, is better than synthetic chelate application in terms of public acceptance of a phytoextraction technology. However, many authors have found lower effectiveness with aliphatic LMWOA, such as citric and oxalic acids, in inducing metal accumulation in plants compared to the effectiveness of synthetic chelates. Natural aromatic compounds, such as vanillic and gallic acids, could be a compromise between fast degradation and efficient enhanced phytoextraction, although information on degradation of vanillic and gallic acids in soils is not available. Both phenolic acids occur naturally in plants and are exuded by living roots. They contain acidic functional groups, such as carboxyl and phenolic groups, which have an important role in the chelation of heavy metals.

2.4 Mechanisms of Heavy Metal Accumulation in Plants in the Presence of Organic Pollutants

Heavy metal hyperaccumulation in plants is a multistep process that includes mobilization from soil into soil solution, transport of the heavy metal across the plasma membrane of root cells, xylem loading and root-to-shoot translocation, detoxification, and sequestration of the heavy metal in the leaf tissue (Figure 2.4). Most of the research on the mechanism of root and plant cell uptake has focused on the metals N, P, S, Fe, Ca, K, and Cl. However, little is known about the mechanisms of mobilization, uptake, and transport of heavy metals into plants. It is clear that for a large proportion of these metal when soil bound, phytoextracting plants mobilize the metals from soil solutions and accumulate them in their harvestable parts (Raskin et al. 1997).

2.4.1 Metal Phytoavailability in the Rhizosphere

Any phytoremediation process starts at the soil–plant interface or the root zone region, termed the rhizosphere, and hence the process of heavy metal transformation occurs mainly in this region.

Phytoavailability of metals in soil is the first step for successful phytoextraction. A major proportion of heavy metals in soil exists as the bound fraction at alkaline pH and needs to be mobilized into the soil solution to be made available for plant uptake. In addition, plant roots and soil microbes and their interactions can improve metal phytoavailability in the rhizosphere through secretion of root exudates, siderophores, protons, amino acids, and enzymes (Yang et al. 2005). Root exudates contain not only organic acids (e.g., lactate, acetate, oxalate, glutamate, salicylic acid, succinate, fumarate, malate, and citrate), sugars, amino acids, and enzymes (e.g., proteins, lectins, proteases, acid phosphatases, peroxidases,

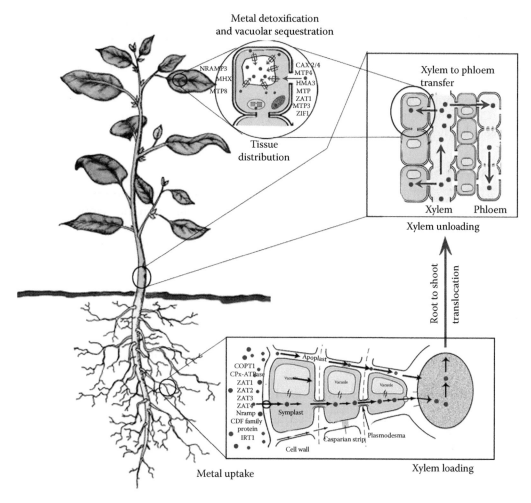

FIGURE 2.4
Mechanisms of heavy metal transport and accumulation in plants.

hydrolases, and lipases) as the main components but also secondary metabolites (e.g., isoprenoids, alkaloids, and flavonoids) which are released to soil as rhizodeposits (Wenzel et al. 2003). Root exudates can be used as an energy source by microorganisms. However, there is no definite answer as to whether and how hyperaccumulator and nonhyperaccumulator plants, or their rhizospheric microbial communities, have different effects on metal availability in their rhizopshere. The heavy metal mobility in soils also depends on many factors, including pH, redox potential, cation exchange capacity, presence of organic matter and the clay component, biological activity (Yang et al. 2005).

2.4.2 Transport of Heavy Metals across Plasma Membranes of Root Cells

Metal ions enter into roots through either an active (symplastic) or a passive (apoplastic) pathway (Xing et al. 2008). In the apoplastic pathway, metal ions or metal–chelate complexes enter roots through intercellular spaces. In contrast, the symplastic pathway is an energy-dependent process that is mediated by specific or generic metal ion carriers or channels. In the symplastic pathway, nonessential metal ions compete for the transmembrane carrier used by essential trace elements. For example, Ni and Cd compete for the transmembrane carrier used by Cu and Zn. Even metal chelates, such as Fe–phytosiderophore complexes, can be transported by the symplastic pathway. Transporter proteins and an intracellular high-affinity binding site mediate the uptake of metals across the root cell plasma membrane. Several classes of protein families, including the cation diffusion facilitator (CDF), zinc-iron permease (ZIP), CPx-type ATPases, and Nramp familes, have been implicated in heavy metal transport in plants (Yang et al. 2005).

2.4.3 Root-to-Shoot Translocation of Heavy Metals

Unlike nonaccumulator plants, which retain most heavy metals taken up from a contaminated site in their root cells, detoxifying the metals via chelation in the cytoplasm or storing them in vacuoles, hyperaccumulators efficiently translocate metal ions from the roots to the shoots via the xylem. Efficient translocation of metal ions to the shoot requires radial symplastic passage and active loading into the xylem, for which metal ions first have to cross the casparian band on the endodermal layer, which is a water-impervious barrier that blocks the apoplastic flux of metal ions from the root cortex to the stele. To cross this barrier and to reach the xylem, metals must move symplastically; this is a rate-limiting step in metal translocation from roots to shoots. Xylem loading is a tightly regulated process mediated by membrane proteins such as the P-type ATPase-HMA (heavy metal transporting ATPase), MATE, and oligopeptide transporter (OPT) proteins.

2.4.4 Detoxification and Sequestration of Heavy Metals in Aerial Parts of Plants

Great efficiency in detoxification and sequestration are key properties of hyperaccumulators, as these properties allow them to concentrate huge amounts of heavy metals in aboveground organs without suffering any phytotoxic effect. Heavy metal detoxification and sequestration occurs in location such as the epidermis, trichome, and even the cuticle (Rascio and Navari-Izzo 2011). A general mechanism for heavy metal detoxification in plants is the distribution of metals to the trichone and cell wall, with chelation of metals in complex with ligands, followed by the sequestration of the metal–ligand complexes into vacuoles. Small ligands such as organic acids, like malate and citrate, play major roles as detoxifying factors. Such ligands may be instrumental in preventing the persistence of heavy metals as free ions in the cytoplasm and even moreso in enabling their entrapment in vacuoles, where metal–organic acid chelates primarily become localized (Figure 2.5). The formation of metal–organic acid complexes is favored in the acidic environment of the vacuole. However, intracellular complexation involves peptide ligands, such as metallothioneins (MTs) and phytochelateins (PCs). PCs form a family of peptides that consist of repetitions of a γ-Glu–Cys dipeptide followed by a terminal Gly residue. The general formula of PCs is $(GluCys)_n$-Gly, where n is generally in the range of 2 to 11. PCs are synthesized enzymatically from glutathione in the presence of certain heavy metals and metalloids, and they are ubiquitous in plants. In hyperaccumulators, PCs are mainly induced in the roots, in particular by Cd, but not by Zn or Ni. However, MTs are low-molecular-weight proteins that are characterized by their high cysteine content and thus often give rise to metal–thiolate clusters. Most MTs have two metal clusters, one containing three and one containing four bivalent metal ions. MTs have been reported to play roles in other cellular processes, including regulation of cell growth and proliferation, DNA damage repair, and scavenging of reactive oxygen species. MTs are able to bind a variety of metals by the

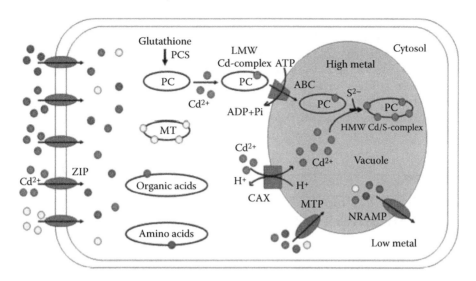

FIGURE 2.5
Role of ligands in heavy metal detoxification and sequestration. *Source*: Chandra, R., Kumar, V., *Environmental Science and Pollution Research* 24(3):2605–2619, 2016.

formation of mercaptide bonds between the numerous Cys residues present in the proteins and the metal, and it is the arrangement of these residues that in part determines the metal-binding properties of the MT proteins. Although MTs are expressed throughout a plant, some have been found to be expressed in a tissue-specific manner. Further, the PC–metal complexes are pumped into vacuoles. Vacuoles are generally considered the main storage site for heavy metals in plant cells, and they often occupy between 60 and 95% of the cell volume in mature parenchymatous and epidermal cells. Several families of transporter proteins, such as the CDF, HMA, ZAT, NRAMP, CAX, MHX, and ABC families, are involved in sequestration of heavy metals in vacuoles.

2.5 Use of Transgenic Plants for Heavy Metal Accumulation

Many of the genes involved in metal uptake, translocation, and sequestration have been identified by using the model plant *Arabidopsis* or naturally hyperaccumulating plants. However, the phytoremediation capacities of these natural hyperaccumulators are limited by their small size, slow growth rates, and limited growth habitats. Therefore, if the genes were transferred to plant species such as poplar and willow, with their high biomass and extensive root systems, significant removal of the heavy metals could be achieved. Genetic engineering of plants for enhanced phytoremediation has obvious environmental benefits, yet some would see potential risks (Linacre et al. 2003). This is especially true when using genetically altered trees. Their long life cycle makes risk assessment more challenging, and so more specific research is needed in this realm. In a commentary on this topic, Nicholas Linacre and colleagues described a risk assessment scenario for enhanced metal remediation (Linacre et al. 2003). They stated that the risk of contamination of food with an engineered metal hyperaccumulator, for example, is low because plants used for phytoextraction would be in isolated, industrial-type areas, not in agricultural areas. Furthermore, crops used for phytoextraction would be harvested before seed set, thus reducing the threat of crossing with other crops intended for food or entering the food supply. Plants engineered to hyperaccumulate toxic metals in foliage could be harmful to wildlife; however, studies have demonstrated that such foliage is not appealing in taste and is avoided. The best way to determine the ecological impact of transgenic plants for phytoremediation is by conducting field trials designed to assess risks (Linacre et al. 2003).

2.6 Anatomical, Physiological, and Morphological Changes in Plants after Heavy Metal Accumulation

Heavy metals can cause severe phytotoxicity, and they may act as powerful factors in the evolution of tolerant plant populations. Phytotoxicity results in chlorosis, weak plant growth, yield depression, and may be accompanied by disorders in plant metabolism, such as reduction of the meristematic zone, plasmolysis, and reduced chlorophyll and carotenoid production. Therefore, plants have developed a complex network of homeostatic mechanisms to modulate the internal concentrations of trace heavy metal elements in the cytosol to maintain metal levels below a certain toxicity threshold. These tolerance strategies include detoxification processes, complexation by organic chelators, and accumulation and compartmentalization of metals in the vacuoles. In addition to the experiments investigating the levels of metal accumulation in the organs of plants, some research has been conducted to study the anatomy, physiology, and morphology of plants in order to better understand the mechanisms by which they adapt to adverse environments. However, little is known about which specific tissues of plants can accumulate metals and whether the heavy metal concentration in the soil influences the organs for metal storage.

Investigations into metal accumulation and histological alterations in plant tissues caused by heavy metal stress will help us better understand the adaptation mechanism(s) of plants and provide more information about phytoremediation. Many phytoremediation studies have reported changes in the anatomical, physiological, and morphological characteristics of plants as a consequence of their adaption to heavy metal-induced environmental stress. Mangabeira et al. (2001) studied the ultrastructure of different organs of tomato plants (root, stem, leaf) which showed visible symptoms of Cr toxicity, and they argued that Cr(VI) induced changes in the ultrastructure of these organs. As stated by Preeti and Tripathi (2011), there is a direct relationship between chemical characteristics of a soil, heavy metal concentrations, and morphological and biochemical responses of plants. Yet, the metabolic and physiological responses of plants to heavy metal concentrations can be viewed as potentially adaptive changes of the plants during stress. Morphological and biochemical responses are probably due to high metal concentrations that damage plant roots and inhibit uptake of nutrients, thus inhibiting normal plant growth. Anatomical and physiological changes in Indian mustard plants (*Brassica juncea*) due to uptake and accumulation of Zn and Cd were investigated by Sridhar et al. (2005). They found that mustard plant accumulated significant amounts of Zn and Cd without exhibiting symptoms of phytotoxicity. Accumulation of Zn and Cd in all parts of the plant increased significantly with an increase in the applied metal concentration. Microscopic studies revealed clotted deposition in roots and stems, breakdown of parenchyma cells, and a decrease in starch content in leaves of plants treated with high concentrations of Zn. Physiological and morphological changes of Zn-treated plants included significant decreases in relative water content, dry weight, and plant height. Cd present at higher concentrations resulted in structural changes but only in stems and roots. Further, similar researchers also reported the effects of Zn and Cd accumulation on anatomical, morphological, and physiological characteristics of barley (*Hordeum vulgare*) plants. First, the plants were exposed to the metals Zn and Cd for 19 and 16 days, respectively. The leaves, stems, and roots were harvested to identify structural changes and analyze metal accumulation levels. Microscopic structural changes, such as decreases in intracellular spaces, breakdown of vascular bundles, and shrinking of palisade and epidermal cells, occurred in leaves, stems, and roots of plants treated with a high concentration of Zn (Figure 2.6). Zn accumulation was also associated with significant decreases in water content, fresh dry weight, and plant height. Cd caused only structural changes in roots at the high concentration. Barley plant was able to accumulate a significant amount of Zn and Cd without exhibiting symptoms of phytotoxicity when the metal concentrations were relatively low (Sridhar et al. 2007).

Daud et al. (2009) also studied Cd-induced alterations in plant leaves of transgenic cotton cultivars, namely, BR001 (herbicide resistant) and GK30 (insect resistant), as well as their wild relative cotton genotype (Coker 312) by using both ultramorphological and physiological indices. Transmission electron microscope (TEM) images of leaf mesophyll cells of BR001, GK30, and Coker 312 grown for 6 days with 1000 µM Cd showed several changes in ultrastructural features of the leaf mesophyll cells (Figure 2.7a to h). The greatest alterations occurred in the plastid regions, followed by changes in the nuclear setup and plasma membrane in all cultivars. The size of chloroplasts decreased and their outer membranes were broken at many places. Thylakoids became swollen, shorter, and broken at different points. The size and number of the plastoglobuli, starch granules, and lipid bodies also subsequently increased. However, the vacuolar compartmentalization was enhanced, and

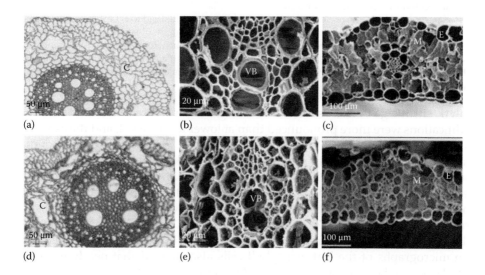

FIGURE 2.6
Microscopic image of a transverse section of *Hordeum vulgare* before and after 19 days of Zn treatment: (a) transverse section of control root; (b) control stem; (c) control leaf; (d) Cd-treated plant, showing breakdown of cells in the cortex region; (e) breakdown and deposition of a vascular bundle in the stem; (f) leaf of the treated plant, showing a decrease in intracellular space and reduction in epidermal and palisade cell sizes. C; cortex, VB; vascular bundle. *Source*: Sridhar, B.B.M. et al., *Brazilian Journal of Plant Physiology* 19(1):15–22, 2007.

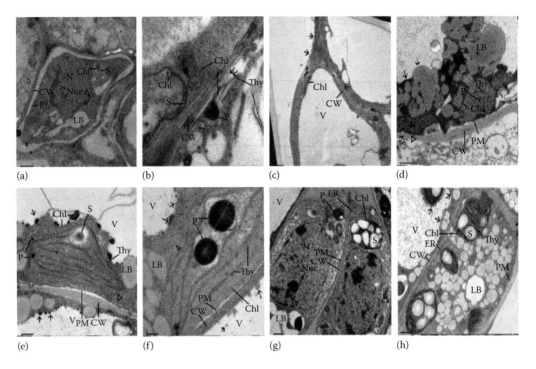

FIGURE 2.7
TEM images of leaf mesophyll cells of germinated seedlings of two transgenic cotton cultivars (BR001 and GK30) and of their wild relative cotton genotype (Coker 312) grown for 6 days in 1000 μM Cd. (a to c) Mesophyll cells of BR001 at low and high magnification, showing a small and irregularly shaped nucleus (N) along with detrimental plasmolytic (PL) shrinkage and enlarged vacuoles (V). Cd deposition (arrow) is apparent as electron-dense granules inside the vacuoles and attached to the cell wall (CW). A disrupted nuclear membrane (S) as well as the outer membrane of a chloroplast (Chl) is shown for BR001. (d to f) Leaf mesophyll cells of GK30 at low and high magnification, showing less irregularly shaped and small-sized nuclei (N) along with less severe plasmolysis (PL) of the plasma membrane (PM) and enlarged vacuoles (V). (g and h) Moreover, the shape of the chloroplast is less irregular than that of BR001, while thylakoids have become swollen and broken (Thy) at many places. The sizes of the lipid bodies (LB) and starch grains (S) increased. *Source*: Daud, M.K. et al., *Journal of Hazardous Materials* 168:614–625, 2009.

plasmolytic shrinkage became severe. At this highest concentration, the ultramorphological modifications were more pronounced than at lower levels of Cd, and the extent of damage was severe for cultivar BR001, followed (i.e., with a less severe extent of damage) by GK30 and Coker 312. Some other prominent changes, like the size of vacuoles, number of lipid bodies, and thickness of cell walls, also increased with increased Cd levels. Moreover, the plasmolytic shrinkage of the plasma membrane at the highest level of Cd (1000 μM) was much more pronounced in BR001 than in GK30, while it was almost absent in the wild relative genotype (Coker 312). The deposition of Cd in the forms of electron-dense granules inside vacuoles and or attached to the cell wall was also found. In addition, the electron micrographs of the leaf mesophyll cells also showed that nearly inconspicuous intercellular spaces became prominent as Cd toxicity increased.

Similarly, Najeeb et al. (2011) studied Cd-induced toxicity stress in the wetland plant *Juncus effusus*, and they explored its potential for Cd phytoextraction via chelators (citric acid and EDTA). *J. effusus* plants were able to survive under Cd stress without showing any visible symptoms of phytotoxicity, even at higher Cd levels. Higher Cd concentrations caused ultrastructural alterations in root and shoot cells (Figure 2.8). Citric acid application

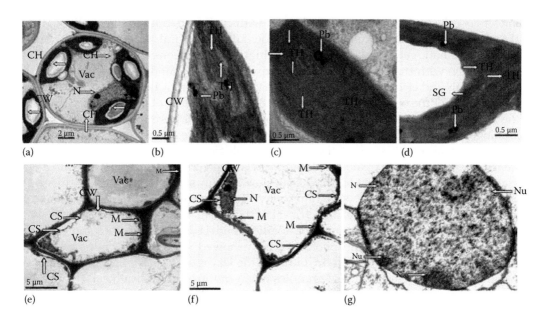

FIGURE 2.8
TEM images of a root and shoot of *Juncus effuses*. (a) Mesophyll cell of a shoot exposed to Cd plus EDTA. (b to d) Magnified views of a chloroplast and thylakoid membrane exposed to 100 µM Cd alone (b), Cd plus citric acid (c), or Cd plus EDTA (d). (e and f) View of a root cell exposed to Cd alone (e) or Cd plus EDTA (f). (g) Nucleus of root tip cells exposed to Cd alone. CH, chloroplast; CW, cell wall; N, nucleus; Mt, mitochondria; SG, starch grain; Vac, vacuole; M, metal deposition; Nu, nucleolus. *Source*: Najeeb, U. et al., *Journal of Hazardous Materials* 186:565–574, 2011.

significantly enhanced Cd accumulation in *J. effusus* plant roots recovering from Cd-induced ultrastructural damage. These findings aided in understanding the tolerance mechanisms of *J. effusus* against Cd stress. Cd altered morphological and physiological attributes of *J. effusus*, as reflected by growth retardation and damaged root cells through cytoplasmic shrinkage and metal deposition. Citric acid restored the structure and shape of root cells and eliminated plasmolysis, whereas EDTA exhibited no positive effect on the cells. Shoot cells remained unaffected under Cd treatment, alone or with citric acid, except for chloroplast swelling. Morphometric observation via electron microscopic images revealed significant reductions in the size and diameter of mesophyll cells of plants exposed to combined treatments of Cd and the chelators citric acid and EDTA. Chloroplast swelling (increased size, length, and width) was also obvious in plants treated either with Cd alone or with EDTA. Nucleus size and diameter of *J. effusus* plants remained statistically unaffected under all these treatments. Mesophyll cells of the control plant had typical, mature cells with a definite cell wall and contained the nucleus and chloroplast. Chloroplast was lens shaped and possessed a well-organized grana and thylakoid membrane system with a few dense plastogobuli inside the stroma (Figure 2.8a). Mesophyll cells of Cd-treated plants showed no change in cellular shape, but there was a substantial increase in the number of plastogobuli inside chloroplasts (Figure 2.8a and b). Application of chelators along with Cd altered cellular and chloroplast shape and structure. These cells were circular in shape with swollen chloroplasts and disintegrated thylakoid membrane systems (Figure 2.8c and d). Higher magnification disclosed a loose thylakoid membrane system under treatment with Cd alone or with chelators (Figure 2.8b to d). Starch granules appeared only in the chloroplast of plant cells exposed to Cd plus EDTA (Figure 2.8d). Data collected from micrographs of root meristematic cells demonstrated no significant changes in the size or diameters of cells under Cd treatment alone or with citric acid, but treatment with Cd and EDTA significantly reduced cell sizes. Numerous ultrastructural alterations were noticed in root meristematic cells of plants exposed to high concentrations of Cd (100 μM) alone or in combination with citric acid or EDTA. A considerable amount of Cd was present in the form of crystals and electron-dense granules in vacuoles and attached to the cell wall along with shrinkage of the cell membrane (Figure 2.8b to d). Swelling of the nucleus, disruption of the nuclear membrane, and increased numbers of nucleoli were some of the other obvious changes observed in the nuclei of Cd-treated cells (Figure 2.8b to d).

Azzarello et al. (2012) studied the ultramorphological and physiological changes in *Paulownia tomentosa* plants treated with high levels of Zn. They observed *P. tomentosa* was tolerant to a high concentration of Zn without showing any symptoms of damage by visual assessment. The TEM analysis of leaf mesophyll cells showed membrane vesiculation in all of the tissue together with the withdrawal of the plasma membrane from the cell wall. Moreover, a large increase in the number of mitochondria and peroxisomes and dramatic changes in the chloroplasts's morphology, which showed a few starch grains and many plastoglobules, were also observed. In addition, the leaves of *P. tomentosa* plants exhibited unusual shapes and electron-dense granules of Zn in the bundle sheath cell (Figure 2.9). The cell cytoplasm was found more vacuolated after treatment with Zn. However, the TEM analysis of *P. tomentosa* root cortical cells showed unusual deposition of electron-dense granules of Zn in the clumps or in the form of globules (Figure 2.9).

A study conducted by Alkhatib et al. (2013) aimed to understand the accumulation pattern of Pb in tobacco plant (*Nicotiana tabacum* var. Turkish) roots and leaves. This study demonstrated

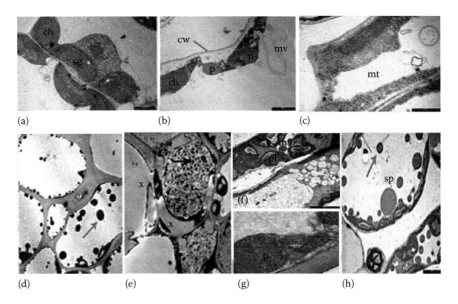

FIGURE 2.9
TEM images of *Paulownia tomentosa*. (a) Leaf mesophyll cells of a control plant with well-developed chloroplast and starch grains. (b) Leaf cell showing membrane vesiculation, withdrawal of the plasma membrane from the cell wall, and an increased number of peroxisomes. (c) Enlarged mitochondrion. (d) Electron-dense globules (arrows) in the spongy parenchyma (e) Vacuolated cytoplasm with large starch grains. (f) Chloroplast with an irregular shape, many plastoglobules, and an increased number of peroxisomes. (g) Increases in the number of peroxisomes. (h) Vacuolated chloroplast and cytoplasm with large starch gains. ch, chloroplast; sg, starch grains; mv, membrane vesiculation; cw, cell wall; p, peroxisome; x, xylem elements; sp, spongy parenchyma; pl, plastoglobules; mt, mitochondria. *Source*: Azzarello, E. et al., *Environmental and Experimental Botany* 81:11–17, 2012.

accumulation of Pb in the form of electron-dense crystalline precipitates in various parts of root tissues relative to control plant tissues. The exposure to a high concentration of Pb also led to distortion of epidermal cell walls and overall disruption in the organization of vascular tissue. The microscopic analysis of plant root sections showed thickening of cell walls in the vascular tissue, enlargement of the intracellular space in the cortex, and severe distortion in the shapes of the cells. The roots of the Pb-treated plants also showed the presence of Pb as dark precipitates in the vascular tissue and to some extent in the cortical and epidermal cells and cell walls (Figure 2.10a). Further, the ultramorphological analysis confirmed the presence of Pb outside the epidermis of roots in the form of electron-dense clusters of fine needles primarily in the cell wall (Figure 2.10b and c). In addition, the cells in the vascular tissues also showed the presence of Pb as electron-dense deposits in the cell walls. The root epidermal cells also showed severe inward invaginations of cell walls (Figure 2.10c).

Chandra and Yadav (2011) studied the ultramorphological changes in three wetland plants (*Phragmites cummunis*, *Typha angustifolia*, and *Cyperus esculentus*) after heavy metal accumulation from an aqueous solution (containing Cd, Cr, Cu, Mn, Fe, Ni, Pb, and Zn) and determined their phytoextraction capacities. This study showed that the given multimetal solution after 56 days of incubation did not lead to any visible morphological adverse effect. The tested plants showed defense mechanism in the multimetal solution. The TEM observation of root sections of *P. cummunis* and *T. angustifolia* showed metal deposits near the cell wall, intracellular spaces, and thinning of the cell wall, while loss of cell shape and thickened cell walls were found in *C. esculentus* due to deposition of metals (Figure 2.11, top row). The TEM observations of leaf tissues of all three plants are shown in Figure 2.11, bottom row; no histological damage was observed in leaves of *P. cummunis* except for changes in cell shape. Similarly, *T. angustifolia* leaves also showed no any apparent deformation. Meanwhile, leaf tissues of *C. esculentus* showed reductions of intracellular spaces, plastids, and irregular cell shape. This study indicated the higher metal tolerance behavior of *P. cummunis* and *T. angustifolia* compared to *C. esculentus*.

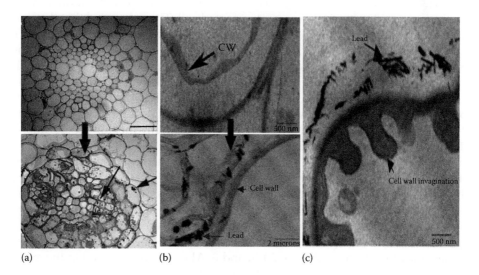

(a)   (b)   (c)

FIGURE 2.10
Microscopic images of *Nicotiana tabacum* var. Turkish after Pb treatment, compared to an untreated plant (a). Disruption and organization of the vascular cortical cells in a Pb-treated plant, showing Pb accumulation in the cell wall (b). (c) Pb outside epidermal cells and in inward invaginations of the cell walls in a root. The arrow, dark spot, and dark lines along the cell wall indicate Pb accumulation. *Source*: Alkhatib, R. et al., *Journal of Microscopy and Ultrastructure* 1(1–2):57–62, 2013.

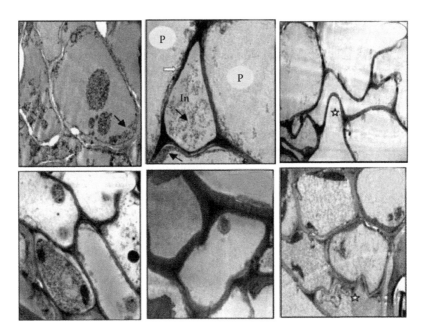

FIGURE 2.11

(Top row) TEM images of roots of *Phragmites cummunis, Typha angustifolia,* and *Cyperus esculentus*. (Bottom row) Leaves of *Phragmites cummunis, Typha angustifolia,* and *Cyperus esculentus* after 56 days of treatment with multiple metals. Metal granules (filled arrows), thinning of cell walls (open arrows), loss of cell shape (star), and intracellular spaces (In) are shown in some of the panels. *Source*: Chandra, R., and Yadav, S., *International Journal of Phytoremediation* 13(6):580–591, 2011.

Guo et al. (2014) studied the histological effects of heavy metal accumulation in *Phragmites australis* grown in acid mine drainage-contaminated soil. They observed that the roots and rhizomes accumulated more metals than other plant parts from contaminated soil. The Fe accumulation in roots was higher than that in the rhizome of *P. australis*. Light microscopic observations of unstained roots and rhizomes of reed prior to staining are shown in Figure 2.12. The roots consisted of an exodermis and inner arenchyma separated from the stele by an endodermis. The exodermis can be an effective barrier to restrict the penetration of Fe into root tissues. The exodermis and endodermis play important roles in the protection of root tissues against heavy metal stress. The rhizome mainly includes epidermis, cortex, central cylinder, and vascular bindles.

The representative images in Figure 2.12f to i show a stained root and rhizome section of reed after Fe and Al accumulation. Fe accumulation (blue hue) but no structural or histological deformities were found in the sample. Most of the Fe was sequestered in the endodermis of the root (Figure 2.12f). Fe was distributed in the epidermis, cortex, and central cylinder of the rhizome. Fe accumulated around the vascular bundles, which are the transport system of the plant (Figure 2.12g). The root and rhizome tissues that contained Al turned magenta after staining (Figure 2.12h and i). Al was mainly stored in the exodermis and endodermis of the root and was also found in the stele of roots (Figure 2.12h). Similar to Fe, Al was also sequestered in the epidermis, cortex, and central cylinder of the rhizome.

Recently, Chandra and Kumar (2017) observed anatomical parts of *Solanum nigrum* and *Ricinus communis* in the presence of heavy metal accumulation under TEM. The TEM observations of root of *S. nigrum*, and *R. communis* showed physiologically and biochemically linked deformities in the plant tissue (Figure 2.13). The TEM observations of roots of

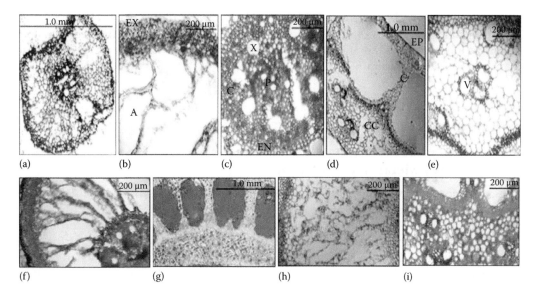

FIGURE 2.12

Light microscopic images of *Phragmites australis* root and rhizomes. (a) Unstained root section of stele. (b) Subsection of root exodermis and arenchyma. (c) Stele of root pith. (d) Section of rhizome epidermis. (e) Vascular bundle of rhizome. (f) Stained section of root. (g) Rhizome after Fe accumulation. (h) Stained section of root. (i) Rhizome after Al accumulation. P, pith; En, endodermis; X, xylem; EP, epidermis; C, cortex; CC, central cylinder; V, vascular bundle. *Source*: Guo, L., Ott, D.W., Cutright, T.J., *Environmental and Experimental Botany* 105:46–54, 2014.

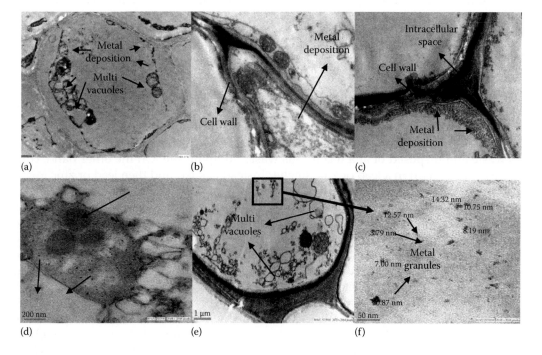

FIGURE 2.13

Microscopic images of native plants root after accumulation of heavy metals. (a to c) *Solanum nigrum*; (d to f) *Ricinus communis*. *Source*: Chandra, R., Kumar, V., *Environmental Science and Pollution Research* 24(3):2605–2619, 2017.

representative native plants revealed that tested hyperaccumulator plants showed formation of multinucleoli, multivacuoles, and deposition of metal granules in their cellular components as a process of adaptation to an environmental site with heavy metal-containing organic waste-rich pollution.

2.7 Management of Hazardous Plant Biomass after Heavy Metal Accumulation

Phytoextraction involves repeated cropping of plants in contaminated soil, sludge, or sediment until the metal concentration(s) drops to an acceptable level. Successful phytoextraction will generate huge quantities of hazardous plant biomass. The problem of contaminated plant and crop disposal after phytoextraction is very important for minimizing environmental health risks. After removal of harvestable metal-enriched biomass from a site, the possible methods of disposal of contaminated plants include landfills, surface impoundments, deep well injection, or incineration. The postharvest treatments (i.e., composting, compacting, thermal treatments) can reduce the volume and/or weight of the biomass for disposal as a hazardous waste (Vassilev et al. 2004). Leaching tests of the composted material have shown, however, that the composting process forms soluble organic compounds that enhance lead solubility. These results document that composting can significantly reduce the volume of harvested biomass; however, lead-contaminated plant biomass would still require treatment prior to disposal (Padmavathiamma and Li 2007). In the case of compaction, care should be taken to collect and dispose the leachate. The leachate produced by pressing contaminated plant biomass will contain high concentrations of heavy metals complexed with chelators (after induced phytoextraction) or Ni, Cd, and/or Zn (after continuous phytoextraction) in soluble and bioavailable forms. A conventional and promising route to utilize hazardous biomass produced by phytoextraction is through pyrolysis, incineration, and gasification. Pyrolysis decomposes material under anaerobic conditions and moderate temperatures. The final products are pyrolytic gas and coke breeze. Heavy metals from contaminated biomass will be contained in the coke breeze. This means that the product should be treated as hazardous waste and disposed of at a hazardous waste dumping site. On the other hand, coke breeze could be used in a lead or zinc smelter instead of coke, and then lead or zinc might be recovered during the smelting process. The process destroys organic matter, releasing metals, mainly as oxides. The liberated metals are entrained in the slag or released into effluent gases. Modern flue gas-cleaning technology ensures effective capture of the metal-containing dust. Plant material also can be destroyed in an incineration plant. To diminish the amount of plant material to be transported to a smelter, desiccation can be conducted. Desiccation also reduces the likelihood of leachate production from the plant material during harvest and transport.

When phytoextraction is combined with biomass generation and the commercial utilization as an energy source, then this type of remediation can be turned into a profitable operation, with the residual ash available to be used as an ore. Phytomining provides a potential means for energy storage that can be utilized to generate thermal energy, which is a cheap and renewable resource. Phytomining has the potential to provide opportunities for sustainable postmining livelihoods, especially in tropical regions. Phytomining exploits plants as renewable and cheap sources, with sunlight as the primary energy resource. As a result, phytomining is considered an esthetic, nondestructive technology that is safe and nonaggressive and is accepted by communities and commerce. This technology involves growing and harvesting metal-accumulating plants on appropriate sites and treating the biomass to recover the metal (Brooks 1998).

2.8 Conclusion

Phytoremediation is a highly efficient means of ameliorating heavy metals and hazardous organic pollutants from contaminated sites through the use of hyperaccumulator and nonhyperaccumulator plants. This green technology is often favored over more conventional methods of cleanup due to its low cost, low impact, and wide public acceptance. The plants used in phytoremediation are generally selected on the basis of their growth rate and biomass, their ability to tolerate and accumulate contaminants, the depth of their root zone, and their potential to transpire groundwater. The plants used in phytoextraction should not only accumulate, degrade, or volatilize the contaminants, but also they should grow quickly in a wide range of environmental conditions. From results obtained in the lab and from field study research, it is clear that phytoextraction could be applied either as the main remediation tool or more likely as a "polishing" technique when combined with other conventional and "biotechniques." Both approaches (using hyperaccumulator or nonhyperaccumulator plants) may yield interesting results, providing that they are used in optimal situations. Indeed, the choice of the plant species (one or several) requires thorough study of the plant's potential and its suitability for the site.

2.9 Recommendations and Future Perspectives

Plants suitable for successful phytoextraction of heavy metals should be tolerant to high concentrations of metals and, at the same time, be able to accumulate high amounts of essential and nonessential elements in harvestable plant organs. Some hyperaccumulators are quite rare and live in very specialized niches, while other hyperaccumulator plants are locally adapted, metal-tolerant populations of species often restricted to one or a few sites, making the broad collection and utilization of these species difficult. Their narrow, highly localized distributions make these species difficult to find; rarity also raises conservation and management issues that need to be immediately addressed if the species are to be used for phytoremediation purposes. However, metal-tolerant species should be carefully examined via laboratory screening, as they could harbor locally adapted, metal-hyperaccumulating plant populations in danger of being extirpated. Thus, for conservation of metal tolerance and accumulating traits to be most effective, it is vital that even isolated plant populations among widespread species be preserved. Consequently, nowadays fast-growing, high-biomass crop plant species that accumulate moderate levels of metals in their shoots are actively being tested for their metal phytoremediation potential. Interestingly, some of these fast-growing, high-biomass crop plant species are known to display significant heavy metal tolerance. These plants can be used for minimization of heavy metals from contaminated soil, but they should be prohibited for use as food and fodder due to the health hazards posed by various heavy metals that may be recovered from specific parts of plants through phytomining.

Hyperaccumulators will continue to have exciting applications if we are able to identify their unique hypertolerance and hyperaccumulation mechanisms, including how they manipulate their rhizospheres. Moreover, the physiological and molecular bases of metal hyperaccumulation in plants are still in the research and development phase. Recent evidence has focused on identification of the molecular mechanisms that may be involved in resistance and tolerance as well as hyperaccumulation of heavy metals. Several findings suggest that in some plants of the ZIP family, genes contribute to metal hyperaccumulation and transport, but their individual functions are yet to be identified and further intensive research is needed in this concern. Identification of novel genes with high-biomass yield characteristics and the subsequent development of transgenic plants with superior remediation capacities are another encouraging area for further research. An in-depth research study is warranted to find out which plants are maximally resistant and best adapted to metallic environments or regions.

Acknowledgments

Financial assistance from the Department of Science and Technology, Science and Engineering Research Board, New Delhi (file no. SB/SO/BB-0042/2013; Professor Ram Chandra) and a Rajiv Gandhi National Senior Research Fellowship from the University Grant Commission, New Delhi, to Vineet Kumar, Ph.D. Scholar, are greatly acknowledged.

References

Abou-Shanab, R.A.I., Angle, J.S., Chaney, R.L. 2006. Bacterial inoculants affecting nickel uptake by *Alyssum murale* from low, moderate and high Ni soils. *Soil Biology and Biochemistry* 38:2882–2889.

Adhikari, T., Kumar, A. 2012. Phytoaccumulation and tolerance of *Riccinus communis* L. to nickel. *International Journal of Phytoremediation* 14:481–492.

Aksoy, A., Oztiirk, M.A. 1997. *Nerium oleander* L. as a biomonitor of lead and other heavy metal pollution in Mediterranean environments. *Science of the Total Environment* 205:145–150.

Algreen, M., Trapp, S., Rein, A. 2014. Phytoscreening and phytoextraction of heavy metals at Danish polluted sites using willow and poplar trees. *Environmental Science and Pollution Research International* 21(15):8992–9001.

Ali, M.H.H., Al-Qahtani, K.M. 2012. Assessment of some heavy metals in vegetables, cereals and fruits in Saudi Arabian markets. *Egyptian Journal of Aquatic Research* 38:31–37.

Alkhatib, R., Bsoul, E., Blom, D.A., Ghoshroy, K., Creamer, R., Ghoshroy, S. 2013. Microscopic analysis of lead accumulation in tobacco (*Nicotiana tabacum* var. Turkish) roots and leaves. *Journal of Microscopy and Ultrastructure* 1(1–2):57–62.

Amira, M.S., Abdul, Q. 2015. Phytoremediation of Pb and Cd by native tree species grown in the kingdom of Saudi Arabia. *Journal of Science Research and Technology* 3(1):22–34.

Ang, L.H., Tang, L.K., Ho, W.M., Hui, T.F., Theseira, G.W. 2010. Phytoremediation of Cd and Pb by four tropical timber species grown on an ex-tin mine in peninsular Malaysia. *International Journal of Environmental Chemical Ecological Geological and Geophysical Engineering* 4(2):70–74.

Azzarello, E., Pandolfia, C., Giordanob, C., Rossia, M., Mugnaia, S., Mancuso, S. 2012. Ultramorphological and physiological modifications induced by high zinc levels in *Paulownia tomentosa*. *Environmental and Experimental Botany* 81:11–17.

Baker, A.J.M., Brooks, R.R. 1989. Terrestrial higher plants which hyperaccumulate metallic elements: A review of their distribution, ecology and phytochemistry. *Biorecovery* 1:81–126.

Baker, A.J.M., Walker, P.L. 1989. Ecophysiology of metal uptake by tolerant plants, pp. 155–177. *Heavy Metal Tolerance in Plants: Evolutionary Aspects*, Shaw, A.J. (ed.). Boca Raton, FL: CRC Press.

Belimov, A.A., Kunakova, A.M., Safronova, V.I., Stepanok, V.V., Yudkin, L.Y., Alekseev, Y.V., Kozhemyakov, A.P. 2004. Employment of rhizobacteria for the inoculation of barley plants cultivated in soil contaminated with lead and cadmium. *Microbiology* 73:99–106.

Bidwell, S. D., Woodrow, I.E., Batianoff, G.N., Sommer-Knudsen, J. 2002. Hyperaccumulation of manganese in the rainforest tree *Austromyrtus bidwillii* (Myrtaceae) from Queensland, Australia. *Functional Plant Biology* 29:899–905.

Blaylock, M.J., Salt, D.E., Dushenkov, S., Zakharova, O., Gussman, C., Kapulnik, Y. et al. 1997. Enhanced accumulation of Pb in Indian mustard by soil-applied chelating agents. *Environmental Science and Technology* 31(3):860–865.

Boyd, R.S. 2007. The defense hypothesis of elemental hyperaccumulation: Status, challenges and new directions. *Plant and Soil* 293:153–76.

Brandt, R., Merkl, N., Schultze-Kraft, R., Infante, C., Broll, G. 2006. Potential of vetiver (*Vetivera zizanioides* (L.) Nash) for phytoremediation of petroleum hydrocarbon-contaminated soils in Venezuela. *International Journal of Phytoremediation* 8(4):273–284.

Broadhurst, C.L., Chaney, R.L., Angle, J.S., Maugel, T.K., Erbe, E.F., Murphy, C.A. 2004. Simultaneous hyperaccumulation of nickel, manganese, and calcium in *Alyssum* leaf trichomes. *Environmental Science and Technology* 38:5797–5802.

Brooks, R. R. (ed.) 1998. *Plants That Hyperaccumulate Heavy Metals*, p. 384. Wallingford: CAB International.

Brooks, R.R., Lee, J., Reeves, R.D., Jaffre, T. 1977. Detection of nickeliferous rocks by analysis of herbarium specimens of indicator plants. *Journal of Geochemical Exploration* 7:49–58.

Bucheli-Witscheli, M., Egli, T. 2001. Environmental fate and microbial degradation of aminopolycarboxylic acids. *FEMS Microbiology Reviews* 25:69–106.

Chandra, R., Yadav, S. 2011. Phytoremediation of Cd, Cr, Cu, Mn, Fe,Ni, Pb and Zn from aqueous solution using *Phragmites cummunis*, *Typha angustifolia* and *Cyperus esculentus*. *International Journal of Phytoremediation* 13(6):580–591.

Chandra, R., Saxena, G., Kumar, V. 2015. Phytoremediation of environmental pollutants: An eco-sustainable green technology to environmental management, pp. 1–29. *Advances in Biodegradation and Bioremediation of Industrial Waste*, Chandra, R. (ed.). Boca Raton, FL: CRC Press.

Chandra, R., Kumar, V. 2017. Phytoextraction of heavy metals by potential native plants and their microscopic observation of root growing on stabilised distillery sludge as a prospective tool for in situ phytoremediation of industrial waste. *Environmental Science and Pollution Research* 24(3):2605–2619.

Chandra, R., Yadav, S., Yadav, S. 2017. Phytoextraction potential of heavy metals by native wetland plants growing on chlorolignin containing sludge of pulp and paper industry. *Ecological Engineering* 98:134–145.

Chang, C.Y., Yu, H.Y., Chen, J.J., Li, F.B., Zhang, H.H., Liu, C.P. 2014. Accumulation of heavy metals in leaf vegetables from agricultural soils and associated potential health risks in the Pearl River Delta, South China. *Environmental Monitoring and Assessment* 186:1547–1560.

Chatterjee, S., Chetia, M., Singh, L., Chattopadhyay, B., Datta, S., Mukhopadhyay, S.K. 2011. A study on the phytoaccumulation of waste elements in wetland plants of a Ramsar site in India. *Environmental Monitoring and Assessment* 178:361–371.

Chen, T., Wei, C., Huang, Z. et al. 2002. Arsenic hyperaccumulator *Pteris vittata* L. and its arsenic accumulation. *Chinese Science Bulletin* 47:902–905.

Chehregani, A., Malayeri, B.E. 2007. Removal of heavy metals by native accumulator plants. *International Journal of Agriculture Biology* 9:462–465.

Chiu, K.K., Ye, Z.H., Wong, M.H. 2005. Enhanced uptake of As, Zn and Cu by *Vetiveria zizanioides* and *Zea mays* using chelating agents. *Chemosphere* 60:1365–1375.

Cobb, G.P., Sands, K., Waters, M., Wixson, B.G., Dorward-King, E. 2000. Accumulation of heavy metals by vegetables grown in mine wastes. *Environmental Toxicology and Chemistry* 19(3):600–607.

Cortez, L.A.S., Ching, J.A. 2014. Heavy metal concentration of dumpsite soil and accumulation in *Zea mays* (corn) growing in a closed dumpsite in Manila, Philippines. *International Journal of Environmental Science and Development* 5(1):77–80.

Cui, Y.S., Wang, Q.R., Dong, Y.T., Li, H.F., Christie, P. 2004. Enhanced uptake of soil Pb and Zn by Indian mustard and winter wheat following combined soil application of elemental sulphur and EDTA. *Plant and Soil* 261:181–188.

Dalun, T., Fan, Z., Wende, Y., Xi, F., Wenhua, X., Xiangwen, D., Guangjun, W., Changhui, P. 2009. Heavy metal accumulation by panicled goldenrain tree (*Koelreuteria paniculata*) and common elaeocarpus (*Elaeocarpus decipens*) in abandoned mine soils in southern China. *Journal of Environmental Sciences* 21:340–345.

Daud, M.K., Variath, M.T., Ali, S., Najeeb, U., Jamil, M., Hayat, Y., Dawood, M., Khand, M.I., Zaffar, M., Cheema, S.A., Tong, X.H., Zhu, S. 2009. Cadmium-induced ultramorphological and physiological changes in leaves of two transgenic cotton cultivars and their wild relative. *Journal of Hazardous Materials* 168:614–625.

de Andrade, S.A.L., da Silveira, A.P.D., Jorge, R.A., de Abreu, M.F. 2008. Cadmium accumulation in sunflower plants influenced by arbuscular mycorrhiza. *International Journal of Phytoremediation* 10:1–14.

de Andrade, S.A.L., Jorge, R.A., da Silveira, A.P.D. 2005. Cadmium effect on the association of jackbean (*Canavalia ensiformis*) and arbuscular mycorrhizal fungi. *Scientia Agricola* 62:389–394.

Deng, H., Ye, Z.H., Wong, M.H. 2004. Accumulation of lead zinc, copper and cadmium by 12 wetland plant species thriving in metal contaminated site in China. *Environmental Pollution* 132:29–40.

de Souza, L.A., de Andrade, S.A.L., de Souza, S.C.R., Schiavinato, M.A. 2011. Tolerance and phytoremediation potential of *Stizolobium aterrimum* associated to the arbuscular mycorrhizal fungi *Glomus etunicatum* in lead-contaminated soil. *Revista Brasileira de Ciência do Solo* 35:1441–1451. (In Portuguese, with English abstract.)

Evangelou, M.W., Ebel, M., Schaeffer, A. 2007. Chelate assisted phytoextraction of heavy metals from soil. Effect, mechanism, toxicity, and fate of chelating agents. *Chemosphere* 68(6):989–1003.

Fontem Lum, A., Ngwa, E.S.A., Chikoye, D., Suh, C.E. 2015. Phytoremediation potential of weeds in heavy metal contaminated soils of the Bassa Industrial Zone of Douala, Cameroon. *International Journal of Phytoremediation* 16(3):302–319.

Galfati, I., Bilal, E., Sassi, A.B., Abdallah, H., Zaïer, A. 2011. Accumulation of heavy metals in native plants growing near the phosphate treatment industry, Tunisia. *Carpathian Journal of Earth and Environmental Sciences* 6(2):85–100.

Garcia-Salgado, S., Garcia-Casillas, D., Quijano-Nieto, M.A., Bonilla-Simon, M.M. 2012. Arsenic and heavy metal uptake and accumulation in native plant species from soil polluted by mining activities. *Water Air and Soil Pollution* 223:559–572.

Gardea-Torresdey, J.L., De la Rosa, G., Peralta-Videa, J.R., Montes, M., Cruz-Jimenez, G., Cano-Aguilera, I. 2005. Differential uptake and transport of trivalent and hexavalent chromium by tumbleweed (*Salsola kali*). *Archives of Environmental Contamination and Toxicology* 48:225–232.

Gaskin, S., Soole, K., Bentham, R. 2008. Screening of Australian native grasses for rhizoremediation of aliphatic hydrocarbon-contaminated soil. *International Journal of Phytoremediation* 10(5):378–389.

Ghosh, M., Singh, S.P. 2005. A review on phytoremediation of heavy metals and utilization of its byproducts. *Applied Ecology and Environmental Research* 3:1–18.

Giordani, C., Cecchi, S., Zanchi, C. 2005. Phytoremediation of soil polluted by nickel using agricultural crops. *Environmental Management* 36:675–681.

Girdhar, M., Sharma, N.R., Rehman, H., Kumar, A., Mohan, A. 2014. Comparative assessment for hyperaccumulatory and phytoremediation capability of three wild weeds. *Biotechnology* 4:579–589.

Grcman, H., Velikonja-Bolta, S., Vodnik, D., Kos, B., Leštan, D. 2001. EDTA enhanced heavy metal phytoextraction: Metal accumulation, leaching, and toxicity. *Plant and Soil* 235:105–114.

Grcman, H., Vodnik, D., Velikonja-Bolta, S., Leštan, D. 2003. Ethylenediaminedissuccinate as a new chelate for environmentally safe enhanced lead phytoextraction. *Journal of Environmental Quality* 32:500–506.

Guo, L., Ott, D.W., Cutright, T.J. 2014. Accumulation and histological location of heavy metals in *Phragmites australis* grown in acid mine drainage contaminated soil with or without citric acid. *Environmental and Experimental Botany* 105:46–54.

Hernandez-Allica, J., Becerril, J.M., Garbisu C. 2008. Assessment of the phytoextraction potential of high biomass crop plants. *Environmental Pollution* 152:32–40.

Ingle, R.A., Smith, J.A., Sweetlove, L.J. 2005. Responses to nickel in the proteome of the hyperaccumulator plant *Alyssum lesbiacum*. *Biometals* 18(6):627–641.

Karimzadeh, L., Heilmeier, H., Merkel, B.J. 2012. Effect of microbial siderophore DFO-B on Cd accumulation by *Thlaspi caerulescens* hyperaccumulator in the presence of zeolite. *Chemosphere* 88:683–687.

Kaya, G., Okumus, N., Yaman, M. 2010. Lead, cadmium and copper concentrations in leaves of *Nerium oleander* L. and *Robinia pseudoacacia* L. as biomonitors of atmospheric pollution. *Fresenius Environmental Bulletin* 19(4A):669–675.

Khan, A.L., Lee, I.J. 2013. Endophytic *Penicillium funiculosum* LHL06 secretes gibberellin that reprograms *Glycine max* L. growth during copper stress. *BMC Plant Biology* 13:86.

Kubota, H., Takenaka, C. 2003. *Arabis gemmifera* is a hyperaccumulator of Cd and Zn. *International Journal of Phytoremediation* 5:197–120.

Kumar, P.B., Dushenkov, V., Motto, H., Raskin, I. 1995. Phytoextraction: The use of plants to remove heavy metals from soils. *Environmental Science and Technology* 29:1232–1239.

Lampis, S., Santi, C., Ciurli, A., Andreolli, M., Vallini, G. 2015. Promotion of arsenic phytoextraction efficiency in the fern *Pteris vittata* by the inoculation of As resistant bacteria: A soil bioremediation perspective. *Frontiers in Plant Science* 6:80.

Li, Y.M., Chaney, R., Brewer, E., Roseberg, R., Angle, J.S., Baker, A., Reeves, R., Nelkin, J. 2003. Development of a technology for commercial phytoextraction of nickel: Economic and technical consideration. *Plant and Soil* 246:107–115.

Linacre, N.A., Whiting, S.N., Baker, A.J.M., Angle, S., Ades, P.K. 2003. Transgenics and phytoremediation: The need for an integrated risk assessment, management, and communication strategy. *International Journal of Phytoremediation* 5:181–185.

Liphadzi, M.S., Kirkham, M.B., Mankin, K.R., Paulsen, G.M. 2003. EDTA-assisted heavy-metal uptake by poplar and sunflower grown at a long-term sewage-sludge farm. *Plant and Soil* 257:171–182.

Liu, W., Wang, Q., Wang, B., Hou, J., Luo, Y., Tang, C., Franks, A.E. 2015. Plant growth-promoting rhizobacteria enhance the growth and Cd uptake of *Sedum plumbizincicola* in a Cd-contaminated soil. *Journal of Soils and Sediments* 15:1191–1199.

Lorestani, B., Cheraghi, M., Yousefi, N. 2011. Phytoremediation potential of native plants growing on a heavy metals contaminated soil of copper mine in Iran. *World Academy of Science Engineering and Technology* 53:377–382.

Luis, R., Rincon, J., Asencio, I., Rodriguez-Castellanos, L. 2007. Capability of selected crop plants for shoot mercury accumulation from polluted soils: Phytoremediation perspectives. *International Journal of Phytoremediation* 9:1–13.

Ma, L.Q., Komar, K.M., Tu, C., Zhang, W., Cai, Y., Kenelly, E.D. 2001. A fern that hyperaccumulates arsenic. *Nature* 409:579–582.

Ma, Y., Prasad, M.N.V., Rajkumar, M., Freitas, H. 2011. Plant growth promoting rhizobacteria and endophytes accelerate phytoremediation of metalliferous soils. *Biotechnology Advances* 29:248–258.

Malik, R.N., Husain, S.Z., Nazir, I. 2010. Heavy metal contamination and accumulation in soil and wild plant species from industrial area of Islamabad. *Pakistan Journal of Botany* 42(1):291–301.

Mander, U., Mitsch, W.J. 2009. Pollution control by wetlands. *Ecological Engineering* 35(2):153–158.

Mangabeira, P., Almeida, A.A., Mielke, M., Gomes, F.P., Mushrifah, I., Escaig, F., Laffray, D., Severo, M.I., Oliveira, A.H., Galle, P. 2001. Ultrastructural investigations and electron probe X-ray microanalysis of chromium-treated plants, p. 55. *Proceedings of the VI International Conference on the Biogeochemistry of Trace Elements.*

Mazumdar, K., Das, S. 2015. Phytoremediation of Pb, Zn, Fe, and Mg with 25 wetland plant species from a paper mill contaminated site in North East India. *Environmental Science and Pollution Research* 22:701–710.

McGrath, S.P., Zhao, F.J. 2003. Phytoextraction of metals and metalloids. *Current Opinion in Biotechnology* 14:277–282.

Meers, E., Samson, R., Tack, F.M.G., Ruttens, A., Vandegehuchte, M., Vangronsveld, J., Veerloo, M.G. 2007. Phytoavailability assessment of heavy metals in soils by single extractions and accumulation by *Phaseolus vulgaris*. *Environmental and Experimental Botany* 60(3):385–396.

Mench, M., Tancogne, J., Gomez, A., Juste, C. 1989. Cadmium bioavailability to *Nicotiana tabacum* L., *Nicotiana rustica* L., and *Zea mays* L. grown in soil amended or not amended with cadmium nitrate. *Biology and Fertility of Soils* 8:48–53.

Mertens, J., Vervaeke, P., De Schrijver, A., Luyssaert, S. 2004. Metal uptake by young trees from dredged brackish sediment: Limitations and possibilities for phytoextraction and phytostabilisation. *Science of the Total Environment* 326(1–3):209–215.

Najeeb, U., Jilani, G., Ali, S., Sarward, M., Xua, L., Zhoua, W. 2011. Insights into cadmium induced physiological and ultra-structural disorders in *Juncus effusus* L. and its remediation through exogenous citric acid. *Journal of Hazardous Materials* 186:565–574.

Nazemi, S. 2012. Concentration of heavy metals in edible vegetables widely consumed in Shahroud, the North East of Iran. *Journal of Applied Environmental and Biological Sciences* 2(8):386–391.

Nouri, J., Lorestani, B., Yousefi, N., Khorasani, N., Hasani, A.H., Seif, F., Cheraghi, M. 2011. Phytoremediation potential of native plants grown in the vicinity of Ahangaran lead–zinc mine (Hamedan, Iran). *Environment Earth Sciences* 62:639–644.

Odjegba, V.J., Fasidi, I.O. 2004. Accumulation of trace elements by *Pistia stratiotes*: Implications for phytoremediation. *Ecotoxicology* 13:637–646.

Oti Wilberforce, J.O., Nwabue, F.I. 2013. Heavy metals effect due to contamination of vegetables from Enyigba lead mine in Ebonyi State, Nigeria. *Environment and Pollution* 12(1). doi:10.5539/ep.v2np19.

Ozen, S.A., Yaman, M. 2016. Phytoextraction of lead and its relationship with histidine in six plant species using ICP-MS and HPLC-MS. *Chemistry and Ecology* 32(4):346–356.

Padmavathiamma, P.K., Li, L.Y. 2007. Phytoremediation technology: Hyperaccumulation metals in plants. *Water Air and Soil Pollution* 184:105–126.

Płociniczak, T., Sinkkonen, A., Romantschuk, M., Piotrowska-Seget, Z. 2013. Characterization of *Enterobacter intermedius* mh8b and its use for the enhancement of heavy metals uptake by *Sinapis alba* L. *Applied Soil Ecology* 63:1–7.

Prasad, M.N.V., Freitas, H.M.D. 2003. Metal hyperaccumulation in plants: Biodiversity prospecting for phytoremediation technology. *Electronic Journal Biotechnology* 93(1):285–321.

Preeti, P., Tripathi, A.K. 2011. Effect of heavy metals on morphological and biochemical characteristics of *Albizia procera* (Roxb.) Benth. seedlings. *International Journal of Environmental Science* 1(5):1009.

Rai, P.K. 2008. Phytoremediation of Hg and Cd from industrial effluent using an aquatic free floating macrophytes *Azolla pinnata*. *International Journal of Phytoremediation* 10:430–439.

Rajkumar, M., Ma, Y., Freitas, H. 2013. Improvement of Ni phytostabilization by inoculation of Ni resistant *Bacillus megaterium* SR28C. *Journal of Environmental Management* 128:973–980.

Rani, A., Souche, Y., Goel, R. 2013. Comparative in situ remediation potential of *Pseudomonas putida* 710A and *Commamonas aquatica* 710B using plant (*Vigna radiata* (L.) wilczek) assay. *Annals of Microbiology* 63:923–928.

Rascio, N., Navari-Izzo, F. 2011. Heavy metal hyperaccumulating plants: How and why do they do it? And what makes them so interesting? *Plant Science* 180:169–181.

Raskin, I., Smith, R.D., Salt, D.E. 1997. Phytoremediation of metals: Using plant to remove pollutants from the environment. *Current Opinion in Biotechnology* 8:221–226.

Reeves, R.D., Brooks, R.R. 1983. Hyperaccumulation of lead and zinc by two metallophytes from mining areas of Central Europe. *Environmental Pollution Series A* 31:277–285.

Reeves, R.D., Baker, A.J.M. 2000. Metal-accumulating plants, pp. 193–229. *Phytoremediation of Toxic Metals: Using Plants To Clean Up the Environment*, Raskin, I., Ensley, B.D. (eds.). New York: Wiley.

Riddell-Black, D. 1993. A review of the potential for the use of trees in the rehabilitation of contaminated land. WRC report CO 3467. Water Research Centre, Medmenham, United Kingdom.

Rosselli, W., Keller, C., Boschi, K. 2003. Phytoextraction capacity of trees growing on metal contaminated soil. *Plant and Soil* 256:265–272.

Saglam, C. 2013. Heavy metal accumulation in the edible parts of some cultivated plants and media samples from a volcanic region in southern Turkey. *Ekoloji* 22(86):1–8.

Salt, D.E., Blaylock, M., Kumar, N.P.B.A., Dushenkov, V., Ensley, B.D., Chat I., Raskin, I. 1995. Phytoremediation: A novel strategy for the removal of toxic metals from the environment using plants. *Nature Biotechnology* 12(5):468–474.

Sarret, G., Saumitou-Laprade, P., Bert, V., Proux, O., Hazemann, J.-L., Traverse, A., Marcus, M.A., Manceau, A. 2002. Forms of zinc accumulated in the hyperaccumulator *Arabidopsis halleri*. *Plant Physiology* 130:1815–1826.

Selvam, A., Wong, J.W. 2008. Phytochelatin synthesis and cadmium uptake of *Brassica napus*. *Environmental Technology* 29:765–773.

Sharma, S., Sharma, P., Mehrotra, P. 2010. Bioaccumulation of heavy metals in *Pisum sativum* L. growing in fly ash amended soil. *Journal of American Science* 6(6):43–50.

Shu, W.S., Ye, Z.H., Zhang, Z.Q., Wong, M.H. 2002. Lead, zinc, and copper accumulation and tolerance in populations of *Paspalum distichum* and *Cynodon dactylon*. *Environmental Pollution* 120(2):445–453.

Singh, A., Prasad, S.M., Singh, S., Singh, M. 2016. Phytoremediation potential of weed plants' oxidative biomarker and antioxidant responses. *Chemistry and Ecology* 32(7):684–706.

Solhi, M., Shareatmadari, H., Hajabbasi, M.A. 2005. Lead and zinc extraction potential of two common crop plants, *Helianthus annuus* and *Brassica napus*. *Water Air and Soil Pollution* 167:59–71.

Sridhar, B.B.M., Diehl, S.V., Hanc, F.X., Monts, D.L., Su, Y. 2005. Anatomical changes due to uptake and accumulation of Zn and Cd in Indian mustard (*Brassica juncea*). *Environmental and Experimental Botany* 54:131–141.

Sridhar, B.B.M., Han, F.X., Diehl, S.V., Monts, D.L., Su, Y. 2007. Effect of Zn and Cd accumulation on structural and physiological characteristics of barley plants. *Brazilian Journal of Plant Physiology* 19(1):15–22.

Srivastava, M., Ma, L.Q., Santos, J.A.G. 2006. Three new arsenic hyperaccumulating ferns. *Science of the Total Environment* 364:24–31.

Stomp, A.M., Han, K.H., Wilbert, S., Gordon, M.P. 1993. Genetic improvement of tree species for remediation of hazardous wastes. *In Vitro Cell Division B* 29:227–232.

Sun, Y., Zhou, Q., Wei, S., Ren, L. 2007. Growth responses of the newly discovered Cd-hyperaccumulator *Rorippa globosa* and its accumulation characteristics of Cd and As under joint stress of Cd and As. *Frontiers of Environmental Science and Engineering* 1:107–113.

Syam, N., Wardiyati, T., Maghfoer, M.D., Handayantoc, E., Ibrahima, B., Muchdar, A. 2016. Effect of accumulator plants on growth and nickel accumulation of soybean on metal-contaminated soil. *Agriculture and Agricultural Science Procedia* 9:13–19.

Szabó, L., Fodor, L. 2006. Uptake of microelements by crops grown on heavy metal-amended soil. *Communications in Soil Science and Plant Analysis* 37:2679–2689.

Taghipour, H., Mosaferi, M. 2013. Heavy metals in the vegetables collected from production sites. *Health Promotion Perspectives* 3(2):185–193.

Tanhuanpää, P., Kalendar, R., Schulman, A.H., Kiviharju, E. 2007. A major gene for grain cadmium accumulation in oat (*Avena sativa* L.). *Genome* 50(6):588–594.

Turgut, C., Pepe, M.K., Cutright, T.J. 2004. The effect of EDTA and citric acid on phytoremediation of Cd, Cr, and Ni from soil using *Helianthus annuus*. *Environmental Pollution* 131:147–154.

Vassilev, A., Schwitzguébel, J.P., Thewys, T., van der Lelie, D., Vangronsveld, J. 2004. The use of plants for remediation of metal-contaminated soils. *Scientific World Journal* 4:9–34.

Verbruggen, N., Hermans, C., Schat, H. 2009. Molecular mechanism of metal hyperaccumulation in plants. *New Phytologist* 181:759–776.

Wang, Y., Stauffer, C., Keller, C. 2005. Changes in Hg fractionation in soil induced by willow. *Plant and Soil* 275:67–75.

Watson, C. 2002. The phytoremediation potential of Salix: Studies of the interaction of heavy metals and willows. Ph.D. thesis. University of Glasgow, Glasgow, Scotland.

Wenzel, W.W., Bunkowski, M., Puschenreiter, M., Horak, O. 2003. Rhizosphere characteristics of indigenously growing nickel hyperaccumulator and excluder plants on serpentine soil. *Environmental Pollution* 123:131–138.

Whiting, S.N., de Souza, M.P., Terry, N. 2001. Rhizosphere bacteria mobilize Zn for hyperaccumulation by *Thlaspi caerulescens*. *Environmental Science and Technology* 35:3144–3150.

Xing J.P., Jiang, R.F., Ueno, D., Ma, J.F., Schat, H., McGrath, S.P., Zhao, F.J. 2008. Variation in root to shoot translocation of cadmium and zinc among different accession of the hyperaccumulators *Thlaspi caerulescens* and *Thlaspi praecox*. *New Phytologist* 178:315–325.

Yadav, S., Chandra, R. 2011. Heavy metals accumulation and ecophysiological effect on *Typha angustifolia* L. and *Cyperus esculentus* L. growing in distillery and tannery effluent polluted natural wetland site, Unnao, India. *Environmental Earth Sciences* 62(6):1235–1243.

Yang, X., Feng, Y., He, Z., Stoffella, P.J. 2005. Molecular mechanism of heavy metal hyperaccumulation and phytoremediation. *Journal of Trace Elements in Medicine and Biology* 18(4):339–353.

Ye, Z.H., Yang, Z.Y., Chan, G.Y.S., Wong, M.H. 2001. Growth response of *Sesbania rostrata* and *S. cannabina* to sludge-amended lead/zinc mine tailings: A greenhouse study. *Environmental International* 26(5):449–455.

Yoon, J., Cao, X., Zhou, Q., Ma, L.Q. 2006. Accumulation of Pb, Cu, and Zn in native plants growing on a contaminated Florida site. *Science of the Total Environment* 368(2):456–464.

Zhang, X. et al. 2012. A nonpathogenic *Fusarium oxysporum* strain enhances phytoextraction of heavy metals by the hyperaccumulator *Sedum alfredii* Hance. *Journal of Hazardous Materials* 229–230:361–370.

3

Adaptation Strategies of Plants against Heavy Metal Stress

Supriya Tiwari and N. K. Dubey

CONTENTS

3.1 Introduction

Anthropogenic interferences in the biosphere, such as the accelerated rates of industrialization and urbanization, changes in agricultural practices, and extensive mining activities, have led to dangerously high levels of contamination of natural resources with subsequent increased hazardous effects on natural biochemical cycles. Of the several human-induced disturbances affecting various global phenomena, accumulation of heavy metals is one of the problems that has drawn serious concern of environmentalists for ecological, nutritional, and environmental reasons. Heavy metals are nonbiodegradable and persistent environmental contaminants which have cytotoxic, genotoxic, and mutagenic effects on living cells. Heavy metals enter the food chain generally via plants, which take up heavy metals through contaminated soil or irrigation waters or through atmospheric deposition on the areal plant parts (Rascio and Navari-Izzo 2011). Heavy metals can be divided into two categories:

1. Essential micronutrients: Some heavy metals are essential for normal growth and development of plants. These essential metals are important structural parts of several enzymes and proteins and are necessary for normal physiological functioning. Plants show deficiency symptoms if they do not get the required levels of the essential heavy metals. Fe, Zn, Cu, Mg, Ni, and Mo are essential heavy metals.

2. Nonessential elements: Nonessential heavy metals do not have any known biological or physiological functions and hence are not essential for plants. Cd, Sb, Cr, Pb, Ar, Co, Ag, Se, and Hg are in this category. Many of these heavy metals are toxic to plants, even at very low concentrations.

The heavy metals, essential or nonessential, play a very significant role in plant developmental processes. Essential heavy metals are necessary for plants, but in very small amounts. On the other hand, nonessential heavy metals may be toxic to plants and can hamper the normal physiological and metabolic processes in the plants. Heavy metals may either disrupt the structure of protein molecules by formation of bonds between the heavy metals and sulfhydryl groups of amino acids (Hall 2002) or may deactivate important enzymes and functional groups of important cellular molecules (Hossain and Piyatida 2012). In addition, heavy metals can also disrupt the structures of plasma membranes, which in turn will have adverse effects on vital plant processes, such as photosynthesis, respiration, and enzyme activities (Hossain et al. 2012). The deleterious effects of heavy metals can be explained by the increased generation of reactive oxygen species (ROS), which disturb the cellular homeostasis of plants (Hossain et al. 2012).

In their efforts to cope with the negative effects of heavy metals, plants have evolved enzymatic and nonenzymatic antioxidants to cope with heavy metal-induced ROS production. Of the different antioxidants that are stimulated under heavy metal stress, the enzymes of the glutathione (GSH) family play a significant role, not only in direct quenching of ROS but also in the detoxification of methylglyoxal (MG), which is produced in excess amounts in response to heavy metal stress (Hossain et al. 2010). In addition to a plant's response at the biochemical level, plants also employ a few physiological strategies to combat heavy metal stress, the most important one being synthesis of certain specialized biomolecules, such as phytochelatins, methallothioneins, proline, etc. These biomolecules provide increased tolerance in plants against heavy metal stress (Hossain et al. 2010). In certain cases, plants also develop morphological adaptations, including association with mycorrhizal fungi, which serve to reduce the heavy metal uptake from the soil. In addition to these adaptations, plants also make use of certain phytoremediation methods to handle heavy metal stresses. This chapter focuses upon the different strategies adopted by plants to survive through the unfavorable conditions created by heavy metal stress. We also discuss the oxidative stress generated due to heavy metals and the different enzymatic components of the defense machinery that are activated to deal under stress conditions, along with different morphological and physiological adaptations made by plants under heavy metal stress.

3.2 Sources of Heavy Metals in the Environment

Heavy metals are introduced into the environment via natural as well as anthropogenic sources. Natural processes that add heavy metals to the environment include geogenic processes, like weathering of parental rocks, volcanic eruptions, natural fires, etc. The heavy metals added through natural processes are limited and remain well below dangerous levels (<1000 mg/kg) and are rarely found to be above toxic limits. Contamination of the environment with heavy metals generally occurs by various anthropogenic activities, which include improper irrigation and agricultural practices, unmanaged disposal of industrial effluents, mining and metallurgical industries, organic waste transport, combustion of fuels, and power generation (Figure 3.1). Industrialization and urbanization are the most important causes of heavy metal contamination in the environment. Improper disposal of municipal solid wastes is also responsible for adding large quantities of heavy metals into aquatic and terrestrial ecosystems (Tiwari et al. 2013). Figure 3.2 illustrates the percentages of heavy metal contamination that occur from different human-based sources.

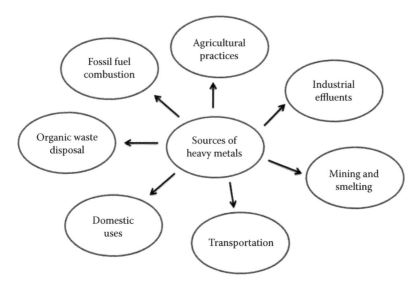

FIGURE 3.1
Anthropogenic sources of heavy metal contamination in the environment.

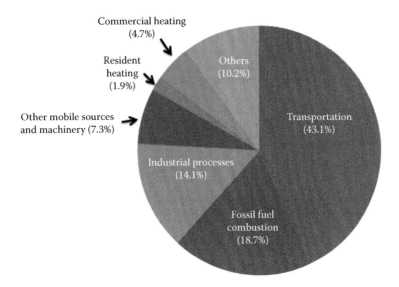

FIGURE 3.2
Heavy metal contamination that occurs due to different human sources.

3.3 Mechanisms of Action of Heavy Metals

Heavy metals at toxic levels disrupt normal metabolic and physiological activities. However, the site of action within a plant can differ with different heavy metals. The most common effects of heavy metal toxicity include chlorosis, necrosis, damaged photosynthetic apparatus, and reduction in plant growth (Dalcorso et al. 2008). Heavy metals can bring about ultrastructural, biochemical, and molecular changes in plant tissues which can speed up senescence processes and ultimately lead to plant death (Gamalero et al. 2009). Heavy metal-induced toxicity can be attributed to the following events occurring in plant cells:

1. Stimulation of production of ROS, which can be free radical species such as superoxide anions and hydroxyl ions or non-free radicals, such as singlet oxygen and hydrogen peroxide. In addition to this, certain cytotoxic compounds like MG generated under heavy metal stress also create a state of oxidative stress by disturbing the cellular homeostasis within the cell (Hossain et al. 2012, Sytar et al. 2013).

2. Direct interaction with cellular macromolecules, such as proteins, due to their high affinity towards thiol, histidyl, and carboxyl groups. Due to this ability, heavy metals can inactivate enzymes by binding to the sulfhydryl group of cysteine residues. For instance, Cd toxicity can lead to misfolding of structural proteins and inhibition of enzyme activity due to Cd binding to sulfhydryl groups on such proteins (Dalcorso et al. 2008, Hall 2002).

3. Displacement of essential metal ions from specific binding sites, which can disrupt important cellular functions (Sharma and Dietz 2009). For example, displacement of Ca^{2+} by Cd^{2+} in calmodulin, an important entity in cell signalling, leads to an inhibition of calmodulin-dependant phosphodiesterase activity in radishes (Rivetta et al. 1997).

Heavy metals can be divided into two categories on the basis of their mechanism of action. One group is redox active; this group contains heavy metals like Fe, Cu, Cr, and Co. The other group is redox inactive (e.g., Cd, Zn, Ni, and Al). The heavy metals belonging to the redox-active group are directly involved in redox reactions in cells, resulting in the formation of ROS (Schutzendubel and Polle 2002). Redox-inactive cells also cause oxidative stress in cells but do so indirectly via interactions with antioxidant defense systems, disruption of the electron transport chain, or induction of lipid peroxidation.

Plasma membranes are considered the primary targets of heavy metal toxicity. Different studies have shown that heavy metal stress triggers the peroxidation of lipid components of cell membranes (Yusuf et al. 2012, Kumar et al. 2012). Lipid peroxidation is a biochemical marker that indicates the state of oxidative stress in plants due to overproduction of ROS under stress conditions. Enhanced lipid peroxidation under heavy metal stress can be attributed to heavy metal-induced lipoxygenase (LOX) activity. Studies have shown that accumulation of Cd, Cu, and Pb in plant tissues enhances lipid peroxidation (Verma and Dubey 2003, Ann et al. 2011). Heavy metals can also cause membrane damage, either by oxidation of or cross-linking with protein thiols, via inhibition of the membrane protein H^+- ATPase or by bringing about changes in the composition and fluidity of membrane lipids (Meharg 1993).

The most important physiological process affected by heavy metal exposure is photosynthesis. Accumulation of heavy metals not only affects the process of biosynthesis of photosynthetic pigments but also targets the photosynthetic machinery (Vesely et al. 2011). Heavy metals alter the membrane structure and function of chloroplasts, along with the electron transport chain, thus affecting the light reaction of photosynthesis (Mysliwa-Kurdziel and Strzalka 2002, Ventrella et al. 2011). Heavy metal ions affect the electron transport chain at multiple sites (Khan et al. 2007), with PS II being more susceptible to stress conditions than PS I (Horcik et al. 2007). However, the activity of PS I was also inhibited by high heavy metal concentration (Horcik et al. 2007). Divalent cations like Co^{2+}, Ni^{2+}, and Zn^{2+} displace Mg^{2+} in ribulose-1,5-bisphosphate carboxylase/oxygenase, adversely affecting the dark reaction of photosynthesis (Van Assche and Clijsters 1986). The negative effects of heavy metals on light as well as dark reactions of photosynthesis were found to be dose dependent (Babu et al. 2010) and increased as the duration of exposure was increased (Chugh and Sawhney 1999). Due to their hazardous effects on both phases of photosynthesis, heavy metals decrease the photosynthetic rate, stomatal conductance, transpiration rate, plant growth, and agricultural productivity.

Heavy metals can bind to the cell nucleus, resulting in DNA-based modifications, inter- and intramolecular cross-linking of DNA and proteins, DNA strand breakage, rearrangement, depurification, and mutagenic damage (Kasprzak 1995). Ni stress inhibits the activities of ribonuclease and protease in rice seedlings, suppressing the hydrolysis of RNA and proteins (Dubey 2011). Heavy metals affect microtubule assembly and disassembly, thereby altering the cell cycle and cell division (Fusconi et al. 2006). Heavy metals also influence other metabolic processes, like carbohydrate metabolism, as the activities of enzymes like α-amylase, β-amylase, and sucrose phosphate synthase were decreased, while those of starch phosphorylase, acid invertase, and sucrose synthase increased under heavy metal stress (Jha and Dubey 2004). Nitrogen metabolism enzymes, such as nitrate reductase, nitrite reductase, and glutamine synthase, were inhibited upon exposure to heavy metals (Sangwan et al. 2014). It has been reported that arsenic may interact with the sulfhydryl group of proteins and can replace a phosphate group of ATP. Similarly, Hg also interacts with the sulfhydryl group that forms the S–Hg–S bridge and affects seed germination and embryo growth (Patra et al. 2004). Cu-induced stress stimulates ethylene synthesis, which promotes the process of senescence (Maksymiec 1997).

3.4 Defense Strategies Adopted by Plants under Heavy Metal Stress

As discussed earlier, heavy metal-induced oxidative stress adversely affects the various cytological events that occur in cells and disturbs the different metabolic functions of plants. Plants have developed a number of strategies to overcome the unfavorable impacts of heavy metal stress. These include direct scavenging of cytotoxic species (ROS and MG), formation of metal–ligand complexes and their sequestration into vacuoles, or immobilization of heavy metals to reduce their uptake by plants. These strategies are discussed in detail in the following sections. Figure 3.3 shows the possible mechanisms for mitigation of heavy metal stress in plants.

3.4.1 Biochemical Strategies: Antioxidant Defense Mechanisms

Antioxidant systems are a plant's intrinsic defense mechanism to regulate ROS levels, which increase abruptly under heavy metal stress. The antioxidant system includes enzymatic and nonenzymatic components. The enzymatic components include superoxide dismutase (SOD), ascorbate peroxidase (APX), monodehydroascorbate reductase, dehydroascorbate reductase (DHAR), glutathione reductase (GR), catalase (CAT), glutathione peroxidase (GPX), and glutathione *S*-transferase (GST), and the nonenzymatic components include ascorbic acid (ASH) and GSH. In addition to direct quenching of ROS, ascorbic acid and GSH function as cofactors for enzymes of the glyoxalase pathway, which is an essential component of cellular metabolism for MG produced in the cytosol under heavy metal stress (Hossain 2012). Heavy metal stress causes increased ROS generation in plants, which in turn stimulates enzyme activities of the antioxidant defense system. It has been shown that Pb treatment increases the activation of SOD, CAT, APX, POD, and LOX (Huang et al. 2012). Increased activities of SOD, CAT, APX, and guaiacol-dependent peroxidase are indicative of oxidative damage induced by Mg, Cu, Zn, and Mn (Candan and Tarhan 2003); these authors reported the potential for these three metals to cause oxidative damage in

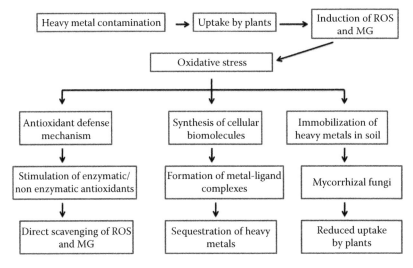

FIGURE 3.3
Possible defense strategies adopted by plants under heavy metal stress.

the following order: Mn < Zn < Cu. It has been shown that the activities of SOD, CAT, APX, and DHAR increase upon Cd treatment in two rice varieties (Iqbal et al. 2010). In the presence of excess Cd ions, the expression of Zn-SOD slightly decreased while that of Mn-SOD increased (Ann et al. 2011). Some of the important components of the antioxidant defense system in plants include GSH, GST, GPX, and GR.

3.4.1.1 Glutathione

GSH is a nonenzymatic thiol-containing antioxidant that has been detected in all cell compartments, including the cytosol, chloroplast, endoplasmic reticulum, vacuole, and mitochondria. The reactive nature of the thiol group along with its relative stability and high solubility in water make it ideal to protect plants against the oxidative stress produced by heavy metals (Millar et al. 2003). GSH not only controls the level of hydrogen peroxide via the ASH-GSH cycle, but also it scavenges various ROS directly. Changes in the ratio of reduced GSH to (oxidized) glutathione disulfide (GSSH) is important in certain redox signaling pathways. It has been suggested that the GSH/GSSH ratio is indicative of the cellular redox balance (Millar et al. 2003). Studies have also suggested that GSH may function as a cellular sensor to ensure maintenance of the NADPH pool (May et al. 1998). Increments in GSH contents have been observed in *Brassica juncea, B. compestris* (Qadir et al. 2004, Anjum et al. 2008), and *Pisum sativum* (Metwally et al. 2003) when grown under heavy metal stress.

GSH also plays a significant role in mitigating heavy metal toxicity by conjugating with excess heavy metals and reducing their subsequent transport to vacuoles, thereby protecting the plants from their harmful effects (Klein et al. 2006). In addition, GSH also acts as a precursor to phytochelatins, which are a group of heavy metal binding proteins. Phytochelatins are synthesized by GSH through activity of the enzyme phytochelatin synthase (PCS). The protective role of phytochelatins in mitigating heavy metal stress is discussed below. Figure 3.4 shows the role of glutathione in different defense mechanisms adopted by plants in mitigating heavy metal stress.

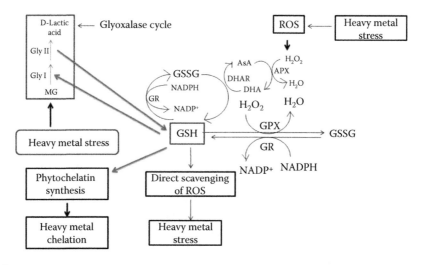

FIGURE 3.4
Role of glutathione in the defense system of plants under heavy metal stress. (Modified from Hossain, A. et al., *Physiology and Molecular Biology of Plants* 16(3):259–272, 2010.)

3.4.1.2 Glutathione S-Transferase

GSTs are a family of multifunctional enzymes known for their role in heavy metal detoxification. GST assists GSH in conjugation with heavy metals, forming specific derivatives which can be secreted from cells, sequestered in the vacuole, or catabolized (Dixon and Edward 2010). In addition, GST also provides protection against heavy metal-induced oxidative stress (Dixon and Edward 2010). An increase in GST activity was reported for pumpkin (*Curbita maxima* Duch) in the presence of high levels of Cd, Cr, Mn, and As stress (Hossain et al. 2006). A similar response was observed in rice (*Oryza sativa* L. cv. N07-63) when it was treated with 50 μM Cd for 7 days (Hu et al. 2009). Zhang and Ge (2008) reported a close relationship between Cd level, GSH content, and GST activity, suggesting utilization of GSH and GST as biomarkers of Cd-induced stress. GST also plays an important role in transportation of phytochelatin–metal complexes (formed under heavy metal stress) to the vacuole (Marrs and Walbot 1997). GST may also be important for elimination of lipid peroxidation induced by heavy metal toxicity (Gullner et al. 1998).

3.4.1.3 Glutathione Peroxidase

The main function of GPX is to catalyze the reduction of H_2O_2, organic hydroperoxides, and lipid hydroperoxides to H_2O and alcohol via GSH and/or other reducing agents (Foyer and Noctor 2011). GPX has been suggested as a potent biomarker for heavy metal activity (Cuypers et al. 2002). Significant increases in GPX activity were recorded in mung bean seedlings exposed to 1 mM $CdCl_2$ for 24 h (Hossain et al. 2010) and in *Arabidopsis thaliana* plants exposed to 1 and 10 μM $CdCl_2$ for 7 days (Semane et al. 2007). In addition to these studies, it has been shown that other heavy metals, like Ni, Hg, As, and Cu, also led to an increase in GPX activity; however, Pb, Co, and Zn had no significant effects on GPX activity (Haluskova et al. 2009). These studies clearly indicated that GPX serves as an intrinsic tool in the antioxidant defense machinery to resist heavy metal-induced oxidative stress in plants. Aravind and Prasad (2005) evaluated the synergistic effects of Zn and Cd in *Ceratophyllum demersum* and found a reduction of 40% in the GPX activity in plants grown under Cd stress, whereas Zn supplementation for this plant while under Cd stress efficiently restored and even increased the GPX activity. These results suggest that Zn has the potential to stimulate detoxification mechanisms, thus controlling the ROS production induced by Cd stress (Aravind and Prasad 2005).

3.4.1.4 Glutathione Reductase

GR is known for its role in catalyzing the reduction of GSSG to GSH. It plays an important role in a cell's defense against ROS by maintaining the reduced states of pools of GSH and ASH, which in turn maintain the cellular redox state under heavy metal stress (Contour-Ansel et al. 2006). Evidence suggests that GR plays a significant role in detoxification of heavy metal-induced ROS through the ASH-GSH cycle. Aravind and Prasad (2005) reported a significant decline in GR activity in *Ceratophyllum demersum* under 10 μM Cd stress, which in turn increased upon Zn supplementation. Exposure of mung bean (1 mM $CdCl_2$ for 24 h) (Hossain et al. 2010) and *Arabidopsis thaliana* (1 and 10 μM for 7 days) (Semane et al. 2007) showed significant declines in GR activity upon exposure to Cd stress. Cultivar-specific responses in GR activity were also observed in mung bean cultivars Pusa 9531 (Cd tolerant) and PS16 (Cd sensitive) (Anjum et al. 2011). Increased GR activity in heavy metal-treated plants suggested a possible role for GR in regenerating GSH from

GSSR to increase the GSH/GSSR ratio and glutathione pool, which is used during phytochelatin synthesis and by DHAR and other GSH-dependent enzymes involved in ROS and MG metabolism (Hossain et al. 2010).

3.4.1.5 Detoxification of Methylglyoxal

The cytotoxic entity MG forms under heavy metal stress and is detoxified via the GSH-based glyoxalase pathway, which includes the enzymes glyoxalase I (Gly I) and glyoxalase II (Gly II). Gly I activity has been reported to increase in many plant species exposed to different heavy metal stresses, such as pumpkins (Cd stress) and tobacco (Zn stress) (Singla-Pareek et al. 2006). However, the activity of this enzyme was found to decrease in onion exposed to Cd stress (Hossain and Fujita 2009). It is well accepted that elevated levels of Gly I activity are required to remove excessive MG produced under stress conditions; however, Gly I activity varies due to differences in genetic backgrounds of plants, the type of heavy metal used, and differential regulation of different Gly I isoforms (Tuomainen et al. 2011). Gly II catalyzes the second step of MG detoxification and forms d-lactate while GSH is regenerated. Gly II activity was found to decrease in onion callus (Hossain and Fujita 2009) and mung bean seedlings (Hossain et al. 2010) subjected to Cd stress. Regeneration of Gly II activity in these experiments could be attributed to the Fe–Zn binuclear metal center that is essential for Gly II activity. It was concluded that Cd interacts with the Fe-Zn binuclear center and decreases the catalytic activity of Gly II. However, Gly II activity increased in transgenic tobacco plants overexpressing both Gly I and Gly II under Zn stress (5 mM $ZnCl_2$ for 24 h) (Singla-Pareek et al. 2006). The glyoxalate cycle helps in upregulation of GSH levels under heavy metal stress; as GSH is utilized in the formation of phytochelatins, it thus provides additional heavy metal tolerance for plants. The other antioxidant enzymes (SOD, APX, CAT, GST, GPX, and GR) also showed significant increases along with Gly I and Gly II, which shows the close association between antioxidant systems and glyoxalate systems of plants exposed to heavy metal stress (Hoque et al. 2008).

3.4.2 Physiological Strategies: Synthesis of Cellular Biomolecules

In addition to the antioxidant defenses, another important strategy adopted by plants for heavy metal detoxification is the synthesis of certain biomolecules that are capable of chelating with heavy metals, thus avoiding the hazardous impacts of the metals on plants. Implementation or upregulation of this mechanism depends on the plant species, the level of its metal tolerance, the plant growth stage, and the metal type (Solanki and Dhankhar 2011). Some of the important cellular biomolecules are phytochelatins, metallothioneins, proline, organic acids, and amino acids.

3.4.2.1 Phytochelatins

PCs are short-chained, thiol-rich repetitious peptides of low molecular weight that are synthesized in plants in response to heavy metal stress. PCs are synthesized from GSH via the enzyme PCS, which has a high affinity to bind to heavy metals when they are at toxic levels (Shukla et al. 2013). The enzyme PCS is activated when metal ions bind to it, resulting in the conversion of GSH to PCs. PCs have been reported to be used as biomarkers for the early detection of heavy metal stress in plants (Saba et al. 2013). PCs are found in the cytosol, where they form metal–phytochelatin complexes and are then transported to vacuoles (Song et al. 2014). This transport is believed to be mediated by a Mg-ATP dependent carrier or ATP binding cassette transporter (Sytar et al. 2013). The sulfhydryl and carboxyl groups of PCs attach to the heavy metals, such as Cd, Cu, Ag, Zn, Pb, Ni, and Ar, that are present in a cationic or anionic form (Gupta et al. 2013). Cd^{2+} ions show the most effective stimulation of PCs synthesis, where they are 4- to 6-fold stronger in inducing PCs synthesis than are Cu^{2+} or Zn^{2+} in cell cultures of *Rauwolfia serpentina* (Kotrba et al. 1999) or *Picea rubens* (Thangavel et al. 2007), respectively. PCs also contribute to homeostasis of Cu and Zn by providing transitory storage for these ions.

It has been found that both heavy metal-resistant and heavy metal-susceptible plants produce PCs, which accumulated in both the aboveground and belowground plant parts. However, studies have suggested that PCs are biosynthesized in the roots. Sunflower exposed to Cd intoxication showed twice the amount of PC in roots compared to leaves (Yurekli and Kucukbay 2003). However, Szalai et al. (2013) reported that long-term treatment of maize plants with Cd led to decreased amounts of PCs in roots and increased levels of the enzyme PCS in leaves. The contradictory results obtained in the two studies were attributed to either the feedback regulation process or substrate reduction.

3.4.2.2 Metallothioneins

MTs are another category of small cysteine-rich, low-molecular-weight, cytoplasmic metal binding polypeptides. Contrary to PCs, which are more effective against Cd stress, MTs are responsive to a number of heavy metals, such as Cu, Zn, Cd, and As (Yang and Chu 2011). MTs help in mitigating heavy metal toxicity in plants through cellular sequestration, homeostasis of intracellular metal ions, and metal transport adjustments (Kohler et al. 2004). In addition, MTs also play a significant role in ROS scavenging (Wong et al. 2004), maintenance of redox levels (Macovei et al. 2010), repair of plasma membranes (Mishra and Dubey 2006), and cell proliferation and DNA repair (Grennan 2011). Based upon the arrangement of cysteine residues, MTs are divided into four categories: type I MTs are mainly expressed in roots, type II MTs are primarily in shoots, type III MTs are induced in leaves and during fruit ripening, and type IV MTs are found abundantly in seeds (Yang and Chu 2011). Studies with *Arabidopsis* have shown that type I and type II MTs are involved in Cu chelation, while type IV has more affinity for Zn (Grennan 2011). In mature seeds of barley, MT III was reported to maintain the homeostasis of Zn and Cu while MT IV was involved in Zn storage (Hegelund et al. 2012), whereas in soybean MT I, MT II and MT III were involved in Cd detoxification while MT IV showed Zn affinity (Pagani et al. 2012). These studies suggest that different types of MTs function differently in maintaining homeostasis and detoxification processes in plants.

3.4.2.3 Proline

Proline is a five-carbon multifunctional amino acid in plants that acts as a metabolic osmolyte, a constituent of cell walls, a free radical scavenger, and an antioxidant (Szabados and Savouré 2010). Under heavy metal stress, proline production in higher plants is elevated, with proline playing a significant role in combating stress via adaptation, recovery, and signaling (Szabados and Savouré 2010). It has been proposed that accumulation of proline under heavy metal stress is actually due to a metal-induced water balance disorder, with proline acting as an osmoregulator (Clemens 2006). ROS scavenging by proline stimulated by heavy metal stress is primarily conducted through detoxifying hydroxyl radicals and quenching singlet oxygen (Tripathi and Gaur 2004). It was reported that proline can act as a metal chelator and metal stabilizer (Mishra and Dubey 2006). Several studies have shown that proline accumulation is more prominent in roots than the ariel parts when treated differentially with heavy metals like Pb, Cd, Cu, Hg, etc. (Fidalgo et al. 2013). Spraying proline on foliar parts of plants grown under heavy metal stress has been shown to be an effective method to reduce metal toxicity and stimulate the protective mechanisms in plants. It was observed that with exogenous application of proline, its endogenous production showed a significant increase (Hayat et al. 2013).

3.4.2.4 Organic Acids and Amino Acids

Organic acids (malate, citrate, oxalate, etc.) present in cells also act as efficient heavy metal chelators. Organic acids confer metal tolerance in two ways: by either external exclusion or internal tolerance. The mechanism of external exclusion involves organic acids being secreted out of the roots, thus hindering metal ions from entering and accumulating in the sensitive sites of the roots (Sharma and Dietz 2009). Citrate, which is synthesized in plants, plays an important role in Fe chelation.

Amino acids also play an important role in heavy metal detoxification in plants. Histidine is considered the most important amino acid in heavy metal metabolism. Carboxyl, amino, and imidazole groups make histidine an efficient metal chelator (Krämer 2010). In the Ni hyperaccumulator plants *Alyssum* spp. and *Thlaspi goesingense*, the concentration of histidine in xylem exudates was higher than in xylem of nonaccumulator species (McNear et al. 2010). Nicotinamine, a low-molecular-weight amino acid, chelates Cu, Fe, and Zn, forming metal complexes which are then stored in vacuoles. In *A. thaliana*, nicotinamine is involved in influx and efflux of Cu, Zn, and Fe via transport of the metals from one cell to another (Klatte et al. 2009).

3.4.3 Morphological Strategies: Symbiotic Associations with Mycorrhizal Fungi

The mutualistic symbiotic relationship that mycorrhizal fungi develop with the plants is one of the best strategies adopted by the plants to mitigate the deleterious effects of heavy metal stresses. The important mechanisms by which mycorrhizal fungi help in amelioration of heavy metal stresses include:

1. Acting as barriers by deporting metals within the cortical cells
2. Binding metals to the cell wall or sequestering them in the vacuoles
3. Releasing heat shock proteins and glutathione
4. Chelating metals in the soil matrix

Mycorrhizal fungi can extend their hyphae up to several meters outside the plant rooting zone and transfer the necessary elements into the plants (Ernst 2006). The metal-tolerant fungi can increase a plant's resistance towards heavy metals by restricting metal transfer into the plants. Arines et al. (1989) found that mycorrhiza *Trifolium pratense* (red clover) growing on acid soils had lower levels of Mn in its roots and shoots compared to nonmycorrhizal plants. In *Lolium perenne* L. (ryegrass) in symbiotic association with mycorrhizal fungi, the translocation of Cd, Ni, and Zn from soil to different parts of the plants was reduced as a result of immobilization of heavy metals from the soil (Takács et al. 2001). Changes in the soil pH due to mycorrhizal activities are one significant factor in immobilization of heavy metals (Bano and Ashfaq 2013). Mycorrhizal fungi even reduce availability of heavy metals like Zn, Cu, and Pb to plants growing in metal-contaminated soils, either through binding the metal to the soil organic matter or absorbing the metals into its own organs (cell wall, vacuoles, etc.), thereby reducing the possibility of metal uptake by the plants (Huang et al. 2005). Mycorrhizal fungi can reduce ROS induction and can stimulate the activity of antioxidative enzymes in plants. Abad and Khara (2007) showed that wheat plants colonized by mycorrhizal fungi that were subjected to toxic levels of Cd had higher levels of protective antioxidants, such as APX and GPX, in their aboveground and belowground plant parts than occurred in nonmycorrhizal plant parts. Reduction in lipid peroxidation and electrolytic leakage and increased activities of SOD and CAT were observed in mycorrhizal pigeon pea (*Cajanus cajan*) plants in Cd- and Pb-contaminated soils (Garg and Aggrawal 2012).

3.4.4 Phytoremediation Strategies

Phytoremediation is the term applied to an eco-friendly process of using plants to combat a stressed environment. The basic concept of phytoremediation is to utilize the potential of the green plants in such a way that they can efficiently clean up the environment without exploiting it. The potential for this technology in the tropics is high due to the prevailing climate conditions, which favor plant growth and stimulate microbial activity (Zhang et al. 2000). It is suitable to use this technique when metal contaminants have spread over a wide region or within the root zones of plants. Phytoremediation is an energy-efficient, aesthetically pleasing method of remediating sites with low to moderate levels of contamination. There are about more than 400 plants which can be exploited for phytoremediation or cleaning up of contaminated sites. Many *Brassica* species, such as *B. juncea* L. Czen, *B. napus* L., and *B. rapa* L., exhibit moderately enhanced Zn and Cd accumulation. They are also found to be most effective in removing Zn from contaminated soils. Phytoremediation of heavy metals is more effective when the contaminants are present in low to medium concentrations, as high metal concentrations inhibit plant and microbial growth and activity. Selection of plant species for phytoremediation is based upon the length of the roots, nature of the contaminants, the soil, and the regional temperature. Suitable plants must possess these abilities in order to be identified for remediation of soil: either high-biomass plants, such as willow (Landberg and Greger 1996) or those that have low biomass but hyperaccumulator characteristics, such as *Thlaspi* and *Arabidopsis* species.

Phytoremediation of heavy metals from the soil involves several processes, like phytoextraction, phytostabilization, phytostimulation, phytovolatization, phytodecomposition, and rhizofiltration, as shown in Figure 3.5. Each process has its own mechanism of minimizing the harmful effects of heavy metals on a plant's productivity. Table 3.1 summarizes the principal operative mechanism of each process. Plant species differ widely in their ability to accumulate heavy metals. Many authors have concluded that concentrations of metals in plants growing in the same soil vary between species and even between genotypes of a species. It has been reported that for phytoremediation, grasses are the most commonly evaluated plants. They have been more preferable in use for phytoremediation because, compared to trees and shrubs, herbaceous plants and especially grasses are characterized by rapid growth, producing a large amount of biomass, strong resistance, effective stabilization to soils, and the ability to remediate different types of soils. The grasses can even adapt to adverse conditions such as low nutrient contents and other environmental stresses. The large surface area of fibrous roots of grasses and their intensive penetration of soil reduce leaching, runoff, and erosion via stabilization of soil and thus offers advantages for phytoremediation.

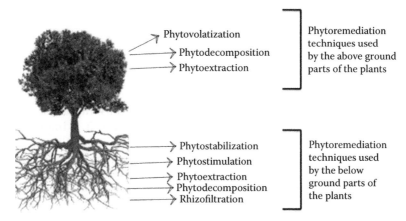

FIGURE 3.5
Different phytoremediation techniques adopted by plants.

TABLE 3.1

Plant Adaptation Strategies against Heavy Metal Stress

Type of Phytoremediation	Mechanism	Heavy Metals Treated
Phytoextraction	Plants tend to accumulate heavy metals in harvestable parts	Cd, Co, Cr, Ni, Hg, Pb, Se, Zn
Phytostabilization	Immobilization of metal through absorption and accumulation by roots via vacuolar sequestration or cell wall binding, precipitation within the root zone (rhizosphere) by formation of complexes	Pb, Zn, Cu
Phytostimulation	Root exudates promote development of microorganisms (bacteria and fungi) within rhizosphere that are capable of degrading heavy metal contaminants into nontoxic products	Petroleum hydrocarbons, polyaromatic benzene, toluene, etc.
Phytovolatilization	Plants take up contaminants, convert them into volatile forms, and release them into atmosphere via transpiration	Hg, Se, chlorinated solvents (carbon tetrachloride and trichloromethane)
Phytodecomposition	Both terrestrial and aquatic plants take up organic compounds and store or degrade them to less toxic or nontoxic by-products	TNT, DNT, nitrobenzene, nitrotoluene, chlorinated solvents
Rhizofiltration	Plant roots absorb, concentrate, or degrade heavy metals from contaminated liquid effluents	Cd, Co, Cr, Ni, Pb, Hg, Se, Zn

Phytoremediation, like other remediation technologies, has a range of both advantages and disadvantages. The most positive aspect of using phytoremediation is as follows: (1) more cost-effective; (2) more environmentally friendly; (3) applicable to a wide range of toxic metals, and (4) more aesthetically pleasing. On the other hand, phytoremediation presents some limitations. It is a lengthy process, and thus may take several years or longer to clean up a site, and it is only applicable for surface soils.

3.5 Conclusion

Increasing heavy metal contamination and the hazardous effects it has on plant and animal life has become an issue of serious concern among environmentalists. The problem of heavy metal contamination intensifies as it easily enters the food chain and can be transferred from one trophic level to another. Plants generally act as important targets or entry points through which the heavy metals enter the food chain. It is important for plants to develop certain defense strategies that help the plants to cope with the heavy metal-induced oxidative stress. A plant's heavy metal tolerance is a multigenic trait that can be controlled at multiple levels. In addition to other mechanisms, plants protect themselves from heavy metal toxicity through elevated levels of enzymatic and nonenzymatic components of antioxidant defense machinery and the glyoxalase defense system. Glutathione is the most important antioxidant involved in ameliorating heavy metal-induced oxidative stress. Glutathione not only works in direct scavenging of heavy metal-generated reactive oxygen species and methylglyoxal, a cytotoxic compound, but also it plays a significant role in heavy metal uptake, transport, and sequestration and formation of specific metal binding ligands, such as phytochelatin. Certain specific biomolecules that are synthesized in the plants in response to heavy metal stress also serve as an important means to overcome heavy metal stress in plants. Phytoremediation strategies also serve to play an important role in this context and therefore, future research should emphasize the development of these phytoremediation strategies for mitigating heavy metal stress in plants.

Acknowledgment

Financial assistance to S.T. by UGC [43-132/2014(SR)] is sincerely acknowledged.

References

Abad, K.J., Khara, J. 2007. Effect of cadmium toxicity on the level of lipid peroxidation and antioxidative enzymes activity in wheat plants colonized by arbuscular mycorrhizal fungi. *Pakistan Journal of Biological Sciences* 10(14):2413–2417.

Anjum, N.A., Umar, S., Ahmad, A., Iqbal, M., Khan, N.A. 2008. Ontogenic variation in response of *Brassica campestris* L. to cadmium toxicity. *Journal of Plant Interactions* 3(3):189–198.

Anjum, N.A., Umar, S., Iqbal, M., Khan, N.A. 2011. Cadmium causes oxidative stress in moongbean [*Vigna radiate* (L.) Wilczek] by affecting antioxidant enzyme systems and ascorbate-glutathione cycle metabolism. *Russian Journal of Plant Physiology* 58(1):92–99.

Ann, C., Karen, S., Jos, R. et al. 2011. The cellular redox state as a modulator in cadmium and copper responses in *Arabidopsis thaliana* seedlings. *Journal of Plant Physiology* 168:309–316.

Aravind, P., Prasad, M.N.V. 2005. Cadmium-zinc interactions in a hydroponic system using *Ceratophyllum demersum* L.: Adaptive ecophysiology, biochemistry and molecular toxicology. *Brazilian Journal of Plant Physiology* 17:3–20.

Arines, J., Vilarino, A., Sainz, M. 1989. Effect of different inocula of vesicular-arbuscular mycorrhizal fungi on manganese content and concentration in red-clover (*Trifolium pratense* L.) plants. *New Phytologist* 112:215–219.

Babu, N.G., Sarma, P.A., Attitalla, I.H., Murthy, S.D.S. 2010. Effect of selected heavy metal ions on the photosynthetic electron transport and energy transfer in the thylakoid membrane of the cyanobacterium, *Spirulina platensis*. *Academic Journal of Plant Sciences* 3:46–49.

Bano, A., Ashfaq, D. 2013. Role of mycorrhiza to reduce heavy metal stress. *Natural Science* 5(12):16–20.

Candan, N., Tarhan, L. 2003. The correlation between antioxidant enzyme activities and lipid peroxidation levels in *Mentha pulegium* organs grown in Ca^{2+}, Mg^{2+}, Cu^{2+}, Zn^{2+} and Mn^{2+} stress conditions. *Plant Science* 165:769–776.

Chugh, L.K., Sawhney, S.K. 1999. Photosynthetic activity of *Pisum sativum* seedling grown in presence of cadmium. *Plant Physiology and Biochemistry* 37:297–303.

Clemens, H. 2006. Toxic metal accumulation, responses to exposure and mechanisms of tolerance in plants. *Biochimie* 88(11):1707–1719.

Contour-Ansel, M.L., Torres-Franklin, M.H., De Carvalho, C., D'Arcy-Lameta, A., Zuily-Fodil, Y. 2006. Glutathione reductase in leaves of cowpea: Cloning of two cDNAs, expression and enzymatic activity under progressive drought stress, desiccation and abscisic acid treatment. *Annals of Botany* 98(6):1279–1287.

Cuypers, A.N.N., Vangronsveld, J., Clijsters, H. 2002. Peroxidases in roots and primary leaves of *Phaseolus vulgaris* copper and zinc phytotoxicity: A comparison. *Journal of Plant Physiology* 159(8):869–876.

DalCorso, G., Farinati, S., Maistri, S., Furini, A. 2008. How plants cope with cadmium: Staking all on metabolism and gene expression. *Journal of Integrative Plant Biology* 50(10):1268–1280.

Dalcorso, S., Farinati, S., Furini, A. 2010. Regulatory networks of cadmium stress in plants. *Plant Signaling and Behavior* 5(6):1–5.

Dixon, D.P., Edwards, R. 2010. Glutathione transferases. *The Arabidopsis Book*, vol. 8. American Society of Plant Biologists, Austin, TX.

Dubey, R.S. 2011. Metal toxicity, oxidative stress and antioxidative defense system in plants, pp. 177–203. *Reactive Oxygen Species and Antioxidants in Higher Plants*, Gupta, S.D. (ed.). Boca Raton, FL: CRC Press.

Ernst, W.H.O. 2006. Evolution of metal tolerance in higher plants. *Forest Snow and Landscape Research* 80:251–274.

Fidalgo, M., Azenha, A., Silva, F. et al. 2013. Copper-induced stress in *Solanum nigrum* L. and antioxidant defense system response. *Food and Energy Security* 2(1):70–80.

Foyer, H., Noctor, G. 2011. Ascorbate and glutathione: The heart of the redox hub. *Plant Physiology* 155(1):2–18.

Fusconi, A., Repetto, O., Bona, E., Massa, N., Gallo, C., Dumas-Gaudot, E., Berta, G. 2006. Effect of cadmium on meristem activity and nucleus ploidy in roots of *Pisum sativa* L. cv. Frisson seedlings. *Environmental and Experimental Botany* 58:253–260.

Gamalero, G., Lingua, G., Berta, A., Glick, B.R. 2009. Beneficial role of plant growth promoting bacteria and arbuscular mycorrhizal fungi on plant responses to heavy metal stress. *Canadian Journal of Microbiology* 55(5):501–514.

Garg, A., Aggarwal, N. 2012. Effect of mycorrhizal inoculations on heavy metal uptake and stress alleviation of *Cajanus cajan* (L.) Millsp. genotypes grown in cadmium and lead contaminated soils. *Plant Growth Regulation* 66(1):9–26.

Grennan, K. 2011. Metallothioneins: A diverse protein family. *Plant Physiology* 155(4):1750–1751.

Gullner, C., Uotila, M., Komives, T. 1998. Responses of glutathione and glutathione *S*-transferase to cadmium and mercury exposure in pedunculate oak (*Quercus robur*) leaf discs. *Botanica Acta* 111(1):62–65.

Gupta, H., Vandenhove, A., Inouhe, M. 2013. Role of phytochelatins in heavy metal stress and detoxification mechanisms in plants, pp. 73–94. *Heavy Metal Stress in Plants*. Berlin: Springer.

Hall, L. 2002. Cellular mechanisms for heavy metal detoxification and tolerance. *Journal of Experimental Botany* 53(366):1–11.

Haluskova, L., Valentovicova, K., Huttová, J., Mistrík, I., Tamas, L. 2009. Effect of abiotic stresses on glutathione peroxidase and glutathione *S*-transferase activity in barley root tips. *Plant Physiology and Biochemistry* 47(11–12):1069–1074.

Hayat, S., Hayat, Q., Alyemeni, M.N., Ahmad, A. 2013. Proline enhances antioxidative enzyme activity, photosynthesis and yield of *Cicer arietinum* L. exposed to cadmium stress. *Acta Botanica Croatica* 72(2):323–335.

Hegelund, N., Schiller, M., Kichey, T. et al. 2012. Barley metallothioneins: MT3 and MT4 are localized in the grain aleurone layer and show differential zinc binding. *Plant Physiology* 159(3):1125–1137.

Hoque, M.A., Banu, M.N.A., Nakamura, Y., Shimoishi, Y., Murata, Y. 2008. Proline and glycine betaine enhance antioxidant defense and methylglyoxal detoxification systems and reduce NaCl-induced damage in cultured tobacco cells. *Journal of Plant Physiology* 165(8):813–824.

Hossain, A., Hasanuzzaman, M., Fujita, M. 2010. Up-regulation of antioxidant and glyoxalase systems by exogenous glycine betaine and proline in mung bean confer tolerance to cadmium stress. *Physiology and Molecular Biology of Plants* 16(3):259–272.

Hossain, A., Piyatida, P., da Silva, J.A.T., Fujita, M. 2012. Molecular mechanism of heavy metal toxicity and tolerance in plants: Central role of glutathione in detoxification of reactive oxygen species and methylglyoxal and in heavy metal chelation. *Journal of Botany* 37:872875.

Hossain, M.A., Fujita, M. 2009. Purification of glyoxalase I from onion bulbs and molecular cloning of its cDNA. *Bioscience Biotechnology and Biochemistry* 73(9):2007–2013.

Hossain, M.Z., Hossain, M.D., Fujita, M. 2006. Induction of pumpkin glutathione *S*-transferases by different stresses and its possible mechanisms. *Biologia Plantarum* 50(2):210–218.

Hu, Y., Ge, Y., Zhang, C., Ju, T., Cheng, W. 2009. Cadmium toxicity and translocation in rice seedlings are reduced by hydrogen peroxide pretreatment. *Plant Growth Regulation* 59(1):51–61.

Huang, H., Gupta, D.K., Tian, S., Yang, X., Li, T. 2012. Lead tolerance and physiological adaptation mechanism in roots of accumulating and non-accumulating ecotypes of *Sedum alfredii*. *Environmental Science and Pollution Research* 19:1640–1651.

Huang, Y., Tao, S., Chen, Y.-J. 2005. The role of arbuscular mycorrhiza on change of heavy metal speciation in rhizosphere of maize in wastewater irrigated agriculture soil. *Journal of Environmental Sciences* 17(2):276–280.

Iqbal, N., Masood, A., Nazar, R., Syeed, S., Khan, N.A. 2010. Photosynthesis, growth and antioxidant metabolism in mustard (*Brassica juncea* L.) cultivars differing in Cd tolerance. *Agriculture Science China* 9:519–527.

Jha, A.B., Dubey, R.S. 2004. Arsenic exposure alters activity behaviour of key nitrogen assimilatory enzymes in growing rice plants. *Plant Growth Regulation* 43(3):259–268.

Kasprzak, K.S. 1995. Possible role of oxidative damage in metal induced carcinogenesis. *Cancer Investigation* 13:411–430.

Khan, N.A., Samiullah, Singh, S., Nazar, R. 2007. Activities of antioxidant enzymes, sulphur assimilation, photosynthetic activity and growth of wheat (*Triticum aestivum*) cultivars different in yield potentials under cadmium stress. *J Agronomy and Crop Science* 193:435–444.

Klatte, M., Schuler, M., Wirtz, M., Fink-Straube, C., Hell, R., Bauer, P. 2009. The analysis of *Arabidopsis* nicotianamine synthase mutants reveals functions for nicotianamine in seed iron loading and iron deficiency responses. *Plant Physiology* 150(1):257–271.

Klein, M., Burla, B., Martinoia, E. 2006. The multidrug resistance-associated protein (MRP/ABCC) subfamily of ATP-binding cassette transporters in plants. *FEBS Letters* 580(4):1112–1122.

Kohler, D., Blaudez, M., Chalot, L., Martin, F. 2004. Cloning and expression of multiple metallothioneins from hybrid poplar. *New Phytologist* 164(1):83–93.

Kotrba, P., Macek, T., Ruml, T. 1999. Heavy metal-binding peptides and proteins in plants. A review. *Collection of Czechoslovak Chemical Communications* 64(7):1057–1086.

Krämer, U. 2010. Metal hyperaccumulation in plants. *Annual Review of Plant Biology* 61:517–534.

Kumar, A., Prasad, M.N.V., Sytar, O. 2012. Lead toxicity, defense strategies and associated indicative biomarkers in *Talinum triangulare* grown hydroponically. *Chemosphere* 89:1056–1165.

Landberg, T., Greger, M. 1996. Differences in uptake and tolerance to heavy metals in *Salix* from unpolluted and polluted areas. *Applied Geochemistry* 11(1–2):175–180.

Macovei, L., Ventura, M., Dona, M., Balestrazzi, F.A., Carbonera, D. 2010. Effects of heavy metal treatments on metallothioneins expression profiles in white poplar (*Populus alba* L.) cell suspension cultures. *Analele Universitatii din Oradea-Fascicula Biologie* 18(2):274–279.

Maksymiec, W. 1997. Effect of copper on cellular processes in higher plants. *Photosynthetica* 34(3):321–342.

Marrs, K.A., Walbot, V. 1997. Expression and RNA splicing of the maize glutathione *S*-transferase Bronze2 gene is regulated by cadmium and other stresses. *Plant Physiology* 113(1):93–102.

May, M.J., Vernoux, T., Leaver, C., Van Montagu, M., Inze, D. 1998. Glutathione homeostasis in plants: Implications for environmental sensing and plant development. *Journal of Experimental Botany* 49(321):649–667.

McNear, D., Chaney, R., Sparks, D. 2010. *The Metal Hyperaccumulator Alyssum murale Uses Nitrogen and Oxygen Donor Ligands for Ni Transport and Storage.* USDA publication no. 239775. Washington, DC: USDA Agricultural Research Service.

Meharg, A.A. 1993. The role of the plasmalemma in metal tolerance in angiosperms. *Physiologia Plantarum* 88(1):191–198.

Metwally, A., Finkemeier, I., Georgi, M., Dietz, K.J. 2003. Salicylic acid alleviates the cadmium toxicity in barley seedlings. *Plant Physiology* 132(1):272–281.

Millar, A.H., Mittova, V., Kiddle, G. et al. 2003. Control of ascorbate synthesis by respiration and its implications for stress responses. *Plant Physiology* 133:443–447.

Mishra, S., Dubey, R.S. 2006. Heavy metal uptake and detoxification mechanisms in plants. *International Journal of Agricultural Research* 1(2):122–141.

Mysliwa-Kurdziel, B., Strzalka, K. 2002. Influence of metals on the biosynthesis of photosynthetic pigments, pp. 201–228. *Physiology and Biochemistry of Metal Toxicity and Tolerance in Plants*, Prasad, M.N.V., Strzalka (eds.). Dordrecht, Netherlands: Springer.

Pagani, M., Tomas, J., Carrillo, C. et al. 2012. The response of the different soybean metallothionein isoforms to cadmium intoxication. *Journal of Inorganic Biochemistry* 117:306–315.

Patra, M., Bhowmik, N., Bandopadhyay, B., Sharma, A. 2004. Comparison of mercury, lead and arsenic with respect to genotoxic effects on plant systems and the development of genetic tolerance. *Environmental and Experimental Botany* 52:199–223.

Qadir, S., Qureshi, M.I., Javed, S., Abdin, M.Z. 2004. Genotypic variation in phytoremediation potential of *Brassica juncea* cultivars exposed to Cd stress. *Plant Science* 167(5):1171–1181.

Rascio, N., Navari-Izzo, F. 2011. Heavy metal hyperaccumulating plants: How and why do they do it? And what makes them so interesting? *Plant Science* 180(2):169–181.

Rivetta, N., Negrini, R., Cocucci, M. 1997. Involvement of Ca^{2+}-calmodulin in Cd^{2+} toxicity during the early phases of radish (*Raphanus sativus* L.) seed germination. *Plant Cell and Environment* 20(5):600–608.

Saba, H., Jyoti, P., Neha, S. 2013. Mycorrhizae and phytochelators as remedy in heavy metal contaminated land remediation. *International Research Journal of Environment Sciences* 2(1):74–78.

Sangwan, P., Kumar, V., Joshi, U.N. 2014. Effect of chromium(VI) toxicity on enzymes of nitrogen metabolism in clusterbean (*Cyamopsis tetragonoloba* L.). *Enzyme Research* 2014:784036.

Schutzendubel, A., Polle, A. 2002. Plant responses to abiotic stresses: Heavy metal-induced oxidative stress and protection by mycorrhization. *Journal of Experimental Botany* 53(372):1351–1365.

Semane, A., Cuypers, K., Smeets, S. et al. 2007. Cadmium responses in *Arabidopsis thaliana*: Glutathione metabolism and antioxidative defence system. *Physiologia Plantarum* 129(3):519–528.

Sharma, S.S., Dietz, K.J. 2009. The relationship between metal toxicity and cellular redox imbalance. *Trends in Plant Science* 14(1):43–50.

Shukla, D., Tiwari, M., Tripathi, R.D., Nath, P., Trivedi, P.K. 2013. Synthetic phytochelatins complement a phytochelatin deficient *Arabidopsis* mutant and enhance the accumulation of heavy metal(loid)s. *Biochemical and Biophysical Research Communications* 434(3):664–669.

Singla-Pareek, S.L., Yadav, S.K., Pareek, A., Reddy, M.K., Sopory, S.K. 2006. Transgenic tobacco over-expressing glyoxalase pathway enzymes grow and set viable seeds in zinc-spiked soils. *Plant Physiology* 140(2):613–623.

Solanki, S., Dhankhar, R. 2011. Biochemical changes and adaptive strategies of plants under heavy metal stress. *Biologia* 66(2):195–204.

Song, W.Y. et al. 2014. Phytochelatin-metal(loid) transport into vacuoles shows different substrate preferences in barley and *Arabidopsis*. *Plant Cell and Environment* 37(5):1192–1201.

Sytar, O., Kumar, A., Latowski, D., Kuczynska, P., Strzałka, K., Prasad, M.N.V. 2013. Heavy metal-induced oxidative damage, defence reactions, and detoxification mechanisms in plants. *Acta Physiologiae Plantarum* 35(4):985–999.

Szabados, L., Savouré, A. 2010. Proline: A multifunctional amino acid. *Trends in Plant Science* 15:89–97.

Szalai, G., Krantev, A., Yordanova, R., Popova, L.P., Janda, T. 2013. Influence of salicylic acid on phytochelatin synthesis in *Zea mays* during Cd stress. *Turkish Journal of Botany* 37(4):708–714.

Takács, B., Biro, I., Voros, I. 2001. Arbuscular mycorrhizal effect on heavy metal uptake of ryegrass (*Lolium perenne* L.) in pot culture with polluted soil, pp. 480–481. *Plant Nutrition: Food Security and Sustainability of Agro-Ecosystems through Basic and Applied Research*, Horst, W.W.J., Scheck, M.K., Burkert, A. et al. (eds.), Dordrecht: Kluwer Academic Publishers.

Thangavel, P., Long, S., Minocha, R. 2007. Changes in phytochelatins and their biosynthetic intermediates in red spruce (*Picea rubens* Sarg.) cell suspension cultures under cadmium and zinc stress. *Plant Cell Tissue and Organ Culture* 88(2):201–216.

Tiwari, K.K., Singh, N.K., Rai, U.N. 2013. Chromium phytotoxicity in radish (*Raphanus sativus*): Effects on metabolism and nutrient uptake. *Bulletin of Environmental Contamination and Toxicology* 91:339–344.

Tripathi, N., Gaur, J.P. 2004. Relationship between copper and zinc-induced oxidative stress and proline accumulation in *Scenedesmus* sp. *Planta* 219(3):397–404.

Tuomainen, M., Ahonen, V., Karenlampi, S.O. et al. 2011. Characterization of the glyoxalase 1 gene *TcGLX1* in the metal hyperaccumulator plant *Thlaspi caerulescens*. *Planta* 233(6):1173–1184.

Van Assche, F., Clijsters, H. 1986. Inhibition of photosynthesis in Phaseolus vulgaris by treatment with toxic concentration of zinc: Effect on ribulose-1,5-biphosphate carboxylase/oxigenase. *Journal of Plant Physiology* 125:355–360.

Ventrella, A. et al. 2011. Interactions between heavy metals and photosynthetic materials studied by optical techniques. *Bioelectrochemistry* 77:19–25.

Verma, S., Dubey, R.S. 2003. Lead toxicity induces lipid peroxidation and alters the activities of antioxidant enzymes in growing rice plants. *Plant Science* 164:645–655.

Vesely, T., Tlustos, P., Szakova, J. 2011. The use of water lettuce (*Pistia stratoites* L.) for rhizofiltration of a highly polluted solution by cadmium and lead. *Journal of Phytoremediation* 13:859–872.

Wong, L., Sakamoto, T., Kawasaki, T., Umemura, K., Shimamoto, K. 2004. Down-regulation of metallothionein, a reactive oxygen scavenger, by the small GTPase OsRac1 in rice. *Plant Physiology* 135(3):1447–1456.

Yang, Z., Chu, C. 2011. Towards understanding plant response to heavy metal stress, pp. 59–78. *Abiotic Stress in Plants: Mechanisms and Adaptations*. Shanghai: InTech.

Yurekli, F., Kucukbay, Z. 2003. Synthesis of phytochelatins in *Helianthus annuus* is enhanced by cadmium nitrate. *Acta Botanica Croatica* 62(1):21–25.

Yusuf, M., Fariduddin, Q., Varshney, P., Ahmad, A. 2012. Salicylic acid minimizes nickel and/or salinity-induced toxicity in Indian mustard (*Brassica juncea*) through an improved antioxidant system. *Environmental Science and Pollution Research* 19:8–18.

Zhang, C.H., Ge, Y. 2008. Response of glutathione and glutathione *S*-transferase in rice seedlings exposed to cadmium stress. *Rice Science* 15(1):73–76.

Zhang, Q., Davis, L.C., Erickson, L.E. 2000. An experimental study of phytoremediation of methyl-tert-butyl ether (MTBE) in groundwater. *Journal of Hazardous Substance Research* 2000:2.

4

Molecular Mechanisms of Heavy Metal Hyperaccumulation in Plants

Anupa Fonia, Preeti Singh, Vijetna Singh, Dhananjay Kumar, and Bhumi Nath Tripathi

CONTENTS

4.1 Heavy Metal Pollution

The term heavy metal has been used frequently in the scientific literature in referring to more than 60 metallic elements of a density greater than 5 g/cm³ (Nies and Silver 1995). Many times, this term is also used collectively in discussing transition metals, some metalloids, lanthanoids, actinides, and organometallic compounds (Gadd 2009). Although low levels of some heavy metals, such as, Cu, Fe, Zn, Mn, Co, Mo, Ni, etc., are essential for the physiological machinery of living organisms, elevated concentrations of all kinds of heavy metals are invariably toxic to biota (Rai et al. 1981, Kotrba et al. 2009). Besides toxicity, heavy metals have bioenrichment tendencies due to their nonbiodegradable nature, and so they are considered more hazardous than other pollutants. The availability of heavy metals to living organisms is often low because the metals remain either immobilized in rocks in the form of ores or are restricted only to some specific locations, such as volcanic soils and hot springs. However, anthropogenic activities, particularly industries, have contributed much to today's enhanced metal concentrations in soils and aquatic systems. Several industries, like smelters and refining, electroplating, metal polishing, explosives, mineral mining, metal finishing, storage batteries, petroleum, welding, and alloy manufacture, have added enormous amounts of various heavy metals, such as cadmium, chromium, copper, nickel, lead, and zinc, to soils and natural water bodies. Certainly, if these toxic metals are allowed to enter the food chain, they will pose serious threats to living organisms, including plants and humans. Thus, developing appropriate processes for cleaning up these contaminants from the environment is an urgent need.

4.2 Conventional Methods for Removal of Heavy Metals

Physico-chemical methods, insofar as they are available for the removal of metal pollutants from soils and sediments, involve extraction of adsorbed metals via acids or chelating agents. Such extraction processes offer the possibility to recover metals through electrolysis. However, they are entirely conventional and comprise several inherent disadvantages. For example, they are inefficient, costlier, and frequently alter the natural property of the soil (Kotrba et al. 2009). Alternatively, some methods, rather than extracting metals from soil, try to stabilize them in the contaminated soil so that their spread to other environments can be minimized. Although this approach is helpful in developing strategies to manage heavy metal pollution to some extent, it does not provide an actual solution to the problem. Moreover, it also involves similar limitations, for example, with acid-based extraction methods. In other words, we can say that conventional methods are not helpful in effectively removing heavy metal pollution. Because of such dissatisfaction, efforts are continuously being made to develop new efficient and viable technologies. Presently, removal of metal ions by plants, that is, phytoremediation, has attracted a great deal of attention due to its cost-effective, efficient, and environmentally friendly features.

4.3 Phytoremediation Technologies

Phytoremediation technology refers to the well-organized use of plants to remove environmental pollutants, including heavy metals, from contaminated sites. In a general scheme of phytoremediation, plants are grown in contaminated soils or sediments until they reach a suitable growth period. During their growth, plants accumulate, degrade, or immobilize the contaminants, and then the plants can be subsequently harvested and processed for necessary disposal. The phytoremediation phenomenon may comprise various processes, such as phytoextraction, rhizofiltration, phytostabilization, phytovolatilization, and phytodegrdation (Figure 4.1). However, only the first four processes are useful for the removal of metal contaminants, and the fifth process is a slightly different strategy for the remediation of organic pollutants.

Phytoextraction is a process of uptake of heavy metals from the soil by plant roots and their translocation to harvestable aboveground parts. Sometimes, workers imprecisely use the term phytoextraction as a synonym for phytoremediation. However, in a true sense, it is only one particular process of phytoremediation technology. Certain plant species, compared to other plants, encompass an inherent capability to absorb extraordinarily large amounts of metals into their aerial parts. Such plants are useful for phytoextraction-based remediation of heavy metals and are widely regarded as metal hyperaccumulators. Interestingly, such plants employ a variety of physiological and molecular mechanisms to detoxify and sequester metal ions. Hyperaccumulator plants can be distinguished from nonhyperaccumulators based on their enhanced capability for metal uptake and faster

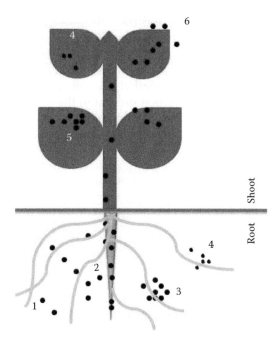

FIGURE 4.1
Diagrammatic presentation of various processes involved in phytoremediation of heavy metals from contaminated soils and sediments. (1) The metal ions; (2) rhizofiltration; (3) phytostabilization; (4) phytodegradation (below- and aboveground); (5) phytoextraction; (6) phytovolatization.

root-to-shoot translocation. However, there are several factors that can affect the phytoextraction performance of metal hyperaccumulator plants. A few of these factors are the extent of soil contamination, bioavailability of metals to roots, capability of aerial parts to accumulate metals, and interaction of plant roots with rhizospheric microbial flora.

Contrary to phytoextraction, rhizofiltration is the capability of plant roots to absorb, concentrate, and precipitate toxic metals from contaminated soils. In this process, metals absorbed from soil are retained within the roots. This is an inexpensive and effective process, and it causes fewer disturbances to the environment. Rhizofiltration is regarded as an *in situ* treatment in which the plants are first exposed to low concentrations of contaminants for acclimatization and thereafter the plants are transferred to contaminated sites for remediation purposes. Rhizofiltration is also an effective technology for removing metal ions from wastewaters. Several plant species, such as sunflower, Indian mustard, tobacco, rye, spinach, and corn, have been widely explored for their rhizofiltration capabilities. These species can efficiently localize a variety of metals, like Pb, Cd, Cu, Ni, Zn, and Cr, in their roots (U.S. EPA 2000).

Phytostabilization, also referred to as phytorestoration, uses plants roots and microbial interactions to limit mobility and bioavailability of contaminants by immobilizing the contaminants. Soil, sediment, and sludges are spread on sites in preparation for inactivation through phytostabilization (U.S. EPA 2000). This process inhibits direct contact with polluted soil by creating a barrier and interferes with soil erosion. It is used to remediate soils, sediments, or sludges contaminated with Cd, Cu, As, Zn, or Cr and has certain advantages, including its not requiring disposal of any material or biomass and its effectiveness for rapid immobilization of ground and surface water contaminants (U.S. EPA 2000).

In phytovolatilization, the absorption and transpiration capabilities of plants are used to release contaminants into the atmosphere in a volatile form (U.S. EPA 2000, Prasad and Freitas 2003). During passage through a plant's vascular system, contaminants are modified. Primarily mercury and selenium are removed through this method. Some plants are able to transform Se into the forms dimethyl selenide and dimethyl diselenide, which are reportedly less toxic forms than Se when present in the soil (Prasad and Frietas 2003). However, the major disadvantage of this technology is that once the volatilized contaminant is released into the atmosphere, no control can be implemented over its migration.

Plants employ both internal and external metabolic processes, and their interactions with microorganisms when breaking down organic contaminants in soil are known as phytodegradation (Prasad and Freitas 2003). Because the contaminants are degraded into a less harmful product under this process, it is also known as phytotransformation. Herbicides, insecticides, and chlorinated solvents are the major organic contaminants that are degraded through this method.

4.4 Metal Hyperaccumulator Plants

Hyperaccumulator plants have the capability to accumulate exceptional concentrations of heavy metals, particularly trace metals, in their aboveground parts without showing any signs of toxicity. However, this definition is not a precise one, as it fails to convey a quantitative sense. Hence, several researchers have suggested threshold concentrations for several metals in order to define the metal hyperaccumulation phenomenon more precisely. These threshold concentrations are 10,000 μg/g for Mn or Zn, 1,000 μg/g for Ni, Cu, or Se, and 100 μg/g for Cd, Cr, Pb, Co, Al, or As (Boyd 2004).

The majority of hyperaccumulators are found in environments where soils naturally contain high concentrations of metals, such as ultramafic soils (Boyd 2004). Recently, several researchers elegantly described the physiological mechanisms through which hyperaccumulator plants readily absorb, process, and sequester toxic metals in their tissues. However, the concept to use plants as a tool to remediate heavy metal contaminants from soil and aquatic environments is not recent. *Thlaspi caerulescens* and *Viola calaminaria* were the first plant species in which heavy metal accumulation was recognized, in 1865 (Assuncão et al. 2003). Baumann (1885) reported hyperaccumulation of Zn in the leaves of these two plant species. Later, Byers (1935) reported Se accumulation by *Astragalus* spp. Likewise, Minguzzi and Vergnano (1948) discovered Ni accumulation in *Alyssum bertolonii* (Brassicaceae). Despite these various reports, the idea to employ plants in bioremediation technology was introduced much later, in 1983 (Chaney 1983), and the term "hyperaccumulation" was coined by Brooks et al. in 1977 (Assuncão et al. 2003). At present, about 450 plants belonging to ~45 families of angiosperms and some pteridophytes have been reported for their capability to hyperaccumulate metal ions (Verbruggen et al. 2009, Rascio and Navari-Izzo 2011). Some important and widely studied metal hyperaccumulator plant species are listed in Table 4.1.

Several plants of the family Brassicaceae have been described for their capacities for hyperaccumulating trace metals (Zn, Ni, Mn, Cu, Co, and Cd), metalloids (As), and nonmetals (Se) (Verbruggen et al. 2009). Genera such as *Alyssum* and *Thlaspi* of the family Brassicaceae are well-studied metal hyperaccumulators. Plants such as *Elsholtzia splendens* and maize have been successfully explored for their Cu and Cd phytoextraction capabilities

TABLE 4.1

Some Important and Widely Reported Metal Hyperaccumulator Plants

Plant Name	Family	Metal
Helianthus annuus	Asteraceae	Ce
Brassica juncea	Brassicaceae	Pb, Zn, Se, Cr, Ag
Avena strigosa	Poaceae	Cd
Tagetes erecta	Asteraceae	Cd
Solanum photeinocarpum	Solanaceae	Cd
Thlaspi caerulescens	Brassicaceae	Cd
Azolla pinnata	Azollaceae	Cd
Vallisneria spiralis	Hydrodictaceae	Cd
Hordeum vulgare	Poaceae	Al
Pistia stratiotes	Araceae	Hg
Larrea tridentata	Zygophyllaceae	Cu
Haumaniastrum robertii	Lamiaceae	Cu
Euphorbia cheiradenia	Euphorbiaceae	Pb
Silene vulgaris	Caryophyllaceae	Zn
Thlaspi caerulescens	Brassicaceae	Mo
Corrigiola telephiifolia	Caryophyllaceae	As
Pteris quadriaurita	Pteridaceae	As
Walsura monophylla	Meliaceae	Ni
Aeolanthus biformifolius	Lamiaceae	Cu
Dicoma niccolifera	Asteraceae	Cr
Alyssum crenulatum	Brassicaceae	Ni
Alyssum heldreichii	Brassicaceae	Ni
Pteris vittata	Pteridaceae	Cr
Alyssum argenteum	Brassicaceae	Ni
Schima superba	Theaceae	Mg

Sources: Prasad, M.N.V., Freitas, H.M.O., *Molecular Biology and Genetics* 6:3–9, 2003; Verbruggen, N. et al., *New Phytologist* 181:759–776, 2009; Kotrba, P. et al., *Biotechnology Advances* 27:799–810, 2009; Krämer, U., *Annual Reviews of Plant Biology* 61:517–534, 2010.

(Verbruggen et al. 2009, Krämer 2010, Rascio and Navari-Izzo 2011). Among various heavy metals, Ni is reportedly the most preferred metal for hyperaccumulation, as >75% of hyperaccumulator plants accumulate this metal in their aboveground tissues (McGrath et al. 2001, Milner and Kochian 2008, Verbruggen et al. 2009). Conversely, Cd is considered the least preferred metal for hyperaccumulation. Cd has a very narrow range of plants for its accumulation, as only five plants species are known to hypeaccumulate it (Kotrba et al. 2009, Rascio and Navari-Izzo 2011). Alpine pennycress, *Thlaspi caerulescens*, is capable of hyperaccumulating Zn^{2+}, but occasionally it also accumulates Cd^{2+} and Ni^{2+} (Milner and Kochian 2008, Assuncão 2010, Rascio and Navari-Izzo 2011). Another potent metal hyperaccumulator plant species is *Brassica juncea*. This plant can accumulate a variety of metals, like Zn, Cd, and Cr, in larger amounts than does *Thlaspi caerulescens* (Gisbert et al. 2006, Kotrba et al. 2009). Likewise, several cereal crops, such as maize (*Zea mays*), sorghum (*Sorghum bicolor*), and alfalfa (*Medicago sativa*), also have high metal accumulation potentials (Milner and Kochian 2008, Assuncão 2010, Rascio and Navari-Izzo 2011). Some other plants, like *Astragalus racemosus* (Leguminosae), *Nicotiana glaucum* (Solanaceae), *Helianthus annuus* (Asteraceae), and *Liliodendron tulipifera* (Magnoliaceae), are also recognized as accumulators of extraordinarily high amounts of heavy metals and metalloids in their tissues (Milner and Kochian 2008, Assuncão 2010, Rascio and Navari-Izzo 2011).

4.5 Functions of Metal Hyperaccumulation in Plants

Although the potential of hyperaccumulators in cleaning up metal-contaminated sites is a beneficial aspect, the role of accumulated metals for these plants is debatable. Why do hyperaccumulators accumulate exceptionally high concentrations of metals in their tissues? What significance do these metals actually have for these plants? Several researchers think that the hyperaccumulation process performs several beneficial functions for hyperaccumulator plants (Boyd 2004). Hanson et al. (2004) reported that Se hyperaccumulated by *Brassica juncea* caused toxicity to invertebrate herbivores and also to fungal pathogens, and the Se thereby provided tolerance to these other environmental factors. Some researchers have reported that perennial hyperaccumulator plants elevate the levels of metals in their canopy-covered soil via production of metal-enriched litter. This process ultimately helps the plants prevent growth of other less tolerant species in their territory and thereby serves to ensure their dominance for obtaining natural resources. Such a phenomenon is also sometimes known as elemental allelopathy (Boyd 2004). In this regard, the effect of metal-enriched litter on decomposer communities and also on rates of nutrient cycling in such soils have not been well elucidated to date, and this topic demands serious efforts. According to another hypothesis, it has been suggested that hyperaccumulated metals foster drought resistance in hyperaccumulator species. However, no strong evidence for this is available to date. Nevertheless, the hypothesis that metals accumulated in hyperaccumulator tissues protect the plants from predators and pathogens has attracted a great deal of attention in the research community, and a number of groups are still focused on elucidating this phenomenon (Boyd 2004).

4.6 Mechanisms of Metal Hyperaccumulation in Plants

The hyperaccumulation of metals by plants involves various steps and mechanisms (Table 4.2 and Figure 4.2). However, the overall process can be described in three subsections: heavy metal uptake by roots, root-to-shoot translocation of absorbed metals, and detoxification and sequestration mechanisms.

4.6.1 Heavy Metal Uptake by Roots

The uptake of metal contaminants by hyperaccumulator plants occurs primarily through the root system, in which the principal mechanisms for preventing metal-induced toxicity are found. The root system provides an enormous surface area that absorbs and accumulates water and nutrients essential for growth, as well as metal contaminants. Plant roots also cause changes at the soil–root interface as they release inorganic and organic compounds (root exudates) in the rhizosphere. These root exudates affect the number and activity of microorganisms, the aggregation and stability of the soil particles around the

TABLE 4.2

Steps and Mechanisms of Heavy Metal Accumulation by Plants

Step	Plant Part(s) Involved	Functional Mechanism(s)
Metal uptake	Root and root exudates	Precipitation, mobilization, chelation, and chemical modification
	Cell wall	Adsorption, chemisorption, microprecipitation, and ion exchange
	Plasma membrane	Metal uptake and efflux by various transporters
Root-to-shoot translocation of absorbed metals	Vascular tissues	Uploading to xylem sap through help of various biological moieties, e.g., histidine, nicotinamine
Detoxification and sequestration	Trichomes	Localization of metals to nonmetabolic cells, such as trichomes
	Vacuole	Accumulation inside vacuole via tonoplast-localized transporters
	Biomolecules	Biomolecules, such as lipids, polypeptides, RNA, and DNA could bind metal ions due to availability of various functional groups
	Specially synthesized biomolecules	Metallothioneins and phytochelatins
	Biomolecules to cope with heavy metal-induced oxidative stress	Enzymes, HSPs, GSH, carotenoids, salicylic acid, proline, polyamines, nitric oxide

root, and the solubility and thereby availability of metal contaminants in the rhizospheric zone.

Metal transporters are an extremely important component of the cellular machinery and play a crucial role in the intracellular uptake of metals, as well as metal accumulation into their vacuoles (Kotrba et al. 2009, Krämer 2010, Rascio and Navari-Izzo 2011). Metal transporters generally have broad substrate specificities and are encoded by genes of different families, like ZIP, HMA, MATE (multidrug and toxin efflux), YSL (yellow strip1-like), and MTP. The enhanced uptake of Zn was noticed in *Thlaspi caerulescens* and *Alyssum halleri* roots after overexpression of some genes that belong to the ZIP (zinc-regulated transporter iron-regulated transporter proteins) family and encode plasma membrane-located cation transporters (Kotrba et al. 2009). Likewise, some workers reported the overexpression of ZIP family genes ZTN1 and ZTN2 in *Thlaspi caerulescens* and ZIP6 and ZIP9 in *Alyssum halleri* under Zn-deficient conditions (Assuncão et. al. 2003, Rascio and Navari-Izzo 2011). The broad substrate specificity of a metal transporter is also important for the hyperaccumulation of nonessential metals. Now, enough evidence is available that reflects that Zn transporters help in the influx of Cd (Rascio and Navari-Izzo 2011). In the case of *Thlaspi caerulescens* and *Alyssum halleri*, Cd uptake is reduced in the presence of Zn. However, in another Cd hyperaccumulator ecotype of *Thlaspi caerulescens*, the Cd uptake remained unaffected, even in the presence of Zn. Some workers also reported the role of Zn transporters in Ni accumulation (Kotrba et al. 2009, Assuncão 2010, Rascio and Navari-Izzo 2011). Similarly, phosphate transporters of root tissues are reported to assist As hyperaccumulation in *Pteris vittata*. In the same way, hyperaccumulation of Se by plants like *Astragalus bisulcatus* and *Stanleya pinnata* involves the active participation of sulfate transporters located in the plasma membranes of root tissues (Rascio and Navari-Izzo 2011). Recently, an aquaporin gene, PvTIP4;1, was reported to mediate As(III) uptake in *Pteris vittata* (He et al. 2016).

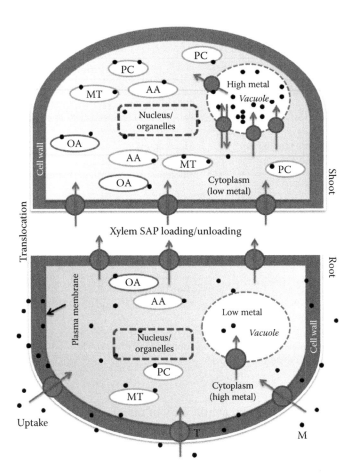

FIGURE 4.2
Diagrammatic presentation of surface binding, uptake, translocation, and intracellular accumulation of metal ions by a metal accumulator plant. A variety of transporters are involved in uptake of metal ions, and the plant cell has numerous intracellular sites for the binding and sequestration of metal ions. AA, amino acids; M, metal ion; MT, metallothioneins; OA, organic acids; PC, phytochelatins; T, transporter.

4.6.2 Root-to-Shoot Translocation of Absorbed Metals

The hyperaccumulator plants have very efficient mechanisms for translocating heavy metals from the root to the aboveground parts. Various reports have indicated enhanced uploading of heavy metals to the xylem due to overexpression of genes which are also present in nonhyperaccumulators but only in their normal or downregulated forms (Kotrba et al. 2009, Rascio and Navari-Izzo 2011). Heavy metal-transporting ATPases play a significant role in metal translocation. Metal hyperaccumulators are reported to have upregulated expression of ATPase genes upon exposure to metal-enriched soils (Rascio and Navari-Izzo 2011). Conversely, in nonhyperaccumulator plants there are downregulated forms of the genes encoding heavy metal-transporting ATPases.

Besides heavy metal-transporting ATPases, another kind of genes that are active in translocation of heavy metals are genes from the MATE families. FDR3, a gene that belongs to the MATE family, is overexpressed in roots of *Thlaspi caerulescens* and *Alyssum halleri*. The product of this gene is located on plasma membranes of root pericycle cells and is involved in xylem loading of citrate. This gene product is also necessary for maintaining homeostasis and transport of heavy metals like Fe and Zn (Krämer 2010, Rascio and Navari-Izzo 2011). The three YSL family genes, namely, TcYSL3, TcYSL5, and YSL7, are upregulated in *Thlaspi caerulescens*. The products of these genes are thought to be involved in xylem loading and translocation of nicotinamine–Ni complexes (Verbruggen et al. 2009, Rascio and Navari-Izzo 2011).

4.6.3 Detoxification and Sequestration of Heavy Metals in Hyperaccumulators

The elevated levels of both essential and nonessential heavy metals elicit adverse effects on the various biological processes of plants. However, both excluders and metal hyperaccumulators have evolved constitutive and adaptive mechanisms to combat the adverse effects of elevated levels of these metals. The various steps and mechanisms involved in metal hyperaccumulation by plants are listed in Table 4.2.

Metal excluders develop hypertolerance for heavy metals via various avoidance mechanisms, such as precipitation of metal ions in soil, active efflux outside the cell, and localized accumulation into root tissues. Contrary to this, hyperaccumulators are known to encourage metal uptake by roots and efficient translocation of them to the aboveground plant parts. Ultimately, after reaching the appropriate sites, these metals either accumulate in various cellular compartments or are transformed to less toxic forms through the help of certain enzymes and metal-chelating ligands (Rascio and Navari-Izzo 2011). In hyperaccumulators, an active loading of heavy metals to xylem is a common phenomenon that may involve upregulation of several genes. The hyperaccumulator plants have constitutive high expression of heavy metal transporter genes in plasma membranes, which results in high heavy metal uptake at the root surface. This particular mechanism is well studied in *Thlaspi caerulescens* for Zn and Cd accumulation (Rascio and Navari-Izzo 2011). The different kinds of chelators are also known to play roles in heavy metal uptake. Moreover, their utility in promoting translocation of metals from root to shoot systems is also well documented (Rascio and Navari-Izzo 2011).

The cell wall is the site at which a metal ion first encounters the cell, but its role in metal hyperaccumulation is controversial (Hall 2002). Next to the cell wall, the first living entity of the cell is the plasma membrane, which is regarded as the major site governing heavy metal uptake by plant cells. It has been found that tolerant cultivars generally have more cation exchangers in their plasma membranes than do sensitive cultivars (Rascio and Navari-Izzo 2011). Once heavy metals enter a plant cell, they are either ejected out of the cytosol or are accumulated into various cellular compartments, like vacuoles. The vacuolar accumulation of metal ions occurs through the help of different transporters located in the tonoplast. Inside the cell, the various organelles, such as chloroplasts and mitochondria, can also provide sites for the binding of metal ions (Kumar et al. 2016). The various probable sites available inside a plant cell for the binding of metal ions is shown in Figure 4.2.

Membrane transporters play important roles in conferring hyperaccumulation properties to the plants. The importance of the AhHMA4 transporter was deciphered in *A. halleri* for the hyperaccumulation of Cd and Zn by using the RNA interference technique (Kotrba et al. 2009, Rascio and Navari-Izzo 2011). This particular transporter belongs to heavy metal P-type ATPases and is located at the plasma membrane of cells. The AhHMA4 transporter is known to have several homologous forms which have only minor differences in their transit peptide. Moreover, the AtHMA4 gene product, along with loading Zn and Cd into xylem, also regulates the metal homeostasis network of plants. In the case of *Noccaea caerulescens* and *Alyssum halleri*, it has been found that alteration of a single key process can result in hypertolerance in these plants for a number of metals (Kotrba et al. 2009, Rascio and

Navari-Izzo 2011). The genes encoding vacuolar metal transporters, such as for Zn/H$^+$ anti-porter, AhMTP1, are capable of sequestering 20-fold-higher amounts of metals in leaves of *Alyssum halleri* than in *Alyssum thaliana* (Hall 2002, Kotrba et al. 2009, Rascio and Navari-Izzo 2011). Zn tolerance is enhanced by heterologous expression of AhMTP1 and AtMTP1 in yeast (Hall 2002) and in *Alyssum thaliana* by overexpression of AtMTP1 and NgMTP1 (Deng et al. 2007). The MTP1 gene increases the Zn metal sink in the shoots via expression of NgMTP1 in *Alyssum thaliana* (Deng et al. 2007). Overexpression of shoot-specific NgMTP1 reportedly triggers enhanced expression of Zn deficiency response genes in the root (Deng et al. 2007). In the case of *A. thaliana*, the overexpression of a fusion of green fluorescent protein (GFP) and AhIRT3, a gene of the ZIP family, caused a considerable increase in Zn and Fe accumulation (Krämer 2010, Rascio and Navari-Izzo 2011).

Metals inside plant cells are either becomes associated with biomolecules such as proteins, carbohydrates, lipids, DNA, etc., or accumulate in vacuoles. Metallothioneins are well-known metal binding proteins of low molecular weight which are synthesized inside cells. These proteins are rich in cysteine residues and also have specific patterns of sulfur-containing amino acids. The metal binding abilities of these proteins are the property of cysteine residues (Turner and Robinson 1995). There are three classes of metallothioneins. The first and the third are mainly reported in animal systems; while all metallothioneins reported in cyanobacteria belong to the second class. Initially, metallothioneins were reported in only three genera of cyanobacteria, but subsequently, their wide distribution among various genera and species of cyanobacteria and other plants has been shown.

Phytochelatins are the other low-molecular-weight peptides which act as metal ion-sequestering ligands (Rauser 1995). Phytochelatins consist of only three amino acids: glutamine, cystine, and glycine. The sequence of these amino acids can be represented as Gly-(Glu-Cys)$_n$-Gly, where $n = 2$ to 5. Phytochelatins are rapidly induced when the plant is exposed to metal ions. Besides metallothioneins and phytochelatins, a wide variety of ligands of carboxylic and amino acids, such as citrate, malate, and oxalate, histidine (His) and nicotianamine (NA), phosphate derivatives (phytate), and phytosiderophores are produced by plants. These ligands are also known to play an essential role in tolerance and detoxification of heavy metals.

Histidines have a high affinity for metal cations such as Zn^{2+}, Co^{2+}, Ni^{2+}, and Cu^{2+}. In Ni-hyperaccumulating plants like *Alyssum* spp., *Noccaea goesingense*, and *Noccaea caerulescens*, the concentration of free amino acids, such as histidine, was found to be considerably higher than in nonhyperaccumulator species (Verbruggen et al. 2009). The roles of histidine in Ni chelation and prompt Ni fluxing to the xylem have been well demonstrated in *Alyssum lesbiacum* (Rascio and Navari-Izzo 2011). Moreover, a number of ligands produced by plants show a strong correlation with exposure to metal ions. For example, a significant increase in production of citrate ligands has been reported in *Phaseolus vulgaris*, *Crotalaria cobalticola*, *Raufolia serpentina*, and *Silene cucubalus* upon their exposure to elevated levels of Ni and Co (Boyd 2007). Likewise, Al exposure enhances the secretion of oxalic acid from the roots of buckwheat (*Fagopyrum esculentum*). This plant also showed enhanced malate synthesis upon exposure to elevated levels of Zn. Besides playing a role in the translocation process, histidine and nicotinic acid also cause chelation of metals within plant cells and xylem sap (Zhao and McGrath 2003). The increased level of histidine is a result of enhanced transcription of genes encoding ATP-phosphoribosyl transferase (Krämer 2010, Assunacao 2010, Rascio and Navari-Izzo 2011). The low level of histidine causes a reduction in chelation and transport of Ni into the root vacuoles of *Thlaspi caserulescens*, while increased citrate synthesis shows positive effects.

Nicotianamine (NA) is a low-molecular-weight nonproteinogenic metal chelator. It is found in high concentrations in *Alyssum halleri* and *Noccaea caerulescens* (Rascio and Navari-Izzo 2011). This metal chelator moiety has a very strong affinity for binding Fe, Zn, Cu, and Ni. The levels of nicotianamine within a cell is determined by the transcript levels of three NA synthase genes, NAS2, NAS3, and NAS4 (Rascio and Navari-Izzo 2011). Other than metal chelation inside the cell, NA also plays important role in maintaining metal homeostasis and long-distance translocation in some solanaceous plants (Krämer 2010). The active role of NA in Zn homeostasis and hyperaccumulation has also been documented for *Alyssum halerii* (Rascio and Navari-Izzo 2011).

Another group of ligands involving phytate also form complexes with heavy metals. Vazquez et al. (1999) reported phytate-Al complex formation in Al-tolerant *Zea mays*. Although phytates are present in both metal-tolerant and metal-sensitive species, their levels are always higher in tolerant ecotypes than in sensitive ones. For example, the level of zinc-phytate globules was considerably higher in tolerant ecotypes of *Deschampsia caespitose* than in sensitive ones (van Steveninck et al. 1987).

Metal hyperaccumulator plants, compared to the nonhyperaccumulator plants, generally have a higher concentration of glutathione (GSH), cysteine, and O-acetylserine (Hall 2002, Kotrba et al. 2009, Rascio and Navari-Izzo 2011). These compounds are part of the antioxidant system of the cellular machinery and help the cell to cope with metal-induced oxidative stress responses. Likewise, upon metal exposure, high expression levels of genes encoding serine acetyltransferase (SAT) and glutathione reductase were reported in *Noccaea goesingense* (Kotrba et al. 2009). GSH possesses the dual activity, as it can play a direct role in metal chelation and also can be a substrate for the biosynthesis of phytochelatins (Rauser 1995). A cytotoxic compound such as methylglyoxal (MG) increases under abiotic stress conditions but is detoxified by the activity of the glyoxalase system in plants (Singla-Pareek et al. 2006). By the action of GSH S-transferases, GSH conjugates with xenobiotics and helps endow heavy metal tolerance to plants.

Heavy metals such as Zn, Cu, Cd, Hg, Al, and Cr are known to induce the formation of heat shock proteins (HSPs). These molecules are also known as molecular chaperones, as they function in normal folding of polypeptides and proteins. Under stress conditions, HSPs help in protection and repair of proteins. HSPs are found in all groups of organisms and are classified in various groups based on their molecular size. Ireland et al. (2004) reported Cd-induced accumulation of HSP70 in marine and freshwater algal species. Root tissues of *Armeria maritima* showed increased HSP17 levels upon exposure to Cu-rich soil (Hall 2002, Assuncão et al. 2003). Al-induced accumulation of HSP25 was reported in *Glycine max* (Zhen et al. 2007). HSPs have also been identified to play an essential role in the protection of metal transporters by minimizing the chance of their proteolytic degradation (Hall 2002).

Chemical modifications of heavy metals by plants into less toxic forms are another interesting adaptation for their detoxification. Several enzymes catalyze assimilation of metal ions, such as selenate, which is metabolized to dimethylselenide. Similarly, reductases can change the oxidation state of metals. For example, root and shoot tissues of *Eichhornia crassipes* showed the capacity to reduce toxic Cr(VI) to the less-toxic Cr(III) via the activity of reductases (Lytle et al. 1998). Similar kinds of enzymes are also reported to reduce Fe(III) and Cu (II) at root cell membranes prior to their uptake (Pilon-Smits and Pilon 2002).

Metal-responsive transcription factor 1 (MTF-1), identified in many plant species, confers tolerance against heavy metal stress by activating expression of genes responsible for metal uptake, transport, and detoxification. Several transcription factors, such as WRKY, basic leucine zipper (bZIP), ethylene-responsive factor (ERF), and myeloblastosis protein

(MYB), are known to play a significant role in controlling the expression of specific stress-related genes in response to Cd stress (Verbruggen et al. 2009).

A number of physiological and biological processes that make a plant resistant to biotic and abiotic stresses are governed by natural signal molecules, such as salicylic acid. Salicylic acid enhances SAT activity and GSH accumulation, which are known to protect plants from metal toxicity by reducing lipid peroxidation (Krämer 2010). The external supply of salicylic acid also enhances heavy metal tolerance in plants. Perhaps salicylic acid acts through regulation of reactive oxygen species levels by way of inducing gene expression and enhancing antioxidant activities inside the cell (Metwally et. al. 2003; Zhou et al., 2009).

Proline, an amino acid that accumulates under the response of stress, is known to provide enhanced protection against heavy metal stress in plants and algae (Tripathi and Gaur 2004). Even in the absence of metal ions, tolerant plants possess elevated levels of proline compared to their nontolerant relatives (Sharma and Dietz 2009). The proline minimizes metal stress by making the cellular environment a reducing one, along with the help of high levels of other antioxidants, such as GSH.

Polyamines are ubiquitous low-molecular-weight organic compounds that play important role in plant growth and development, and also in scavenging free radicals (Hall 2002). Spermidine, spermine, and their diamine precursor, putrescine, are the common polyamines found in plants. Polyamines are known to confer essential roles in the regulation of transcriptional and translational activities of processes which are crucial for plants as they lessen the negative effects of various biotic and abiotic stresses, including heavy metal stress (Hall 2002, Krämer 2010, Rasico and Navari-Izzo 2011).

Nitric oxide, a ubiquitous bioactive signaling molecule, protects plants from oxidative damage. It acts either as an antioxidant or as a stimulant of GSH synthesis. Nitric oxide can also enhance the activities of H_2O_2-scavenging enzymes. During the last decade, it has been shown that the exogenous supply of nitric oxide can also help plants alleviate heavy metal toxicity in a similar manner as endogenous nitric oxide does (Xiong 2009). Xu et al. (2011) showed an increase in nitric oxide in *Solanum nigrum* upon its exposure to elevated levels of Cd.

4.7 Use of Transgenic Plants for Metal Hyperaccumulation

It is now well established that several biochemical and physiological processes play roles in the metal hyperaccumulation capabilities of certain plants. Since all such processes are ultimately genetically regulated, it is quite possible to modify the metal accumulation capabilities of plants by using the tools of molecular biology. Recently, various genetically modified plants have been raised by transferring genes from one plant to another (Table 4.3). These transgenic plants have been found useful in deciphering the various mechanisms of metal hyperaccumulators. However, as ecological impacts of such plants are largely unknown, their real field application is still controversial. Thus, serious efforts are needed to resolve the issues related to their application in the field.

Gene transfer of heavy metal hyperaccumulators to nonhyperaccumulators enhances accumulation of heavy metals. The transfer of the phytochelatin synthetase (PCS) gene from *Cynodon dactylon* to tobacco enhanced 3.88-fold the amount of phytochelation and subsequently enhanced Cd accumulation about 3-fold (Li and Chen 2006). The expression of the phytochelatin synthase gene of *Arabidopsis* in a transgenic version of *Brassica juncea*

TABLE 4.3

Some Examples of Genetically Modified Plants Developed for Phytoremediation of Heavy Metals in Soil

Transgenic Plant	Gene	Source Organism	Remark	Active Mechanism
N. tabacum	MT-I	*Mus musculus*	Developed tolerance against Cd	Synthesis of metal-binding polypeptides and proteins
	MT-II	*Homo sapiens*	Developed tolerance against Cd	
	CUP1	*S. cerevisiae*	Enhanced Cu accumulation capability	
	HisCUP1	*Recombinant fusion*	Enhanced Cd accumulation capability	
A. thaliana	merP	*Bacillus megaterium*	Developed tolerance against mercury exposure and increased accumulation in young seedlings	
N. tabacum	AtPCS1	*A. thaliana*	Enhanced Cd accumulation capability	Overproduction of phytochelatin synthase
N. glauca	TaPCS1	*T. aestivum*	Improved Pb, Cd, Zn, Cu, and Ni accumulation	
A. thaliana	GSH1	*S. cerevisiae*	Increased accumulation of Cd and As	Overproduction of enzymes involved in synthesis of GSH and utilization of sulfur
Populus sp.	gshI	*E. coli*	Increased Cd accumulation ability from media and soil	
N. tabacum	OAS-TL	*S. oleracea*	Improved Cd accumulation in shoots but reduced in roots	
B. juncea	SMT and APS1	*A. bisulcatus* and *A. thaliana*	Increased Se accumulation	
A. thaliana	ZAT	*A. thaliana*	Enhanced Zn accumulation in root	Overproduction of metal transporters
	YCF1	*S. cerevisiae*	Improved Cd and Pb accumulation capability	
	NtCBP4	*N. tabacum*	Improved Ni and Pb accumulation capability	

Source: Kotrba, P. et al., *Biotechnology Advances* 27:799–810, 2009.

increased Cd and As accumulation (Kotrba et al. 2009). The overexpression of PCS in garlic (*Allium sativum*) (Zhang et al. 2005) and the marine alga *Dunaliela tertiolecta* (Tsuji et al. 2002) help in stress mitigation caused by heavy metals. *Arabidopsis thaliana* PCR (AtPCS 1) expression enhanced by 2-fold Cd tolerance by increasing phytochelatin content up to 15%. As plant systems have several enzymes that can change the oxidation state of metal ions to less toxic states, the overexpression of genes encoding such enzymes could also help in improving the metal accumulation capability of plants. In this regard, the overexpression of the *merA* gene, the product of which plays a role in conversion of highly toxic Hg^{2+} to the less toxic elemental Hg, was studied in yellow poplars. The results suggested that transgenic plants performed metal transformation 10 times more efficiently compared to the wild type (Doty 2008, Kotrba et al. 2009).

Comparative transcriptome analysis of transgenic and wild-type plants showed that metal sequestration traits of hyperaccumulators depend on the constitutive overexpression of genes belonging to the CDF (cation diffusion facilitator) family (Rascio and Navari-Izzo 2011). Since genes of this family encode metal transporters, it is also named the MTP family. The metal transporters are known to have an immense role in compartmentation of intracellular metals. The gene HMA3 encodes a vacuolar P1B-ATPase that is involved in Zn compartmentation, and CAX genes, which encode members of a cation exchanger family, are known to mediate Cd sequestration (Kotrba et al. 2009). The overexpression of genes related to the antioxidant machinery of the cell also presents potential uses in improving metal accumulation in plants based on enhanced synthesis of antioxidants, resulting in a better capability for the transgenic plant to manage the oxidative stress that stems from increased accumulation of metals inside a plant (Rascio and Navari-Izzo 2011).

Integration of metallothioneins into transgenic plants was reported for the first time in 1989. Now, metallothioneins have been widely exploited to increase Cd tolerance or accumulation in *Escherichia coli* cells and also in plants. Metallothionein genes, such as the MT-II gene and the MT-II fused to the glucuronidase gene, when introduced into tobacco under control of the CaMV 35S promoter with a double enhancer (35S2) was shown to have enhanced Cd^{2+} tolerance. Transgenic lines expressing the 35S2-hMT-II gene that were grown in a greenhouse and a field with a negligible amount of Cd accumulation showed a significant modification in Cd distribution (Kotrba et al. 2009). MT-expressing plants show a reduced level of Cd translocation to leaves compared to control plants and accumulate a higher amount of it in stems or roots. The overexpression of the *Salmonella enterica* serovar Typhimurium ATP phosphoribosyl transferase gene (StHisG) in transgenic plants in the presence of Ni in the growth medium resulted in 2-fold-higher histidine levels compared to the wild type and also increased biomass production (Wycisk 2004).

Transfer of polyamine biosynthetic genes to transgenic plants also confers tolerance to heavy metals (Franchin et al. 2007). Specialized protein transporters present at the tonoplasts of plants and yeast cells, such as the ATP-binding cassette (ABC), also help in metal sequestration within vacuoles (Krämer 2010). Constitutive overexpression of genes encoding transport systems, such as P1B-type ATPases, is known to play an important role in xylem loading for fast and efficient root-to-shoot translocation. The increased expression of HMA4 genes plays a crucial role in upregulating the expression of FDR3, TcYSL3, TcYSL5, YSL7, aquaglyceroporins, and MATE genes, which are known to intervene in the hyperaccumulation of metals in shoots (Kotrba et al. 2009, Krämer 2010).

In roots of Ni hyperaccumulator plants, such as *Alyssum* spp. and *Thlaspi* spp., histidine is known to be an important metal-chelating ligand (Verbuggen et al. 2009, Kotrba et al. 2009, Krämer 2010). The level of histidine inside the cell is decided by the level of expression of the TP-PRT1 gene, which encodes ATP-phosphoribosyl transferase, an enzyme required in the first step of the histidine biosynthesis pathway. The constitutive overexpression of the TP-PRT1 gene causes an increase in the histidine pool of the cell, and thus it promotes metal xylem loading in the form of metal–His complexes (Verbuggen et al. 2009, Kotrba et al. 2009, Krämer 2010). Likewise, the overexpression of genes encoding enzymes of the nicotinamine biosynthesis pathway results in a greater accumulation of nicotinamide–metal complexes in root tissues of the Ni hyperaccumulators *Thlaspi caerulescens* and *Alyssum halleri*, compared to nonhyperaccumulators (Verbruggen et al. 2009, Kotrba et al. 2009, Krämer 2010).

4.8 Conclusions and Future Perspectives

Hyperaccumulator plants have attracted biologists because of the plants' unique traits of extraordinary metal accumulation in aboveground parts. These plants have opened new dimensions in developing techniques to remediate metal-contaminated soils and also for the mining of precious metals. Some metal hyperaccumulator plant species have shown specificity in accumulating certain metals, such as Pb and Ni. Although the exact reason(s) behind such specificities is yet to be elucidated, the traits might be beneficial in remediating soils contaminated by these metals. Moreover, there is a great possibility for transfer of genes of hyperaccumulator species into high-growth biomass-producing plants, and thus more efficient metal hyperaccumulators might be raised. Some positive reports are now also available in this regard, but before their application to real fields, a lot remains to be done, particularly determination of their ecological impacts on other plant species. As accumulated metals can cause toxicity to predators, such as insects, the applied uses of hyperaccumulators in phytoremediation processes may also raise environmental concerns. Moreover, serious efforts are also essential to find the environmental requirements of such transgenic plants so that the best possible growth and metal accumulation can simultaneously be ensured upon their transfer to real metal-contaminated soils.

Acknowledgments

We thank the Head of the Department of Botany, School of Life Sciences, H.N.B. Garhwal University, Srinagar, Garhwal, for use of their necessary facilities. A.F. and P.S. acknowledge the University Grants Commission, New Delhi, for the award of university research fellowships. D.K. and B.N.T are grateful to Professor N. K. Dubey, Department of Botany, Institute of Science, Banaras Hindu University, Varanasi, for encouragement and support.

References

Assuncão, A.G.L. 2010. *Arabidopsis thaliana* transcription factors bZIP19 and bZIP23 regulate the adaptation to zinc deficiency. *Proceeding of the National Academy of Sciences USA* 107:10296–10301.

Assuncão, A.G.L., Schat, H., Aarts, M.G.M. 2003. *Thlaspi caerulescens*, an attractive model species to study heavy metal hyperaccumulation in plants. *New Phytologist* 159:351–360.

Baumann, A. 1885. Das Verhalten von Zinksatzen gegen Pflanzen und Imboden. *Landwirtsch* 3:1–53.

Boyd, R.S. 2004. Ecology of metal hyperaccumulation. *New Phytologist* 162:563–567.

Boyd, R.S. 2007. The defense hypothesis of elemental hyperaccumulation: Status, challenges and new directions. *Plant and Soil* 293:153–176.

Byers H.G. 1935. Selenium occurrence in certain soils in the United States, with a discussion of the related topics. *USDA Technical Bulletin* 482:1–47.

Chaney R.L. 1983. Plant uptake of inorganic waste constitutes, pp. 50–76. *Land Treatment of Hazardous Wastes*, J.F. Parr, P.B. Marsh, J.M. Kla (eds.). Park Ridge, IL: Noyes Data Corp.

Deng, D.M., Shu, W.S., Zhang, J., Zou, H.L., Ye, Z.H., Wong, M.H. 2007. Zinc and cadmium accumulation and tolerance in populations of *Sedum alfredii*. *Environmental Pollution* 147:381–386.

Doty, S.L. 2008. Enhancing phytoremediation through the use of transgenics and endophytes. *New Phytologist* 179:318–333.

Franchin, C., Fossati, T., Pasquini, E., Lingua, G., Castiglione, Torrigiani, P., Biondi, S. 2007. High concentrations of zinc and copper induce differential polyamine responses in micropropagated white poplar (*Populus alba*). *Physiologia Plantarum* 130:77–90.

Gadd, G.M. 2009. Heavy metal pollutants: Environmental and biotechnological aspects, pp. 321–334. *Encyclopedia of Microbiology*, M. Schaechter (ed.). Oxford: Elsevier.

Gisbert, C., Clemente, R., Navarro-Aviñó, J., Baixauli, C., Ginér, A., Serrano, R., Walker, D.J., Bernal, M.P. 2006. Tolerance and accumulation of heavy metals by Brassicaceae species grown in contaminated soils from Mediterranean regions of Spain. *Environmental and Experimental Botany* 56:19–27.

Hall, J.L. 2002. Cellular mechanisms for heavy metal detoxification and tolerance. *Journal of Experimental Botany* 53:1–11.

Hanson, B., Lindblom, S.D., Loeffler, M.L., Pilon-Smits, E.A.H. 2004. Selenium protects plants from phloem-feeding aphids due to both deterrence and toxicity. *New Phytologist* 162:655–662.

He, Z., Yan, H., Chen, Y., Shen, H., Xu, W., Zhang, H., Shi, L., Zhu, Y.G., Ma, M. 2016. An aquaporin PvTIP4;1 from *Pteris vittata* may mediate arsenite uptake. *New Phytologist* 209:746–761.

Ireland, H.E., Harding, S.J., Bonwick, G.A., Jones, M., Smith, C.J., Williams, J.H.H. 2004. Evaluation of heat shock protein 70 as a biomarker of environmental stress in *Fucus serratus* and *Lemna minor*. *Biomarkers* 9:139–155.

Kotrba, P., Najmanova, J., Macek, T., Ruml, T., Mackova, M. 2009. Genetically modified plants in phytoremediation of heavy metal and metalloid soil and sediment pollution. *Biotechnology Advances* 27:799–810.

Krämer, U. 2010. Metal hyperaccumulation in plants. *Annual Reviews of Plant Biology* 61:517–534.

Kumar, D., Pandey, L.K., Gaur, J.P. 2016. Metal sorption by algal biomass: From batch to continuous system. *Algal Research* 18:95–109.

Li, H.Y., Chen, Z.S. 2006. The influence of EDTA application on the interactions of cadmium, zinc, and lead and their uptake of rainbow pink (*Dianthus chinensis*). *Journal of Hazardous Materials* 137:1710–1718.

Lytle, C., Lytle P., Yang N. 1998. Reduction of Cr(VI) to Cr(III) by wetland plants: Potential for in situ heavy metal detoxification. *Environmental Science and Technology* 32:3087–3093.

McGrath, S.P., Lombi, E., Zhao, F.J. 2001. What's new about cadmium hyperaccumulation? *New Phytologist* 149:2–3.

Metwally, A., Finkemeier, I., Georgi, M., Dietz, K. 2003. Salicylic acid alleviates the cadmium toxicity in barley seedlings. *Plant Physiology* 132:272–281.

Milner, M.J., Kochian, L.V. 2008. Investigating heavy-metal hyperaccumulation using *Thlaspi caerulescens* as a model system. *Annals of Botany* 102:3–13.

Minguzzi, C., Vergnano, O. 1948. Il contenuto in nickel nelle ceneri di Alyssum bertolonii. *Atti della Societa Toscana di Scienze* A55:49–77.

Nies, D.H., Silver, S. 1995. Ion efflux systems involved in bacterial metal resistances. *Journal of Industrial Microbiology* 14:186–199.

Pilon-Smits, E., Pilon, M. 2002. Phytoremediation of metals using transgenic plants. *Critical Reviews in Plant Sciences* 21:439–456.

Prasad, M.N.V., Freitas, H.M.O. 2003. Metal hyperaccumulation in plants: Biodiversity prospecting for phytoremediation technology. *Molecular Biology and Genetics* 6:3–9.

Rai, L.C., Gaur, J.P., Kumar, H.D. 1981. Phycology and heavy-metal pollution. *Biological Reviews of the Cambridge Philosophical Society* 56:99–151.

Rascio, N., Navari-Izzo, F. 2011. Heavy metal hyperaccumulating plants: How and why do they do it? And what makes them so interesting. *Plant Science* 180:169–181.

Rauser, W.E. 1995. Phytochelatins and related peptides: Structure, biosynthesis and function. *Plant Physiology* 109:1141–1149.

Sharma, S.S., Dietz, K.J. 2009. The relationship between metal toxicity and cellular redox imbalance. *Trends in Plant Science* 14:43–50.

Singla-Pareek, S.L., Yadav, S.K., Pareek, A., Reddy, M.A., Sopory, S.K. 2006. Transgenic tobacco overexpressing glyoxalase pathway enzymes grow and set viable seeds in zinc-spiked soils. *Plant Physiology* 140:613–623.

Tripathi, B.N., Gaur, J.P. 2004. Relationship between copper- and zinc-induced oxidative stress and proline accumulation in *Scenedesmus* sp. *Planta* 219:397–404.

Tsuji, N., Hirayanagi, N., Okada, M., Miyasaka, H., Hirata, K., Zenk, M.H., Miyamoto, K. 2002. Enhancement of tolerance to heavy metals and oxidative stress in *Dunaliella tertiolecta* by Zn-induced phytochelatin synthesis. *Biochemical and Biophysical Research Communications* 293:653–659.

Turner, J.S., Robinson, N.J. 1995. Cyanobacterial metallothioneins: Biochemistry and molecular genetics. *Journal of Industrial Microbiology* 14:119–125.

U.S. Environmental Protection Agency. 2000. *Introduction to Phytoremediation*. EPA report 600/R-99/107. U.S. EPA Office of Research and Development, Cincinnati, OH.

van Steveninck, R.F.M., van Steveninck, M.E., Fernando, D.R., Horst, W.J., Marschner, H. 1987. Deposition of zinc phytate in globular bodies in roots of *Deschampsia caespitosa*: A detoxification mechanism? *Journal of Plant Physiology* 131:247–257.

Vazquez, M.D., Poschenrieder, C., Corrales, I., Barcelo, J. 1999. Change in apoplastic aluminum during the initial growth response to aluminum by roots of a tolerant maize variety. *Plant Physiology* 119:435–444.

Verbruggen, N., Hermans, C., Schat, H. 2009. Molecular mechanisms of metal hyperaccumulation in plants. *New Phytologist* 181:759–776.

Wycisk, K., Kim, E.J., Schroeder, J.I., Krämer, U. 2004. Enhancing the first enzymatic step in the histidine biosynthesis pathway increases the free histidine pool and nickel tolerance in *Arabidopsis thaliana*. *Federation of European Biochemical Societies Letters* 578:128–134.

Xiong, L. 2009. Deficient SUMO attachment to Flp recombinase leads to homologous recombination-dependent hyperamplification of yeast 2µ circle plasmid. *Molecular Biology of the Cell* 20:1241–1251.

Xu, J., Wang, W., Sun, J., Zhang, Y., Ge, Q., Du, L., Yin, H., Xiaojin, L. 2011. Involvement of auxin and nitric oxide in plant Cd-stress responses. *Plant and Soil* 346:107–119.

Zhang, H., Xu, W., Guo, J., He, Z., Ma, M. 2005. Coordinated responses of phytochelatins and metallothioneins to heavy metals in garlic seedlings. *Plant Science* 69:1059–1065.

Zhao, F.J., McGrath S. 2003. Phytoextraction of metals and metalloids from contaminated soils. *Current Opinion in Biotechnology* 14:277–282.

Zhen, Y., Qi, J.L., Wang, S.S., Su, J., Xu, G.H., Zhan, M.S., Miao, L., Peng, X.X., Tian, D., Yang, Y.H. 2007. Comparative proteome analysis of differentially expressed proteins induced by Al toxicity in soybean. *Physiologia Plantarum* 131:542–554.

Zhou, Z.S., Guo, K., Elbaz, A.A., Yang, Z.M. 2009. Salicylic acid alleviates mercury toxicity by preventing oxidative stress in roots of *Medicago sativa*. *Environmental and Experimental Botany* 65:27–34.

5

Effects of Heavy-Metal Accumulation on Plant Internal Structure and Physiological Adaptation

B. B. Maruthi Sridhar, Fengxiang X. Han, and Yi Su

CONTENTS

5.1 Introduction

Heavy-metal contamination increases in the environment as a result of several anthropogenic activities such as mining, smelting, electroplating, energy and fuel production, land application of sewage sludge, pesticide and herbicide applications, and melting operations (Welch 1995, Samarghandi et al. 2007, Nouri et al. 2009). Heavy-metal contaminants at higher concentrations are regarded as environmental pollutants. Heavy metals are natural constituents of the earth's crust and their principal characteristics are an atomic density of more than 5 g/cm^3 and an atomic number of greater than 20 (Alloway and Ayres 1997, Adriano 2001). The most common heavy-metal contaminants found in the environment are Cd, Cr, Cu, Hg, Pb, and Zn. The source of heavy metals in soils can be either natural or anthropogenic. Heavy metals in the soil from anthropogenic sources tend to be more mobile and can be taken up by and accumulated in living organisms. Out of all the heavy metals that are bioavailable to plants, Fe, Mn, Zn, Cu, Mg, Mo, and Ni are required in smaller quantities and are referred to as essential micronutrients. Cd, Sb, Cr, Pb, As, Co, Ag, Se, and Hg are the nonessential heavy metals, which are not required for plant growth and have no significant biological and physiological functions (Schutzendubel and Polle 2002, Rascio and Navari-Izzo 2011). Based on their physicochemical properties, bioavailable heavy metals are further divided into two groups: namely, redox-active metals such as Cr, Cu, Mn, and Fe and non-redox-active metals such as Cd, Ni, Hg, Zn, and Al (Jozefczak et al. 2012).

Based on the levels of metals in aboveground plant parts, plants can be distinguished as accumulators and hyperaccumulators. The hyperaccumulator plants can accumulate and sequester the metals at very high concentrations. Some of the known metal uptake concentrations in these hyperaccumulator plants are above 100 mg/kg for Cd; above 500 mg/kg for Co, Cu, and Ni; and above 1000 mg/kg for Mn and Zn (Brooks 1998). High tolerances to Cd, Co, Cu, Ni, and Zn have evolved in the plant orders of Brassicales, Caryophyllales, Plumbaginales, and Poales around the world, whereas tolerances to Co, Cu, and Ni are found in Asterales, Commelinales, Cyperales, Ericales, Fabales, Lamiales, and Liliales (Ernst 2006).

Heavy metals at toxic levels have the capability to interact with several vital cellular biomolecules, such as nuclear proteins and DNA, leading to excessive augmentation of reactive oxygen species. Heavy metals also cause severe morphological, metabolic, and physiological anomalies in plants ranging from chlorosis of the shoot to lipid peroxidation and protein degradation (Emamverdian et al. 2015). The cell layers of endodermis and exodermis in the root tissues and the cell walls of the xylem and cortical parenchyma thicken as plants are exposed to high concentrations of metal contaminants. When penetrating the roots, heavy metals are predominantly accumulated and translocated in the cell wall system (MacFarlane and Burchett 2000, Gomes et al. 2011), with the exodermis and endodermis constituting an effective barrier to the movement of these ions (Ederli et al. 2004, Lux et al. 2004, Wójcik et al. 2005, Gomes et al. 2011). In the leaf tissues, the adaxial and abaxial epidermis show increased thickness while the leaf blade shows reduced thickness as contamination increases with consequent change in the root growth rate.

Soil and water contamination by toxic heavy metals is a serious environmental concern. The magnitude of this problem is aggravated due to increased mining, industrial usage, and anthropogenic activity. Cadmium is abundantly used in surface coatings, pigment formulations, and manufacture of batteries, automobiles, and military equipment; Zn is used in the paint, rubber, dye, and wood preservative industries. In the waste disposed by most manufacturing industries, Zn and Cd often exist as by-products. When released into the environment, both Zn and Cd tend to concentrate at toxic levels by accumulating in soils and sediments. Severe contamination of soil and groundwater by Zn and Cd affects vegetative growth and also poses significant risks to human and animal health.

Many efforts and several methods have been directed to the cleanup of soils contaminated with heavy metals. Phytoremediation is a multidisciplinary approach where plants are used to extract or sequester metal contaminants during the process of environmental restoration (Raskin and Ensley 2000, Lasat 2002). Phytoremediation is a cost-effective and environmentally friendly alternative to engineering approaches of excavation and landfill techniques for hazardous materials cleanup (Raskin and Ensley 2000); it is also considered to be a cost-effective way for large-scale cleanups of metal-contaminated sites. Successful phytoremediation requires efficient uptake of heavy metals without diminishing the plant growth and yield. Hence, it is important to monitor the status of plant health for any stress-related changes. Spectral reflectance is a nondestructive technique that has been used to monitor the effects of disease infestation, mineral nutrition, and pollution in plants (Myers and Allen 1968, Horler and Barber 1980, Collins et al. 1983, Labovitz et al. 1983, Milton et al. 1989, Kraft et al. 1996, Merzlyank et al. 2003).

5.2 Uptake and Accumulation of Zn and Cd in Mustard

Mustard seeds used in an experiment were a commercial variety obtained locally; the seeds were sown in plastic pots, each containing approximately 2 kg of potting mix

(Miracle Grow Lawn Products Inc., Marysville, OH). Two plants per pot were allowed to grow for 5 weeks, and thereafter the plants were supplied with metal treatments at the rate of 100 mL pot^{-1} day^{-1} for 16 days in Zn-treated plants and 17 days in Cd-treated plants. The metal treatments were the following: 2 mM (ZnT1), 10 mM (ZnT2), 50 mM (ZnT3), and 100 mM (ZnT4) of Zn applied in the form of $ZnSO_4 \cdot 7H_2O$; 50 µM (CdT1), 500 µM (CdT2), 2.5 mM (CdT3), and 10 mM (CdT4) of Cd applied in the form of $CdCl_2$. All the treatment groups along with the control (T0) or untreated group were arranged in a completely randomized design with six replicates in each group. The plants were supplied with nutrient solution throughout the plant growth (Sridhar et al. 2004, 2005).

Mustard plants in all the Zn- and Cd-treated groups grew steadily, and chlorosis was not visually observed during the treatment process except in ZnT3- and ZnT4-treated groups toward the end of the metal treatment process. Metal accumulation in plant tissues are given in Figure 5.1. Metal accumulation in plant shoots and roots increased significantly ($p < 0.05$),

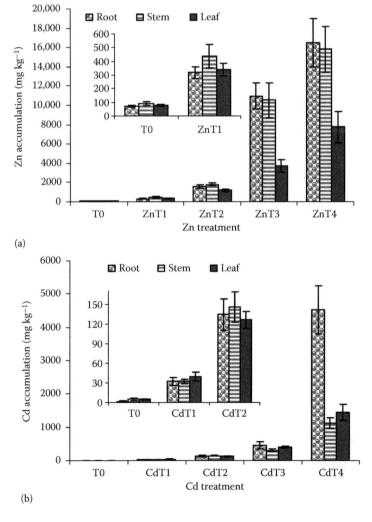

(a)

(b)

FIGURE 5.1
Accumulation of Zn and Cd concentrations in the root, stem, and leaves of the mustard plant.

with increase in applied metal solution concentration for both Zn- and Cd-treated groups (Figure 5.1). The dry weight of the plants showed a decrease in both ZnT3- and ZnT4-treated groups when compared to the control (T0) group (Sridhar et al. 2004, 2005). The Cd-treated groups showed no significant changes in relative water content (RWC) of the plants, while the Zn-treated groups showed a significant decrease ($p < 0.05$) in RWC (Sridhar et al. 2004, 2005).

5.3 Uptake and Accumulation of Zn and Cd in Barley

The barley seeds in the experiment were a commercial variety obtained locally. To simulate heavy-metal contamination in soil, plant groups were supplied with 2 mM (ZnT1), 10 mM (ZnT2), 50 mM (ZnT3), and 150 mM (ZnT4) of Zn solution, supplied as $ZnSO_4.7H_2O$, and 1 mM (CdT1) and 10 mM (CdT2) of Cd solution, supplied as $CdCl_2$. All the treatment groups along with the control (T0) were arranged in a completely randomized design with eight replicates in each Zn group and five replicates in each Cd group. The metal treatment was started at 5 weeks after sowing at the rate of 100 mL pot^{-1} day^{-1} for Zn and 6 weeks after sowing for Cd at the same rate. The metal treatments were supplemented with nutrient solution to avoid any nutrient deficiencies.

The metal accumulation in plant shoots increased with increase in applied metal solution concentration in both the Zn- and Cd-treated groups (Figure 5.2). The plant biomass decreased with increase in metal treatment. Chlorosis was not visually observed during the treatment process except in the higher metal-treated groups toward the end of the experiment. General effects of metal treatment included stunted growth, a decrease in RWC, and an increase in metal concentration. Please note that Zn acts as a growth-promoting micronutrient at lower concentrations; hence, the ZnT1 and ZnT2 groups recorded a slight increase in biomass and RWC over the control group (T0) (Sridhar et al. 2007a).

5.4 Chemical Analysis

In both the mustard and barley studies, the roots and shoots of the plants were harvested at the end of the experiment. The harvested leaves and roots were dried at 80°C in an oven for 48 hours. Dry shoots were ground and weighed. Plant samples (approximately 0.5 g) were digested with concentrated HNO_3 and H_2O_2 (Jackson 1958, Han and Banin 1997). The digested solution was filtered and then analyzed for Zn and Cd concentration using inductively coupled plasma-atomic emission spectrometry (ICP-AES).

5.5 Internal Structural Changes of Mustard

Leaf samples from 24 mustard plants consisting of plants treated with T0, ZnT3, ZnT4, and CdT4 were collected and prepared for light microscopy (LM). Leaves of 5 mm in length were excised on the last day of metal treatment. LM samples were immediately fixed in formaldehyde-acetic acid (FAA). The plant samples were alcohol dehydrated,

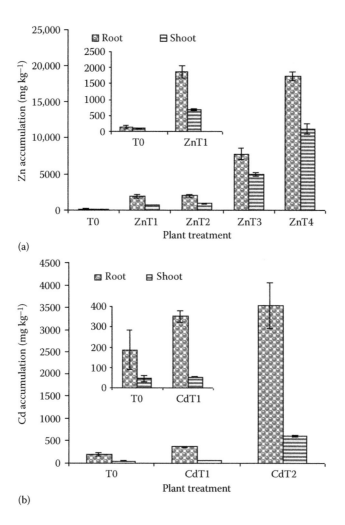

FIGURE 5.2
Accumulation of Zn and Cd concentrations in the root and shoot of the barley plant.

paraffin embedded, and ultramicrotomed. Leaf samples were stained with safranin (0.1%)–fastgreen (0.2%) to observe the structural changes (Sass 1958). The anatomical changes of leaf, stem, and root samples were also studied using light microscopy, scanning, and transmission electron microscopy. The detailed procedures of the microscopic studies and the chemical analysis were reported in Sridhar et al. (2005).

The ZnT4-treated plants (Figure 5.3b,d) showed significant foliar structural changes compared to the control group, T0 (Figure 5.3a,c). The leaves of ZnT4-treated mustard plants showed a reduction in the number of chloroplasts and breakdown of the palisade and epidermal cells compared to the control (T0) plants. In CdT4-treated plants, there were no significant changes in the internal structure of leaves compared to the control group (data not shown). The detailed structural changes due to accumulation of Zn and Cd were published in Sridhar et al. (2005).

The Zn accumulation resulting in significant changes in the structural characteristics of the mustard leaves was also revealed by our scanning electron microscopic results (Figure 5.4).

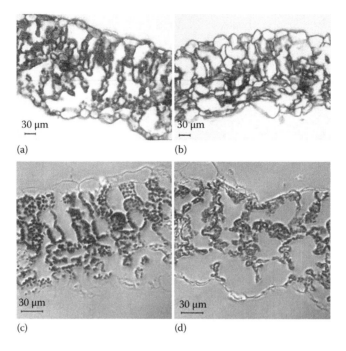

(a)   (b)

(c)   (d)

FIGURE 5.3
Light micrographs showing transverse section of control (a, c) and ZnT4-treated (b, d) leaves. The plants treated with ZnT4 (b, d) show decreases in numbers of chloroplasts and breakdown of epidermal and palisade cells. The leaf samples were collected the last (16th) day of metal treatment.

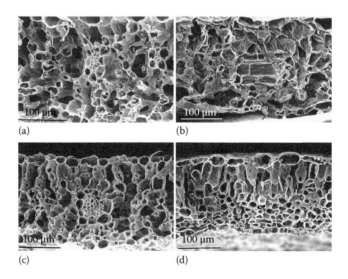

(a)   (b)

(c)   (d)

FIGURE 5.4
Scanning electron micrographs (SEMs) showing transverse section of control (a, c), and ZnT4- (b) and CdT4-treated leaves. The plants treated with ZnT4 (b) show decreases in sizes of epidermal and palisade cells and loss of cell turgidity compared to control leaves. The leaf samples were collected the last day (16th day) of metal treatment.

The significant structural changes in the mustard leaf of ZnT4-treated plants (Figure 5.4b) showed that Zn at higher concentrations causes severe damage to the palisade cells of mustard leaves. Barcelo et al. (1986) and Vazquez et al. (1987) reported similar foliar structural changes caused by Cr in *Phaseolous vulgaris*. The application of Zn at higher concentrations also resulted in decrease in fresh weight, dry weight, and RWC of Zn-treated plants compared to the control (T0) group (Sridhar et al. 2005). No significant structural changes were observed in CdT4-treated plants (Figure 5.4d) compared to the control plants (Figure 5.4a,c).

5.6 Heavy Metal-Induced Structural Changes in Barley

The leaf samples were excised 2 cm above the leaf–stem intersection of the plants for LM. The LM samples were immediately fixed in FAA and the plant samples were alcohol dehydrated, ultramicrotomed, and subjected to 1% toluidine blue staining for further observation (Sass 1958).

The Zn-treated plants showed gradual changes in leaf structure with an increase in metal concentration. The ZnT3 and ZnT4 groups showed significant foliar structural changes compared to the control group (T0). The light micrographs obtained from the leaf samples showed significant reduction in number of chloroplasts in the palisade and epidermal cells of ZnT3 and ZnT4 groups (Figure 5.5b,c) compared to the control group (Figure 5.5a). The epidermis is a single layer of cells present on either side of a plant leaf, and the palisade mesophyll is the cells present on the upper side of the leaf that contains most of the leaf chloroplasts. The chloroplasts are the cellular organelles in which photosynthesis occurs. The chloroplasts of the palisade and epidermal cells were dilated, showing an irregular outline due to metal stress. The chloroplasts were less intensely stained and fewer in number, with an increase in concentration from ZnT3 to ZnT4 compared to the control group

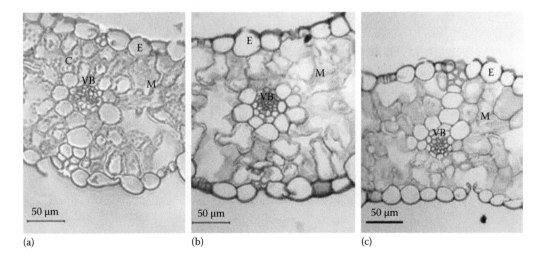

(a)   (b)   (c)

FIGURE 5.5
Light micrographs showing transverse section of control (a), ZnT3-treated (b), and ZnT4-treated (c) leaves. Note the decrease in the chloroplast, cell size, and leaf thickness with increase in Zn treatment. E = epidermal cells, M = palisade mesophyll cells, VB = vascular bundles, and C = chloroplast.

(Figure 5.5). The cellular changes are consistent in both the upper and lower leaves with an increase in metal concentration. The decrease in cellular size and intercellular spaces with an increase in metal concentration affected the thickness of the leaf in Zn-treated plants. This is confirmed from the light micrographs obtained at regular intervals during the process of metal treatment. The changes in chloroplasts and cellular components in ZnT1 and ZnT2 are not distinct from the control group (T0). This may be due to fact that Zn acts as a micronutrient at lower concentrations. In Cd-treated plants, the changes in the chloroplast number were significant over those in the control group (T0), while the epidermal and mesophyll cellular changes were not significant. The magnitude of structural changes increased with an increase in metal accumulation among the Zn-treated groups. The light microscopic results were also confirmed by detailed scanning and transmission electron microscopic studies (Sridhar 2004, 2007a).

The Zn-treated plants showed structural changes, such as decreases in intercellular spaces, thickening of vascular bundles, and shrinkage of palisade cells, which resulted in refractive index discontinuities within the leaf (Figure 5.6). In the higher concentration

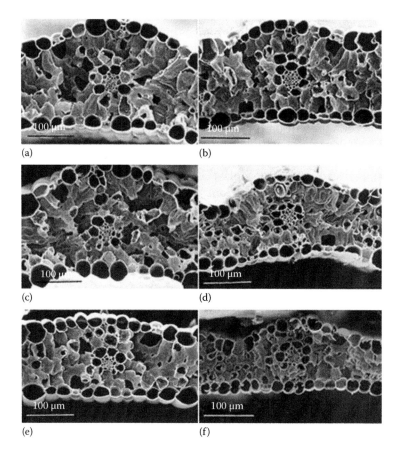

(a)   (b)

(c)   (d)

(e)   (f)

FIGURE 5.6
SEMs showing transverse section of control (a, b), and ZnT3-treated (c, d) and ZnT4-treated (e, f) upper (a, c, e) and lower (b, d, f) leaves of the plant. The upper (e) and lower (f) leaves of ZnT4-treated plants and upper (c) and lower (d) leaves of ZnT3-treated plants show decrease in sizes of epidermal and palisade cells and loss of cell turgidity compared to the upper (a) and lower (b) leaves of control or untreated plants. The leaf samples were collected the last day of metal treatment.

Zn-treated plants (ZnT3 and ZnT4), the breakdown of chloroplasts started with the upper palisade cells and gradually proceeded toward the lower mesophyll spongy cells (Figure 5.6). In the case of Cd-treated plants, no significant leaf structural changes were observed even though Cd accumulation in plant shoots was confirmed by analytical results (Figure 5.2).

5.7 Plant Spectral Reflectance

Plant reflectance is governed by leaf surface properties and internal structure, as well as by the concentration and distribution of foliar pigments and biochemical components (Vazquez et al. 1995, Penuelas and Filella 1998, Raun et al. 1998). Analysis of remotely sensed reflected light can be used to assess both the biomass and the physiological status of plants (Raun et al. 1998). Spectral reflection of solar radiation from leaves and canopy at given wavelengths has been discussed by Myers and Allen (1968), Gausman (1977), Sinclair et al. (1971), and Boyer et al. (1988). Briefly, reflectance of leaves is relatively low in the visible region (400–700 nm), where light absorption by leaf pigments (primarily due to chlorophyll) is the determining factor. The absorption maxima occur in the blue and red regions at 470 and 680 nm, respectively, while the familiar green reflectance peak occurs at 550 nm. In the near- and middle infrared (IR) regions, these pigments are transparent and internal leaf structure and biochemical composition control reflectance. The reflectance spectrum of principal biological interest occurs in the near-IR between 700 and 1,300 nm, where reflectance is high and absorption is minimal (with two minor water absorption bands at 975 and 1,175 nm); beyond 1,300 nm, major water absorption bands (at 1,450 and 1,950 nm) become significant.

In geobotanical remote sensing, vegetation reflectance anomalies have been used to locate mineral deposits (Ustin et al. 1999). Collins et al. (1983) were the first to show narrowband spectral changes in airborne data. They associated mineral deposits with a blue shift of the "red edge," which is defined as the point of maximum slope (about 700 nm) on a vegetation reflectance spectrum between the red and the near-IR regions. The value range of the red edge position is 690–740 nm and high-resolution sensors are necessary to obtain an accurate value (Goetz 2001). The conclusions of Collins et al. (1983) were supported by a subsequent leaf-level laboratory experimental study of mineral-stressed plants (Chang and Collins 1983). Other laboratory, field, and airborne studies have also produced evidence of spectral changes in plants growing in substrates enriched in metallic elements (Horler and Braber 1980, Labovitz et al. 1983, Milton et al. 1989). Most of these studies focused on the position shift of the "red edge" due to metal-induced stress on plants. However, as Goetz and Rock (1983) pointed out, the blue shift of the red edge represents a universal response to certain, but not all forms of geochemical stress. Furthermore, the blue shift of the red edge is also related to vegetation stress caused by deficiencies of N, P, Fe, S, Mg, etc. Most vegetation indices in remote sensing applications use visible and near-infrared spectral bands around the red edge (Bannari et al. 1995). According to Carter (1993), the changes in visible reflectance of plants were spectrally similar due to different agents of stress in several plant species. According to Chapin (1991), all plants respond to environmental stress in basically the same way, through the decline in growth rate and acquisition of nutrient resources. Several studies have indicated that visible chlorosis is a common stress response, resulting in an increased visible reflectance affecting the red edge of a spectral curve compared to normal plants. However, the infrared region beyond the red edge has not been studied in as much detail as the visible and the "red edge" regions. As the earlier studies on leaf reflectance and spectral properties of plants indicate, changes in the internal structure of leaves cause changes in the 700–1,300 nm spectral region (Myers and Allen 1968, Woolley 1971, Sinclair et al. 1973, Gausman 1974, Grant 1987, Boyer et al. 1988, Kim et al. 2001). Barcelo et al. (1988) and Torresdey et al. (1996) suggested that accumulation of metals in the plant leaf lead to chlorosis and deformation of cells. These structural changes or breakdown of cellular components result in refractive index discontinuities within the cell and cause the change in reflectance (Sinclair et al. 1973, Gausmann 1977, Grant 1987). Slaton et al. (2001) estimated NIR leaf reflectance at 800 nm from leaf structural characteristics and found that the leaf trichome density, leaf thickness, and mesophyll proportion filled with air spaces were not considered as effective predictors at this wavelength for the alpine angiosperms.

Photosynthetic pigments, leaf internal structure, and water content of plants have distinct light absorption and reflectance characteristics that can affect the amount of leaf reflectance in the respective regions of the spectrum (Gates et al. 1965, Sinclair et al. 1971, Curran et al. 1992, Slaton et al. 2001, Davids and Tyler 2003). The absorption maxima of leaf pigments occur in the blue and red regions at 470 and 680 nm, respectively; the familiar green reflectance peak occurs at 550 nm. In the NIR region, these pigments are transparent and internal leaf structure and biochemical composition control reflectance. The reflectance spectra of principal biological interest occur in the NIR between 700 and 1300 nm, where reflectance is high and absorption is minimal (with two minor water absorption bands at 975 and 1,175 nm). Beyond 1,300 nm, major water absorption bands (at 1,450 and 1,950 nm) become significant. Thus, analysis of remotely sensed reflected light could be used to assess both biomass and physiological status of plants (Raun et al. 1998).

Increase in metal accumulation results in foliar structural changes, such as breakdown of chloroplasts and decline in chlorophyll synthesis (Barcelo et al. 1988), and reduction in size of mesophyll cells (Zhao et al. 2000). Most of the previous studies on plant reflectance have been focused on the red-edge area (about 700 nm), which is predominantly influenced by pigment concentration. The red edge is defined as the point of maximum slope on a vegetation reflectance spectrum between the red and the NIR regions (Goetz 2001). However, red-edge or red-edge-based indices cannot differentiate changes due to metal accumulation from other stresses, because loss of chlorophyll is a general symptom of many biological and physiological stresses in plants.

Plant spectral reflectance can be used to monitor the process of metal accumulation and to develop a non-red-edge spectral index, sensitive only to leaf structural changes caused by metal accumulation during the process of phytoremediation. Two studies using Indian mustard (*Brassica juncea*) and barley (*Hordeum vulgare*) were summarized earlier in this chapter where changes in spectral reflectance were evaluated with Zn and Cd accumulations and the associated plant structural changes. Spectral reflectance can be used for long-term monitoring of mobility and bioavailability of certain heavy-metal contaminants in soils. It will also allow rapid assessment of metal accumulation during the process of metal phytoremediation and hence expedite plant harvest, optimizing the rate of soil remediation.

5.8 Spectral Reflectance of Mustard

A Fieldspec Pro spectroradiometer (350–2,500 nm) from Analytical Spectral Devices Inc. (Boulder, CO) was used to collect reflectance spectra with solar radiation in outdoor experiments and with quartz–tungsten–halogen (QTH) lamps for laboratory measurements. Diffused light from two 100-W Lowell Pro-Lights were used to illuminate the plant canopy at 45° angles, when spectra were collected in the laboratory. The foreoptics were aligned vertically and the height of the foreoptics from the plant canopy level was adjusted and kept constant so that the whole canopy filled the field of view (FOV) of the instrument. The experimental setup was put on top of a four-wheel wagon and a black cloth was used as a nonreflective background for spectral collection. The same experimental setup was used to collect both indoor and outdoor spectra except that the lamps were turned off for outdoor measurements.

Calibration spectra using a white Labsphere Spectralon panel were acquired before canopy spectra were recorded. Diffuse reflectance spectra (350 to 2,500 nm) of the plant canopies were collected regularly with laboratory illumination and with solar radiation during cloud-free days. Spectra collection was started prior to the beginning of the metal treatment procedure. Both outdoor and laboratory spectra were recorded throughout the metal treatment process for all the mustard plants, while only laboratory spectra were collected on cloudy days. The spectral measurements of canopies of each treatment group ($n = 6$) were averaged to overcome individual variations and the spectral indices of each treatment group were calculated throughout the metal treatment process. Only laboratory spectral results are shown in this chapter; the outdoor results are consistent with laboratory data except that there is more noise (resulting from atmospheric interferences) in outdoor data.

The spectral measurements of canopies of each treatment group of mustard were averaged to overcome individual variations. The canopy spectra of Zn-treated plants (Figure 5.7b) changed gradually over time and showed a consistent and systematic difference from untreated plants. The ZnT3- and ZnT4-treated plants showed a decrease in chlorophyll absorption around 680 nm and thus showed an increased reflectance in the visible red region of the spectrum and a significant decrease in reflectance in the NIR region (Figure 5.7b) compared to the control plants. No significant spectral changes were observed in Cd-treated plants (Figure 5.7a).

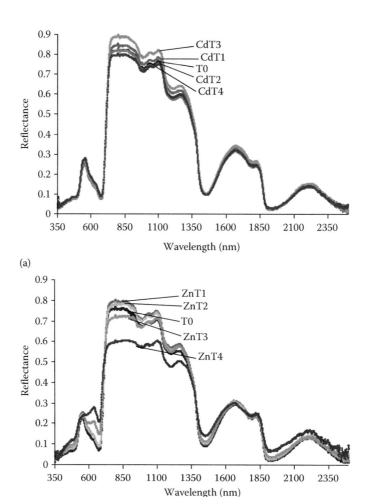

(a)

(b)

FIGURE 5.7
Averaged ($n = 5$) canopy reflectance spectra of control (T0), Zn- and Cd-treated groups of mustard plants on the last day of metal treatment. All the spectra were obtained inside the lab.

5.9 Spectral Reflectance of Barley

The spectral measurements for leaves and canopies of each treatment group of barley were averaged to overcome individual variations. The canopy spectra of Zn-treated plants changed gradually over time and showed a consistent and systematic difference from untreated plants. The ZnT3- and ZnT4-treated plants showed a decrease in chlorophyll absorption around 680 nm (Figure 5.8a). The blue shift of the red edge is also evident for these two groups. The reflectance spectra of leaves from the ZnT1 and ZnT2 groups were almost indistinguishable from those of untreated plants. This is due to the fact that the applied Zn concentrations—2 mM (130.8 mgL^{-1} and 10 mM (654 mg L^{-1}) for these two groups—were relatively low.

For Cd-treated plants, the blue shift of the red edge and an overall increase of spectral reflectance in the visible region were due to decrease in chlorophyll absorption

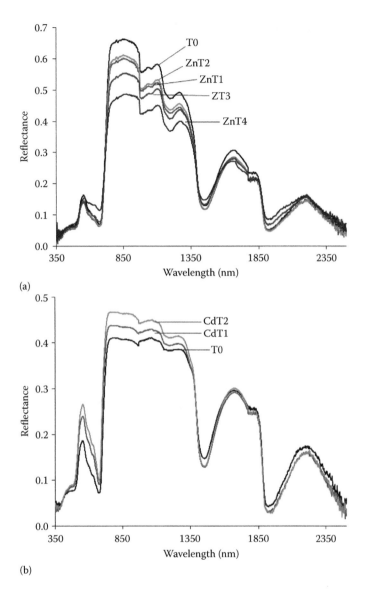

(a)

(b)

FIGURE 5.8
Averaged canopy-level spectral reflectance of barley plants treated with Zn on the 19th day (a) and Cd-treated plants on the 16th day (b) after starting the treatment.

(Figure 5.7b). For Zn-treated plants, the plant spectra showed a decrease in the 800–1,300 nm, 1,470–1,850 nm, and 2,000–2,400 nm regions, especially for the ZnT3 and ZnT4 groups (Figure 5.8a). For Cd-treated plants, an increased spectral reflectance in the 800–1,300 nm region was observed. The canopy reflectance spectra of Zn- and Cd-treated plants (Figure 5.8a, b, respectively) were consistent with the leaf spectra throughout the experimental period.

The dilation of chloroplasts, disintegration of nucleus, and possible accumulation of metal in the vacuoles resulted in increased scattering of incident light rays within the leaf. Both the refractive index discontinuities and scattering of light rays contributed to the decrease in leaf reflectance in the near-IR region (800–1,300 nm). The closure of stomata with higher Zn accumulation (ZnT4) further increased the leaf internal scattering, leading to a further decrease in the near-IR (800–1,300 nm) spectral reflectance. The disintegration and decrease in the number of chloroplasts caused the increase in reflectance in the visible region, particularly around the chlorophyll absorption region.

5.10 Mustard Spectral Analysis

The reflectance spectra of canopies of each treatment group were averaged to overcome individual variations. Spectral indices, normalized difference vegetative index (NDVI), and R_{1110}/R_{810} were calculated with the averaged spectra ($n = 6$) for each treatment. The spectral indices are calculated based on the formula NDVI = $R_{810} - R_{680}/R_{810} + R_{680}$ (Rouse et al. 1974). All the spectral analyses were performed using Microsoft Excel spreadsheets. The spectral measurements of canopies of each treatment group ($n = 6$) were averaged to overcome individual variations. Changes in spectral reflectance along with metal accumulation in plant tissues were used to characterize the phytoextraction process.

Despite the fact that all the treatment groups were arranged randomly, there were still some slight differences between groups before the metal treatment started (Figure 5.9).

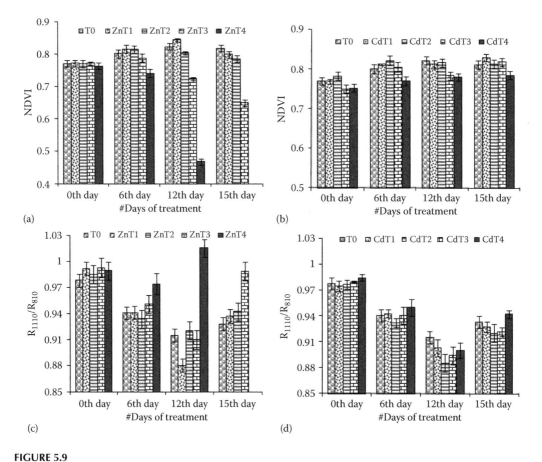

FIGURE 5.9
Changes in NDVI (a, b) and in R_{1110}/R_{810} (c, d) of plants during the Zn and Cd treatment processes. Bars are plus/minus standard error of six replicates.

These summarized trends of spectral indices for the metal treatment period are presented for NDVI in Figure 5.9(a, b) and for R_{1110}/R_{810} in Figure 5.9c,d. As seen in Figure 5.9, the groups with higher Zn treatment (ZnT3 and ZnT4) can be distinguished from other groups from the 12th day of the metal treatment process (Figure 5.9). The stress caused by Zn accumulation increased with the progress of metal treatment as indicated by decrease in NDVI values (Figure 5.9). As Indian mustard is an accumulator of Zn, the applied ZnT1 and ZnT2 concentrations are within the tolerance limit. The Cd-treated groups were not significantly different from the control (T0) group, while the CdT4-treated group showed stress starting from the 16th day of metal treatment.

The R_{1110}/R_{810} value of the ZnT3- and ZnT4-treated groups (Figure 5.9c,d) became higher than that of the control group (T0), starting from the 12th day of the metal-treatment process. This index can be used to delineate groups with significant Zn accumulation. For the Cd-treated groups, the R_{1110}/R_{810} values showed no significant difference from the control (T0) group (Figure 5.9) throughout the phytoextraction process. This agrees with the microscopic observations of no significant leaf structural changes in Cd-treated plants (Figure 5.4). The Zn results imply that the ratio index R_{1110}/R_{810} is only sensitive to leaf structural changes.

5.11 Barley Spectral Analysis

The NDVI results of all metal-treated groups are shown in Figure 5.10a,b, where NDVI = $(R_{810} - R_{680})/(R_{810} + R_{680})$. The groups with higher Zn treatment (ZnT3 and ZnT4) can be distinguished from other groups (Figure 5.10), from the seventh or eighth day onward. The stress caused by Zn accumulation increased along with the progress of metal treatment. Since Zn is a natural micronutrient and low concentrations were applied to the ZnT1 and ZnT2 groups, these two groups were not distinct from the control (T0) group. Pierzynski and Schwab (1992) reported that the metal concentration of 1,165 mg kg^{-1} Zn was suspected to have phytotoxic effects on soybean. Nandakumar et al. (1995) suggested that 100 mg L^{-1} of Zn concentration was not phytotoxic to *Brassica juncea* when added to a soil mixture. Two Cd-treated groups, CdT1 and CdT2, also showed stress starting on the eighth day of the treatment (Figure 5.10b).

The R_{1110}/R_{810} value (Figure 5.10c) of the ZnT3 and ZnT4 group became higher than that of the control group (T0) during the later stages of the metal-treatment process. This index can be used to delineate groups with significant Zn accumulation, starting on the 13th day of treatment. Once again the ZnT1 and ZnT2 groups could not be distinguished. For the two Cd-treated groups, the R_{1110}/R_{810} value did not differ from the T0 group (Figure 5.10d). This agrees with our observations of no significant leaf structural changes in Cd-treated plants. We believe that the stress indicated by NDVI is primarily due to the decrease in numbers of chloroplasts as observed in light micrographs (Sridhar et al. 2007a).

NDVI analysis revealed the stress caused by Zn and Cd accumulation prior to visual detection. NDVI is sensitive to chlorophyll concentration, but it is also sensitive to changes in the internal structure of the cell (Adams et al. 1999). Decrease of chlorophyll concentration can also be the end result of a number of other stresses, such as moisture loss and nutrient deficiency. To detect the leaf structural changes due to Zn accumulation, we found that the ratio index of R_{1110}/R_{810} closely correlated to the leaf structural changes and the

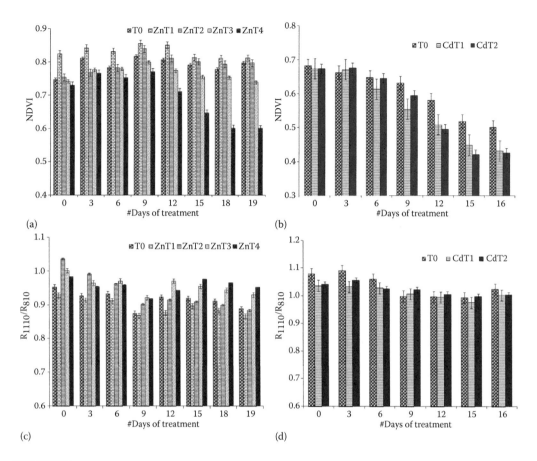

FIGURE 5.10
Changes in NDVI (a, b) and in R_{1110}/R_{810} (c, d) of plants during the Zn and Cd treatment processes. Bars are plus/minus standard error of six replicates.

content of Zn in the leaf. This index can be used to monitor the accumulation and concentration of Zn. It is not sensitive to chlorophyll concentration because both wavelengths are away from the red edge and are within the spectral region that is sensitive to internal leaf structure.

The stress caused by deformation and decrease in chloroplasts was revealed by NDVI analysis. The ratio index R_{1110}/R_{810} did not distinguish the Cd-treated groups from the control group (T0). The Cd results imply that the ratio index R_{1110}/R_{810} is only sensitive to leaf structural changes, but not to chlorophyll concentration. The increased reflectance in the 800 to 1,300 nm and 1,450 to 1,900 nm regions of Cd-treated plants is possibly due to a decrease in the leaf water content due to accumulation of phytotoxic metal in plant shoots. Carter (1991) reported that a decrease in leaf water content generally increased reflectance throughout the 400–2,500 nm wavelength region. The RWC of plants from CdT1 and CdT2 groups was lower than in the T0 group. The RWC values were not significantly different in both the Cd- and Zn-treated groups, except for CdT2 and ZnT4 compared to the control.

5.12 Conclusion

Our chemical, structural, and spectral analysis indicates that Zn at higher concentrations has a profound impact on physiology and internal structure of mustard and barley plants. The Zn and Cd uptake and accumulation at higher concentrations causes severe morphological, metabolic, and physiological changes in the plants. High metal concentrations will affect the cortical cells, reduce the size of xylem vessels, and further cause reduced root growth of the plant. High metal uptake into the leaves results in loss of pigment concentration, shrinkage of palisade, and spongy parenchyma cells. Apart from the structural changes, morphological changes such as decrease in water content of the plants and decline in plant growth rate have been observed. The combination of remote sensing indices, such as NDVI, and a spectral ratio index in the 800–1,300 nm region, such as R_{1110}/R_{810}, can provide a nonintrusive and continuous monitoring method for the impact and content of certain heavy metals in the canopies of living plants grown in contaminated soils. Our results, presented here, and earlier studies suggest that it is feasible to use spectral reflectance for monitoring the physiological status of plants during the process of phytoremediation and hence the process of uptake and accumulation of certain heavy-metal contaminants.

In summary, the ratio index R_{1110}/R_{810} is sensitive to foliar structural changes in metal accumulator plants. This index closely correlates to the magnitude of leaf structural changes and hence the concentration of a specific accumulated metal (e.g., Zn) that caused the structural changes in the leaf. The accumulation of metals such as Cd does not result in significant leaf structural changes. In the Cd case, the ratio index R_{1110}/R_{810} was not sensitive to metal accumulation in the leaf. We also found that even though NDVI can be used to detect the general stresses caused by accumulation of different metals, it can neither distinguish between two metal species nor the stress caused by different physiological disturbances. The combination of remote sensing indices, such as NDVI and a special ratio index in the 800–1,300 nm region such as R_{1110}/R_{810}, can provide a nonintrusive and continuous monitoring method for the impact and content of certain heavy metals in the canopies of living plants grown on contaminated soils. We have applied similar methodology to monitor the phytoextraction of other metals by other plant species (Sridhar et al. 2007b, Su et al. 2007).

Acknowledgments

We acknowledge Mr. Dharmendra K. Singh, Mr. Cheng Wang, Mr. Shyam S. Balasubramaniam, and Mr. Thomas W. Hallmark, for their contribution in data collection and plant culture activities. The authors are thankful to Ms. Yunju Xia, Mr. Dean Patterson, and Dr. Thomas Meaker of DIAL for their help in chemical analysis. We also thank Mr. Richard F. Kuklinski, Mr. William A. Monroe, and Ms. Kay N. Milam for expert assistance in microscopy; and Ms. Amanda M. Lawrence for help in sample processing and providing microtomes and other accessories. This work was supported by funding from the US Department of Energy through Cooperative Agreement DE-FC26-98FT-40395.

References

Adams, J.L., Philpot, W.D., Norvell, W.A. 1999. Yellowness index: An application of spectral second derivatives to estimate chlorosis of leaves in stressed vegetation. *International Journal of Remote Sensing* 20:3663–3675.

Adriano, D.C. 2001. *Trace elements in terrestrial environments: Biochemistry, bioavailability and risks of metals.* 2nd Edition, Springer–Verlag, New York.

Alloway, B.J., Ayres, D.C. 1997. *Chemical principles of environmental pollution.* Blackie Academic, London.

Bannari, A., Morin, D., Bonn, F. 1995. A review of vegetation indices. *Remote Sensing Reviews* 13:95–120.

Barcelo, J., Poschenrieder, Ch., Gunse, B. 1986. Water relations in chromium VI treated bush bean plants (*Phaseolous vulgaris* L. cv. Contender) under both normal and water stress conditions. *Journal of Experimental Botany* 37:178–187.

Barcelo, J., Vazquez, M.D., Poschenrieder, C. 1988. Cadmium induced structural and ultrastructural changes in the vascular system of bush bean stems. *Botanica Acta* 101:254–261.

Boyer, M., Miller, J., Belanger, M., Hare, E. 1988. Senescence and spectral reflectance in leaves of northern pin oak (*Quercus palustris Muenchh*). *Remote Sensing of Environment* 25:71–87.

Brooks, R.R. (ed) 1998. *Plant that hyperaccumulate heavy metals.* Wallingford, CAB.

Carter, G.A. 1991. Primary and secondary effects of water content on the spectral reflectance of leaves. *American Journal of Botany* 78:916–924.

Carter, G.A. 1993. Responses of leaf spectral reflectance to plant stress. *American Journal of Botany* 80:239–243.

Chang, S.H., Collins, W. 1983. Confirmation of the airborne biogeophysical mineral exploration technique using laboratory methods. *Economic Geology* 78:723–736.

Chapin, F.S. 1991. Integrated responses of plants to stress. *Bio Science* 41:29–36.

Collins, W., Chang, S.H., Raines, G., Canney, F., Ashley, R. 1983. Airborne biogeophysical mapping of hidden mineral deposits. *Economic Geology* 78:737–749.

Curran, P.J., Dungan, J.L., Macler, B.A., Plummer, S.E., Peterson, D.L. 1992. Reflectance spectroscopy of fresh whole leaves for the estimation of chemical concentration. *Remote Sensing of Environment* 39:153–166.

Davids, C., Tyler, A.N. 2003. Detecting contamination-induced tree stress within the Chernobyl exclusion zone. *Remote Sensing of Environment* 85:30–38.

Ederli, L., Reale, L., Ferranti, F., Pasqualini, S. 2004. Responses induced by high concentration of cadmium in *Phragmites australis* roots. *Physiologia Plantarum* 121:66–74.

Emamverdian, A., Ding, Y., Mokhberdoran, F., Xie, Y. 2015. Heavy metal stress and some mechanisms of plant defense response. *The Scientific World Journal*, Article ID 756120, 1–18.

Ernst, W.H.O. 2006. Evolution of metal tolerance in higher plants. *Forest Snow Landscape Research* 80(3):251–274.

Gates, D.M., Keegan, H.J., Schleter, J.C., Wiedner, V.R. 1965. Spectral properties of plants. *Applied Optics* 4:11–20.

Gausman, H.W. 1974. Leaf reflectance of near-infrared radiation, *Photogrammetric Engineering & Remote Sensing* 40:183–191.

Gausman, H.W. 1977. Reflectance of leaf components. *Remote Sensing of Environment* 6:1–9.

Goetz, A.F.H. 2001. Progress in hyperspectral imaging of vegetation, In *Optics in agriculture: 1990–2000* (pp. 20–36), Deshazer, J.A., Meyer, G.E. (eds.). SPIE Optical Engineering Press.

Goetz, A.F.H., Rock, B.N. 1983. Remote sensing for exploration: An overview. *EconomicGeology* 78:573–590.

Gomes, M.P., Marques, T.C.L.L.S.M., Oliveira, M.D., Nogueira, G., Castro, E.M.D., Soares, A.M. 2011. Ecophysiological and anatomical changes due to uptake and accumulation of heavy metal in *Brachiaria decumbens*. *Scientific Agriculture* 68(5):566–573.

Grant, L. 1987. Diffuse and specular characteristics of leaf reflectance. *Remote Sensing of Environment* 22:309–322.

Han, F.X., Banin, A. 1997. Long-term transformations and redistribution of potentially toxic heavy metals in arid zone soils. I. Under saturated conditions. *Water Air Soil Pollution* 95:399–423.

Horler, D.N.H., Barber J. 1980. Effects of heavy metals on the absorbance and reflectance spectra of plants. *International Journal of Remote Sensing* 1:121–136.

Jackson, M.L. 1958. *Soil chemical analysis*. 1st ed. Prentice Hall, New Jersey, USA.

Jozefczak, M., Remans, T., Vangronsveld, J., Cuypers, A. 2012. Glutathione is a key player in metal-induced oxidative stress defenses. *International Journal of Molecular Sciences* 13(3):3145–3175.

Kim, M.S., McMutrey, J.E., Charles, L.M., Daughtry, C.S.T., Chappelle, E.W., Chen, Y.R. 2001. Steady state multi spectral fluorescence imaging system for plant leaves. *Applied Optics* 40:157–166.

Kraft, M., Weigel, H., Mejer, G., Brandes, F. 1996. Reflectance measurements of leaves for detecting visible and non-visible ozone damage to crops. *Journal of Plant Physiology* 148:148–154.

Labovitz, M.L., Masuoka, E.J., Bell, R., Siegrist, A.W., Nelson, R.F. 1983. The application of remote sensing to geobotanical exploration for metal sulfides—Results from the 1980 field season at mineral, Virginia. *Economic Geology* 78:750–760.

Lasat, M.M. 2002. Phytoextraction of toxic metals: A review of biological mechanisms. *Journal of Environmental Quality* 31:109–120.

Lux, A.A., Sottniková, A., Opatrná, J., Greger, M. 2004. Differences in structure of adventitious roots in *Salix* clones with contrasting characteristics of cadmium accumulation and sensitivity. *Physiologia Plantarum* 120:537–545.

MacFarlane, G.R., Burchett, M.D. 2000. Cellular distribution of copper, lead and zinc in the grey mangrove, *Avicennia marina* (Forsk.) Vierh. *Aquatic Botany* 68:45–59.

Merzlyak, M.N., Gitelson, A.A., Chivkunova, O.B., Solovchenko, A.E., Pogosyan, S.I. 2003. Application of reflectance spectroscopy for analysis of higher plant pigments. *Russian Journal of Plant Physiology* 50:704–710.

Milton, N.M., Ager, C.M., Eiswerth, B.A., Power, M.S. 1989. Arsenic and selenium induced changes in spectral reflectance and morphology of soybean plants. *Remote Sensing of Environment* 30:263–269.

Myers, V.I., Allen, W.A. 1968. Electro optical remote sensing methods as nondestructive testing and measuring techniques in agriculture. *Applied Optics* 7:1819–1838.

Nandakumar, P.B.A., Dushenkov, V., Motto, H., Raskin, I. 1995. Phytoextraction: The use of plants to remove heavy metals from soils. *Environmental Science & Technology* 29:1232–1238.

Nouri, J., Khorasani, N., Lorestani, B., Karami, M., Hassani, A.H., Yousefi, N. 2009. Accumulation of heavy metals in soil and uptake by plant species with phytoremediation potential. *Environmental Earth Science* 59:315–323.

Penuelas, J., Filella, I. 1998. Visible and near infrared reflectance techniques for diagnosing plant physiological status. *Trends in Plant Science* 3:151–156.

Pierzynski, G.M., Schwab, A.P. 1992. Metal availability to Soybeans grown on a Metal contaminated soil. *Proceedings of Hazardous Waste Research Conference* (pp. 543–553) Boulder, CO.

Rascio, N., Navari-Izzo, F. 2011. Heavy metal hyperaccumulating plants: How and why do they do it? And what makes them so interesting? *Plant Science* 180(2):169–181.

Raskin, I., Ensley, B.D. 2000. *Phytoremediation of Toxic Metals: Using plants to clean the environment*. 1st ed. John Wiley & Sons. Inc, New York, USA.

Raun, W.R., Johnson, G.V., Sembiring, H., Lukina, E.V., Laruffa, J.M., Thomason, W.E., Phillips, S.B., Solie, J.B., Stone, M.L., Whitney, R.W. 1998. Indirect measures of plant nutrients. *Communications in Soil Science and. Plant Analysis* 29:1571–1581.

Rouse, J.W., Haas, R.H., Schelle, J.A., Deering, D.W., Harlan, J.C. 1974. Monitoring the vernal advancement and retrogradation (green wave effect) of natural vegetation. Type III Final Report, NASA Goddard Space Flight Center, Green belt, Maryland, MD.

Samarghandi, M.R., Nouri, J., Mesdaghinia, A.R., Mahvi, A.H., Nasseri, S., Vaezi, F. 2007. Efficiency removal of phenol, lead and cadmium by means of UV/TiO2/H2O2 processes. *International Journal of Environmental Science and Technology* 4(1):19–25.

Sass, J.E. 1958. *Botanical microtechniques.* 3rd ed. Iowa St. Univ Press, Ames, USA.

Schutzendubel, A., Polle, A. 2002. Plant responses to abiotic stresses: Heavy metal-induced oxidative stress and protection by mycorrhization. *The Journal of Experimental Botany* 53(372):1351–1365.

Sinclair, T.R., Schreiber, M.M., Hoffer, R.M. 1973. Diffuse reflectance hypothesis for the pathway of solar radiation through leaves. *Agronomy Journal* 65:276–283.

Sinclair, T.R., Hoffer, R.M., Schreiber, M.M. 1971. Reflectance and internal structure of leaves from several crops during a growing season. *Agronomy Journal* 63:864–868.

Slaton, M.R., Hunt, E.R., Smith, W.K. 2001. Estimating near-infrared leaf reflectance from leaf structural characteristics. *American Journal of Botany* 88:278–284.

Sridhar, B.B.M. 2004. Monitoring plant spectral reflectance and internal structure changes during phytoremediation processes for selected heavy metals, Ph.D. Dissertation, Mississippi, Mississippi State University.

Sridhar, B.B.M., Diehl, S.V., Han, F.X., Monts, D.L., Su, Y. 2005. Changes in plant anatomy due to uptake and accumulation of Zn and Cd in Indian mustard (*Brassica juncea*). *Environmental and Experimental Botany* 54:131–141.

Sridhar, B.B.M., Han, F.X., Monts, D.L., Diehl, S.V., Su, Y. 2007a. Spectral reflectance and leaf internal structure changes of barley plants due to phytoextraction of zinc and cadmium. *International Journal of Remote Sensing* 28(5):1041–1054.

Sridhar, B.B.M., Han, F.X., Diehl, S.V., Monts, D.L., Su, Y. 2007b. Monitoring the effects of arsenic- and chromium-accumulation in Chinese brake fern (*Pteris vittata*) using microscopy and near infrared spectral reflectance. *International Journal of Remote Sensing* 28(5):1055–1067.

Su, Y., Sridhar, B.B.M., Han, F.X., Diehl, S.V., Monts, D.L. 2007. Effect of bioaccumulation of Cs and Sr natural nuclides and impact on foliar structure and plant spectral reflectance of Indian mustard (*Brassica juncea*). *Water Air and Soil Pollution* 180:65–74.

Torresdey, J.L.G., Polette, L., Arteaga, S., Tiemann, K.J., Bibb, J., Gonzalez, J.H. 1996. Determination of the content of hazardous heavy metals on *Larrea tridentata* grown around a contaminated area, *Proceedings of HSRC/WERC Conference*, Albuquerque, NM.

Ustin, S.L., Smith, M.O., Jacquemoun, S., Verstraete, M., Govaerts, Y. 1999. Geobotany: Vegetation mapping for earth sciences, *Remote sensing for earth sciences* (pp. 189–248), edited by A.N. Rencz. John Wiley & Sons. Inc, New York, USA.

Vazquez de Aldana, B.R., Garcia crisdo, B., Garcia ciudad, A., Perez corona, M.E. 1995. Estimation of mineral content in natural grasslands by near infrared reflectance spectroscopy, *Communications in Soil Science and Plant analysis* 26:1383–1396.

Welch, R.M. 1995. Micronutrient nutrition of plant. *Critical Reviews of Plant Science* 14(11):49–82.

Wójcik, M., Vangronsveld. J., D´Haen, J., Tukiendorf, A. 2005. Cadmium tolerance in *Thlaspi caerulescens. Environmental and Experimental Botany* 53:163–171.

Woolley, J.T. 1971. Reflectance and transmittance of light by leaves. *Plant Physiology* 148:515–522.

Zhao, F.J., Lombi, E., Breedon, T., McGrath, S.P. 2000. Zinc hyperaccumulation and cellular distribution in *Arabidopsis halleri. Plant Cell Environment* 23:507–514.

6

Role of Rhizospheric Mycobiota in Remediation of Arsenic Metalloids

Manvi Singh, Pankaj Kumar Srivastava, and Ravindra Nath Kharwar

CONTENTS

6.1 Introduction

Arsenic (As) is a naturally occurring toxic metalloid that is widely distributed in the earth's crust at an average concentration of 5–7 mg/kg (Stroud et al. 2011). In natural system, it exists in four oxidation states: 0 (elemental), −3 (arsine), +3 (arsenite), and +5 (arsenate). Inorganic As(V) and As(III) are found to be more toxic, predominant, and phytoavailable forms than organic forms of As, such as DMA (dimethylarsenic), MMA (monomethyl-arsenic), TMAO (trimethylarsine oxide), and TMA (trimethylarsine) (Gillispie et al. 2015, Chakraborti et al. 2016). Arsenic is found in soil and water due to various geochemical and anthropogenic sources such as volcanic eruptions, weathering of minerals and rocks, ore processing, pesticide application for agricultural purposes, and mining and production of energy from fossil fuels (Panda et al. 2010) (Figure 6.1).

Groundwater is more vulnerable to As contamination than surface water because of its interaction with As-bearing minerals in the aquifer (Smedley 2013) and, as the major source irrigation, contaminates agricultural soils with As in the Ganga Meghna

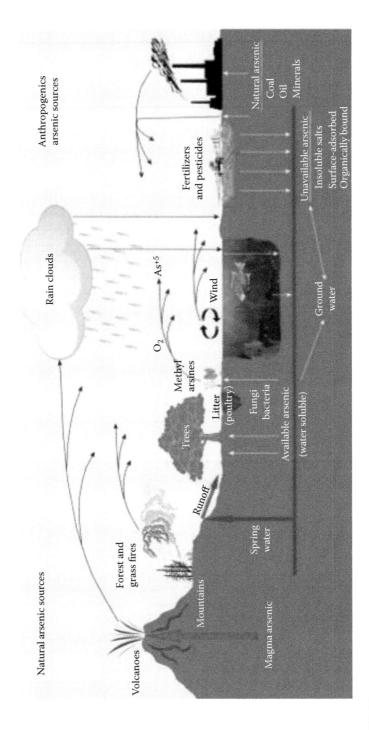

FIGURE 6.1
An overview of As sources and spread to ecosystem.

Brahamaputra (GMB) plain. Indian states, such as West Bengal, Bihar, Uttar Pradesh, Assam, Punjab, Haryana, Jharkhand, and Andhra Pradesh, have been severely affected by As poisoning (Figure 6.2). The amount of arsenic in the groundwater of the Bengal Delta Plain (West Bengal, India) ranges from 50 to 3200 µg/L, a value that is significantly higher than national drinking water standards of 50 µg/L set by the Bureau of Indian Standards (BIS 2009) and 10 µg/L set by the World Health Organization (WHO 2004). In an arsenic monitoring study of districts (Ballia, Bahraich, Gorakhpur, Lakhimpur-kheri and Gazipur) of Uttar Pradesh, India (238,000 km^2 located in the upper and middle Ganga plain), Srivastava et al. (2015) revealed the arsenic status in irrigation groundwater and agricultural soils. It was found that the Ballia and Bahraich districts were highly contaminated with arsenic metalloid.

More than 25 countries worldwide, such as Bangladesh, India, China, the United States, Chile, Mexico, Japan, Argentina, Taiwan, etc., are severely affected by arsenic toxicity (Bundschuh et al. 2012). The As metalloid can be retained for a long time in the soil matrix, from where it migrates to crops and enters the food chain, endangering people's health. In India and Bangladesh, approximately 70 million people are at risk of As toxicity (WHO 2008) by consuming contaminated drinking water and various food products. Due to exposure to such high As concentrations, various arsenic-induced chronic human diseases have been reported: namely, arsenicosis, black foot, skin cancer, lung cancer, diseases of blood vessels, and reproductive disorders (WHO 2011). Also, cattle are being exposed to As by consuming contaminated paddy straw, which in turn increases As content in meat and milk products (Figure 6.3).

It has been reported that Southeast Asia is the major producer of rice within the world. Approximately 90% production of rice comes from the area, which is heavily contaminated by As (Williams et al. 2007, Dixit et al. 2015). Rice is the major staple food of this area and it has been reported to accumulate high As content (Zhao et al. 2010). Rice is more efficient in accumulating As compared to other crops, as it is generally grown under flooded conditions, where arsenic mobility is high (Meharg et al. 2009). A significant amount of As (up to 2 mg/kg) has been found to accumulate in rice grains and in rice straw (up to 92 mg/kg) (Abedin et al. 2002). In rice, *NIP2* (nodulin 26-like intrinsic proteins) (also named *OsLsi1*), a silicon (Si) influx transporter, is involved in a major pathway responsible for As(III) uptake by roots. *OsNIP2* is localized at the distal side of both exodermis and endodermis cells of rice to mediate As(III) influx sequentially from roots to shoots (Ma et al. 2008). The native people of these areas are more prone to As-associated health issues by consuming these As-contaminated paddy grains (Hua et al. 2011).

Many physical and chemical treatments such as co-precipitation, adsorption onto coagulated flocs, and lime treatment, as well as membrane techniques such as permeable reactive barriers (PRBs), have been reported to remove arsenic from aquifers. However, application of these methods has been limited because of massive inputs of equipment, energy, reagents/chemicals, manpower, inefficiency in treating lower concentrations of As, and generation of highly toxic wastes and their disposal.

Hence, this has led to the exploration of alternative methods for arsenic bioremediation. Microbial remediation of arsenic-polluted sites can be considered as an important approach for restricting arsenic contamination to crops/plants during cultivation. Like other microbes (bacteria, algae, cyanobacteria), fungi are also capable of removing heavy metals from contaminated sites (Volesky and Holan 1995). The use of fungi to combat environmental pollution is a well-known strategy known as *mycoremediation*. It is an escalating field of research that can provide novel methods of eco-friendly approach for cleaning up pollution in soils.

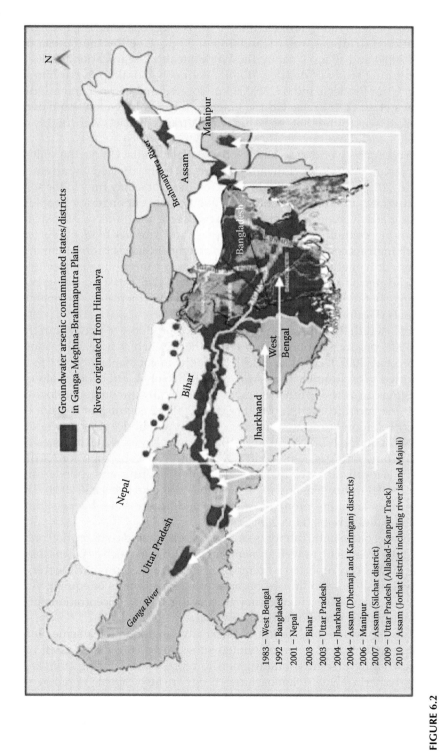

FIGURE 6.2
Major As-contaminated states of India. (Adapted from Chakraborti et al. 2013.)

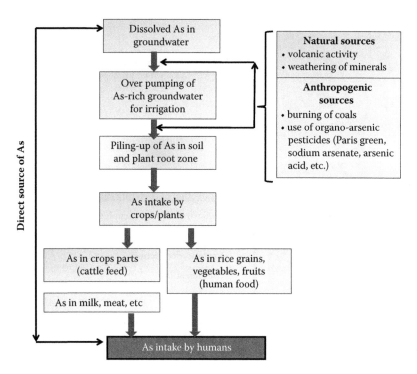

FIGURE 6.3
Various routes of arsenic exposure to humans.

Fungi are unique organisms due to their physiological, morphological, and genetic features and are a well known dominating living biomass in the soil; they may sequester metals or metalloids (Massaccesi et al. 2002, Amin et al. 2013, Batista et al. 2016). They also stimulate soil enzyme activity and promote plant growth (Srivastava et al. 2012, Singh et al. 2015). Fungal cellular structure has well established functioning and is capable of As detoxification within the rhizopheric zone of plants by employing various mechanisms: (1) biosorption of metal ions by negative groups present on the cell walls of fungi; (2) sequestration of metal–thiol complexes into vacuoles; (3) chelation in the cytosol by metallothioneins, organic acids, amino acids, and compound-specific chaperones; (3) reduction in mobilization in plants due to absortion of As by mycorrhizal roots; (4) complexation; (5) oxidation/reduction; and (6) biomethylation. Because of their versatility in growth conditions and ability to tolerate adverse physical and chemical conditions (Haferburg and Kothe 2010), fungi can be used to remediate heavy metals not just in soils and groundwater, but also in industrial effluents and wastewater treatment facilities and on a variety of synthetic materials (Gadd et al. 2012) (Figure 6.4).

Arsenic detoxification by many fungal species, such as *Fusarium oxysporum* (Su et al. 2010), *Trichoderma* sp. (Srivastava et al. 2011), *Aspergillus oryzae* (Singh et al. 2015), *Scopulariopsis brevicaulis* (Boriova et al. 2014), *Aspergillus foetidus* (Chakraborty et al. 2014), *Rhizopus nigricans* (Bai and Abraham 2003), *Aspergillus niger* (Dhankhar and Hooda 2011), and *Penicillium janthinellum* (Zeng et al. 2010), has been reported for potential As bioremediation.

Mycorrhizae are symbioses between plant roots and fungi and another way of protecting plants' exposure to As. Arbuscular mycorrhizal fungi (AMF) are considered the most important type of mycorrhizae for phytoremediation. AMF occur broadly in contaminated

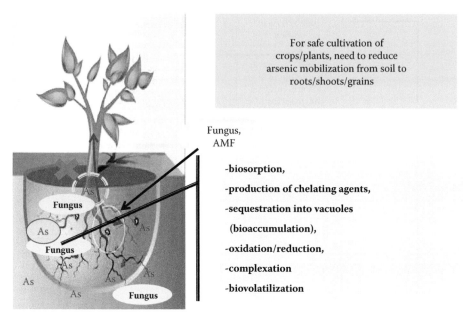

FIGURE 6.4
Role of fungi in rhizospheric zone of plant/crop roots.

soils and are reported to form symbiotic relationships with 80% of terrestrial plant species (Brundrett 2009). They are well reported in improving plant tolerance to environmental stress (Nadeem et al. 2014), help in mobilizing nutrients from organic sources, and appear to possess well-developed saprotrophic capabilities (i.e., oxidative and hydrolytic enzymes). These pioneer fungi grow rapidly and have a short exploitative phase and a high competitive ability. However, other AMF species can protect plants by improving As resistance (Gomes et al. 2014), allowing for discrimination between As and Pi, which reduces As uptake via the AM pathway and results in lower toxicity (Christophersen et al. 2012, Zhang et al. 2016).

Therefore, application of fungi can be beneficial for modulating the rhizospheric zones of crops/plants grown in As-rich soils by restricting the mobilization of arsenic into crops/plants. The mass application of highly arsenic-tolerant strains into fields/agricultural land will sequester the arsenic into their vacuoles by the process of bioaccumulation/biosorption of arsenic (Stolz et al. 2006, Srivastava et al. 2011, Singh et al. 2015, Batista et al. 2016).

6.2 Arsenic Toxicity

Inorganic arsenate As(V) is the dominant species in aerobic soils and readily enters plant roots via phosphate (Pi) transporters, as the oxyanion chemical structure of As(V) is structurally analogous to that of Pi. Phosphorous transporters have been classified in the Pi transporter 1 family (Pht1; TC 2.A.1.9, the PHS family), some of which are also involved in As(V) transport in plants (Meharg 2004). Therefore, any processes requiring phosphate can be interrupted or terminated by the presence of the analogous As(V) (Garg and Singla

2011). Due to arsenate toxicity, cellular processes such as the formation of the cellular fuel source ATP are negatively affected (Gbaruko et al. 2008), resulting in the formation of the adenosine diphosphate–As(V) compound instead of the necessary adenosine triphosphate. Other problems caused by the arsenate analogs to phosphate include blockage of oxidative phosphorylation via inhibition of phosphate-utilizing enzymes (Yang et al. 2012), induction of detrimental cross-links between protein–DNA and DNA–DNA, and replacement of phosphorus in bone (Gbaruko et al. 2008). In the case of inorganic arsenite As(III), flooded and anaerobic conditions are favorable for its existence. As(III) has high affinity to bind with sulfhydryl (–SH) groups present on cystein residues and disrupts the protein metabolism. It also stimulates generation of reactive oxygen species (ROS) to damage proteins, lipids, and DNA.

6.3 Mycoremediation over Phytoremediation

Remediation of contaminated sites is a group of practices and techniques aimed at alleviating impacts caused by contaminants such as arsenic metalloid. Among the available techniques for protecting natural ecosystems, bioremediation (use of microbes) and phytoremediation (use of plants) are generally considered good choices. However, there are some limitations to applying phytoremediation techniques. Some important facts to be considered include:

- Plants take longer to complete the As bioremediation process in contrast to fungi, which have faster life cycles.
- In shallow soils, the phytoremediation technique is limited due to a plant's roots' length. Fungi are present at all depths of soil and can extend their network with the help of their hyphae.
- Use of plants produces a huge biomass containing toxic As metalloids for disposal; further composting or incineration causes release of toxic gases.
- Reduction in yield of vegetative crops and degradation of nutrient contents due to high contamination of As can be overcome by application of fungal consortia into fields to quickly take up As and nullify its mobilization from soil to roots to shoot.

6.3.1 Mycoremediation

The term *mycoremediation* was coined by the American mycologist Paul Stamets, who has studied many potential uses of mushrooms. According to Stamets (2005), mycoremediation methods are preferable to conventional remediation methods due to low-cost input of approximately $50 per ton of contaminated soil compared to $1,000 per ton for conventional methods. This lower cost can be attributed to fungi cheaply producing high yields of biomass in a short time with minimal input (Gadd 1993).

6.3.1.1 Fungi

Fungi are well known dominating living biomass in the soil, may sequester metals or metalloids, and, in parallel, stimulate soil enzyme activity and plant growth promotion

(Srivastava et al. 2012, Tripathi et al. 2012, Singh et al. 2015). By different estimates, it is believed that there are about 20 million fungal cells and 5 m of fungal hyphae in every gram of soil (Stamets 2005). This remarkable ability to build such a living network is shown by fungi residues in the soil.

Fungi exhibit a high range of tolerance within adverse conditions that are often present in contaminated environments. They are able to thrive in high pH (Cernansky et al. 2009) as well as low pH (Siokwi and Anyanwu 2012) conditions, the latter of which are common when As mobilization through an acidic environment is a concern. Extreme temperatures, lack of adequate nutrients, and combinations of organic and inorganic contaminants with multiple adverse physicochemical conditions present (Haferburg and Kothe 2010) are also well tolerated by various types of fungi. In general, fungi have demonstrated higher prevalence and dominance over other microbes (i.e., bacteria) in heavy metal–contaminated habitats (Gadd 1993, Cernansky et al. 2009) due to their especially enhanced capacities for tolerating, adsorbing, accumulating, and detoxifying metals in adverse conditions (Gadd 1993, Stamets 2005, Ahluwalia and Goyal 2010, Mukherjee et al. 2010, Srivastava et al. 2011, Singh et al. 2015, Batista et al. 2016).

6.3.1.2 Mycorrhiza

Mycorrhiza is a widespread mutual association between agriculturally relevant plant species and arbuscular mycorrhizal (AM) fungi. AM fungi were found to belong to the phylum *Glomeromycota* and to be present in all heavy metal–polluted soils (Gamalero et al. 2009). *Glomus* species are reported as the most abundant AM fungi (Larrocea et al. 2010). Some of the important species belonging to this genus are *G. fasciculatum*, *G. intraradices*, *G. etunicatum*, and *G. mosseae*. Recently, Sun et al. (2016) investigated AMF biodiversity by sequencing the nuclear small subunit ribosomal RNA (SSU rRNA) gene fragments using a 454-pyrosequencing technique in rhizosphere soils contaminated with arsenic. They found a total of 11 AMF genera: namely, *Rhizophagus, Glomus, Funneliformis, Acaulospora, Diversispora, Claroideoglomus, Scutellopora, Gigaspora, Ambispora, Praglomus,* and *Archaeospora,* among which *Glomus and Rhizophagus* were dominant AMF genera in the Realgar mining area. When the symbiosis is established between the plant root and fungus, the fungus grows within the cells of the roots, forming arbuscules, which are the main site of nutrient exchange between plant root and fungus. Moreover, the fungus develops an extensive extraradical mycelium that enhances the absorption ability of the plant root system. AM fungi may stimulate metal phytoextraction by essentially improving plant growth and consequent increase in total metal uptake. Li et al. (2016) reported *Rhizophagus intraradices* (AMF) transformed inorganic arsenic to less toxic organic dimethylarsinic acid (DMA) when inoculated into a rice rhizospheric zone amended with 60 mg of As per kilogram of soil. Li et al. (2011) showed that rice/AMF combinations had beneficial effects on lowering grain As concentration, improving grain yield and grain P uptake.

6.3.2 Mechanism Adopted by Fungi in Response to Arsenic Stress

Wysocki et al. (1997) reported that, in yeast, resistance toward As was conferred by contiguous three-gene clusters ACR1, ACR2, and ACR3 located on the arsenic resistance (*ars*) operon, where ACR1 encodes a putative transcription factor, ACR2 encodes arsenate reductase, and ACR3 encodes a plasma membrane As(III) efflux transporter (Tripathi et al. 2007). In a similar way, fungi may employ a different mechanism or combination of mechanisms for tolerating and responding to arsenic in their environment that relies on

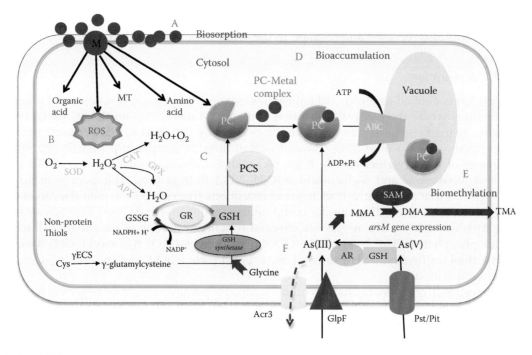

FIGURE 6.5
Schematic representation of fungal mechanism to As exposure includes: (a) *biosorption*: binding of metal ions to the functional groups present on fungal cell wall, (b) ROS defense mechanism (antioxidant enzymes [SOD: superoxide dismutase; CAT: catalase; GR: glutathione reductase; APX: ascorbate peroxidase]), (c) *bioaccumulation*: induction of phytochelatin (PC) synthesis by As stress using GSH as substrate, (d) formation and sequestration of As–PC complexes into vacuoles, (e) *biomethylation:* reduction of As(V) to As(III) by *arsenate reductase* (AR) enzyme and As biomethylation catalyzed by S-adenosylmethionine (SAM) and arsenite methyltransferase (*arsM*), (f) arsenite efflux.

their activities to reduce, mobilize, or immobilize arsenic through bioaccumulation, biosorption, biomethylation, complexation, and oxidation–reduction processes. Schematic representation of fungal responses to arsenic can be seen in Figure 6.5.

6.3.2.1 Bioprecipitation

Precipitation is a result of the bioactive reaction between metal(s) and metabolites produced by microbes as a result of their defense system against a heavy-metal pollutant. In this case, microorganisms sense the toxicity of metals and start producing metabolites that favor the precipitation process. During this process, a toxic form of arsenic is converted to less toxic forms outside the cell, through reactions like complexation, crystallization, and extracellular precipitation. Fungi accomplish this by secreting extracellular compounds like oxalic acids and citric acids (Baldrian 2003) into the environment that complex and precipitate with arsenic (Gadd 1993). The metabolite oxalic acid is shuttled outside the cell, where it scavenges and forms insoluble oxalate crystals with nearby metal cations (Gadd et al. 2012). Recently, Kaewdoung et al. (2016) reported two fungal strains, *Fomitopsis meliae* and *Ganoderma steyaertanum*, to form metal oxylates (zinc oxylate, cadmium oxylate, lead oxylate) by precipitation. The organic acid-producing soil fungus *Aspergillus niger* was able to solubilize mimetite insoluble mineral with simultaneous precipitation of lead oxalate as a new mycogenic biomineral (Ceci et al. 2015). The production of H_2S by *Saccharomyces cerevisiae* was also reported and resulted in the precipitation of metals (Gadd 1993).

6.3.2.2 Biosorption

Biosorption is defined as a physiochemical interaction between metal species and microbial biomass where the passive adsorption of dissolved metal ions on the microbial biomass due to their chemical activities takes place (Volesky and Holan 1995, Cernansky et al. 2009, Haferburg and Kothe 2010). The cell wall of fungi is composed primarily of chitin, a polymer of *N*-acetylglucosamine ($C_8H_{15}NO_6$), and chitosan, a mixed polymer of both *N*-acetylglucosamine and D-glucosamine ($C_6H_{13}NO_5$). Both of these polymers have functional groups that are negatively charged in conditions from pH 3 to 10 (Seh-Bardan et al. 2013), which serve as binding sites for metal cations (Ahluwalia and Goyal 2010, Srivastava et al. 2011, Gadd 2012). The negatively charged functional groups of the fungal cell wall include phosphate ($-PO4^{3-}$), carboxyl ($-COOH$), carbonyl ($-C=O$), hydroxyl ($-OH$), amino ($-NH_2$), thiol/sulfhydryl ($-SH$), sulfate ($-SO4^{2-}$), sulfonate ($-SO_3$), and thioesters ($-C-S-C$) (Simonescu et al. 2012).

Other cell wall components such as proteins can also possess numerous negative charges and help in binding As metal ions (Srivastava et al. 2011). Melanins (fungal pigments found interior or exterior to cell walls) also play an important role in binding with toxic metals as they contain phenolic groups, peptides, carbohydrates, aliphatic hydrocarbons, and fatty acids that possess many metal-binding sites (Gadd 1993). Fungi can increase their production of melanin and even secrete it into the environment under heavy-metal stress. Melanized soil fungi such as *Cladosporium* sp. and *Alternaria* sp. are often isolated from toxic metal-contaminated soils (Gadd et al. 2012). *Aspergillus clavatus* and *Aspergillus niger* species have been reported in efficient adsoption of arsenic onto their cell walls (Cernansky et al. 2009) and cadmium biosorption by *Aspergillus foetidus* (Chakraborty et al. 2014). The biomass of the fungus *Penicillium purpurogenum* has been found effective in biosorption of As (Say et al. 2003).

Sathish Kumar et al. (2008) have used live and pretreated biomass of *Aspergillus fumigatus* for the biosorption of As(III) from aqueous solution. Cernansky et al. (2007) found that a heat-resistant strain of *Neosartorya fischeri* isolated from soil was able to bioabsorb 15.2% to 31.5% of As in its medium (originally at 50 mg/L) within just 1 hour. The valence state of the As (trivalent or pentavalent) may also impact the pH level at which biosorption is maximized; Seh-Bardan et al. (2013) found that maximum As(V) was biosorbed by *Aspergillus niger* at pH 3, while pH 6 was best for As(III). The efficient biosorption potency of chromium (VI) was also observed in *Aspergillus niger* (Srivastava and Thakur 2006). The binding of heavy metals exterior to cell walls of fungi has been documented by scanning and transmission electron microscopy (SEM/TEM) and by Fourier transform infrared (FTIR) spectroscopy, x-ray diffraction analysis (XRD), and nuclear magnetic resonance (NMR) (Seh-Bardan et al. 2013), and studies have found that the cell wall is indeed the primary site for metal accumulation.

6.3.2.3 Bioaccumulation

Another mechanism known to be adopted for detoxification of arsenic is *bioaccumulation*. The bioaccumulation of heavy metals by fungi has received more attention in recent years due its advantages over bioprecipitation and biosorption. Many studies are taking place because of concern with environmental protection and recovery of metals by using the hidden potential of diverse fungi. This pathway involves the intracellular chelation of arsenic metal ions by glutathione (GSH) and other nonprotein thiols (phytochelatins and metallothioneins) through arsenic–thiol complexes (Cobbett and Goldsbrough 2002, Seith et al. 2012) that sequester into the vacuoles through ABC transporters (Figure 6.5).

In order to transport a metal into the cell, the fungus first mobilizes it outside the cell. Cells must secrete compounds to free the metals from their mineralized state. For example, in iron-limiting conditions, fungi will secrete a class of low molecular weight metal chelators called siderophores whose function is to scavenge, solubilize, and chelate the critical Fe^{3+} in the environment (Hastrup et al. 2013). Chelation occurs when two or more functional groups of the siderophore chelator bind to a metal cation to form a coordination complex. Once bound to a siderophore chelator, the metal-containing complex can then be transported into the cell (Srivastava et al. 2011) and subsequent transformations can take place.

The main way for metal cations to get into the cell is via plasma membrane proteins functioning as cation transporters. Because it is the analog of phosphorus, pentavalent arsenate, specifically, chooses the existing phosphate transport systems to gain entry into the cell (Rosen 1999, Tripathi et al. 2007, Yang et al. 2012). After successful transportation, As(V) gets reduced to As(III) by the enzyme *arsenate reductase*, which is then either extruded from the cells or sequestered in intracellular components (either as free As(III) or as conjugates with glutathione or other thiols) (Figure 6.5). Thiol–arsenic complexes have been found to be formed and stored into vacuoles of *Aspergillus* sp. P37 (Canovas et al. 2004). Zeng et al. (2010) studied the bioaccumulation potential of their *Penicillium janthinellum* isolate, able to accumulate 87.0 mg/kg of dry tissue of As(V) at 50 mg/L. Su et al. (2010) reported that fungi (*Penicillin janthinellum, Fusarium oxysporum*, and *Trichoderma asperellum*) bioaccumulate As(V) under laboratory conditions. *Aspergillus* species have been shown to grow in liquid medium with 20 mg/L arsenate (Urik et al. 2007). Strains of *Penicillium* sp., *Neocosmospora* sp., and *Trichoderma* sp. isolated from As-contaminated agricultural soils in West Bengal, India, have been reported to accumulate 58.4%, 57.8%, and 56.2%, respectively, of the total environmental As when grown in media containing 10 mg/L As (Srivastava et al. 2011). In a study, a fungal strain, *Aspergillus oryzae*, was identified as the most efficient among all 54 isolates for removing arsenic from liquid medium by 82% through maximum bioaccumulation (0.259 g/kg) (Singh et al. 2015). Batista et al. (2016) isolated and identified filamentous fungi from a rice rhizospheric zone. Most of them were found to belong to genus *Penicillium* sp., *Aspergillus* sp., *Trichoderma* sp., *Cladosporium* sp., *Rhizopus* sp., and *Westerdykella* sp. In their arsenic-tolerance test (0–50 mg/L As^{3+}), the author showed that the most arsenic-tolerant genera were *Penicillium* sp. and *Aspergillus* sp.

As soon as arsenic enters the cell, the intracellular mechanism starts immediately to deal with the metal toxic effects. To quickly take up the newly imported metal, the internal environment of the cell starts multiple responses toward As stress, such as induction of antioxidant enzymes against ROS, increase in GSH level, phytochelatin synthesis, and expression of the *arsM* gene (Canovas et al. 2004, Gadd et al. 2012). Details on intracellular mechanisms initiated on reaching arsenic metalloid into the cell are discussed below.

6.3.2.3.1 ROS Generation

Metal-induced oxidative damage and generation of reactive oxygen species (ROS) are the most common responses of fungi (Gratao et al. 2005). When the cellular redox balance is compromised, the generation of ROS, such as superoxide (O_2^-) and hydrogen peroxide (H_2O_2), promotes the oxidation of membrane lipids, proteins, and nucleic acids, affecting plant metabolism. To counteract this oxidative stress, fungi have a built-in antioxidative system, which helps in scavenging ROS and the by-products generated by arsenic (Chou et al. 2004). Superoxide dismutase (SOD, EC 1.15.1.1) is a metalloenzyme that catalyzes the dismutation of superoxides (O_2^-) into oxygen and hydrogen peroxide. Such enzymes provide a defense system for the survival of aerobic organisms (Beyer et al. 1991). The hydrogen peroxide generated due to SOD activity is also very toxic. This may be converted to a hydroxyl free radical (OH·) by the catalase (CAT, EC 1.11.1.6) enzyme, which detoxifies the H_2O_2 by converting it into water and oxygen in aerobic organisms. The role of thiol compounds in dealing with oxidative stress caused by As is also significant. The most common thiol compound is glutathione, which has been found to be efficient in counteracting As stress (Riccillo et al. 2000). Within the cell, the enzyme glutathione reductase (GR) is involved in generating a more reduced form of glutathione (GSH) from its oxidized form (GSSG). Todorova et al. (2008) showed cadmium-induced ROS generation in *Aspergillus niger* B77 and that tolerance of the strain was highly correlated to the efficiency of its antioxidative defense system. *Trichosporon asahii* yeast isolated from industrial wastewater was exposed for 2 days to 100 mg/L sodium arsenite (NaAsO$_2$) and increased GR activity was reported (Ilyas et al. 2014). Activities of SOD and GR were found to sharply increase in an *Aspergillus niger* strain on arsenate stress up to 100 mg/L (Mukherjee et al. 2010).

6.3.2.3.2 Nonprotein Thiols

Thiols are a class of organic compounds that contain sulfhydryl groups (–SH) composed of a sulfur atom and a hydrogen atom attached to a carbon atom. The sum of nonprotein thiols is within a cell composed of mainly GSH, GSSG, PCs, and cystein (Seith et al. 2012), where cysteine was found to be rapidly converted into other compounds. Thiols are essential agents in cellular redox signaling in animals, plants, and fungi (Pócsi et al. 2004). Mukherjee et al. (2010) found that when a chelating agent, EDTA, was used in the extraction buffer, arsenic was chelated, making more thiols free to react with thiol-specific reagent DTNB and resulting in an increase of thiol. This would indicate that conjugation of thiols and arsenate forms a thiol–arsenate complex, which may have a role in detoxification. The level of nonprotein thiols was also found to be significantly increased by 134% in Cd-treated yeast cells (Rehman and Anjum 2011).

6.3.2.3.2.1 Glutathione Glutathione (GSH) is widely recognized as a relevant nonprotein thiol (NPT) present in most microorganisms (Dickinson and Forman 2002). GSH is a tripeptide (γ-glutamyl-cysteinyl-glycine) (Figure 6.6) and is a predominant nonprotein thiol involved in the chelation and detoxification of free metal (Anjum et al. 2014). The synthesis of GSH occurs in two ATP-dependent enzymatic steps: The first is catalyzed

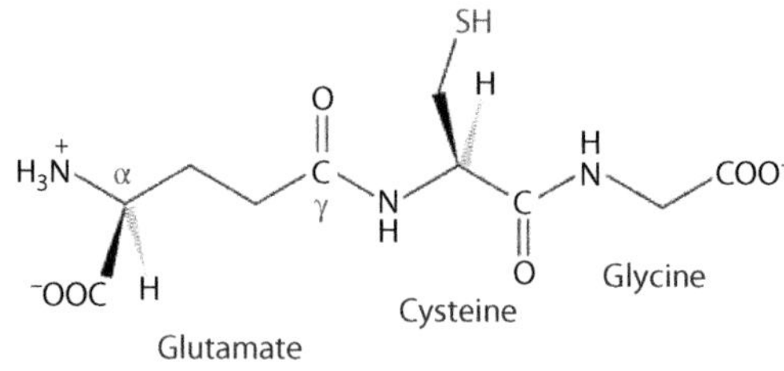

FIGURE 6.6
General structure of glutathione (GSH).

by γ-glutamylcysteine synthetase (γ-ECS, *AtGSH1* gene) to bind glutamine and cysteine in chloroplasts synthesizing γ-EC (Zechmann 2014). Glutathione synthetase (GS, *AtGSH2* gene) catalyzes the second step, where glycine is added to γ-EC (Noctor et al. 2014). GSH overproduction up to 10 mM has been reported in fungi (Pocsi et al. 2004).

GSH plays a dual role as an antioxidant in quenching ROS and chelation of heavy-metal ions (Hossain et al. 2012). The role of GSH in metal chelation lies behind its significance as a precursor for the synthesis of phytochelatins (PCs, a family of peptides structurally related to GSH). In higher eukaryotes, the reduction of As(V) to As(III) was also mediated by GSH (Bleeker et al. 2006). A disulfide redox couples with GSH and reduces arsenate to arsenite *in vitro* and further forms arsenotriglutathione (Equations 6.1 and 6.2). The –H group of the GSH molecule undergoes arsenolysis, where anhydride bond formation between As(V) and the Cys residue occurs; thus, it might play a favored role in reduction of As(V) coupled to the GS-catalyzed arsenolysis of GSH (Hayakawa et al. 2005, Watanabe and Hiano 2013).

$$H_3As^5O_4 + 2GSH \rightarrow H_3As^3O_3 + H_2O + GS\text{–}SG \tag{6.1}$$

$$H_3As^3O_3 + 3GSH \rightarrow As(SG)_3 + 3H_2O \tag{6.2}$$

Hayakawa et al (2005) first reported that GSH forms conjugates with trivalent forms of arsenic. The As(SG)$_3$ complex formed from As(III) and GSH has a pyramidal configuration with a molar ratio of As(III):GSH = 1:3 (Han et al. 2007).

6.3.2.3.2.2 Phytochelatins (PCs) Phytochelatins (PCs) are short, cysteine-rich, thiol-containing nonprotein peptides with the general structure of (γ-Glu-Cys)$_n$Gly$_{(n = 2–11)}$ that are present in plants, fungi, and other organisms (Figure 6.7) (Minocha et al. 2008). They were first discovered in a yeast, but have been named after a plant that they were isolated from rather than the fungus (hence, *phyto* instead of *myco*). These PCs start making complexes with arsenic metal and sequester into vacuoles through ABC transporters (Yang et al. 2012). This has been well studied by Canovas (2004) in *Aspergillus* sp. P37 (arsenate hypertolerant fungus). Under high arsenate exposure, *Aspergillus* sp. P37 implies vacuolar sequestration of As(GS)$_3$. In addition, *Saccharomyces cerevisiae* species have been found to sequester As(GS)$_3$ conjugate into the vacuole by the ABC transporters Ycf1 (Maciaszczyk-Dziubinska et al. 2012).

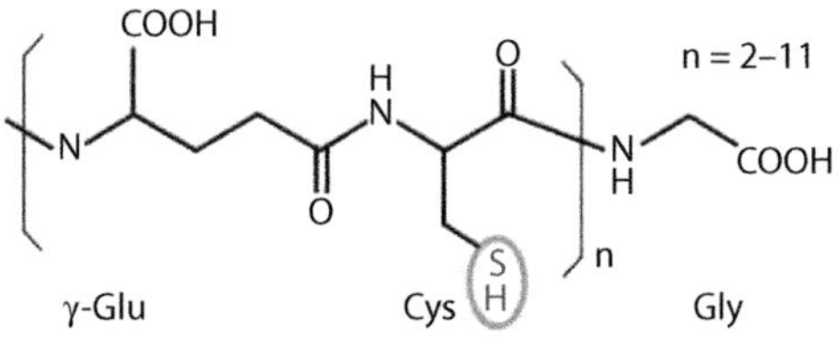

FIGURE 6.7
General structure of phytochelatins (PCs).

6.3.2.4 Biomethylation/Biovolatilization

The final mechanism of fungal remediation of toxic metals to be discussed is biomethylation/ biovolatilization. This classical mechanism involves a series oxidation, reduction, and methylation, where inorganic arsenic species get methylated into less toxic pentavalent mono-, di-, and trimethylated arsenicals catalyzed by a S-adenosylmethionine-dependent methyltransferase enzyme known as *Challenger Pathway* (Figure 6.8). In this pathway, the resulting compound, TMA, is volatile and 1,000 times less toxic than As(III) species (Hirano et al. 2004) and hence easily escapes from the cell. The pathway was proposed in 1951 by Frederick Challenger, who studied formation of the toxic methylarsines by the plant pathogen *Scopulariopsis brevicaulis* (Challenger 2006). It is reported that sequencing of multiple fungal genomes identified genes encoding homologs of As(III) methyltransferase (*arsM*) in fungi. This *arsM* gene has been previously isolated from the soil bacterium *Rhodopseudomonas palustris* and has shown As(III) resistance when expressed in an As-sensitive strain of *Escherichia coli* (Qin et al. 2006). Moreover, phylogenetic analysis has shown that bacterial *arsM* is more closely related to fungal *arsM* (Meng et al. 2011).

The fungus actively converts either intracellular inorganic As(III) or As(V) to an organometallic compound by the addition of one or more methyl groups. Similarly, several fungal species (i.e., *Candida humicola, Gliocladium roseum, and Penicillium* sp.) have been established to methylate arsenic (Tamaki and Frankenberger 1992). Resistant fungi *Pennicillum janthinellum, Trichoderma asperellum,* and *Fusarium oxysporum* also accumulate and volatilize As (ranging from 100 to 304.06 μg) from culture medium (Su et al. 2010). *Aspergillus clavatus, Aspergillus niger, Trichoderma viride,* and *Penicillium glabrum,* also isolated from As-contaminated soils, could biovolatilize 9%–27% of the pentavalent As provided in media within 30 days (Urík et al. 2007). In 2015, Singh et al. reported that fungal strains FNBR_L-35 (*Aspergillus* sp.) and FNBR_L-82 (fungal strain) biovolatilized higher amounts of arsenic with 6.4 and 6.15 mg/kg, respectively. A fungal strain, *Westerdykella*

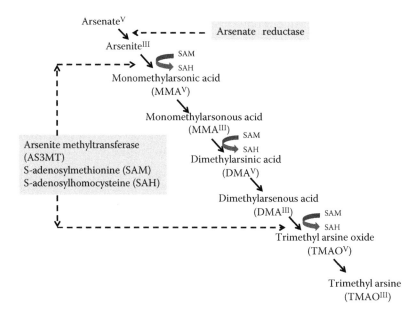

FIGURE 6.8
Biomethylation of arsenic in fungi using S-adenosylmethionine (SAM) as methyl group donor.

aurantiaca, isolated from arsenic-contaminated (9.45–15.63 mg/kg) agricultural soils from West Bengal, India, was shown to have high capability for As-methylation against arsenic stress (Srivastava et al. 2011, Verma et al. 2016).

Once the biomethylated As diffuses out of the cell (Srivastava et al. 2011), the gases will dissipate into the surrounding soil or water and be released into the air. This reduces toxicity for the exposed organisms by decreasing the concentrated dose of As in their local environment.

6.3.3 Contribution of Fungi/Mycorrhiza in Rhizosphere

6.3.3.1 Plant Growth–Promoting Fungi (PGPF)

The rhizospheric zone may be defined as the "heart of soil" as it is the zone of plant roots in direct contact with various active soil microbes. Soil microorganisms including fungi, bacteria, algae, etc. are responsible for producing physiologically active compounds that exert physiological effects on plant growth and development (Duca 2015). These compounds are known as plant growth regulators such as auxin, gibberlic acid, etc. Kingdom fungi are playing a significantly important role in plant growth promotion and are involved in production of a large number of degradative enzymes, including proteases, glucanases, and chitinases converting complex structures into simple forms. Fungi are also involved in production of antibiotics, destruction of fungal pathogens, elicitation of defense responses, etc. *Trichoderma* sp. are well known for bearing tremendous abilities to stimulate the plant growth promotion through root colonization, pathogenic resistance to crop cultivars defending their ecological significance (Hermosa et al. 2012, Tripathi et al. 2013). Many soil fungi have been shown to possess the ability to solubilize sparingly soluble nutrients, like phosphates *in vitro* by secreting inorganic or organic acids. The fungal phytases may increase hydrolysis of inositol phosphates in soil environments (Singh et al. 2015). According to Srivastava et al. (2012), *Westerdykella* and *Trichoderma* strains isolated from arsenic contaminated soils showed better stimulatory plant growth promotion and soil nutrient availability. *Penicillium simplicissimum* is a plant-growth–promoting fungus isolated from the rhizosphere of zoysiagrass (*Zoysia tenuifolia*) and has been shown to induce resistance in cucumber (Koike et al. 2001). In a greenhouse experiment, conducted by Gogoi and Singh (2011), six AM fungal species (*Glomus fasciculatum*, *Glomus versiforme*, *Glomus clarum*, *Glomus* sp. 2, *Glomus mosseae* and *Glomus etunicatum*) were the most promising in plant growth promotion of *Piper longum* L.

6.3.3.2 Phosphate Solubilization

In the rhizospheric zone of plants, microorganisms play an essential role in phosphate solubilization. Phosphate solubilizing microorganisms (PSMs) release organic acids that dissolve the phosphate in soil so that plant roots can easily utilize it. Phosphate solubilizing fungi constitute approximately 0.1%–0.5% of total fungal populations within the soil. Moreover, these fungi are able to travel long distances more easily in soils and produce more acids than the bacteria, showing effective phosphate solubilization (Kucey 1983). The contributions of PSMs to rhizospheric soils includes (i) siderophore production, (ii) antibiotic production, (iii) production of plant growth hormones, and (iv) biocontrol of plant pathogens, etc.

The genera *Aspergillus, Mucor,* and *Penicillium* are reported as most effective and powerful in phosphate solubilizing activity (Saxena et al. 2013), although strains of *Trichoderma*

(Li et al. 2015) and *Rhizoctonia solani* (Jacobs et al. 2002) have also been reported as phosphate solubilizers. A nemato fungus, *Arthrobotrys oligospora*, is reported in solubilizing phosphate *in vivo* as well as *in vitro* (Duponnois et al. 2006). Among yeasts, only a few studies have been conducted to assess their ability in solubilizing phosphate; these include *Schizosaccharomyces pombe* and *Pichia fermentans*.

6.3.3.3 Biocontrol Fungi

Biological control is the use of beneficial microbes to reduce the numbers of other pathogenic organisms. Biocontrol fungi can increase resistance to plants by root colonization from pathogen attack by suppressing the growth of phytopathogens in a variety of ways, such as competing for nutrients and space and limiting the supply of available nutrients to pathogens (Marasco et al. 2012). A range of biopesticides is now available commercially, most of which are based on the fungal genus *Trichoderma* (Woo et al. 2006). Many strains of *Trichoderma* are strong opportunistic invaders (Kumar et al. 2014) and are reported for their efficiency against different phytopathogens, such as *Rhizoctonia, Alternaria, Colletotrichum, Diaporthe, Fusarium, Phytophthora, Phythium, Botrytis, Sclerotinia, Monilinia,* and *Verticillium* (Begum et al. 2008), thus restricting growth and possible spore development. Root colonization by the endophytic basidiomycete fungus *Piriformospora indica* is also reported in protecting the barley plant from the pathogen *Blumeria graminis* sp. *hordei* (Molitor et al. 2011). AMF fungi belonging to genus *Glomus* also act as biocontrols against pathogens like *Rhizoctonia solani* and *Magnaporthe oryzae* in maize and rice, respectively (Song et al. 2011, Campos-Soriano et al. 2012).

6.4 Conclusions and Future Prospects

Bioremediation of arsenic-contaminated soils offers a great opportunity for sustainable future development due to its eco-friendly and cost-effective nature. The persistence of arsenic metalloid in agricultural soil leads to the emergence of new methods for its biological treatment. It relies on microbial activity to remove, mobilize, and contain arsenic through sorption, biomethylation, co-precipitation, complexation, and oxidation-reduction processes. The soil microorganisms are key players in restricting heavy metals' entry into the food chain through root to shoot to grain translocation. Among various available biological methods, mycoremediation is one of the unique and innovative technologies of using the potential of fungi. The use of native soil fungi for biosorption, bioaccumulation, and biomethylation of As metalloid from arsenic-contaminated agricultural soils is one of the potential bioremediation approaches. Fungi sequester arsenic metalloids into their vacuoles as thiol (phytochelatins and metallothionines) conjugates for removing arsenic from contaminated matrices (soil).

Further studies should focus on the factors involved in improving *in situ* bioremediation strategies using genetically engineered microorganisms (GEMs) and also the applicability and adaptability of these GEMs in all the possible adverse/stressed conditions and multiple heavy metal–polluted conditions. An efficient and potent engineered fungal strain can be generated using biotechnological resources, which may help in remediating a variety of heavy metals and are more selective for heavy-metal biotransformation. GEMs have improved heavy-metal biosorption, bioaccumulation, and biotransformation abilities, and their application could be used for bioremediation of heavy metals from contaminated soil.

Acknowledgments

The authors are thankful to the head of the Botany Department, Benares Hindu University, Varanasi, and the director, CSIR-NBRI, Lucknow, for institutional support. Mrs. Manvi Singh, UGC-SRF (ref. no. 3185/NET-DEC-2011) is thankful to UGC for financial support to her as SRF. Prof. R. N. Kharwar is thankful to DST, New Delhi (SB/EMEQ-121/2014), and DST-PURSE, BHU, Varanasi, for financial help.

References

Abedin, M.J., Feldmann, J., Meharg, A.A. 2002. Uptake kinetics of arsenic species in rice plants. *Plant Physiology* 128(3):1120–8.

Amin, A., Latif, Z. 2013. Detoxification of mercury pollutant by Immobilized yeast strain *Candida xylopsoci*. *Pakistan Journal of Botany* 45(4):1437–43.

Anjum, N.A., Aref, I.M., Duarte, A.C., Pereira, E., Ahmad, I., Iqbal, M. 2014. Glutathione and proline can coordinately make plants withstand the joint attack of metal(loid) and salinity stresses. *Frontiers in Plant Science* 5:662.

Batista, B.C., Souza, J.M., Paulelli, A.C., Rocha, B.A., de Oliveira, A.R., Segura, F.R., Braga, G.Ú., Tonani, L., von Zeska-Kress, M.R., Barbosa, F. 2016. A low-cost and environmentally-friendly potential procedure for inorganic-As remediation based on the use of fungi isolated from rice rhizosphere. *Journal of Environmental Chemical Engineering* 4(1):891–8.

Bai, R.S., Abraham, T.E. 2003. Studies on chromium(VI) adsorption–desorption using immobilized fungal biomass. *Bioresource Technology* 87(1):17–26.

Baldrian, P. 2003. Interactions of heavy metals with white-rot fungi. *Enzyme and Microbial Technology* 32(1):78–91.

Begum, M.M., Meon, S., Ahmad, M., Abidin, Z., Puteh, A.A., Rahman, M. 2008. Antagonistic potential of selected fungal and bacterial biocontrol agents against *Colletotrichum truncatum* of soybean seeds. *Pertanika Journal of Tropical Agricultural Science* 31(1):45–53.

Beyer, W., Fridovich, I. 1991. Superoxide dismutases. *Progress in Nucleic Acid Research and Molecular Biology* 40:221–53.

Bleeker, P.M., Hakvoort, H.W.J., Bliek, M., Souer, E., Schat, H. 2006. Enhanced arsenate reduction by a CDC25-like tyrosine phosphatase explains increased phytochelatin accumulation in arsenate-tolerant *Holcus lanatus*. *Plant Journal* 45(6):917–29.

BOI Standards 10500:1991, Second Revision ICS No. ICS No. 13.060.20. 2009: http://bis.org.in/sf/fad/FAD25(2047)C.pdf

Boriova, K., Cernansky, S., Matus, P., Bujdos, M., Simonovicova, A. 2014. Bioaccumulation and biovolatilization of various elements using filamentous fungus *Scopulariopsis brevicaulis*. *Letters in Applied Microbiology* 59(2):217–23.

Brundrett, M.C. 2009. Mycorrhizal associations and other means of nutrition of vascular plants: Understanding the global diversity of host plants by resolving conflicting information and developing reliable means of diagnosis. *Plant and Soil* 320(1–2):37–77.

Bundschuh, J., Litter, M.I., Parvez, F., Roman-Ross, G., Nicolli, H.B., Jean, J.S. et al. 2012. One century of arsenic exposure in Latin America: A review of history and occurrence from 14 countries. *Science of the Total Environment* 429:2–35.

Campos-Soriano, Li., García-Martínez, Jo., Segundo, Bs. 2012. The arbuscular mycorrhizal symbiosis promotes the systemic induction of regulatory defence-related genes in rice leaves and confers resistance to pathogen infection. *Molecular Plant Pathology* 13(6):579–92.

Canovas, D., Vooijs, R., Schat, H., de Lorenzo, V. 2004. The role of thiol species in the hypertolerance of Aspergillus sp P37 to arsenic. *Journal of Biological Chemistry* 279(49):51234–40.

Ceci, A., Kierans, M., Hillier, S., Persiani, A.M., Gadd, G.M. 2015. Fungal bioweathering of mimetite and a general geomycological model for lead apatite mineral biotransformations. *Applied and Environmental Microbiology* 81(15):4955–64.

Cernansky, S., Kolencik, M., Sevc, J., Urik, M., Hiller, E. 2009. Fungal volatilization of trivalent and pentavalent arsenic under laboratory conditions. *Bioresource Technology* 100(2):1037–40.

Cernansky, S., Urik, M., Sevc, J., Khun, M. 2007. Biosorption and biovolatilization of arsenic by heat-resistant fungi. *Environmental Science and Pollution Research—International* 14(1):31–5.

Chakraborti, D., Rahman, M.M., Ahamed, S., Dutta, R.N., Pati, S., Mukherjee, S.C. 2016. Arsenic groundwater contamination and its health effects in Patna district (capital of Bihar) in the middle Ganga plain, India. *Chemosphere* 152:520–9.

Chakraborty, S., Mukherjee, A., Khuda-Bukhsh, A.R., Das, T.K. 2014. Cadmium-induced oxidative stress tolerance in cadmium resistant *Aspergillus foetidus*: Its possible role in cadmium bioremediation. *Ecotoxicology and Environmental Safety* 106:46–53.

Challenger, F. 2006. Biological methylation. *Advances in Enzymology and Related Areas of Molecular Biology* DOI: 10.1002/9780470122709.ch8.

Chou, W.C., Jie, C.F., Kenedy, A.A., Jones, R.J., Trush, M.A., Dang, C.V. 2004. Role of NADPH oxidase in arsenic-induced reactive oxygen species formation and cytotoxicity in myeloid leukemia cells. *Proceedings of the National Academy of Sciences of the United States of America* 101(13):4578–83.

Christophersen, H.M., Smith, F.A., Smith, S.E. 2012. Unraveling the influence of arbuscular mycorrhizal colonization on arsenic tolerance in Medicago: *Glomus mosseae* is more effective than *G. intraradices*, associated with lower expression of root epidermal Pi transporter genes. *Frontiers in Physiology* 3.

Cobbett, C., Goldsbrough, P. 2002. Phytochelatins and metallothioneins: Roles in heavy metal detoxification and homeostasis. *Annual Review of Plant Biology* 53:159–82.

del Pilar Ortega-Larrocea, M., Xoconostle-Cázares, B., Maldonado-Mendoza, I.E., Carrillo-Gonzalez, R., Hernández-Hernández, J., Garduño, M.D., López-Meyer, M., Gómez-Flores, L., González-Chávez, M.D. 2010. Plant and fungal biodiversity from metal mine wastes under remediation at Zimapan, Hidalgo, Mexico. *Environmental Pollution* 158(5):1922–31.

Dhankhar, R., Hooda, A. 2011. Fungal biosorption—An alternative to meet the challenges of heavy metal pollution in aqueous solutions. *Environmental Technology* 32(5):467–91.

Dickinson, D.A., Forman, H.J. 2002. Cellular glutathione and thiols metabolism. *Biochemical Pharmacology* 64(5–6):1019–26.

Dixit, G., Singh, A.P., Kumar, A., Singh, P.K., Kurnar, S., Dwivedi, S. et al. 2015. Sulfur mediated reduction of arsenic toxicity involves efficient thiol metabolism and the antioxidant defense system in rice. *Journal of Hazardous Materials* 298:241–51.

Duca, M. 2015. Plant Growth and Development. In *Plant Physiology* (pp. 187–229). Springer International Publishing.

Duponnois, R., Kisa, M., Plenchette, C. 2006. Phosphate-solubilizing potential of the nematophagous fungus *Arthrobotrys oligospora*. *Journal of Plant Nutrition and Soil Science* 169(2):280–2.

Gadd, G.M. 1993. Interactions of fungi with toxic metals. *New Phytologist* 124(1):25–60.

Gadd, G.M., Rhee, Y.J., Stephenson, K., Wei, Z. 2012. Geomycology: Metals, actinides and biominerals. *Environmental Microbiology Reports* 4(3):270–96.

Gamalero, E., Lingua, G., Berta, G., Glick, B.R. 2009. Beneficial role of plant growth promoting bacteria and arbuscular mycorrhizal fungi on plant responses to heavy metal stress. *Canadian Journal of Microbiology* 55(5):501–14.

Garelick, H., Jones, H., Dybowska, A., Valsami-Jones, E. 2008. Arsenic pollution sources reviews of environmental contamination and toxicology, Vol 197: *Arsenic Pollution and Remediation: An International Perspective* 197:17–60.

Garg, N., Singla, P. 2011. Arsenic toxicity in crop plants: Physiological effects and tolerance mechanisms. *Environmental Chemistry Letters* 9(3):303–21.

Gbaruko, B.C., Ana, G., Nwachukwu, J.K. 2008. Ecotoxicology of arsenic in the hydrosphere: Implications for public health. *African Journal of Biotechnology* 7(25):4737–42.

Gillispie, E.C., Duckworth, O.W., Polizzotto, M.L. 2015. Soil pollution due to irrigation with arsenic-contaminated groundwater: Current state of science. *Current Pollution Reports* 1(1):1–2.

Gomes, M.P., Andrade, M.L., Nascentes, C.C., Scotti, M.R. 2014. Arsenic root sequestration by a tropical woody legume as affected by arbuscular mycorrhizal fungi and organic matter: Implications for land reclamation. *Water, Air, & Soil Pollution* 225(4):1–2.

Gogoi, P., Singh, R.K. 2011. Differential effect of some arbuscular mycorrhizal fungi on growth of *Piper longum* L. (Piperaceae). *Indian Journal of Science and Technology* 4(2):119–25.

Gratao, P.L., Polle, A., Lea, P.J., Azevedo, R.A. 2005. Making the life of heavy metal-stressed plants a little easier. *Functional Plant Biology* 32(6):481–94.

Haferburg, G., Kothe, E. 2010. Metallomics: Lessons for metalliferous soil remediation. *Applied Microbiology and Biotechnology* 87(4):1271–80.

Hastrup, A.C.S., Jensen, B., Jellison, J. 2014. Fungal accumulation of metals from building materials during brown rot wood decay. *Archives of Microbiology* 196(8):565–74.

Hayakawa, T., Kobayashi, Y., Cui, X., Hirano, S. 2005. A new metabolic pathway of arsenite: Arsenic-glutathione complexes are substrates for human arsenic methyltransferase Cyt19. *Archives of Toxicology* 79(4):183–91.

Hermosa, R., Viterbo, A., Chet, I., Monte E. 2012. Plant-beneficial effects of Trichoderma and of its genes. *Microbiology* 158(1):17–25.

Hirano, S., Cui, X., Kanno, S., Hayakawa, T., Shraim, A. 2004. The accumulation and toxicity of methylated arsenicals in endothelial cells: Important roles of thiol compounds. *Toxicology and Applied Pharmacology* 198(3):458–67.

Hossain, M.A., Piyatida, P., da Silva, J.A., Fujita, M. 2012. Molecular mechanism of heavy metal toxicity and tolerance in plants: Central role of glutathione in detoxification of reactive oxygen species and methylglyoxal and in heavy metal chelation. *Journal of Botany* 2:872875.

Hua, B., Yan, W.G., Wang, J.M., Deng, B.L., Yang, J. 2011. Arsenic accumulation in rice grains: Effects of cultivars and water management practices. *Environmental Engineering Science* 28(8):591–6.

Ilyas, S., Rehman, A., Varela, A.C., Sheehan, D. 2014. Redox proteomics changes in the fungal pathogen *Trichosporon asahii* on arsenic exposure: Identification of protein responses to metal-induced oxidative stress in an environmentally-sampled isolate. *Plos One* 9(7):e102340.

Jacobs, H., Boswell, G.P., Harper, F.A., Karl, R.I., Davidson, F.A. 2002. Solubilization of metal phosphates by *Rhizoctonia solani*. *Mycological Research* 106(12):1468–79.

Kaewdoung, B., Sutjaritvorakul, T., Gadd, G.M., Whalley, A.J.S., Sihanonth, P. 2016. Heavy metal tolerance and biotransformation of toxic metal compounds by new isolates of wood-rotting fungi from Thailand. *Geomicrobiology Journal* 33(3–4):283–8.

Koike, N., Hyakumachi, M., Kageyama, K., Tsuyumu, S., Doke, N. 2001. Induction of systemic resistance in cucumber against several diseases by plant growth-promoting fungi: Lignification and superoxide generation. *European Journal of Plant Pathology* 107(5):523–33.

Kucey, R.M. 1983. Phosphate-solubilizing bacteria and fungi in various cultivated and virgin Alberta soils. *Canadian Journal of Soil Science* 63(4):671–8.

Kumar, R., Kumari, A., Singh, M. 2014. *Trichoderma*: A most powerful bio-control agent-a review. *Trends in Biosciences* 7(24):4055–8.

Li, H., Chen, X.W., Wong, M.H. 2016. Arbuscular mycorrhizal fungi reduced the ratios of inorganic/organic arsenic in rice grains. *Chemosphere* 145:224–30.

Li, H., Ye, Z.H., Chan, W.F., Chen, X.W., Wu, F.Y., Wu, S.C., Wong, M.H. 2011. Can arbuscular mycorrhizal fungi improve grain yield, As uptake and tolerance of rice grown under aerobic conditions? *Environmental Pollution* 159(10):2537–45.

Li, R.X., Cai, F., Pang, G., Shen, Q.R., Li, R., Chen, W. 2015. Solubilisation of phosphate and micronutrients by *Trichoderma harzianum* and its relationship with the promotion of tomato plant growth. *PloS One* 10(6):e0130081.

Luef, E., Prey, T., Kubicek, C.P. 1991. Biosorption of zinc by fungal mycelial wastes. *Applied Microbiology and Biotechnology* 34(5):688–92.

Ma, J.F., Yamaji, N., Mitani, N., Xu, X.Y., Su, Y.H., McGrath, S.P. et al. 2008. Transporters of arsenite in rice and their role in arsenic accumulation in rice grain. *Proceedings of the National Academy of Sciences of the United States of America* 105(29):9931–5

Maciaszczyk-Dziubinska, E., Wysocki, R. 2012. Arsenic and antimony transporters in eukaryotes. *International Journal of Molecular Sciences* 13(3):3527–48.

Marasco, R., Rolli, E., Ettoumi, B., Vigani, G., Mapelli, F., Borin, S., Abou-Hadid, A.F., El-Behairy, U.A., Sorlini, C., Cherif, A., Zocchi, G. 2012. A drought resistance-promoting microbiome is selected by root system under desert farming. *PLoS One* 7(10):e48479.

Massaccesi, G., Romero, M.C., Cazau, M.C., Bucsinszky, A.M. 2002. Cadmium removal capacities of filamentous soil fungi isolated from industrially polluted sediments, in La Plata (Argentina). *World Journal of Microbiology & Biotechnology* 18(9):817–20.

Meharg, A.A. 2004. Arsenic in rice—Understanding a new disaster for South-East Asia. *Trends in Plant Science* 9(9):415–7.

Meharg, A.A., Williams, P.N., Adomako, E., Lawgali, Y.Y., Deacon, C., Villada, A. et al. 2009. Geographical variation in total and inorganic arsenic content of polished (white) rice. *Environmental Science & Technology* 43(5):1612–7.

Meng, X.Y., Qin, J., Wang, L.H., Duan, G.L., Sun, G.X., Wu, H.L. et al. 2011. Arsenic biotransformation and volatilization in transgenic rice. *New Phytologist* 191(1):49–56.

Minocha, R., Thangavel, P., Dhankher, O.P., Long, S. 2008. Separation and quantification of monothiols and phytochelatins from a wide variety of cell cultures and tissues of trees and other plants using high performance liquid chromatography. *Journal of Chromatography A* 1207(1–2):72–83.

Molitor, A., Zajic, D., Voll, L.M., Pons-Kühnemann, J., Samans, B., Kogel, K.H., Waller, F. 2011. Barley leaf transcriptome and metabolite analysis reveals new aspects of compatibility and *Piriformospora indica*-mediated systemic induced resistance to powdery mildew. *Molecular Plant–Microbe Interactions* (12):1427–39.

Nadeem, S.M., Ahmad, M., Zahir, Z.A., Javaid, A., Ashraf, M. 2014. The role of mycorrhizae and plant growth promoting rhizobacteria (PGPR) in improving crop productivity under stressful environments. *Biotechnology Advances* 32(2):429–48.

Noctor, G., Mhamdi, A., Foyer, C.H. 2014. The roles of reactive oxygen metabolism in drought: Not so cut and dried. *Plant Physiology* 164(4):1636–48.

Panda, S.K., Upadhyay, R.K., Nath, S. 2010. Arsenic stress in plants. *Journal of Agronomy and Crop Science* 196(3):161–74.

Pócsil, P.R., Penninckx, M.J. 2004. Glutathione, altruistic metabolite in fungi. *Advances in Microbial Physiology* 49:1–76.

Qin, J., Rosen, B.P., Zhang, Y., Wang, G.J., Franke, S., Rensing, C. 2006. Arsenic detoxification and evolution of trimethylarsine gas by a microbial arsenite S-adenosylmethionine methyltransferase. *Proceedings of the National Academy of Sciences of the United States of America* 103(7):2075–80.

Rehman, A., Anjum, M.S. 2011. Multiple metal tolerance and biosorption of cadmium by Candida tropicalis isolated from industrial effluents: Glutathione as detoxifying agent. *Environmental Monitoring and Assessment* 174(1–4):585–95.

Riccillo, P.M., De Bruijn, F.J., Roe, A.J., Booth, I.R., Aguilar, O.M. 2000. Glutathione is involved in environmental stress responses in *Rhizobium tropici*, including acid tolerance. *Journal of Bacteriology* 182(6):1748–53.

Sathishkumar, M., Binupriya, A.R., Swaminathan, K., Choi, J.G., Yun, S.E. 2008. Arsenite sorption in liquid-phase by *Aspergillus fumigatus*: Adsorption rates and isotherm studies. *World Journal of Microbiology & Biotechnology* 24(9):1813–22.

Saxena, J., Basu, P., Jaligam, V. 2013. Phosphate solubilization by a few fungal strains belonging to the genera *Aspergillus* and *Penicillium*. *African Journal of Microbiology Research* 7(41):4862–9.

Say, R., Yilmaz, N., Denizli, A. 2003. Biosorption of cadmium, lead, mercury, and arsenic ions by the fungus *Penicillium purpurogenum*. *Separation Science and Technology* 38(9):2039–53.

Seh-Bardan, B.J., Othman, R., Abd Wahidm, S., Sadegh-Zadeh, F., Husin, A. 2013. Biosorption of heavy metals in leachate derived from gold mine tailings using *Aspergillus fumigatus*. *Clean-Soil Air Water* 41(4):356–64.

Seth, C.S., Remans, T., Keunen, E., Jozefczak, M., Gielen, H., Opdenakker, K. et al. 2012. Phytoextraction of toxic metals: A central role for glutathione. *Plant Cell and Environment* 35(2):334–46.

Simonescu, C.M. 2012. *Application of FTIR spectroscopy in environmental studies*. INTECH Open Access Publisher.

Singh, M., Srivastava, P.K., Verma, P.C., Kharwar, R.N., Singh, N., Tripathi, R.D. 2015. Soil fungi for mycoremediation of arsenic pollution in agriculture soils. *Journal of Applied Microbiology* 119(5):1278–90.

Smedley, P.L. 2013. Arsenic in groundwater and the environment. *Essentials of Medical Geology* 279–310.

Song, Y.Y., Cao, M., Xie, L.J., Liang, X.T., Zeng, R.S., Su, Y.J., Huang, J.H., Wang, R.L., Luo, S.M. 2011. Induction of DIMBOA accumulation and systemic defense responses as a mechanism of enhanced resistance of mycorrhizal corn (*Zea mays* L.) to sheath blight. *Mycorrhiza* 21(8):721–31.

Srivastava, P.K., Shenoy, B.D., Gupta, M., Vaish, A., Mannan, S., Singh, N. et al. 2012. Stimulatory effects of arsenic-tolerant soil fungi on plant growth promotion and soil properties. *Microbes and Environments* 27(4):477–82.

Srivastava, P.K., Singh, M., Gupta, M., Singh, N., Kharwar, R.N., Tripathi, R.D. et al. 2015. Mapping of arsenic pollution with reference to paddy cultivation in the middle Indo-Gangetic Plains. *Environmental Monitoring and Assessment* 187(4).

Srivastava, P.K., Vaish, A., Dwivedi, S., Chakrabarty, D., Singh, N., Tripathi, R.D. 2011. Biological removal of arsenic pollution by soil fungi. *Science of the Total Environment* 409(12):2430–42.

Srivastava, S., Thakur, I.S. 2006. Biosorption potency of *Aspergillus niger* for removal of chromium (VI). *Current Microbiology* 53(3):232–7.

Stamets, P. 2005. Mycelium running: How mushrooms can help save the world. Random House Digital, Inc. http://er.uwpress.org/content/27/2/228.extract.

Stolz, J.E., Basu, P., Santini, J.M., Oremland, R.S. 2006. Arsenic and selenium in microbial metabolism. *Annual Review of Microbiology* 60:107–30.

Stroud, J.L., Norton, G.J., Islam, M.R., Dasgupta, T., White, R.P., Price, A.H. et al. 2011. The dynamics of arsenic in four paddy fields in the Bengal delta. *Environmental Pollution* 159(4):947–53.

Su, Y.H., McGrath, S.P., Zhao, F.J. 2010. Rice is more efficient in arsenite uptake and translocation than wheat and barley. *Plant and Soil* 328(1–2):27–34.

Sun, Y., Zhang, X., Wu, Z., Hu, Y., Wu, S., Chen, B. 2016. The molecular diversity of arbuscular mycorrhizal fungi in the arsenic mining impacted sites in Hunan Province of China. *Journal of Environmental Sciences* 39:110–8.

Tamaki, S., Frankenberger, W.T. 1992. Environmental biochemistry of arsenic. *Reviews of Environmental Contamination and Toxicology* 124:79–110.

Todorova, D., Nedeva, D., Abrashev, R., Tsekova, K. 2008. Cd (II) stress response during the growth of *Aspergillus niger* B 77. *Journal of Applied Microbiology* 104(1):178–84.

Tripathi, P., Singh, P.C., Mishra, A., Chauhan, P.S., Dwivedi, S., Bais, R.T., Tripathi, R.D. 2013. Trichoderma: A potential bioremediator for environmental cleanup. *Clean Technologies and Environmental Policy* 15(4):541–50.

Tripathi, R.D., Srivastava, S., Mishra, S., Singh, N., Tuli, R., Gupta, D.K. et al. 2007. Arsenic hazards: Strategies for tolerance and remediation by plants. *Trends in Biotechnology* 25(4):158–65.

Urik, M., Cernansky, S., Sevc, J., Simonovicova, A., Littera, P. 2007. Biovolatilization of arsenic by different fungal strains. *Water Air and Soil Pollution* 186(1–4):337–42.

Verma, S., Verma, P.K., Meher, A.K., Dwivedi, S., Bansiwal, A.K., Pande, V. et al. 2016. A novel arsenic methyltransferase gene of *Westerdykella aurantiaca* isolated from arsenic contaminated soil: Phylogenetic, physiological, and biochemical studies and its role in arsenic bioremediation. *Metallomics* 8(3):344–53.

Volesky, B. 1995. Biosorption of heavy metals. *Biotechnology Progress* 11(3):235–50.

Watanabe, T., Hirano, S. 2013. Metabolism of arsenic and its toxicological relevance. *Archives of Toxicology* 87(6):969–79.

WHO. 2004. Guidelines for drinking water qualit. 1:3rded: http://www.who.int/water sanitation health/dwq/gdwq3/en/

WHO. 2008. Guidelines for drinking water quality. http://www.who.int/water_sanitation_health /dwq/fulltext.pdf

WHO. 2011. Guidelines for drinking-water quality. *WHO chronicle* 38:104–8.

Williams, P.N., Villada, A., Deacon, C., Raab, A., Figuerola, J., Green, A.J. et al. 2007. Greatly enhanced arsenic shoot assimilation in rice leads to elevated grain levels compared to wheat and barley. *Environmental Science & Technology* 41(19):6854–9.

Woo, S.L., Scala, F., Ruocco, M., Lorito, M. 2006. The molecular biology of the interactions between *Trichoderma* spp., phytopathogenic fungi, and plants. *Phytopathology* 96(2):181–5.

Wysocki, R., Bobrowicz, P., Ulaszewski, S. 1997. The *Saccharomyces cerevisiae* ACR3 gene encodes a putative membrane protein involved in arsenite transport. *Journal of Biological Chemistry* 272(48):30061–6.

Xu, X.Y., McGrath, S.P., Meharg, A.A., Zhao, F.J. 2008. Growing rice aerobically markedly decreases arsenic accumulation. *Environmental Science & Technology* 42(15):5574–9.

Zechmann, B. 2014. Compartment-specific importance of glutathione during abiotic and biotic stress. *Frontiers in Plant Science* 5:566.

Zeng, X.B., Su, S.M., Jiang, X.L., Li, L.F., Bai, L.Y., Zhang, Y.R. 2010. Capability of pentavalent arsenic bioaccumulation and biovolatilization of three fungal strains under laboratory conditions. *Clean-Soil Air Water* 38(3):238–41.

Zhang, X., Wu, S., Ren, B., Chen, B. 2016. Water management, rice varieties and mycorrhizal inoculation influence arsenic concentration and speciation in rice grains. *Mycorrhiza* 26(4):299–309.

Zhao, F.J., McGrath, S.P., Meharg, A.A. 2010. Arsenic as a food chain contaminant: Mechanisms of plant uptake and metabolism and mitigation strategies. *Annual Review of Plant Biology* 61:535–59.

7

Bacteria-Assisted Phytoremediation of Industrial Waste Pollutants and Ecorestoration

Vineet Kumar and Ram Chandra

CONTENTS

7.1 Introduction

With the coming of the industrial revolution, humans were able to advance farther into the 21st century. Technology developed rapidly, science advanced, and the manufacturing age came into view. With all of these came one more effect: industrial pollution in the form of liquid, solid, or gas. Moreover, any form of pollution that can trace its immediate source to industrial practices is known as *industrial pollution*. Generally, industrial pollutants may include organic compounds, such as petroleum hydrocarbons (i.e., benzene, pyrene, toluene, and xylene); polycyclic aromatic hydrocarbons (PAHs); polychlorinated biphenyls (PCBs); phenols; pesticides; herbicides; insecticides; chlorinated solvents (i.e., trichloroethylene [TCE]); and explosives such as 2,4,6-trinitrotoluene

(TNT), hexahydro-1,3,5-trinitro-1,3,5-triazene (RDX), and octahydro-1,3,5,7-tetranitro-1,3,5-tetrazocine (HMX). Inorganic compounds may include heavy metals such as lead (Pb), zinc (Zn), cadmium (Cd), chromium (Cr), cobalt (Co), copper (Cu), nickel (Ni), mercury (Hg), sodium (Na^+), nitrate $\left(NO_3^-\right)$, ammonical nitrogen $\left(N\text{-}NH_4^+\right)$, sulfides ($S^{2-}$), and phosphate $\left(PO_4^{3-}\right)$. If the processing of waste is a cost-prohibitive one, then the industrialist throws the waste into the environment without any adequate treatment. The gases are usually released into the atmosphere; the liquids are discharged into aquatic bodies like canals, rivers, or the sea; and solid wastes are either dumped on the land or in aquatic bodies. In all cases, the air, water, or land is polluted due to dumping of wastes. Industrial pollutants accumulate in the food chain and threaten human health. In wealthy industrialized countries, contamination is often highly localized, and the pressure to use contaminated land and water for agricultural food production or for human consumption, respectively, is minimal. However, soil and water contamination is widespread in Eastern Europe, and is increasingly recognized as dramatic in large parts of the developing world, primarily in India and China. The clean-up of polluted soils and waters is very costly, and for many pollutants no feasible technologies are yet available. Therefore, a search for alternative methods to restore industrial waste–polluted sites in a less expensive, less labor-intensive, safe, and environmental friendly way is required.

In view of sustainability issues, bioremediation, the exploitation of biological process for the cleanup of contaminated site, is a promising benign and ecologically sound alternative to other expensive methods (USEPA 2012). Among different bioremediation approaches, phytoremediation appears to be an ideal approach for the treatment or removal of pollutants from the environment or to render them harmless (Khan et al. 2014; Chandra et al. 2015). It is an emerging green technology where plants are grown in the presence of contaminated soil, sediment, surface, and groundwater to enhance the decomposition or removal of inorganic and organic pollutants *in planta* as well as *ex planta* (Doty et al. 2008). Phytoremediation is more admired because of its solar-driven nature, cost effectiveness, aesthetic advantages, and long-standing applicability as it can be directly employed at polluted sites, compared to other expensive methods of treatment (Mench et al. 2009). For phytoremediation of recalcitrant organic pollutatns (ROPs) in soil, plants are used to take up or enhance the decomposition into non- or less-toxic forms. Additionally, plants can improve the soil structure by increasing aeration, humidity, and also by promoting microbial growth. However, in the rhizosphere, some ROPs can be completely degraded and mineralized by plant enzymes through the process of phytodegradation. This process occurs because many plants secrete enzyme such as peroxidases, dioxygenases, P450 monooxygenases, laccases, phosphatases, dehalogenases, nitrilases, and nitroreductases in rhizosphere that can degrade a wide range of toxic organic pollutants. Unfortunately, inorganic pollutants, mainly heavy metals, cannot readily be degraded. They must either be stabilized in the soil, which makes them less bioavailable and thereby reduces their spread in the environment; extracted; transported; accumulated and concentrated from the soil, sediment, or sludge into plant roots or shoots (phytoextraction); removed from the liquid effluent via the uses of plant root (rhizofiltration); or transformed into a volatile form (phytovolatilization). However, plants possess highly efficient systems that acquire and concentrate nutrients as well as numerous metabolic activities, all of which are ultimately powered by photosynthesis but the major constraints of this technology is that even plants that are tolerant to the presence of these contaminants often remain relatively small, due to the toxicity of the pollutants that they are accumulating or the toxic end-products of their degradation (Glick 2003). In addition, it is time-consuming, with slow degradation, limited uptake, and evotransporation of volatile contaminants and phytotoxicity. This toxicity can be reduced using bacteria-assisted phytoremediation. The combinatorial systems of plant and their associated bacteria (rhizosphere and endophytic bacteria) have been shown to contribute to biodegradation of toxic organic compounds in polluted soil and could have potential for improving phytoremediation. It could be a new and more effective approach the removal or degradation of organic and inorganic contaminants from polluted site (Glick 2010; Khandare et al. 2015). It is an *in situ*, solar-powered remediation technology that requires minimal site disturbance and maintenance. Restoration of habitats and in situ cleanup of contaminants from an industrial waste–polluted site can be achieved with this technology. Owing to its green approach, it is gaining a significant amount of public acceptance. In fact, the real idea of bacteria-assisted phytoremediation is to achieve the enhanced degradation of pollutants by using plants and their associated bacteria (pollutant-degrading or plant growth–promoting) already present in the soil and with bacteria deliberately inoculated by seed inoculation, has been investigated in the laboratory, green house, and field. Many rhizosphere and endophytic bacteria can directly or indirectly improve plant growth because they exhibit plant growth-promoting activities, *viz.* 1-aminocyclopropane-1-carboxylate (ACC)-deaminase activity, siderophore production, and nutrient solubilization (Chandra and Kumar 2015). These plant growth–promoting activities of root-associated bacteria enhance the plant's adaptation and growth in polluted environments and, ultimately, its phytoremediation efficiency (Glick 2010). Plants can provide favorable conditions for the colonization and proliferation of bacteria by offering nutrients and residency while allowing them to feed upon pollutants in the rhizosphere as well as in the endosphere. This chapter presents an overview of the potential role of plant-associated bacteria in the bioremediation of organic compounds and heavy metal–polluted environment, and provides some discussion of how bacteria could be exploited to enhance phytoremediation.

7.2 Bacteria

Bacteria are found almost everywhere in the environment. They grow in and tolerate soil, air, water, and desert conditions, as well as extreme temperatures. The plant-associated bacteria are from nature, present allover (on and inside) the plant. These bacteria can promote plant growth and development and might even be able to degrade or detoxify the organic and inorganic pollutants (Glick 2010). Here, we focus on the role of rhizosphere and endosphere bacterial communities in degradation and detoxification of industrial waste pollutants.

7.2.1 Rhizosphere Bacteria

Plant rhizosphere microorganisms play a significant role in the ecological restoration of industrial waste contaminated sites. The majority of microbial populations are concentrated in nutrient-rich niches like the rhizosphere that have a constant supply of easily utilizable nutrients. The *rhizosphere* term was originally introduced by Lorenz Hiltner in 1904 to illustrate the particular zone of soil surrounding plant roots in which microbe populations are stimulated by root exudates (Haldar and Sengupta 2015). The rhizosphere is the complex and highly dynamic interface between plant roots and surrounding soil (see Figure 7.1) that has a massive amount of soil microorganisms and is the "hot spot"

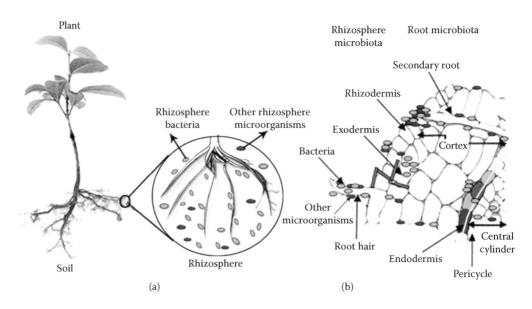

FIGURE 7.1
Plants and their associated bacteria: (a) Rhizosphere bacteria and (b) endophytic bacteria.

for microbial colonization and activity (Bakker et al. 2013). In this environment, almost 5 percent to 21 percent of photosynthetic substances are released by plant roots, which stimulate soil microbial activity (Chaparro et al. 2014). Plant roots exert strong effects on the rhizosphere through *rhizodeposition*, (root exudation, production of mucilages, and release of sloughed-off root cells) and by providing suitable ecological niches for microbial growth. The interactions between plant roots, soil, and microbes significantly change soil physical and chemical properties, which in turn alter the microbial population in the rhizosphere. The rhizosphere microflora includes bacteria, fungi, nematodes, protozoa, algae, and microarthrops. Among them, bacteria are the most abundant microorganisms in the rhizosphere, occupying one gram of soil with up to half a billion individual cells. Bacterial diversity in the rhizosphere is also high, with at least 2000–5000 bacterial species inhabiting a single gram (Dey et al. 2012). Up to 15 percent of the total plant root surface may be covered by a variety of bacterial strains (Van Loon 2007). The bacteria inhabiting the rhizosphere are called *rhizobacteria*. The term *rhizobacteria* was coined by Kloepper and Schroth in 1978, based on their experiments with radishes. They defined these bacteria as a community that competitively colonizes plant roots and enhances their growth and also reduces plant diseases (Shrivastava et al. 2015). The most common genera of bacteria that have been reported in the rhizosphere are *Pseudomonas, Bacillus, Arthrobacter, Rhizobia, Agrobacterium, Alcaligenes, Azotobacter, Mycobacterium, Flavobacter, Cellulomonas* and *Micrococcus* spp. Accordingly, studies suggest that proteobacteria, actinobacteria, pseudomonads, and actinomycetes are the most dominating populations of bacteria (>1 percent, usually much more) found in the rhizosphere of many different plant species (Sylvia et al. 2005). They are the most studied of the rhizobacteria, and as such, contain the majority of the organisms investigated. Rhizobacteria can be classified into beneficial, deleterious, and neutral groups on the basis of their effects on plant growth. The rhizobacteria that stimulate the growth and health of the plant are referred to as *plant growth promoting rhizobacteria* (PGPR; Kloepper 1989; Glick 2012). It is well established that only 1 to 2 percent of bacteria promote plant growth in the rhizosphere (Antoun and Kloepper, 2001). The PGPR may be three general types: (1) Those that form a symbiotic relationship, which involves formation of specialized structures or nodules on host plant roots; (2) those that are endo-phytic and colonize the inner tissues of the plant without being pathogenic; and (3) those that are able to competitively colonize the rhizosphere and plant root surface (Glick et al. 2012). However, PGPR are common in the rhizosphere and perform various mechanisms, including: (1) The production of phytohormones, such as indole acetic acid (IAA), gibberel-lic acid, or cytokinins; (2) antifungal metabolites or lytic enzymes; (3) production of ACC deaminase to reduce the level of ethylene in the roots of developing plants; (4) solubiliza-tion of minerals such as phosphorus and potassium; (5) production of exopolysaccharides and osmoprotectants; and (6) immobilization of heavy metals (Ahmed and Kibret 2014; Glick 2014). Interaction of some PGPR with plant roots can result in plant resistance against some pathogenic bacteria, fungi, and viruses. This phenomenon is called induced sys-temic resistance (ISR; Lugtenberg and Kamilova 2009). Moreover, ISR involves jasmonate and ethylene signaling within the plant and these hormones stimulate the host plant's defense responses against a variety of plant pathogens (Glick 2012). Some important genera of PGPR include *Serratia, Bacillus, Pseudomonas, Burkholderia, Enterobacter, Erwinia, Klebsiella, Beijerinckia, Flavobacterium,* and *Gluconacetobacter* (Ahmed and Kibret 2014; Glick et al. 2012). Some PGPR can also help plants with stand abiotic stresses including contami-nation by heavy metals or ROPs. Therefore, utilizing PGPR is a promising approach for improving the success of phytoremediation of contaminated soil.

7.2.2 Endophytic Bacteria

Bacteria on plant roots and in the rhizosphere benefit from root exudates, but some bac-teria, fungi, and actinomycetes are capable of entering the plant tissues as endophytes that do not cause harm and could establish a mutualistic association (Weyens et al. 2009b; Hardoim et al. 2015). The term *endophyte* was first introduced in 1886 by De Bary for micro-organisms (i.e., fungi, yeast, and bacteria) colonizing internal plant tissues (De Bary 1884). Endophytic bacteria is defined as bacteria colonizing the internal tissues of plants without causing symptoms of infection or negative effects on their host plants or environment (Compant et al. 2005; de Oliveira et al. 2012). In general, endophytic bacteria occur at lower population densities than rhizosphere bacteria or bacterial pathogens (Rosenblueth and Martínez-Romero 2004). They have established harmonious associations with host plants during symbiotic, mutualistic, commensalistic, and trophobiotic relationships over a long evolutionary process (Lodewyckx et al. 2002). Endophytic bacteria in a single plant host are not restricted to a single species but comprise several genera and species. It is well established that the soil environment is the main source of endophytes. However, some other studies have also reported that endophytes can originate from the phyllosphere (Ali et al. 2012). Endophytic bacteria can be facultative, obligate, or passive, depending on the genotype of the host plant and life strategy. Facultative endophytes are free-living, present in the soil, and colonize plants opportunistically. They are also the species most commonly cultivated from plants. Facultative endophytic bacteria can survive and colo-nize outside the plant during a period of their life cycle. In general, bacteria in the family *Pseudomonaceae, Burkholderiaceae,* and *Enterobacteriaceae* are the most frequently isolated, although recent biotechnological advances have exposed the presence of additional taxa, including non-culturable endophytes (Lundberg et al. 2012; Nair and Padmavathy 2014). Species from these families, which are abundant in the soil, first colonize the rhizosphere of the host plant before penetrating the root epidermis and establishing within the host plant (Lodewyckx et al. 2002). However, obligate endophytic bacteria depend on the host plant for their survival and metabolic activities and may be transmitted from one generation to the next through seeds or vegetative plant tissues (Hamilton et al. 2012). The third group, the passive endophytes, do not actively seek to colonize the plant, but do so as a result of stochastic events, such as an open wound along root hairs. This passive life strategy may cause the endophytes to be less competitive since the cellular machinery required for plant colonization is lacking, and therefore may be less appropriate as plant growth promoters. Several studies revealed that endophytic bacteria mainly reside in the intercellular apoplast and in dead or dying cells. They also found in xylem vessels, within which they may be translocated from the root to the aerial parts (Turner et al. 2013). It has been reported that substantial numbers of endophytes (10^3–10^6 cells) can colonize the vascular system (phloem, xylem) (Frommel et al. 1991). The highest densities of endophytic bacteria usually are observed in the roots and decrease progressively from the stem to the leaves. Endophytic bacteria, which have been found in numerous plant species, often belong to genera commonly found in soil, including *Pseudomonas*, *Burkholderia*, *Bacillus*, and *Azospirillum* (Mastretta et al. 2006). In addition, most of the endophytic bacteria (26 percent) could be assigned to the *Gammaproteobacteria*, including 56 recognized and others unidentified genera as well as the *Candidatus Portiera* genus. It should be noted that *Gammaproteobacteria* also comprise a large number of genera and species which are known as phytopathogens. Endophytic *Gammaproteobacteria* are largely represented by a few genera, namely *Pseudomonas*, *Enterobacter*, *Pantoea*, *Stenotrophomonas*, *Acinetobacter*, and *Serratia* (>50 sequences each). Among Gram-positive endophytic bacteria, the class *Actinobacteria* (20 percent) comprises diverse endophytes belonging to 107 recognized and 15 unidentified genera. This assortment of bacteria have a wide variety of functions, including the stimulation of plant growth, the promotion of biological nitrogen fixation, the protection of plants from harsh external environments, and the control of pathogen activities. It has not been resolved whether plants benefit more from an endophyte than from a rhizospheric bacterium or if it is more advantageous for bacteria to become endophytic compared with rhizospheric. It is still not always clear which population of microorganisms (endophytes or rhizospheric bacteria) promotes plant growth; nevertheless, benefits conferred by endophytes are well recognized and will be presented here. Endophytic populations, like rhizospheric populations, are conditioned by biotic and abiotic factors (Hallmann et al. 1999; Seghers et al. 2004), but endophytic bacteria could be better protected from biotic and abiotic stresses than rhizospheric bacteria (Hallmann et al. 1997). Moreover, many endophytic bacteria, particularly those inhabiting plants growing in a polluted environment, produce degradative enzymes and contribute to the degradation of several types of organic compounds present in the rhizosphere and endosphere.

7.3 Interactions in the Rhizosphere: Plant–Bacteria, Pollutants–Bacteria, Plant–Bacteria–Soil

Potential for phytoremediation depends upon the interactions in the rhizosphere among soil, pollutants, bacteria, and plants (Figure 7.2). In the rhizosphere, root-based interactions between plants and organisms are highly influenced by edaphic factors. However, the below-ground biological interactions that are driven by root exudates are more complex than those occurring above the soil surface.

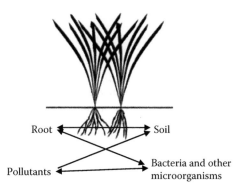

FIGURE 7.2
Possible interaction occurs in the rhizosphere of a plant growing on a contaminated site.

7.3.1 Plant–Bacteria Interaction

The rhizosphere is a dynamic and complex environment where a multitude of biotic and abiotic factors interact, both with one another and with the root of plants, thereby often profoundly impacting on plant–bacteria interactions (Cheng et al. 2010). Plants and bacteria can interact with one another in a variety of different ways to absorb, degrade, or remove toxic contaminants from polluted soil and water. Bacteria that reside in different compartments of plants can synthesize several compounds that assist the plants to overcome stress, providing essential nutrients required for plant growth and development, improving plant defense systems against pathogens, and stimulating contaminants degradation. The interaction between bacteria and plant roots may be beneficial, harmful, or neutral for the plant, and sometimes the effect of a particular bacterium may vary as the soil and environmental conditions change. The first step of the plant–bacteria interaction is the recognition of plant root exudates by the bacteria. Root exudates are the most important part of rhizodeposit and play an active and relatively well-documented role in the regulation of symbiotic and protective interactions with tremendous diversity of microorganisms. Plants release up to 40 percent of their photosynthetic fixed carbon through their roots into the surrounding area due to this so-called *rhizodeposition*. Generally, root exudates are transported across the plasma membrane and secreted into the surrounding rhizosphere by plant roots via either passive (diffusates) or active (secretions) mechanisms. The active process of exudation is mediated by highly specific membrane transporters requiring ATP. Root exudates have been grouped into low molecular weight (LMW) compounds and high molecular weight (HMW) compounds. LMW compounds, that is, amino acids, organic acids, sugars, phenolics, and various other secondary metabolites are believed to comprise the majority of root exudates, whereas HMW exudates primarily include mucilage (high-molecular weight polysaccharides) and proteins (Table 7.1). HMW exudates are more significant in terms of total mass of the root exudates but have comparatively less variety than the first class (Bais et al. 2006).

Exudates may also be classified as active and passive exudates on the basis of their role and mode of secretion from the roots (Bais et al. 2006). Passive exudates have unknown functions and are diffused from the roots as basal exudation (output of waste materials) depending on the gradient (Bais et al. 2006). The active exudates are secreted through open plasma membrane pores of the plants and have a specific functions such as lubrication and defense. The quantity and composition of root exudates fluctuates with plant

TABLE 7.1

Organic Compounds Identified in Root Exudates of Different Plants

Components	Identified Compounds
Sugars	Arabinose, fructose, galactose, glucose, maltose, mannose, mucilages of various compositions, oligosaccharides, raffinose, rhamnose, ribose, sucrose, xylose, deoxyribose
Amino acids	a-Alanine, b-alanine, g-aminobutyric, a-aminoadipic, arginine, asparagine, aspartic, citrulline, cystathionine, cysteine, cystine, deoxymugineic, 3-epihydroxymugineic, glutamine, glutamic, glycine, histidine, homoserine, isoleucine, leucine, lysine, methionine, mugineic, ornithine, phenylalanine, praline, proline, serine, threonine, tryptophan, tyrosine, valine
Organic acids	Acetic, aconitic, ascorbic, aldonic, benzoic, butyric, caffeic, citric, pcoumaric, erythronic, ferulic, formic, fumaric, glutaric, glycolic, lactic, glyoxilic, malic, malonic, oxalacetic, oxalic, p-hydroxybenzoic, piscidic, propionic, pyruvic, succinic, syringic, tartaric, tetronic, valeric, vanillic
Fatty acids	Linoleic, linolenic, oleic, palmitic, stearic
Sterol	Campesterol, cholesterol, sitosterol, stigmasterol
Growth factors and vitamins	p-Amino benzoic acid, biotin, choline, N-methyl nicotinic acid, niacin, pathothenic, thiamine, riboflavin, pyridoxine, pantothenate
Enzymes	Amylase, invertase, peroxidase, phenolase, acid/alkaline phosphatase, polygalacturonase, protease
Flavonones and purine nucleotides	Adenine, flavonone, guanine, uridine/cytidine
Miscellaneous	Auxins, scopoletin, hydrocyanic acid, glucosides, unidentified ninhydrinpositive compounds, unidentifiable soluble proteins, reducing compounds, ethanol, glycinebetaine, inositol, and myo-inositol-like compounds, Al-induced polypeptides, dihydroquinone, sorgoleone, isothiocyanates, inorganic ions and gaseous molecules (e.g., CO_2, H_2, H, OH, HCO_3), some alcohols, fatty acids, and alkyl sulfides.

Source: Dennis, P.G., Miller, A.J., Hirsch, P.R., *FEMS Microbiology Ecology*, 72, 313–327, 2010.

developmental stage and the proximity to neighboring species. Root exudates possibly influence host recognition, biofilm formation, and colonization by bacteria. The components of root exudates such as carbohydrates, organic acids, and amino acids have been demonstrated to stimulate positive chemotactic responses in bacteria, and the abilities of bacterial cells to move toward roots in response to chemical signals (chemotaxis) and grow rapidly are important traits that enable a bacterial species to be competitive in the rhizosphere. Chemotactic motility of bacteria is an energy-dependent process; therefore, colonization ability can be decreased if either ATP or flagella production is disrupted. Another trait that is important in effective root colonization by bacteria is their growth rate, which is partly dependent on their ability to obtain components that are essential for growth or maintenance. The O-antigen of lipopolysaccharide plays a role in root colonization and may be implicated in regulating growth rates. Moreover, bacteria perform a wide range of functions in the rhizosphere, including mediation of biogeochemical cycles, acquisition of nutrients, protection of the host plant from antagonistic microbial attacks, and maintenance of plant populations. Bacteria can also positively interact with plants by producing protective biofilms or antibiotics operating as biocontrols against potential pathogens, or by degrading plant and microbe-produced compounds in the soil that would otherwise be allelopathic or even autotoxic. However, rhizosphere bacteria can also have detrimental effects on plant health and survival through pathogen or parasite infection. Secreted chemical signals from both plants and bacteria mediate these complex exchanges and determine whether an interaction will be malevolent or benign. Chemotaxis towards root exudates is suggested to be the first step of bacterial colonization. This is generally achieved by flagella and adhesion to plant cells via curli fibers and pili. Furthermore, biofilm formation by bacteria on plant roots is a visualized performance of effective colonization (Li et al. 2013). *Bacillus amyloliquefaciens* SQR9, isolated from cucumber rhizosphere, could protect the host from invasion by *Fusarium oxysporum* f. sp. cucumerinum J. H. Owen (FOC) through competition for nutrients and space (Qiu et al. 2012; Xu et al. 2013). On the other hand, *B. subtilis* N11, originated from banana rhizosphere, could successfully colonize the banana rhizosphere and suppress wilt disease caused by *Fusarium oxysporum* f. sp. cubense race 4 (FOC4) (Zhang et al. 2011). Citric acid detected exclusively in cucumber exudates could both attract SQR9 and induce its biofilm formation, whereas only chemotactic response but not biofilm formation was induced in *B. subtilis* N11. Fumaric acid that was only detected in banana root exudates revealed both significant roles on chemotaxis and biofilm formation of *B. subtilis* N11, while showing only effects on biofilm formation but not chemotaxis of SQR9. Colonization of *B. amyloliquefaciens* SQR9 and *B. subtilis* N11 of their original host was found to be more effective when compared to the colonization of the non-host plant (Zhang et al. 2014).

7.3.1.1 Interaction of Endophytic Bacteria with Plants

The endophytic bacteria enter into plant tissue through the root, flower, leaf, stem, and cotyledon and they may either become localized at the point of entry or spread throughout the plant (Kobayashi and Palumbo 2000). Among them, the root is the main organ where endophytic bacteria get entry into plants. Before entering the plant, endophytes have to establish themselves in the rhizosphere and attach themselves to the root surface. The cell wall degrading enzymes facilitate the penetration of such bacteria into plants (Reinhold-Hurek et al. 2006). By using gnotobiotic conditions and with the help of microscopic tools, which allow the detection of *gfp* or *gusA* labeled strains or of strains by immune markers or by fluorescence in situ hybridization (FISH), it has been demonstrated that bacterial cells first colonize the rhizosphere following soil inoculation (Gamalero et al. 2003). Then, bacterial cells have been visualized as single cells attached to the root surfaces, and subsequently as doublets on the rhizodermis, forming a string of bacteria (Figure 7.3; Hansen et al. 1997). Colonization may then occur on the whole surface of some rhizodermal cells (Figure 7.3) and bacteria can even establish as microcolonies or biofilms (Benizri et al. 2001).

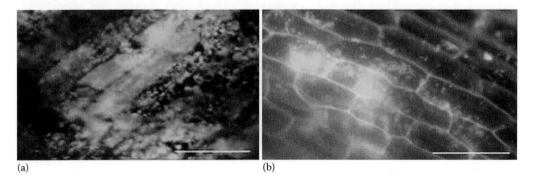

(a) (b)

FIGURE 7.3
Rhizoplane colonization of a beneficial bacterium, *Burkholderia phytofirmans* strain PsJN, tagged with gfp: (a) Under gnotobiotic or (b) nonsterile conditions.

Compant et al. (2007) determined the root colonization by bacterium, *Burkholderia phyto-firmans* strain PsJN tagged with *gfp*. Strain PsJN was found mainly in the root hair zone, lateral root emergence sites, and root tips. The root hair zone was more highly colonized than other root zones with PsJN<gfp2x (Figure 7.4a–c). Other rhizodermal cells were colonized differentially: Some rhizoplane cells were filled with bacteria (Figure 7.4d) whereas *gfp*-tagged cells were found closely attached to cell walls in other cells (Figure 7.4d–e).

During the transition from the rhizosphere to the plant endosphere, bacteria must have the capacity for quick adaptation to a highly different environment (i.e., pH, osmotic pressure, carbon source, availability of oxygen). They also have to overcome plant defense response to the invasion, that is, production of reactive oxygen species (ROS) causing stress to invading bacteria (Zeidler et al. 2004). Thus, bacterial ability to establish endophytic populations is likely to depend on the recognition of signal molecules (e.g., two-component systems or extracytoplasmatic function sigma factors), mobility, penetration capability, and capacity for adjustment of metabolism and behavior. After entering the plant, endophytic bacteria may remain localized at the entrance point or spread throughout the plant. In these very close plant–endophyte interaction, plants provide nutrients and residency for bacteria, which in exchange can directly or indirectly improve plant growth and health. Direct plant growth promoting mechanisms may involve production of plant growth regulators such as auxins, cytokinins, and gibberellins, suppression of stress ethylene production by ACC deaminase activity, nitrogen fixation, and the mobilization of unavailable nutrients such as phosphorus and other mineral nutrients. Endophytic bacteria can indirectly benefit plant growth by preventing the growth or activity of plant pathogens through competition for space and nutrients, production of hydrolytic enzymes, antibiosis, induction of plant defense mechanisms, and inhibition of pathogen-produced enzymes or toxins. In addition to their beneficial effects on plant growth, endophytes have considerable biotechnological potential to improve the applicability and efficiency

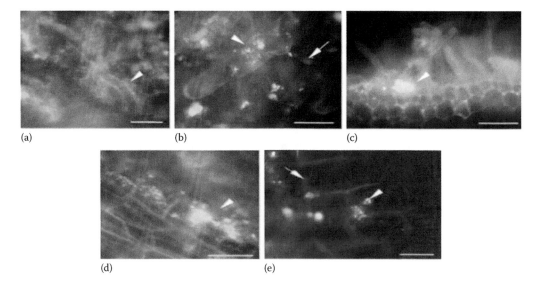

(a)   (b)   (c)

(d)   (e)

FIGURE 7.4
Images under an epifluorescence microscope of roots of grapevine fruiting cuttings inoculated with PsJN<gfp2x 1–4 weeks after soil inoculation. Gfp-tagged bacteria (arrows) were visualized at the root hair zone of colonized root hairs (a–c), on other rhizodermal cells (d and e).

of phytoremediation. Endophytic populations, like rhizospheric populations, are conditioned by biotic and abiotic factors but endophytic bacteria could be better protected from biotic and abiotic stresses than rhizospheric bacteria (Hallmann et al. 1997).

7.3.1.2 Quorum Sensing and Rhizosphere Communication

The rhizosphere is a crossroads for nutrient exchange between plants and soil microbes, and this exchange is likely mediated by bacterial signaling, or quorum sensing (QS). QS is a density-dependent regulatory mechanism used by many bacteria to regulate gene expression in a coordinated manner. It was first described in the aquatic bacterium, *Vibrio fischeri* as the signal mediated induction of the *lux* genes responsible for bioluminescence. Its activation is mediated by small autoinducer molecules, which are responsible for cell–cell communication, and the coordinated action of many bacteria, including plant-associated bacteria. The most commonly reported type of autoinducer signals in Gram-positive bacteria are N-acyl homoserine lactones (AHLs) and 2-heptyl-3-hydroxy-4-quinoline, although half a dozen other molecules, including diketopiperazines in several Gram-negative bacteria, a furanosyl borate diester in *Vibrio harveyi*, and γ-butyrolactone in *Streptomyces*, have also been implicated in density-dependent signaling.

When a root passes through soil and activates the indigenous microbial community, bacteria interact with each other through the release and detection of organic molecules (Fray 2002). Plant addition of resources to the rhizosphere permits increased cell density and activity, which is reflected in the increased abundance of the QS signal AHL, regulating QS and QS-controlled behaviors. AHL-mediated QS is likely to be important in the rhizosphere due to the apparent dominance of Proteobacteria among rhizosphere microbial communities. Plants release plenty of secondary metabolites such as terpenes, flavonoids, glucosinolates, and phenylpropanoids into the rhizosphere. Flavonoids, for instance, are able to mimic QS molecules and thereby influence the bacterial metabolism (Hassan and Mathesius 2012). The production of 2-aminoacetophenone, a volatile metabolite produced by *Pseudomonas aeruginosa*, *Streptomyces* sp., and *Burkholderia ambifaria* is known to be regulated QS. Thus, the QS system that controls basic processes of the bacterial life, for example, biofilm formation and motility and is likely to also affect the quality and quantity of volatiles. This is particularly important in highly competitive situations between different bacterial organisms that benefit from nutrient-rich conditions in the rhizosphere. Biofilms are assemblages of cells embedded in a matrix composed of exopolysaccharides (EPSs), proteins, and sometimes DNA.

Root biofilm initiation and development is complex and not well understood due to the dynamic nature of plant root surfaces. Bacterial species have adapted to these ever-changing conditions and are capable of starting colonization by forming microcolonies on different parts of the roots from tip to elongation zone. Such microcolonies eventually grow into large population sizes on roots to form mature biofilms. Root exudates serve as a major plant-derived factor responsible for triggering root colonization and biofilm associations (Walker et al. 2004). The plant growth–promoting pseudomonads have been reported to discontinuously colonize the root surface, developing as small biofilms along epidermal fissures. *Pseudomonas putida* can respond rapidly to the presence of root exudates in soils, converging at root colonization sites and establishing stable biofilms. A variety of specific functions are relevant to colonization and biofilm formation on plant roots. Motility via flagella or type IV pili is required for competitive colonization of roots by *Pseudomonas putida*. Motility mutants typically demonstrate only modest deficiencies in noncompetitive root colonization, emphasizing the efficacy of passive rhizodeposition. Surface structures such as lipopolysaccharide (LPS) and outer membrane proteins are also important in biofilm formation on roots. Production of

EPS is generally important in biofilm formation, and likewise can affect the interaction of bacteria with roots and root appendages. EPSs were shown to be important for root colonization in many bacterial species, such as for *Azospirillum brasilense*, *Gluconacetobacter diazotrophicus*, *Herbaspirillum seropedicae*, *Agrobacterium tumefaciens*, *Sinorhizobium meliloti*, and *Pseudomonas fluorescens*. Germaine et al. (2006) demonstrated that plants inoculated with *Pseudomonas putida* VM1450 in the presence of 2,4-D (2,4-dichlorophenoxyacetic acid) showed efficient colonization around the plant root and form biofilm on their surfaces, compared to uninoculated plant (Figure 7.5). The number of cells that were visualized increased with the increase in the 2,4-D levels applied. The strain also showed reduction of phytotoxic effect of 2,4-D on plant. *Pseudomonas putida* VM1450 cells could also be seen inside the root and stems, residing as discrete microcolonies or inhabiting the intercellular spaces of these tissues.

Similarly, *Bacillus subtilis*, a plant growth–promoting Gram-positive bacterium widely used as a biofertilizer, exhibits differential biofilm formations on *Arabidopsis thaliana* root surfaces. It is widely known as a soil-dwelling bacterium and also found in association with roots of many different plants where it protects plants from infection. One of the key requirements for biofilm formation is the secretion of an extracellular matrix that holds the cells together, as both the EPS and the protein component of the matrix (TasA) are necessary to form robust biofilms in vitro. The major EPS component required for each biofilm type is synthesized by the products of the *epsA-epsO* operon. *B. subtilis* biofilms; this matrix is produced by a subpopulation of cells that are genetically identical but phenotypically distinct from the rest of the population. The plant polysaccharides present in root exudates act as an environmental cue for biofilm formation. Some other plant polysaccharides, such as arabinogalactan, pectin, or xylan derived from plants' cell walls trigger biofilm formation by *B. subtilis*. All three of the plant polysaccharides tested were capable of inducing pellicle formation after 24 hours in vitro. However, plant polysaccharides and other polysaccharides are used as a carbon source by *B. subtilis* for the synthesis of matrix EPS. Arabinogalactan proteins present in the *A. thaliana* cell wall are potent inducers of *B. subtilis* biofilm formation and thus may be able to induce the expression of matrix genes when the bacteria are in contact with *A. thaliana* roots. These plant polysaccharides can also be digested, converted to UDP-galactose, and incorporated into the matrix EPS. Xylan is a polymer of xylose, a pentose, to which side chains are added in *A. thaliana*. It is present mostly in fiber cells, contrary to pectin and arabinogalactans, which are generally found throughout the plant and thus are more likely to be present on the roots. For surface colonization of *A. thaliana*, pectin and arabinogalactans may be the major players.

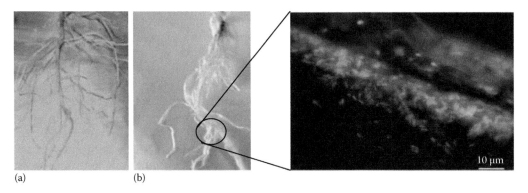

(a)   (b)

FIGURE 7.5
Roots of pea plants (a) noninoculated plant at 0 mg 2,4-dichlorophenoxyacetic acid (2,4-D) level; (b) *Pseudomonas putida* VM1450 biofilm within the rhizosphere of inoculated plants exposed to 54 mg of 2,4-D. (From Germaine, K. J. et al., *FEMS Microbiology Ecology*, 57, 302–310, 2006.)

However, *B. subtilis* is very often found growing associated with decaying plant material, in which case xylan would more likely be exposed. It is possible that xylan serves as a signal for colonization and biofilm formation on decaying plants, which also serve as a rich source of carbon. Pectin is a complex polysaccharide composed mainly of galacturonic acid, branched arabinan, and 1,4-β-galactans. It constitutes about 40 percent of the *A. thaliana* primary cell wall. Finally, arabinogalactans are attached to the arabinogalactan proteins (AGPs), which play important roles in the development of the plant and its interaction with microorganisms. Moreover, *B. subtilis* colonizations of *A. thaliana* roots also require the production of surfactin, a lipopeptide antimicrobial that is also important for biofilm formation in vitro. The production of surfactin and other lipopeptides by *Bacillus* cells is one of main mechanisms for plant biocontrol since these molecules can induce systemic resistance as well as strongly inhibit the growth of common plant pathogens such as *Pseudomonas syringae*. To recruit *B. subtilis*, plants secrete small molecules. For example, when *A. thaliana* is infected with *P. syringae*, malic acid is secreted and this enhances *B. subtilis* biofilm formation on the root. However, one of the key requirements for biofilm formation is the secretion of an extracellular matrix that holds the cells together. Furthermore, root exudates from *P. syringae*–infected plants or purified malic acid induce matrix gene expression in *B. subtilis*. This phenomenon is not specific to *A. thaliana*; malic acid is also found in tomato root exudates and, at high concentrations, can stimulate matrix gene expression and biofilm formation in vitro. A general schematic depicting the different stages of *B. subtilis* biofilm formation is shown in Figure 7.6.

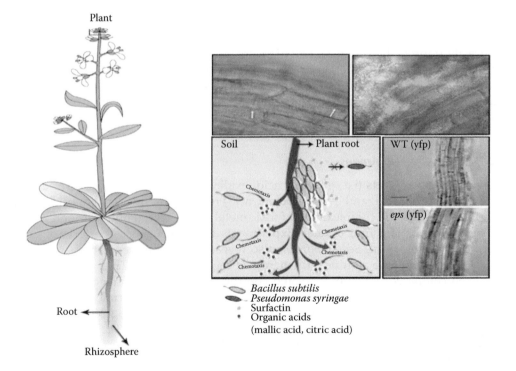

FIGURE 7.6

Bacillus subtilis forms biofilms on *Arabidopsis thaliana* root. Wild-type (WT) cells constitutively expressing YFP and various eps mutant strains constitutively expressing YFP were coincubated with six-day-old seedlings of *A. thaliana*. *B. subtilis* cells colonizing *A. thaliana* roots express matrix genes. Colonization of the root was observed after 24 hours. (a, b) *B. subtilis* forms biofilms on plant roots. Overlays of fluorescence (false-colored green for) and transmitted light images (gray) are shown. Arows point toward some of the nonfluorescent cells. (Scale bars: 10 μm.)

7.3.2 Heavy Metal–Bacteria Interaction and Its Mobilization in Rhizosphere

The problem of heavy metals in the environment is growing worldwide and currently has reached an alarming rate. There are various sources of heavy metal pollution in the environment but tanneries, distilleries, and pulp and paper industries are major sources of Cr, Cu, Mn, Fe, Ni, Cd, Pb, and Zn pollution in the environment. Moreover, mining, metallurgical activities, smelting of metal ores, and fertilizers have contributed to a high level of metal concentration in the environment. They are harmful to humans and animals, and they tend to accumulate in the food chain. The threat that heavy metals pose to human and animal health is aggravated by their long-term persistence in environment. Among 90 naturally occurring elements, 53 are heavy metals, but only 17 are bioavailable and important for ecosystem. Plants growing in a heavy metals–contaminated site harbor a diverse group of microorganisms that are capable of tolerating a high concentration of metals and providing a number of benefits to both the soil and plants. The availability of heavy metals for plant metabolisms is related to the chemical forms of metals, soil–plant or water–plant interaction, and microbial activities, because a large proportion of heavy metals are generally bound to various organic and inorganic constituents in industrial waste or polluted soil and their phytoavailability is closely related to their chemical speciation. The soil microorganisms play a significant role to affect heavy metals' mobility and availability to the plant through release of chelating agents, acidification, phosphate solubilization, and oxidation/redox changes (Abou-Shanab et al. 2003, 2008). The various processes, such as chemical transformation, chelation, and protonation, will lead to mobilization of heavy metals, whereas precipitation or sorption decrease metals availability. Among the microorganisms, rhizosphere bacteria deserve special attention because they can directly improve the phytoremediation process by altering the solubility, availability, and transport of heavy metals through altering the soil pH, release of chelators, and oxidation/reduction reactions in the rhizosphere (Abou-Shanab et al. 2003). To date, two groups of bacterially produced natural chelators are known. These are carboxylic acid anions and siderophores.

7.3.2.1 Carboxylic Acid Anions

Different carboxylic acids (i.e., gluconic acid, oxalic acid, tartaric acid, citric acid, acetic acid, 2-ketogluconic acid, and 5-ketogluconic acid) released by rhizosphere bacteria play an important role in the complexation of toxic metal and essential nutrients and increase their mobility for plant root uptake. Experiments with rhizosphere bacteria of Cd/Zn hyperaccumulating plant, *Sedum alfredii*, also revealed that the inoculation of soil with Cd/Zn–resistant bacteria significantly increased the water soluble Zn and Cd concentrations when compared with uninoculated controls. In this case, enhanced heavy metal mobilization could be correlated with the increased production of formic acid, acetic acid, tartaric acid, succinic acid, and oxalic acid (Li et al. 2010). Wani et al. (2007) also studied the mobilization of Pb and Zn by inoculating three metal-resistant *Bacillus* strains, PSB1, PSB7, PSB10, and found that the *Bacillus* sp. SPSB1 solubilizing phosphate was able to mobilize large amount of Pb and Zn. Where organic acids improve P acquisition, microbial siderophores chelate and solubilize Fe^{3+} in soil and improve iron acquisition by plants (Rajkumar et al. 2010).

7.3.2.2 Siderophores

Siderophores play a significant role in metal mobilization and accumulation in plants. They are low molecular weight (200–2000 Da) chelating agents produced by bacteria, fungi, and plants under iron-limiting conditions. Iron is a necessary cofactor for many enzymatic reactions and thus is an essential nutrient for virtually all organisms. It acts as a catalyst in some of the most fundamental enzymatic processes, including oxygen metabolism, electron transfer, and DNA and RNA synthesis. Generally, iron occurs mainly as Fe^{3+} and forms insoluble hydroxides and oxyhydroxides, thus, it is not easily available to both plants and microorganisms (Ahemad and Kibret 2013b). Siderophores have a strong ferric iron (Fe^{3+}) binding affinity in the rhizosphere (Schalk et al. 2011). In addition to iron, siderophores can also form stable complexes with other metals such as Al, Cd, Cr, Mn, Mg, As, Cu, Ga, Hg, In, Pb, Tb, Tl, Eu, Co, Ag, Zn, and radionuclides, including plutonium (Pu), and uranium (U), with variable affinities (Nair et al. 2007; Rajkumar et al. 2010; Schalk et al. 2011). Supply of iron to growing plants under heavy metal stress becomes more important as bacterial siderophores help to minimize the stress imposed by metal contaminants (Gamalero and Glick 2012). It has been found that competition for iron in the rhizosphere is controlled by the affinity of the siderophores for iron. Interestingly, the binding affinity of phyto-siderophores for iron is less than the affinity of microbial siderophores, but plants require a lower iron concentration for normal growth than do microbes (Meyer 2000). For example, Braud et al. (2009) found that siderophores produced by *P. aeruginosa* significantly enhanced the concentration of bioavailable Pb and Cr in the rhizosphere and made them available for uptake by maize. Similarly, the siderophores produced by *Streptomyces tendae* F4 significantly enhanced Cd uptake by sunflower plants (Dimpka et al. 2009). For instance, siderophore overproducing mutant NBRI K28 SD1 of phosphate-solubilizing bacterial strain *Enterobacter* sp. NBRI K28 not only increased plant biomass but also enhanced phytoextraction of Ni, Zn, and Cr by *Brassica juncea* (Kumar et al. 2009). Siderophore-producing rhizosphere bacteria have been shown to enhance chlorophyll content and growth of various crop plants in contaminated soil by selectively supporting iron uptake from the pool of trace element cations competing for import (Burd et al. 1998, 2000; Dimkpa et al. 2009). Moreover, complexation of trace elements by bacterial siderophores in the rhizosphere likely prevents the generation of free radicals and oxidative stress. Thus, the resulting increase in heavy metal uptake by plants caused by bacterial siderophores might enhance the phytoextraction process of an industrial waste–contaminated site.

7.3.2.3 Other Important Bacterial Chelators

Phosphate-solubilizing microorganisms are believed to increase plant biomass by supporting plant health and mobilizing heavy metals, making them an attractive strategy for improving phytoextraction. Compared to uninoculated controls, the inoculation of *Brassica juncea* with a phosphate solubilizing *Bacillus* spp. induced a 349 percent increase in plant dry weight after 8 weeks and a 148 percent increase in Cd concentration (Jeong et al. 2013). Some plant-associated bacteria can produce biosurfactants that can enhance the bioavailability of hydrocarbons and may be useful for phytoremediation application organic pollutants moving into the rhizosphere through transpiration stream. Biosurfactants are amphiphilic compounds produced in living spaces or excreted extracellular hydrophobic and hydrophilic moieties that confer on the organism the ability to accumulate between fluid phases, thus reducing surface and interfacial tension. Biosurfactants are produced by several microorganisms, which include *Acinetobacter* sp., *Bacillus* sp., *Candida antartica*, and *Pseudomonas aeruginosa*.

7.4 Tolerance Mechanism of Bacteria in Heavy Metal Stress

Generally, bacterial cells uptake the heavy metal cations of similar size, structure, and valence with the same mechanism (Nies 1999). Bacteria generally possess two types of uptake system for heavy-metal ions: One is fast and unspecific and driven by the chemiosmotic gradient across the plasma membrane, and another type is slower, exhibits high substrate specificity, and is coupled with ATP hydrolysis (Nies and Silver 1995). For survival under a metal-stressed environment, rhizosphere bacteria have evolved several mechanisms by which they can immobilize, mobilize, or transform metals, rendering them inactive to tolerate the uptake of heavy metal ions (Nies 1999). These mechanisms include the following: (1) Exclusion, in which the metal ions are kept away from the target sites; (2) extrusion, in which the metals are pushed out of the cell through chromosomal/plasmid mediated events; (3) accommodation, in which metals form complex with metal-binding proteins (e.g., metallothienins or other cell components); (4) bio-transformation, in which toxic metal is reduced to less-toxic forms; (5) methylation and demethylation.

One or more of these defense mechanisms allows the bacteria to function metabolically in heavy metal–polluted environment. Numerous bacteria exhibit efflux transporters (e.g., ATPase pumps or chemiosmotic ion/proton pumps) with high substrate affinity by which they expel a high concentration of toxic metals outside the cell (Haferburg and Kothe 2007; Ahemad 2012). For instance, plasmid-encoded and energy-dependent metal efflux systems involving ATPases and chemiosmotic ion/proton pumps are also reported for As, Cr, and Cd resistance in other bacteria (Roane and Pepper 2000). Many metals bind the anionic functional groups (e.g., sulfhydryl, carboxyl, hydroxyl, sulfonate, amine, and amide groups) of extracellular material present on bacterial cell surfaces and prevent its intake into bacterial cell. Similarly, bacterial extracellular polymers, such as polysaccharides, proteins, and humic substances, also competently bind heavy metals through biosorption (Ahemad and Kibret 2013). These substances thus detoxify metals simply by complex formation or by forming an effective barrier surrounding the cell (Rajkumar et al. 2010). Besides, siderophores secreted by rhizosphere bacteria too have an important role in the acquisition of several heavy metals. Siderophores can also diminish metal bioavailability and, in turn, its toxicity by binding metal ions that have chemistry akin to that of iron (Dimkpa et al. 2008; Rajkumar et al. 2010). Sometimes, crystallization and precipitation of heavy metals takes place because of bacteria–mediate reactions or due to the production of specific metabolites (Diels et al. 2003). In addition, several bacteria have developed a cytosolic sequestration mechanism for defense from heavy metal toxicity. In this process, metal ions might also become compartmentalized or converted into more innocuous forms after entering inside the bacterial cell. This process of the detoxification mechanism in bacteria facilitates metal accumulation in high concentrations (Ahemad 2012). For this, a marvelous example is the synthesis of LMW metallothioneins, which are small cysteine-rich proteins that bind and sequester multiple metal ions. The production of these novel metal detoxifying proteins is

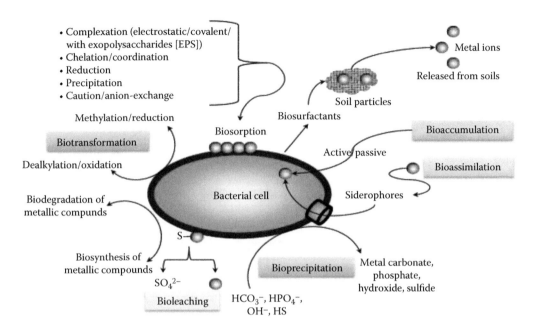

FIGURE 7.7
Various types of bacterial interaction with heavy metals in contaminated soil. (From Ahemad, M., *Arabian Journal of Chemistry*, dx.doi.org/10.1016/j.arabjc.2014.11.020.)

induced in the presence of metal stress. In addition, certain bacteria utilize methylation as an alternative for metal resistance or as a detoxification mechanism. It involves the transfer of methyl (CH_3) groups to metals and metalloids. However, limitation of the application of this methylation-related metal detoxification is that only some metals can be methylated (Ranjard et al. 2003). In addition, several bacteria can eliminate several heavy metals from the metal-polluted soils by reducing them to a lower redox state (Jing et al. 2007). Bacterial species that catalyze such reducing reactions are referred to as *dissimilatory metal-reducing bacteria* and exploit metals as terminal electron acceptors in anaerobic respiration, even though most of them use Fe^{3+} and S0 as terminal electron acceptors (Lovley et al. 1997; Jing et al. 2007). For example, the anaerobic or aerobic reduction of Cr^{6+} to Cr^{3+} by an array of bacterial isolates is an effective means of Cr detoxification (Jing et al. 2007). The various mechanisms how microorganisms may interact with plants in relation to trace elements accumulation or resistance are summarized in Figure 7.7.

7.4.1 Plant–Bacteria–Soil Interaction

The soil represents a favorable habitat for microorganisms and is inhabited by a wide range of microorganisms, including bacteria, fungi, algae, viruses, and protozoa. The soil environment consists of a variety of physical, biological, and chemical factors that affect the abundance, diversity, and interaction of a microorganism with plant root rot. However, the specificity of the plant–bacteria interaction is dependent upon soil conditions, including organic matter, pH, temperature, nutrients, and pollutants level, which can alter contaminant bioavailability, root exudate composition, and nutrient levels as shown schematically in Figure 7.8.

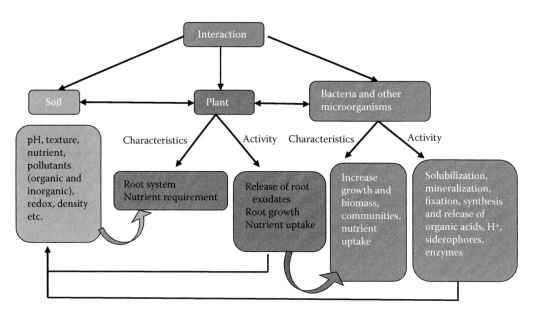

FIGURE 7.8
Plant–soil–bacteria interactions in the rhizosphere.

7.5 Rhizosphere Bacteria–Assisted Phytoremediation of Heavy Metals

Accumulation of heavy metals in the soil environment and their uptake by both rhizosphere bacteria and plants is a matter of growing environmental concern. A variety of plants growing on metalliferous soils and sludge is accumulate metals in their harvestable parts and have the potential to be used for phytoremediation of heavy metal–polluted site. Successful phytoextraction may not only depend on the plant itself but also on the interaction of the plant roots with the rhizosphere bacterial communities (Whiting et al. 2001). The discovery of rhizosphere bacteria that are heavy metal–resistant and able to promote plant growth in highly polluted environments have raised high hopes for reclamation of heavy metal–polluted soil. Plant growth-promoting effects by rhizosphere bacteria can greatly improve plant performance and also result in higher amounts of accumulated trace elements. Rhizosphere bacterial immobilization may lead to the reduction of heavy metal uptake in plants, while bacterial activity that enhances the mobility of heavy metals may cause the increase of heavy metal accumulation by plants (Figure 7.9). For example, rhizosphere bacteria were shown to promote the accumulation of Se/Hg (de Souza et al. 1999), Zn (Whiting et al. 2001), Ni (Abou-Shanab et al. 2003), Cu (Chen et al. 2005), and Cr (Abou-Shanab et al. 2007) in wetland plants *B. juncea, T. caerulescens, A. murale, Elsholtzia splendens*, and *Eichhornia crassipes*, respectively. Prapagdee and Khonsue (2015) demonstrated bacteria-assisted phytoremediation of Cd by aromatic crop plant *Ocimum gratissimum* L. grown in polluted agricultural soil. They found that Cd-resistant bacteria strains *Ralstonia* sp. TISTR 2219 and *Arthrobacter* sp. TISTR 2220 promote Cd accumulation in *O. gratissimum* L. planted in soil with Cd concentrations till 65.2 mg kg^{-1}. After transplantation in contaminated soil for 2 months, soil inoculation with *Arthrobacter* sp. enhanced Cd accumulation in the roots, above-ground tissues, and

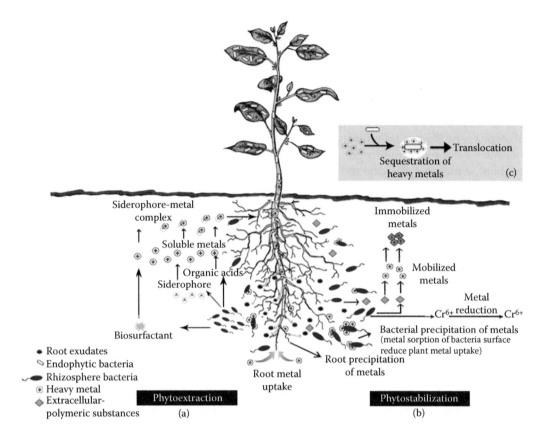

FIGURE 7.9
Action mechanism of rhizosphere bacteria accelerate the phytoremediation process in metal-contaminated soil by enhancing metal mobilization/immobilization: (a) Rhizosphere bacteria improve plant metal uptake (phytoextraction) by producing metal mobilizing chelators. It reduces plant metal uptake or translocation through (b) producing exoplusaccharide and (c) bisorption.

whole plant of *O. gratissimum* L. by 1.2-fold, 1.4-fold, and 1.1- fold. Abou-Shanab et al. (2008) isolated four bacterial isolates, that is, *Bacillus subtilis*, *Bacillus pumilus*, *Pseudomonas pseudoalcaligenes*, and *Brevibacterium halotolerans* and examined for their ability to increase the availability of Cu, Cr, Pb, and Zn in soils and for their effect on metal uptake by *Zea mays* and *Sorghum bicolor* grown in Cr and Cu rich soils. Mixed inoculum of these four bacterial strains increased the Zn, Cr, Cu, and Pb accumulation in *Z. mays* shoots grown in tannery effluent–polluted soil by 3.9, 2.7, 16, and 1.9 times respectively when compared with uninoculated plants. Several bacterial strains with resistance to Cr, isolated from the rhizosphere and endosphere of *Prosopis juliflora* plants grown on tannery effluent contaminated soil also showed tolerance toward other heavy metals such as Cd, Cu, Pb, Zn, and salt. The inoculation of ryegrass (*Lolium multiflorum* L.) with three isolates, that is, *Bacillus*, *Staphylococcus*, and *Aerococcus* promoted plant growth and the removal of toxic metals from polluted soil, demonstrating that the interaction between plants and bacterial strains identified in contaminated areas could improve plant growth and the efficiency of phytoremediation (Khan et al. 2004). A bacterial strain *Brevibacterium casei* MH8a isolated from the heavy metal–contaminated rhizosphere soil of *Sinapisalba* L. is able to promote plant growth and enhance Cd, Zn, and Cu uptake by white mustard (*Sinapis alba* L.) under laboratory conditions (Płociniczak et al. 2016).

However, in spite of the increasing knowledge of metal–microorganism interactions, few studies have attempted to characterize the rhizosphere soil bacterial communities of metallophyte plants under metal stress. These communities might also have important functions in relation to plant growth under adverse conditions as well as trace element uptake. Kunito et al. (2001) compared the characteristics of bacterial communities in the rhizosphere of *Phragmites* with those of nonrhizosphere soil in a highly Cu-contaminated area near a copper mine in Japan. Higher bacterial numbers were detected in the rhizosphere, which may be due to the lower Cu concentrations or due to the availability of root exudates that can serve as a carbon source for bacteria. Nevertheless, the percentage of highly resistant strains was higher in the rhizosphere than in nonrhizosphere soil. Similarly, Bennisse et al. (2004) characterized rhizosphere bacterial populations of metallophyte plants growing in Pb, Zn, and Cu contaminated soils from mining areas in semiarid climate. There, Gram-negative bacteria in the rhizosphere soil were dominant compared to the nonrhizosphere soils. Many reports have shown that Gram-negative bacteria are more tolerant of heavy metal stress than Gram-positive bacteria. The dominant Gram-negative generic groups isolated were *Pseudomonas, Alcaligenes, Flavobacterium, Xanthomonas,* and *Acinetobacter,* whereas Gram-positive bacteria were less abundant and were represented mainly in the *Bacillus* genera. Navarro-Noya et al. (2010) characterized bacterial communities by PCR-DGGE and 16S rRNA gene library analysis of metagenomic DNA extracted from the rhizosphere of pioneer plants *Bahia xylopoda* and *Viguiera linearis* to grow on silver mine tailings with high concentrations of heavy metals in Zacatecas, Mexico. Moderate bacterial diversity and twelve major phylogenetic groups, including *Proteobacteria, Acidobacteria, Bacteroidetes, Gemmatimonadetes, Chloroflexi, Firmicutes, Verrucomicrobia, Nitrospirae,* and *Actinobacteria* phyla, and divisions TM7, OP10, and OD1 were recognized in the rhizospheres. The most abundant groups were members of the phyla *Acidobacteria* and *Betaproteobacteria* (*Thiobacillus* spp., *Nitrosomonadaceae*). As an example, Ni-resistant rhizosphere bacteria have been shown to increase Ni uptake into the shoots of hyperaccumulator *Alyssum murale* (Abou-Shanab et al. 2003). Table 7.2 summarizes the published studies on the rhizosphere bacteria effect on phytoremediation of heavy metal–contaminated soil.

7.6 Rhizosphere Bacteria–Assisted Phytoremediation of Organic Pollutants

Industries are important pollution sources and the discharged wastewaters may contain very diverse organic pollutants that are potentially harmful to human health and wildlife animals. In addition, organic pollutants are toxic to plants and microorganisms; the presence of organic pollutants in soil and water decreases plant growth and its phytoremediation efficacy. However, some organic pollutants can be directly degraded and mineralized by plant enzymes through phytodegradation; many plants produce, and often secrete into the environment, enzymes that can be degraded to a wide range of organic compounds. Plants can accumulate organic pollutants from contaminated sites and detoxify them through their metabolic activities. From this point of view, green plants can be regarded as a "green liver" for the biosphere (Sandermann 1992).

TABLE 7.2

Examples of Successful Remediation of Heavy Metals by Using Plant–Rhizosphere Bacteria Partnerships

Bacteria Species	Host Plant Species	Pollutants	Reference
Microbacterium saperdae, Pseudomonas monteilii, and *Enterobacter cancerogenes,*	*Thlaspi caerulescens*	Zn	Whiting et al. 2001
Sphingomonas macrogoltabidus, Microbacterium liquefaciens, and *Microbacterium arabinogalactanolyticum*	*Alyssum murale*	Ni	Abou-Shanab et al. 2003
Pseudomonas, Bacillus, and *Cupriavidus*	*Rinorea bengalensis* (Wall.) O. K. and *Dichapetalum gelonioides* sp. *andamanicum* (King) Leenh	Ni	Pal et al. 2007
Burkholderia cepacia	*Sedum alfredii*	Cd and Zn	Li et al. 2007
Pseudomonas, Janthinobacterium, Serratia, Flavobacterium, Streptomyces, and *Agromyces*	*Salix caprea.*	Zn and Cd	Kuffner et al. 2008
Bradyrhizobium sp. *Rhizobium* sp., *Sphingomonas* sp., *Variovorax* sp., *Burkholderia* sp., *Janthinobacterium* sp.3, *Pseudomonas* sp., *Streptomyces* sp., *Nocardia* sp. *Microbacterium* sp., *Flavobacterium* sp., *Pedobacter* sp., *Chryseobacterium* sp., *Mucilaginibacter* sp.	*Salix caprea.*	Zn and Cd	Kuffner et al. 2010
Acinetobacter, Alcaligens, Listeri, Staphylococcus. Acinetobacter, Alcaligens, and *Listeria*	*Nymphaea pubescens*	Cu	Kabeer et al. 2014
Ralstonia sp. TISTR 2219 and *Arthrobacter* sp. TISTR 2220	*Ocimum gratissimum*	Cd	Prapagdee and Khonsue, 2015
Arthrobacter sp., *Pseudomonas aeruginosa, Bacillus licheniformis, Bacillus licheniformis, Pseudomonas stutzeri*	*Prosopis juliflora*	Cr	Khan et al. 2014
Brevibacterium casei MH8a	*Sinapis alba* L.	Cd, Zn, Cu	Płociniczak et al. 2016
Bacillus pumilus and *Micrococcus* sp.	*Noccaea caerulescens*	Ni	Aboudrar, 2013

Phytoremediation of organic pollutants may also occur by phytostabilization (the use of certain plants to reduce the mobility and bioavailability of pollutants in the environment, thus preventing their migration to groundwater or their entry into the food chain) or by phytostimulation (the stimulation of microbial degradation of pollutants in the plant rhizosphere, sometime called rhizodegradation). Moreover, plants have certain limits with respect to their capabilities to remove organic pollutants from the environment (Carvalho et al. 2014).

Although some organic pollutants can be metabolized by bacteria that may either be found in or added to the soil in the absence of plants, this process is usually slow and inefficient, in part as a consequence of the relatively low number of these degradative bacteria in soil. Several plants growing in polluted soil and water host different types of rhizosphere bacteria able to degrade organic pollutants, have been isolated, and degradation pathways and genes involved in organic pollutants degradation have been identified (Fatima et al. 2015; Nicoară et al. 2014). Even though these bacteria showed high potential to degrade different POPs, these are unable to survive and proliferate in the contaminated soils (Pandey et al. 2009). Therefore, effective mineralization and degradation of the organic pollutants can be achieved by employing the combined use of plants and their root-associated rhizosphere bacteria (Glick 2010; Haslmayr et al. 2014). In such a relationship, rhizosphere bacteria having catabolic genes feed upon the organic pollutants as a sole source of carbon for their metabolism and cell functioning, whereas plants facilitate the survival of bacteria by adjusting the rhizosphere environment through production of root exudates, rhizosphere oxidation, co-metabolite induction, H^+/OH^- ion excretion, organic acid production, and release of biogenic surfactants (Figure 7.10).

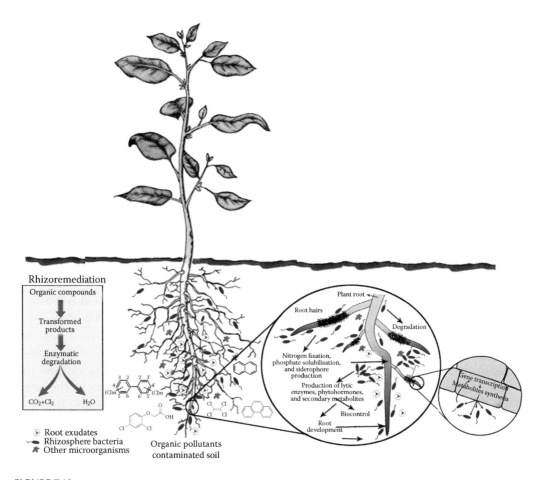

FIGURE 7.10
Degradation of organic pollutants in rhizosphere with plant–rhizosphere bacteria partnership.

Plant–rhizosphere bacteria interactions enhance the abundance and expression of catabolic genes in the rhizosphere, leading to an increase in mineralization, degradation, stabilization, or sequestration of a variety of organic pollutants (Jha and Jha 2015; Passatore et al. 2014). Recently, several studies have been conducted to explore the potential of plant–rhizosphere bacteria partnerships for the remediation of organic pollutant–contaminated soil and water (Jha and Jha 2015; Qin et al. 2014). The partnerships of *Medicago sativa* (alfalfa) with *Pseudomonas fluorescens* sp. strain F113 and *Arabidopsis* with *Pseudomonas putida* strain Flav1-1 enhanced the degradation of a variety of PCBs (Villacieros et al. 2005; Narasimhan et al. 2003). Similarly, enhanced biotransformation of a number of aroclor compounds (e.g., 1242, 1248, 1254, and 1260) by *Medicago sativa* inoculated with a symbiotic N_2-fixing rhizosphere bacteria, *Sinorhizobium meliloti* strain A-025, has been reported (Mehmannavaz et al. 2002). Chaturvedi et al. (2006) isolated 15 bacteria from the rhizosphere soil of wetland plant *Phragmites australis* L. growing in a distillery waste–contaminated site. On the basis of 16S rRNA gene sequence analysis, these bacteria were identified as *Microbacterium hydrocarbonoxydans, Achromobacter xylosoxidans, Bacillus subtilis, B. megaterium, B. anthracis from upper zone, B. licheniformis, Achromobacter sp., B. thuringiensis, B. licheniformis, Staphylococcus epidermidis, Pseudomonas migulae, Alcaligens faecalis,* and *B. cereus.* The study also showed that the bacteria population present in rhizosphere soils of *P. australis* play a major role for the bioremediation of various pollution parameters such biological oxygen demand (BOD), chemical oxygen demand (COD), phenol, sulfate, heavy metals, and melanoidins (a Maillard product of amino carbonyl compound) present in the sugarcane molasses that were present in a post-methanated distillery effluent (PMDE). Similarly, Chandra et al. (2012) characterized rhizosphere bacterial communities of *P. cummunis* growing in constructed wetland during the sequential treatment of PMDE. The PCR–RFLP analysis showed the presence of *Stenotrophomonas, Enterobacter, Pantoea, Acinetobacter,* and *Klebsiella* sp., as dominant rhizosphere bacterial communities play an important role in the degradation and decolorization of PMDE in the wetland treatment system. Guo et al. (2015) has also observed the presence of bacteria in *Typha* rhizosphere, enhancing the phytoremediation of reclaimed water in wetlands and their relation to water quality properties between upstream and downstream of wetlands. The 16S rRNA gene clone library revealed the diversity and structural characteristics of *Typha* rhizosphere bacterial communities in the wetland supplied with reclaimed water in Bai River, Beijing. These bacteria were related to *Betaproteobacteria, Gammaproteobacteria, Deltaproteobacteria, Alphaproteobacteria, Bacteroidales, Gemmatimonadetes, Chlorobi, Chloroflexi, Nitrospira, Acidobacteria,* and *Actinobacteria.* Approximately 47.10 percent of the bacterial community groups showed a close relationship with nitrogen recycling in wetlands, and 15.40 percent of the bacterial community groups were closely related to phosphorus recycling in wetlands. This study showed that 5.90 percent of the bacterial community groups were closely associated with the wetland antibiotic-containing environments, with dominant groups of *Comamonas* sp. and *Rhodocyclus* sp. in *Betaproteobacteria* and the *Alcaligenaceae, Steroidobacter* sp., and *Nitrospira* sp. of *Betaproteobacteria* accounted for 4.60 percent of the clone library groups that had a close relationship with the halogenated hydrocarbon-containing environment, the characteristics of which are one of the marked differences between reclaimed water wetlands and freshwater wetlands (Guo et al. 2015). Many other studies also reported the enhanced degradation of organic pollutants by plant–rhizosphere bacteria partnerships as shown in Table 7.3.

TABLE 7.3

Example of Successful Degradation of Organic Pollutants by Plant–Rhizosphere Bacteria Partnership

Bacteria Species	Host Plant Species	Target Pollutants	Reference
Rhodococcus sp. *Arthrobacter, Oxydans,*	*Robinia pseudoacacia,*	PCBs	Schell, 1985
Rhodococcus erythreus type strain	*Betula pendula, Fraxinus excelsior*		
Pseudomonas fluorescens	*Medicago sativa*	PCBs	Brazil et al. 1995
Burkholderia cepacia	*Hordeum sativum* L.	2,4-D	Jacobsen et al. 1997
Pseudomonas fluorescens	*Triticum* sp.	TCE	Yee et al. 1998
Microbacterium oxydans type strain	*Pinus nigra*	PCBs	Siciliano and Germida 1998
Pseudomonas putida Flav1-1, *Pseudomonas putida* PML2	*Arabidopsis*	PCBs	Narasimhan et al. 2003
Pseudomonas putida PCL1444	*Lolium multiflorum*	Naphthalene	Kuiper et al. 2004
Pseudomonas	*Hordeum sativum* L.	Phenanthrene	Ankohina et al. 2004
Azospirillum lipoferum spp	*Triticum* sp.	Crude oil	Muratova et al. 2005
Pseudomonas fluorescens	*Medicago sativa*	PCBs	Villacieros et al. 2005
Achromobacter sp.	*Salix caprea*	PCBs	Leigh et al. 2006
Microbacterium hydrocarbonoxydans, Achromobacter xylosoxidans, Bacillus subtilis, B. megaterium, B. anthracis, B. licheniformis, Achromobacter sp., B. thuringiensis, B. licheniformis, Staphylococcus epidermidis, Pseudomonas migulae, Alcaligens faecalis and *B. cereus.*	*Phragmites australis* L.	Distillery effluent	Chaturvedi et al. 2006
Pseudomonas mendocina, Pseudomonas fluorescens	*Solanum nigrum*	PCBs	Ionescu et al. 2009
Bacillus pumilus	*Armoracia rusticana*	PCBs	Ionescu et al. 2009
Sphingobacterium mizutae, Burkholderia cepacia	*Salix caprea*	PCBs	Ionescu et al. 2009
Achromobacter sp.	*Nicotiana tabacum*	PCBs	Ionescu et al. 2009
Pseudomonas, Rhodococcus, Rhizobium, and	*Medicago sativa*	PCBs	Ionescu et al. 2009
Exiguobacterium aestuarii strain ZaK	*Zinnia angustifolia*	Remazol Black B dye	Khandare et al. 2012
Microbacterium foliorum, Gordonia, alkanivorans, and *Mesorhizobium*	*Sesbania cannabina*	TPH	Maqbool et al. 2012
Stenotrophomonas, Enterobacter, Pantoea, Acinetobacter, and *Klebsiella* sp.,	*Phragmites cummunis*	Distillery effluent	Chandra et al. 2012
Sphingomonas herbicidovorans, AB042233, *Sphingomonas* sp. DS3-1, *Sphingomonas taejonensis, Sphingomonas, Herbicidovorans, Sphingomonas* sp. D12	*Zea mays*	α, β, γ, δ-hexachlorocyclohexane	Abhilash et al. 2013 Böltner et al. 2008

Notes: PCBs: Polychlorinated biphenyls; TPH: Total petroleum hydrocarbon.

7.7 Endophyte–Assisted Phytoremediation of Organic Pollutants

The presence of organic pollutants in soil, such as hydrocarbons, PAHs, PCBs, phenols, chlorophenols, toluene, trinitrotoluene, benzene, herbicides, and pesticides, inhibit growth and metabolic activities of plants and microorganisms even at very low concentrations. Phytoremediation of these highly water soluble and volatile organic compounds (VOCs) are often inefficient because plants do not completely degrade these compounds through their rhizospheres. The phytotoxicity or volatilization of these toxic chemicals through the leaves causes additional environmental problems. However, several plants grown in polluted environments host different types of associated bacteria able to degrade organic pollutants. As plants can take up organic pollutants from soil, and water through their root transport into shoots and leaves, endophytic bacteria seem to be the best potential candidates for their degradation *in planta*. Bacteria-degrading ROPs are more abundant among endophytic populations than in the rhizosphere of plants in contaminated sites (Siciliano et al. 2001), which could mean that endophytes have a role in metabolizing these substances. Endophytic bacteria are able to utilize several organic compounds as the sole carbon source for their growth and metabolism (Figure 7.11). Plant growth-promoting activities, such as nitrogen fixation, siderophore production, and phosphorous solubilization of endophytic bacteria play an important role during the growth of plants in a polluted environment (Glick 2010; Shehzadi et al. 2015).

The fate of organic pollutants in the root–rhizosphere system largely depends on their physicochemical properties. The octanol–water partition coefficient (K_{ow}) is an important physico–chemical characteristic widely used to describe hydrophobic/hydrophilic

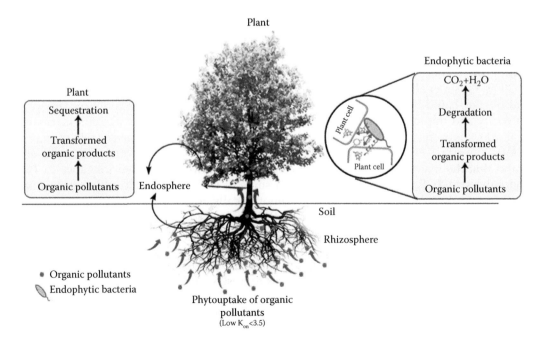

FIGURE 7.11
Mechanism of plants and their associated endophytic bacteria to degradation/detoxification of organic pollutants.

properties of chemical compounds. It is related to the transfer free energy of a compound from water to octanol. In case of constant plant and environmental features, the K_{ow} was shown to be the determining factor for root entry and translocation. Organic pollutants with a log K_{ow} < 1 are considered to be very water-soluble, and plant roots do not generally accumulate them at a rate surpassing passive influx into the transpiration stream, while organic pollutants with a log K_{ow} > 3.5 show high sorption to the roots but slow or no translocation to the stems and leaves. However, plants readily take up organic contaminants with a log K_{ow} between 0.5 and 3.5, as well as weak electrolytes (weak acids and bases or amphoteres as herbicides). These compounds seem to enter the xylem faster than the soil and rhizosphere microflora can degrade them, even if the latter is enriched with degradative bacteria (Trapp et al. 2000). Once these contaminants are taken up, plants metabolize these pollutants, although some of them or their metabolites can be toxic (Doucette et al. 1998). For example, TCE can be transformed into a toxic compound, trichloro acetic acid (TCA). Alternatively, some plants preferentially release VOCs such as TCE and BTEX or their metabolites into the environment by evapotranspiration via the leaves. After uptake and translocation, to avoid the toxicity associated with absorbed pollutants, plants usually follow one of two procedures, evapotranspiration or phytodegradation. For most of the pollutants, evapotranspiration is the major mechanism in which plants release pollutants in the atmosphere through their leaves. Regarding phytodegradation, partial degradation of pollutants within plants takes place through *in planta* detoxification mechanisms, that is, transformation (phase 1), conjugation (phase 2), and compartmentalization (phase 3).

Endophytic bacteria have been shown to contribute to biodegradation of toxic compounds and that the plant–endophyte partnership can be exploited for the remediation of contaminated soil and ground water (Weyens et al. 2009c). They are equipped with pollutant degradation pathways and metabolic activities can diminish both phytotoxicity and evapotranspiration of VOCs (Weyens et al. 2009c; Shehzadi et al. 2015). Endophytic bacteria, able to degrade organic pollutants, were isolated from the internal tissues of tree plants planted in polluted soil. Recently, many studies showed that endophytic bacteria isolated from the plants grown in soil or water contaminated with organic pollutants improved plant growth and pollutant degradation/extraction during phytoremediation (Weyens et al. 2009; Yousaf et al. 2011). Organic pollutants, such as TCE, naphthalene, BTEX, catechol, and phenol, could be degraded by endophytes, which decreased the contaminants phytotoxicity and improved plant growth (Weyens et al. 2010, 2011; Ho et al. 2009; Germaine et al. 2009). Weyens et al. (2009c) isolated endophytic bacteria from *Quercus robur* (English oak) and *Fraxinus excelsior* (Common ash) plants grown on a TCE-contaminated site and found that the majority of the bacterial community showed increased tolerance to TCE and enhanced TCE degradation capacity in some of the strains. Cultivable bacteria isolated from bulk root, stem, and leaf were genotypically characterized by amplified rDNA restriction analysis (ARDRA) of their 16S rRNA gene and identified by 16S rRNA gene sequencing. The cultivable endophytic bacterial community associated with English oak was also dominated by *Actinobacteria* (65.1 percent), with *Frigobacterium* spp. (45.0 percent) and *Okibacterium* spp. (13.0 percent) forming the majority of the group. *Arthrobacter* (3.7 percent) and *Streptomyces* (2.0 percent) were much less represented. *Proteobacteria* represented 23.1 percent of the endophytic collection associated with English oak and were dominated by *Gammaproteobacteria* (17.9 percent), including *Pseudomonas* spp. (9 percent), *Xanthomonas* spp. (4.6 percent), *Enterobacter* spp. (3.4 percent), and *Erwinia* (0.8 percent). The remaining parts of the endophytic community associated with English oak were *Firmicutes* (11.8 percent) with 8.8 percent *Bacillaceae* and 3.0 percent *Paenibacillaceae*. The community of cultivable soil and rhizosphere bacteria associated with common ash was dominated by *Bacteroidetes*, more specifically species of *Flavobacterium* (36.1 percent), and by *Firmicutes* (35.0 percent), including 29.4 percent *Bacillaceae* and 5.6 percent *Paenibacillaceae*. *Actinobacteria* made up 18.9 percent of the community (9.5 percent *Arthrobacter* spp. and 9.4 percent *Streptomyces* spp.) and *Proteobacteria* represented 10.1 percent (5.1 percent *Pseudomonas* spp. and 5.0 percent *Collimonas*). *Proteobacteria* accounted for 67.4 percent of the common ash endophyte isolates. Nearly all of these (67.3 percent) were *Gammaproteobacteria* of the genus *Pseudomonas*; the remaining 0.1 percent was identified as *Alphaproteobacteria Sinorhizobium*. Further, *Actinobacteria* made up 22.1 percent of the endophytic community, including a majority of 19.9 percent *Streptomyces* and a minority of 2.2 percent *Arthrobacter*.

A study done by van Aken et al. (2004) indicated that *Methylobacterium populum* sp. nov., strain BJ001, isolated from poplar trees is able to degrade explosive compounds such as TNT, RDX, and HMX. Mineralization of about 60 percent of RDX to CO_2 was observed within two months' time. Oliveira et al. (2012) have isolated three strains from Cerrado plants exhibiting the capacity for degradation of different fractions of petroleum, diesel oil, and gasoline. Kang et al. (2012) reported a novel endophyte *Enterobacter* sp. strain PDN3 that was isolated from a hybrid poplar (*Populus deltoides* × *Populus nigra*) and showed high tolerance to TCE. Without the addition of inducers, such as toluene or phenol, PDN3 rapidly metabolized TCE to chloride ion, and nearly 80 percent TCE (55.3 µM) was dechlorinated to chloride ions within five days. Guo et al. (2010) showed that the endophytic bacterium *Bacillus* sp. reduced Cd to approximately 94 percent in the presence of industrially used metabolic inhibitors N, N'- dicyclohexylcarbodiimide (specific ATPase inhibitor, DCC) or 2,4-dinitrophenol (DNP). Sheng et al. (2008) isolated an endophytic pyrene-degrading bacterium *Enterobacter* sp. 12J1 from *Allium macrostemon* Bunge grown in PAH-contaminated soils and found that the bacterium increased plant resistance to pyrene by increasing plant biomass (from 13 percent to 56 percent) and promoted pyrene removal from pyrene-amended soils.

Most of the isolated endophytic bacteria exhibited the potential to degrade both alkane and aromatic hydrocarbons. The endophytic microbial community may also assist in phytoremediation of petroleum. Preference for petroleum-degrading bacteria in the root interior has been illustrated with an example of plants growing in petroleum-contaminated soil (Siciliano et al. 2001). The bioremediation potential during degradation of xenobiotic compounds by three strains of *Pseudomonas* sp. isolated from xylem sap of poplar trees was tested by Germaine et al. (2004). *Pisum sativum* has been reported to remove 2,4-dichlorophenoxyacetic acid from soil when inoculated with an endophyte capable of degrading the compound without its accumulation in plant's tissues (Germaine et al. 2006; Figure 7.12). Endophytic bacteria were also isolated from different wetland plants growing on polluted sites (Chen et al. 2012; Ho et al. 2012; Zhang et al. 2013). These endophytes showed the potential to degrade different hydrocarbons and pesticides and belonged mainly to *Gammaproteobacteria, Bacilli, Alphaproteobacteria, Flavobacteria*, and *Actinobacteria*. Hydrocarbon degrading endophytic bacteria were also isolated from different salt marsh plant species (Oliveira et al. 2014). These bacteria might be particularly important for plant colonization in urban estuarine areas exposed to petroleum hydrocarbon contamination. Recently, it has pollutant-degrading endophytic bacteria that may show preferential colonization for a specific plant tissue. For instance, alkane-degrading root endophytic bacteria showed better colonization in the root, whereas shoot endophytic bacteria showed better colonization in the shoot (Andria et al. 2009; Yousaf et al. 2011). However, higher numbers of pollutant-degrading endophytic bacteria were observed in roots of most plants when compared to shoot and leaves (Rosenblueth and Martínez-Romero, 2006). Moreover, lower densities of alkane-degrading endophytic bacteria were observed at the initial stages of plant growth, indicating that first bacteria have to establish in the plant

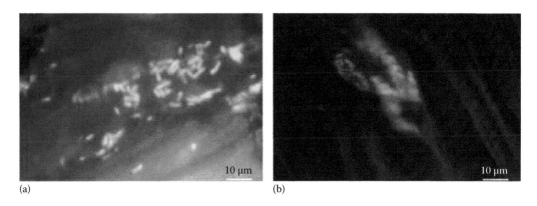

(a)　　　　　　　　　　　　　　　　　　　(b)

FIGURE 7.12
(a) *Pseudomonas putida* VM1450 microcolony within the root of inoculated plants exposed to 54 mg of 2,4-D; (b) microcolony of *Pseudomonas putida* VM1450 within the stem of plants exposed to 27 mg of 2,4-D. (From Germaine, K. J. et al., *FEMS Microbiology Ecology*, 57, 302–310, 2006.)

environment and later become important for the degradation of organic pollutants (Afzal et al. 2011, 2012). Further, the potential of a wetland plant, *Typha domingensis*, to restore quality of water contaminated with textile effluent was evaluated in combination with endophytic bacteria, *Microbacterium arborescens* TYSI04 and *Bacillus pumilus* PIRI30 (Shehzadi et al. 2015). The introduction of endophytic bacterial strains not only improved plant health but enhanced pollutant degradation and reduced mutagenic effects. Natural endophytes, which can use pollutants as a sole carbon source, were reported as most effective in preliminary degradation. Chen et al. (2012) demonstrated that culturable endophytes in aquatic plants have the potential to enhance in situ phytoremediation. This was one of the first studies aimed at isolation and comparison of culturable endophytic bacteria among different aquatic plants showing great diversity of microorganisms dominated by *Gammaproteobacteria*. Ho et al. (2012) isolated endophytic bacteria tolerating aromatic compounds from plants predominantly occurring in constructed wetlands, including reed (*Phragmites australis*) and water spinach (*Ipomoea aquatica*). *Achromobacter xylosoxidans* strain F3B was chosen for *in planta* studies using *Arabidopsis thaliana* as a model plant. It promoted removal of catechol or phenol pollutants (Ho et al. 2012). Recently, Khan et al. (2014a) reported the in vitro phytoremediation of phenanthrene by endophyte bacteria strain *Pseudomonas putida* PD1, inoculation of two different willow clones and a grass with *P. putida* PD1 was found to promote root and shoot growth and protect the plants against the phytotoxic effects of phenanthrene. There was an additional 25–40 percent removal of phenanthrene from soil by the willow and grasses, respectively inoculated with *P. putida* PD1 when compared to the uninoculated controls. The plant type was shown to strongly influence the diversity of their associated endophytic bacteria (Phillips et al., 2008). Moreover, different types of endophytic bacteria were isolated from the root and shoot of different plants planted in a soil contaminated with organic pollutants (Yousaf et al., 2010a).

A bacterial strain of *Burkholderia fungorum* DB was isolated from an oil refinery discharge and capable of transforming dibenzothiophene, phenanthrene, naphthalene, and fluorine in laboratory conditions. This bacterial strain was inoculated with hybrid poplar. It improved the phytoremediation efficiency of PAHs (naphthalene, phenanthrene, fluorine, and dibenzothiophene fluorine). On the other hand, plants inoculated with the strain DBT1 resulted in higher tolerance against the toxic effects of PAHs, in terms of root dry weight (Andreolli et al. 2013). A number of toxic organic compounds in soil have been successfully remediated using endophytic bacteria, as shown in Table 7.4. Assessing the

TABLE 7.4

Examples of Successful Degradation of Organic Pollutants by Using Plant–Endophytic Bacteria Partnerships

Bacteria Species	Host Plant Species	Target Pollutants	Reference
Pseudomonas aeruginosa R75, *Pseudomonas savastanoi* CB35	*Elymus dauricus*	Chlorobenzoic acids	Siciliano et al. 1998
Triticum spp.	*Herbaspirillum* sp. *K1*	PCBs and TCP	Mannisto et al. 2001
Methylobacterium populi BJ001	*Populus alba*	2,4,6-Trinitrotoluene, hexahydro-1,3, 5-trinitro-1,3,5-triazine, octahydro-1,3,5,7-tetranitro-1,3,5-tetrazocine	van Aken et al. 2004
Pseudomonas tolaasii, P. Jessenii, P. Rhodesiae, P. Plecoglossicida, P. veronii, P. Fulva, P. Oryzihabitans, Acinetobacter lwoffi, A. nicotianae, Bacillus megaterium, Paenibacillus amylolyticus	*Populus cv. Hazendans*	BTEX, TCE	Moore et al. 2006
Burkholderia macroides	*Populus cv. Hoogvorst*	BTEX, TCE	Moore et al. 2006
Pseudomonas putida strain POPHV6/VM1450	*Pisum sativum*	2,4-Dichlorophenoxyacetic acid	Germaine et al. 2006
Pseudomonas sp., *Brevundimonas, Pseudomonas rhodesiae*	*Medicago sativa, Puccinellia nuttaalliana, Festuca altaica, Lolium perenne, Thinopyrum ponticum*	n-Hexadecane	Phillips et al. 2008
Enterobacter sp.	*Allium macrostemon*	Pyrene	Sheng et al. 2008a
Pseudomonas sp.	*Lolium multiflorum*	Alkanes	Andria et al. 2009
Achromobacter xylosoxidans etc.	*Phragmites australis, Ipomoea aquatica, and Vetiveria zizanioides*	Catechol and phenol	Ho et al. 2009
Pseudomonas putida	*Pisum sativum*	Naphthalene	Germaine et al. 2009
Pseudomonas sp. strain ITRI53, *Rhodococcus* sp. strain ITRH43	*Lolium perenne*	Hydrocarbons	Andria et al. 2009
Achromobacter xylosoxidans	*Ipomoea aquatica, Chrysopogon zizanioides, Phragmites australis*	Catechol and phenol	Ho et al. 2009
Pseudomonas putida W619-TCE	*Populus alba*	TCE	Weyens et al. 2009a
Burkholderia cepacia	*Zea mays*	Phenol, toluene	Wang et al. 2010
Populus trichocarpa	*Pseudomonas putida* W619-TCE	TCE	Weyens et al. 2010
Burkholderia cepacia VM1468	*Lupinus luteus*	TCE	Weyens et al. 2010a
Pseudomonas putida W619-TCE	*Populus alba*	TCE	Weyens et al. 2010b
Pseudomonas sp. strain ITRI15, *Pseudomonas putida* W619-TCE	*L. multiflorum* var. Taurus and L. corniculatus var. Leo	TCE degradation	Yousaf et al. 2010
Pseudomonas sp. strain ITRI53, *Pseudomonas* sp. strain MixRI75	*Lolium multiflorum*	Diesel	Afzal et al. 2011, 2012

(Continued)

TABLE 7.4 (CONTINUED)

Examples of Successful Degradation of Organic Pollutants by Using Plant–Endophytic Bacteria Partnerships

Bacteria Species	Host Plant Species	Target Pollutants	Reference
Enterobacter ludwigii strains	*Lolium multiflorum, Lotus corniculatus, Medicago sativa*	Diesel	Yousaf et al. 2011
Burkholderia cepacia VM1468	*Populus trichocarpa*	BTEX and TCE	Taghavi et al. 2011
Methylobacterium populi	*Populus trichocarpa*	TNT, RDX and HMX	van Aken et al.2011
Populus trichocarpa	*Enterobacter* sp. strain PDN3	TCE	Kang et al. 2012
Bacillus, Pseudomonas, Staphylococcus, Rhizobium, Lysinibacillus, Caulobacter, Paenibaccillus, Bacillus, Aeromonas, Flavobacterium, Stenotrophomonas, Pantoea, Achromobacter, Sphingomonas, Rahnellas	*Alopecurus aequalis* and *Oxalis corniculata* L.	PAH	Peng et al. 2013
Burkholderia fungorum DBT1	*Hhybrid poplar* (*Populus deltoides* × *Populus nigra*)	Dibenzothiophene, phenanthrene, naphthalene, and fluorene	Andreolli et al. 2013
Rhodococcus erythropolis ET54b, *Sphingomonas* sp. D4	*Cytisus striatus*	Hexachlorocyclohexane	Becerra-Castro et al. 2013
Consortium CAP9	Agrostis	2,4,6-Trinitrotoluene	Thijs et al. 2014
Bacillus sp., *Microbacterium* and *Halomonas* sp.	*Typha domingensis, Pistia stratiotes* and *Eichhornia crassipes,*	Textile effluent	Shehzadi et al. 2015
Pseudomonas and *Acinetobacter,Microbacterium, Enterobacter, Pseudomonas* and *Rahnella*	*Miscanthus giganteus* and *Iris pseudacorus*	Brewery effluent	Dunne et al. 2015
Pseudomonas putida PD1	Willow and grass	Phenanthrene	Khan et al. 2014

Notes: PCBs: Polychlorinated biphenyls; TCP: Trichlorophenol; BTEX: Benzene toluene, ethlybenzene and xylene; TCE: Trichloroethylene; TNT: Trinitrotolene; RDX and HMX.

diversity, distribution, physiology, and ecology of endophytic bacteria in plants is prerequisite for isolating organic contaminant-degrading endophytic bacteria and using them to eliminate organic pollution in plants (Moore et al. 2006; Tang et al. 2010). There have been some reports regarding endophytic bacterial populations in plants grown in soils polluted with different contaminants (Nogales et al. 2001). Moore et al. (2006) investigated endophytic bacterial populations in poplar trees and found that a number of isolates had the ability to degrade BTEX compounds or to grow in the presence of TCE. Ho et al. (2009) isolated 188 endophytic strains from 3 plants and found that among these strains, 29 not only grew well in the presence of naphthalene, catechol, and phenol, but also were able to utilize the pollutant as a sole carbon source for growth. *Burkholderia cepacia* L.S.2.4 bacteria genetically modified by introduction of a pTOM (toluene degradation plasmid) of *B. cepacia* G4, a natural endophyte of yellow lupine, were used for phytoremediation of toluene (Barac et al. 2004).

7.8 Endophyte–Assisted Phytoremediation of Heavy Metals

During phytoremediation of metal-contaminated sites, heavy metal resistance or sequestration systems of bacterial endophytes lower the metal phtotoxicity in hosts and increase
metal accumulation and translocation to aboveground plant tissues; thus, they play a significant role in the adaptation of plants to a polluted environment (Sessitsch et al. 2013).
Bacterial endophytes might function more effectively than bacteria added to the soil
because they participate in a process known as *bioaugmentation* (Newman and Reynol,
2005). Bioremediation of heavy metals involving endophytic bacteria L14 (EBL14) isolated
from a Cd hyperaccumulator *Solanum nigrum* L. has been described by Chen et al. (2012).
Ma et al. (2011) isolated Ni-resistant endophytic bacteria from tissues of *Alyssum serpyllifo-
lium* grown in serpentine soils. Inoculation of *Brassica juncea* seeds with these strains significantly increased the plant biomass. Zhang et al. (2011) isolated Pb-resistant endophytic
bacteria *Acinetobacter* sp. Q2BJ2 and *Bacillus* sp. Q2BG1 from metal-tolerant *Commelina communis* plants grown on Pb and Zn mine tailing. The strains increased the dry weight of
above-ground tissues and roots of *Brassica napus* grown in quartz sand containing 100 mg
kg^{-1} of Pb, compared with the uninoculated control. The Pb content in plants increased
from 58 percent to 62 percent in inoculated rape plants compared with the uninoculated
control. It has been reported recently that bacterial endophytes, *Pantoea stewartii* ASI11,
Microbacterium arborescens HU33, and *Enterobacter* HU38, isolated from *Prosopis juliflora*
vegetated in tannery effluent–contaminated soil, have extremely high levels of Cr, Cd,
Cu, Pb, and Zn, and not only improved plant growth significantly but also enhanced the
uptake and accumulation of Cr in plant tissues of inoculated ryegrass (*Lolium multiflorum* L.)
These bacteria could enhance the establishment of the plant in contaminated soil and also
improve the efficiency of phytoremediation of heavy metal–degraded soils (Khan et al.
2014b). Endophytic bacterium *Neotyphodium* enhanced Zn tolerance and uptake in two
grass species, *Festuca rundinacea* and *Lolium perenne* (Zamani et al. 2015). The symbiosis
with endophyte helped these grasses to take up and translocate more Zn, hence enabling
them to carry out phytoremediation of contaminated sites. The Ni hyperaccumulator
T. goesingense hosted a range of different *Methylobacterium* strains, mostly belonging to the
M. extorquens and *M. mesophilicum* (Idris et al. 2004). In addition, one strain was isolated,
characterized, and described as a novel species, *Methylobacterium goesingense* (Idris et al.
2006). *Methylobacterium* strains in the rhizosphere and inside plants belonged to the same
species, but consisted of different strains indicating that these plant habitats provide different conditions for bacteria (Idris et al. 2004).

Root endophytic bacteria strains *P. putida* and *Rhodopseudomonas* sp. isolated from perennial helophyte *Circuta virosa* L. increased root growth, shoot growth, and Zn uptake by
roots compared with uninoculated plants (Nagata et al. 2014). The heavy metal–resistant
strain *P. putida* CZ1 isolated from the rhizosphere of *Elsholtzia splendens* increased plant
growth under Cu stress condition, whereas Cu concentration in the shoots was increased
up to 211 percent. The strain was also able to stimulate the root system, which improved
nutrient utilization from soil. Similar observations were reported by Kolbas et al. (2015),
where endophytic bacteria from roots of *Agrostis capillaris* grown in Cu contaminated soil
(13–1020 mg kg^{-1}) increased shoot and root dry weight and also Cu concentration in sunflower plants by 1.3- to 2.2-fold in the 13–416 mg kg^{-1} soil condition. Table 7.5 summarizes
the role of endophytic bacteria in heavy metals remediation of contaminated soil.

TABLE 7.5

Examples of Successful Remediation of Heavy Metals by Using Plant–Endophyte Partnerships

Bacteria Species	Host Plant Species	Pollutants	Reference
Burkholderia sp. HU001, *Pseudomonas* sp. HU002	*Salix alba*	Cd	Weyens et al. 2013
Sphingomonas SaMR12	*Sedum alfredii*	Zn	Chen et al. 2014
Rhizobium leguminasarum	*Brassica juncea*	Zn	Adediran et al. 2015
Neotyphodium coenophialum	*Festuca arundinacea*	Cd	Soleimani et al. 2010
Enterobacter aerogenes	*Brassica juncea*	Ni and Cr	Kumar et al. 2009
Pseudomonas putida PD1	*Salix alba*	Cd	Khan et al. 2014a
Pseudomonas sp. M6, *Pseudomonas jessenii* M15	*Ricinus communis*	Ni, Cu, and Zn	Rajkumar and Freitas 2008
Pseudomonas fluorescens G10, *Microbacterium* sp. G16	*Brassica napus*	Pb and Cd	Sheng et al. 2008
Flavobacterium sp.	*Orychophragmus violaceus*	Zn	He et al. 2010
Methylobacterium oryzae strain CBMB20	*Lycopersicon esculentum*	Cd	Madhaiyan et al. 2007
Methylobacterium extorquens, Methylobacterium mesophilicum, Sphingomonas sp.	*Thlaspi goesingense*	Ni	Idris et al. 2004
Serratia sp. RSC-14	*Solanum nigrum*	Cd	Khan et al.2015
Staphylococcus saprophyticus, Massilia sp., *Ochrobactrum intermedium, Bacillus* sp., *Bacillus pumilus, Staphylococcus* sp., *Aerococcus* sp., *Staphylococcus epidermidis, Aerococcus* sp., *Pantoea stewartii, Ochrobactrum* sp., *Bacillus aerophilus, Staphylococcus* sp., *Microbacterium arborescens, Aerococcus viridans, Brevundimonas vesicularis, Enterobacter* sp., *Bacillus aquimaris, Pseudomonas* sp.	*Prosopis juliflora*	Cr	Khan et al. 2014b

7.9 Concluding Remarks and Future Perspectives

Bacteria-assisted phytoremediation is a promising technology for the ecorestoration of industrial waste–contaminated sites whose time has nearly arrived. In the past sixteen years, researchers and scientists have developed a much better understanding of precisely how various rhizosphere and endophytic bacteria contribute to various phytoremediation strategies and the efficacy of these approaches has been demonstrated under laboratory conditions. The phytoremediation process may be influenced by various biotic and abiotic factors, and more work should be carried out to document the role of naturally adapted indigenous rhizosphere and endophytic bacteria. Although plants and their associated bacteria degrade a wide range of hazardous organic pollutants, many compounds are degraded slowly or are not degraded at all. On the other hand, to efficient phytoremediation of metal contaminated site, it is first necessary to address the problem of metal bioavailability to plant. In order to help this technology become rapidly adopted and put into widespread use, more genetically engineered rhizosphere and endophytic bacteria with appropriate biodegradative capabilities to facilitate the bioavailability of metals should be produced. Finally, to employ bacteria-assisted phytoremediation of either organic or metal contaminants on a large scale in the environment, it will be necessary to convince regulatory bodies in various jurisdictions that the deliberate release of selected, or engineered, bacteria to the environment should be viewed not only as benign, but in fact as beneficial. This will require scientists to ensure, based on a thorough understanding of the processes and interactions between plants, pollutants, soil, and bacteria, that bacteria-assisted phytoremediation is a highly reproducible and dependable process. Moreover, to maximize the outcomes of plant–endophyte interactions, high throughput strategies are needed, in which all or most of the genes involved in organic pollutant degradation, proteins, or even metabolites in an organism are subjected to functional analysis. It is expected that the interactions between plants and endophytic bacteria will be explored more efficiently by using latest molecular biology techniques and that endophyte-assisted phytoremediation will be effectively applied in the field for the clean-up of soil and water polluted with organic compounds.

Acknowledgments

We acknowledge financial assistance from the Department of Science & Technology (DST), Science and Engineering Research Board (SERB), New Delhi, to Professor Ram Chandra and Rajeev Gandhi. National Senior Research Fellowship (RGNSRF) from University Grant Commission (UGC), New Delhi, to Mr. Vineet Kumar, PhD scholar, is also highly acknowledged.

References

Abhilash, P.C., Singh, B., Srivastava, P., Schaeffer, A., Singh, N. 2013. Remediation of lindane by Jatropha curcas L: Utilization of multipurpose species for rhizoremediation. *Biomass Bioenergy* 51:189–193.

Aboudrar, W., Schwartz, C., Morel, J.L., Boularbah, A. 2013. Effect of nickel-resistant rhizosphere bacteria on the uptake of nickel by the hyperaccumulator Noccaea caerulescens under controlled conditions. *Journal of Soils and Sediments* 13:501–507.

Abou-Shanab, R.A.I., Angle, J.S., van Berkum, P. 2007. Chromatetolerant bacteria for enhanced metal uptake by *Eichhornia crassipes* (Mart.). *International Journal of Phytoremediation* 9:91–105.

Abou-Shanab, R.A., Ghanem, K., Ghanem, N., Al-Kolaibe, A. 2008. The role of bacteria on heavy-metal extraction and uptake by plants growing on multi-metal-contaminated soils. *World Journal of Microbiology and Biotechnology* 24:253–262.

Abou-Shanab R.A., Angle, J.S., Delorme, T.A., Chaney, R.L., van Berkum, P., Moawad, H., Ghanem, K., Ghozlan, H.A. 2003. Rhizobacterial effects on nickel extraction from soil and uptake by *Alyssum murale*. *New Phytologist* 158:219–224.

Adediran, G.A., Ngwenya, B.T., Mosselmans, J.F.W., Heal, K.V., Harvie, B.A. 2015. Mechanisms behind bacteria induced plant growth promotion and Zn accumulation in *Brassica juncea*. *Journal of Hazardous Materials* 283:490–499.

Afzal, M., Yousaf, S., Reichenauer, T.G., Kuffner, M., Sessitsch, A. 2011. Soil type affects plant colonization, activity and catabolic gene expression of inoculated bacterial strains during phytoremediation of diesel. *Journal of Hazardous Materials* 186:1568–1575.

Afzal, M., Yousaf, S., Reichenauer, T.G., Sessitsch, A. 2012. The inoculation method affects colonization and performance of bacterial inoculant strains in the phytoremediation of soil contaminated with diesel oil. *International Journal of Phytoremediation* 14:35–47.

Ahemad, M. 2014. Remediation of metalliferous soils through the heavy metal resistant plant growth promoting bacteria: Paradigms and prospects. *Arabian Journal of Chemistry* dx.doi .org/10.1016/j.arabjc.2014.11.020

Ahemad, M., Kibret, M. 2014. Mechanisms and applications of plant growth promoting rhizobacteria: Current perspective. *Journal of King Saud University—Science* 26:1–20.

Ali, N., Sorkhoh, N., Salamah, S., Eliyas, M., Radwan, S. 2012. The potential of epiphytic hydrocarbon-utilizing bacteria on legume leaves for attenuation of atmospheric hydrocarbon pollutants. *Journal of Environmental Management* 93:113–120.

Andreolli, M., Lampis, S., Poli, M., Gullner, G., Biróc, B., Vallini, G. 2013. Endophytic *Burkholderia fungorum* DBT1 can improve phytoremediation efficiency of polycyclic aromatic hydrocarbons. *Chemosphere* 92:688–694.

Andria, V., Reichenauer, T.G., Sessitsch, A. 2009. Expression of alkane monooxygenase (alkB) genes by plant-associated bacteria in the rhizosphere and endosphere of Italian ryegrass (*Lolium multiflorum* L.) grown in diesel contaminated soil. *Environmental Pollution* 157:3347–3350.

Ankohina, T., Kochetkov, V., Zelenkova, N., Balakshina, V., Boronin, A. 2004. Biodegradation of phenanthrene by *Pseudomonas* bacteria bearing rhizospheric plasmids in model plant-microbial associations. *Applied Biochemistry and Microbiology* 40:568–572.

Antoun, H., Kloepper, J.W. 2001. Plant growth promoting rhizobacteria, in *Encyclopedia of Genetics*, eds. S. Brenner, J.H. Miller, pp. 1477–1480. New York: Academic.

Badri, D.V., Vivanco, J.M. 2009. Regulation and function of root exudates. *Plant, Cell & Environment* 32:666–681.

Bais, H.P., Weir, T.L., Perry, L.G., Gilroy, S., Vivanco, J.M. 2006. The role of root exudates in rhizosphere interactions with plants and other organisms. *Annual Reviews of Plant Biology* 57:233–266.

Barac, T., Taghavi, S., Borremans, B., Provoost, A., Oeyen, L., Colpaert, J.V., Vangronsveld, J., van der Lelie, D. 2004. Engineered endophytic bacteria improve phytoremediation of water-soluble, volatile, organic pollutants. *Nature Biotechnology* 22:583–588.

Becerra-Castro, C., Prieto-Fernández, Á., Kidd, P.S., Weyens, N., Rodríguez-Garrido, B., Touceda-González, M., Acea, M.J., Vangronsveld, J. 2013. Improving performance of *Cytisus striatus* on substrates contaminated with hexachlorocyclohexane (HCH) isomers using bacterial inoculants: Developing a phytoremediation strategy. *Plant and Soil* 362:247–260.

Bennisse, R., Labat, M., ElAsli, A., Brhada, F., Chandad, F., Lorquin, J., Liegbott, P., Hibti, M., Qatibi, A. 2004. Rhizosphere bacterial populations of metallophyte plants in heavy metal-contaminated soils from mining areas in semiarid climate. *World Journal of Microbiology and Biotechnology* 20:759–766.

Böltner, D., Godoy, P., Muñoz-Rojas, J., Duque, E., Moreno-Morillas, S., Sánchez, L., Ramos, J.L. 2008. Rhizoremediation of lindane by root colonizing *Sphingomonas*. *Microbiology and Biotechnology* 1:87–93.

Brazil, G.M., Kenefick, L., Callanan, M., Haro, A., De Lorenzo, V., Dowling, D.N., O'Gara, F. 1995. Construction of a rhizosphere pseudomonad with potential to degrade polychlorinated biphenyls and detection of bph gene expression in the rhizosphere. *Applied and Environmental Microbiology* 61:1946–1952.

Chandra, R., Bharagava, R.N., Kapley, A., Purohit, H.J. 2012. Characterization of *Phragmites cummunis* rhizosphere bacterial communities and metabolic products during the two stage sequential treatment of post methanated distillery effluent by bacteria and wetland plants. *Bioresource Technology*. doi:10.1016/j.biortech.09.132.

Chandra, R., Kumar, V. 2015. Mechanism of wetland plant rhizosphere bacteria for bioremediation of pollutants in an aquatic ecosystem, in *Advances in Biodegradation and Bioremediation of Industrial Waste*, ed. R. Chandra, pp. 329–379. Boca Raton: CRC Press.

Chandra, R., Saxena, G., Kumar, V. 2015. Phytoremediation of environmental pollutants: An eco-sustainable green technology to environmental management, in *Advances in Biodegradation and Bioremediation of Industrial Waste*, ed. R. Chandra, pp. 1–29. Boca Raton: CRC Press.

Chaparro, J.M., Badri, D.V., Vivanco, J.M. 2014. Rhizosphere microbiome assemblage is affected by plant development. *ISME Journal* 8(4):790–803.

Chaturvedi, S., Chandra, R. 2006. Isolation and characterization of *Phragmites australis* (L) rhizo-sphere bacteria from contaminated site for bioremediation or coloured distillery effluent. *Ecological Engineering* 27:202–207.

Chen, B., Shen, J., Zhang, X., Pan, F., Yang, X., Feng, Y. 2014. The Endophytic bacterium, *Sphingomonas* SaMR12, improves the potential for Zinc phytoremediation by its host, Sedum alfredii. *PLoS One* 9:e106826. doi:10.1371/journal.pone. 0106826.

Chen, L., Luo, S., Xiao, X., Guo, H., Chen, J., Wan, Y., Li, B., Xu, T., Xi, Q., Rao, C., Liu, C., Zeng, G. 2010. Application of plant growth-promoting endophytes (PGPE) isolated from *Solanum nigrum* L. for phytoextraction of Cd-polluted soils. *Applied Soil Ecology* 46:383–389.

Chen, W.M., Tang, Y.Q., Mori, K., Wu, X.L. 2012. Distribution of culturable endophytic bacteria in aquatic plants and their potential for bioremediation in polluted waters. *Aquatic Biology* 15:99–110.

Chen, Y.X., Wang, Y.P., Lin, Q., Luo, Y.M. 2005. Effect of copper-tolerant rhizosphere bacteria on mobil-ity of copper in soil and copper accumulation by *Elsholtzia splendens*. *Environment International* 31:861–866.

Compant, S., Duffy, B., Nowak, J., Clement, C., Barka, E.A. 2005. Use of plant growth-promoting bac-teria for biocontrol of plant diseases: Principles, mechanisms of action, and future prospects. *Applied and Environmental Microbiology* 71(9):4951–4959.

Dams, R.I., Paton, G.I., Killham, K. 2007. Rhizoremediation of pentachlorophenol by *Sphingobium chlorophenolicum* ATCC 39723. *Chemosphere* 68:864–870.

De Bary, H.A. 1884. *Vergleichende Morphologie und Biologie der pilze Mycetozoen und Bacterien*. Leipzig: Verlag von Wilhelm Engelmann.

De Oliveira, N.C., Rodrigues, A.A., Alves, M.I.R., Filho, N.R.A., Sadoyama, G., Vieira, J.D.G. 2012. Endophytic bacteria with potential for bioremediation of petroleum hydrocarbons and deriva-tives. *African Journal of Biotechnology* 11:2977–2984.

De Souza, M.P., Huang, C.P.A., Chee, N., Terry, N. 1999. Rhizosphere bacteria enhance that accumu-lation of selenium and mercury in wetland plants. *Planta* 209:259–263.

Dennis, P.G., Miller, A.J., Hirsch, P.R. 2010. Are root exudates more important than other sources of rhizodeposits in structuring rhizosphere bacterial communities? *FEMS Microbiology Ecology* 72:313–327.

Dey, R., Pal, K.K., Tilak, K.V.B.R. 2012. Influence of soil and plant types on diversity of rhizobacteria. *Proceedings of National Academy of Science, India, Section B Biological Science* 82(3):341–352.

Doty, S.L. 2008. Enhancing phytoremediation through the use of transgenics and endophytes. *New Phytologist* 179:318–333.

Doucette, W.J., Bugbee, B., Hayhurst, S., Plaehn, W.A., Downey, D.C., Taffinder, S.A., Edwards, R. 1998. Phytoremediation of dissolved phase trichloroethylene using mature vegetation, in *Bioremediation and Phytoremediation: Chlorinated and Recalcitrant Compounds*, eds. G.B. Wickramanayake, H.E. Hinchee, pp. 251–256. Columbus, OH: Batelle Press.

Dunne, G.G.C., Keogh, E., Menton, C., Brazil, D., Ryan, D. 2015. Bacterial communities associated with monocotyledonous plants used for effluent treatment. *Envirocore*. doi: 10.13140/RG.2.1.3270.0249.

Ekman, D.R., Lorenz, W.W., Przybyla, A.E., Wolfe, N.L., Dean, J.F. 2003. SAGE analysis of tran-scriptome responses in Arabidopsis roots exposed to 2,4,6-trinitrotoluene. *Plant Physiology* 133:1397–1406.

Frommel, M.I., Nowak, J., Lazarovits, G. 1991. Growth enhancement and developmental modifica-tions of *in vitro* grown potato (*Solanum tuberosum* ssp. *Tuberosum*) as affected by a nonfluores-cent *Pseudomonas* sp. *Plant Physiology* 96:928–936.

Germaine, K., Keogh, E., Garcia-Cabellos, G. et al 2004. Colonisation of poplar trees by gfp express-ing bacterial endophytes. *FEMS Microbiology Ecology* 48:109–118.

Germaine, K.J., Keogh, E., Ryan, D., Dowling, D. 2009. Bacterial endophyte-mediated naphthalene phytoprotection and phytoremediation. *FEMS Microbiology Letters* 296:226–234.

Germaine, K.J., Liu, X., Cabellos, G.G., Hogan, J.P., Ryan, D., Dowling, D.N. 2006. Bacterial endophyte-enhanced phytoremediation of the organochlorine herbicide 2, 4-dichlorophenoxyacetic acid. *FEMS Microbiology Ecology* 57:302–310.

Glick, B.R. 2003. Phytoremediation: Synergistic use of plants and bacteria to clean up the environment. *Biotechnology Advances* 21:383–393.

Glick, B.R. 2010. Using soil bacteria to facilitate phytoremediation. *Biotechnology Advances* 28: 367–374.

Glick, B.R. 2012. Plant growth-promoting bacteria: Mechanisms and applications. *Scientifica Article ID* 963401, 15 pp.

Glick, B.R. 2014. Bacteria with ACC deaminase can promote plant growth and help to feed the world. *Microbiology Research* 69(1):30–39.

Guo, H., Luo, S., Chen, L., Xiao, X., Xi, Q., Wei, W., Zeng, G., Liu, C., Wan, Y., Chen, J., He, Y. 2010. Bioremediation of heavy metals by growing hyperaccumulaor endophytic bacterium *Bacillus* sp. L14. *Bioresour Technology* 101:8599–8605.

Guo, Y., Gong, H., Guo, X. 2015. Rhizosphere bacterial community of *Typha angustifolia* L. and water quality in a river wetland supplied with reclaimed water. *Applied Microbiology and Biotechnology* 99:2883–2893.

Haldar, S., Sengupta, S. 2015. Plant-microbe cross-talk in the rhizosphere: Insight and biotechnological potential. *Open Microbiol Journal* 9:1–7.

Hallmann, J., Quadt-Hallmann, A., Mahaffee, W.F., Kloepper, J.W. 1997. Bacterial endophytes in agricultural crops. *Canadian Journal of Microbiology* 43:895–914.

Hallmann, J., Rodriguez-Kabana, R., Kloepper, J.W. 1999. Chitinmediated changes in bacterial communities of the soil, rhizosphere and within roots of cotton in relation to nematode control. *Soil Biology and Biochemistry* 31:551–560.

Hamilton, C.E., Gundel, P.E., Helander, M., Saikkonen, K. 2012. Endophytic mediation of reactive oxygen species and antioxidant activity in plants: A review. *Fungal Divers.* doi:10.1007 /s13225-012-0158-9.

Hardoim, P.R., van Overbeek, L.S., Berg, G., Pirttilä, A.M., Compant, S., Campisano, A., Döring, M., Sessitsch, A. 2015. The hidden world within plants: Ecological and evolutionary considerations for defining functioning of microbial endophytes. *Microbiology and Molecular Biology Reviews.* doi:10.1128/MMBR.00050-14.

He, C.Q., Tan, G., Liang, X., Du, W., Chen, Y., Zhi, G., Zhu, Y. 2010. Effect of Zn-tolerant bacterial strains on growth and Zn accumulation in *Orychophragmus violaceus. Applied Soil Ecology* 44:1–5.

Ho, Y.N., Mathew, D.C., Hsiao, S.C., Shih, C.H., Chien, M.F., Chiang, H.M., Huang, C.C. 2012. Selection and application of endophytic bacterium *Achromobacter xylosoxidans* strain F3B for improving phytoremediation of phenolic pollutants. *Journal of Hazardous Materials* 15:43–49.

Ho, Y.N., Shih, C.H., Hsiao, S.C., Huang, C.C. 2009. A novel endophytic bacterium, *Achromobacter xylosoxidans*, helps plants against pollutant stress and improves phytoremediation. *Journal of Bioscience and Bioengineering* 108:S75–S95; 108:9–16.

Idris, R., Trifonova, R., Puschenreiter, M., Wenzel, W.W., Sessitsch, A. 2004. Bacterial communities associated with flowering plants of the Ni hyperaccumulator *Thlaspi goesingense. Applied and Environmental Microbiology* 70(5):2667–2677.

Idris, R., Kuffner, M., Bodrossy, L., Puschenreiter, M., Monchy, S., Wenzel, W.W., Sessitsch, A. 2006. Characterization of Ni-tolerant methylobacteria associated with the hyperaccumulating plant *Thlaspi goesingense* and description of *Methylobacterium goesingense* sp. nov. *Systematic and Applied Microbiology* 29:634–644.

Ionescu, M., Beranova, K., Dudkova, V., Kochankova, L., Demnerova, K., Macek, T., Mackova, M. 2009. Isolation and characterization of different plant associated bacteria and their potential to degrade polychlorinated biphenyls. *International Biodeterioration & Biodegradation* 63:667–672.

Jacobsen, C. 1997. Plant protection and rhizosphere colonization of barley by seed inoculated herbicide degrading *Burkholderia* (*Pseudomonas*) *cepacia* DBO1 (pRO101) in 2,4-D contaminated soil. *Plant and Soil* 189:139–144.

Kabeer, R., Varghese R., Kannan, V.M., Thomas, J.R., Poulose, S.V. 2014. Rhizosphere bacterial diversity and heavy metal accumulation in *Nymphaea pubescens* in aid of phytoremediation potential. *Journal of Bioscience and Biotechnology* 3(1):89–95.

Kang, J.W., Khan, Z., Doty, S.L. 2012. Biodegradation of trichloroethylene (TCE) by an endophyte of hybrid popular. *Applied and Environmental Microbiology* 78:3504–3507.

Kesarwani, M., Hazan, R., He, J., Que, Y., Apidianakis, Y., Lesic, B. et al. 2011. A quorum sensing regulated small volatile molecule reduce sacute virulence and promotes chronicin fection phenotypes. *PLoS Pathogens* 7:e1002192. doi: 10.1371/journal.ppat.1002192.

Khan, A.R., Ullah, I., Khan, A.L., Park, G.S., Waqas, M., Hong, S.J., Jung, B.K., Kwak, Y., Lee, I.J., Shin, J.H. 2015. Improvement in phytoremediation potential of *Solanum nigrum* under cadmium contamination through endophytic-assisted *Serratia* sp. RSC-14 inoculation. *Environmental Science and Pollution Research* 2(18):14032–14042.

Khan, M.U., Sessitsch, A., Harris, M., Fatima, K., Imran, A., Arslan, M., Shabir, G., Khan, Q.M., Afzal, M. 2014. Cr-resistant rhizo- and endophytic bacteria associated with *Prosopis juliflora* and their potential as phytoremediation enhancing agents in metal-degraded soils. *Frontier in Plant Science* 1(755):1–10.

Khan, Z., Roman, D., Kintz, T., delas Alas, M., Yap, R., Doty, S. 2014. Degradation, phytoprotection and phytoremediation of phenanthrene by endophyte *Pseudomonas putida*, PD1. *Environmental Science and Technology* 48:12221–12228.

Khandare, R.V., Rane, N.R., Waghmode, T.R., Govindwar, S.P. 2012. Bacterial assisted phytoremediation for enhanced degradation of highly sulfonated diazo reactive dye. *Environmental Science and Pollution Research* 19:1709–1718.

Kloepper, J.W., Lifshitz, R., Zablotowicz, R.M. 1989. Free-living bacterial inocula for enhancing crop productity. *Trends in Biotechnology* 7:39–43.

Kolbas, A., Kidd, P., Guinberteau, J., Jaunatre, R., Herzig, R., Mench, M. 2015. Endophytic bacteria take the challenge to improve Cu phytoextraction by sunflower. *Environmental Science and Pollution Research International* 22(7):5370–53782.

Kuffner, M., De Maria, S., Puschenreiter, M., Fallmann, K., Wieshammer, G., Gorfer, M., Strauss, J., Rivelli, A.R., Sessitsch, A. 2010. Culturable bacteria from Zn- and Cd-accumulating *Salix caprea* with differential effects on plant growth and heavy metal availability. *Journal of Applied Microbiology* 108:1471–1484.

Kuffner, M., Puschenreiter, M., Wieshammer, G., Gorfer, M., Sessitsch A. 2008. Rhizosphere bacteria affect growth and metal uptake of heavy metal accumulating willows. *Plant and Soil* 304:35–44.

Kuiper, I., Lagendijk, E.L., Bloemberg, G.V., Lugtenberg, B.J.J. 2004. Rhizoremediation, a beneficial plant-microbe interaction. *Molecular Plant Microbe Interactions* 17:6–15.

Kumar, K.V., Srivastava, S., Singh, N., Behl, H. 2009. Role of metal resistant plant growth promoting bacteria in ameliorating fly ash to the growth of *Brassica juncea*. *Journal of Hazardous Materials* 170:51–57.

Leigh, M.B., Prouzová, P., Macková, M., Macek, T., Nagle, D.P., Fletcher, J.S. 2006. Polychlorinated biphenyl (PCB)-degrading bacteria associated with trees in a PCB-contaminated site. *Applied and Environmental Microbiology* 72:2331–2342.

Li, W.C., Ye, Z.H., Wong, M.H. 2007. Effects of bacteria on enhanced metal uptake of the Cd/Zn-hyperaccumulating plant, *Sedum alfredii*. *Journal of Experimental Botany* 58(15/16):4173–4182.

Lodewyckx, C., Vangronsveld, J., Porteous, F., Moorea Edward, R.B., Taghavi, S. et al. 2002. Endophytic bacteria and their potential applications. *Critical Review in Plant Science* 21(6):583–606.

Lugtenberg, B., Kamilova, F. 2009. Plant-growth-promoting rhizobacteria. *Annual Review in Microbiology* 63:541–556.

Ma, Y., Prasad, M.N.V., Rajkumar, M., Freitas, H. 2011. Plant growth promoting rhizobacteria and endophytes accelerate phytoremediation of metalliferous soils. *Biotechnology Advances* 29:248–258.

Madhaiyan, M., Poonguzhali, S., Sa, T. 2007. Metal tolerating methylotrophic bacteria reduces nickel and cadmium toxicity and promotes plant growth of tomato (*Lycopersicon esculentum* L.). *Chemosphere* 69:220–228.

Mannisto, M.K., Tiirola, M.A., Puhakka, J.A. 2001. Degradation of 2, 3,4,6-tetraclorophenol at low temperature and low dioxygen concentrations by phylogenetically different groundwater and bioreactor bacteria. *Biodegradation* 12:291–301.

Maqbool, F., Wang, Z., Xu, Y., Zhao, J., Gao, D., Zhao, Y.G., Bhatti, Z.A., Xing, B. 2012. Rhizodegradation of petroleum hydrocarbons by *Sesbania cannabina* in bioaugmented soil with free and immobilized consortium. *Journal of Hazardous Materials* 237–238:262–269.

Mastretta, C., Barac, T., Vangronsveld, J., Newman, L., Taghavi, S., van der Lelie, D. 2006. Endophytic bacteria and their potential application to improve the phytoremediation of contaminated environments. *Biotechnology & Genetics Engineering Reviews* 23:175–207.

Mench, M., Schwitzguébel, J.P., Schroeder, P., Bert, V., Gawronski, S., Gupta, S. 2009. Assessment of successful experiments and limitations of phytotechnologies: Contaminant uptake, detoxification and sequestration, and consequences for food safety. *Environmental Science and Pollution Research* 16:876–900.

Moore, F.P., Barac, T., Borremans, B., Oeyen, L., Vangronsveld, J., Lelie, D., Campbell, C.D., Moore, E.R. 2006. Endophytic bacterial diversity in poplar trees growing on a BTEX-contaminated site: The characterization of isolates with potential to enhance phytoremediation. *Systematic and Applied Microbiology* 29:539–556.

Muratova, A.Y., Turkovskaya, O.V., Antonyuk, L.P., Makarov, O.E., Pozdnyakova, L.I., Ignatov, V.V. 2005. Oil-oxidizing potential of associative rhizobacteria of the genus *Azospirillum*. *Microbiology* 74:210–215.

Nagata, S., Yamaji, K., Nomura, N., Ishimoto, H. 2014. Root endophytes enhance stress-tolerance of *Cicuta virosa* L. growing in a mining pond of eastern Japan: Endophyte promote Zn uptake and tolerance. *Plant Species Biology* 30(2).

Narasimhan, K., Basheer, C., Bajic, V.B., Swarup, S. 2003. Enhancement of plant-microbe interactions using a rhizosphere metabolomics-driven approach and its application in the removal of polychlorinated biphenyls. *Plant Physiology* 132:146–153.

Newman, L.A., Reynol, C.M. 2005. Bacteria and phytoremediation: New uses for endophytic bacteria in plants. *Trends in Biotechnology* 23:6–8.

Nogales, B., Moore, E.R.B., Llobet-Brossa, E., Rossello-Mora, R., Amann, R. et al. 2001. Combined use of 16S ribosomal DNA and 16S rRNA to study the bacterial community of polychlorinated biphenyl-polluted soil. *Applied and Environmental Microbiology* 67(4):1874–1884.

Oliveira, V., Gomes, N.C.M., Almeida, A., Silva, A.M.S., Simões, M.M.Q., Smalla, K., Cunha, Â. 2014. Hydrocarbon contamination and plant species determine the phylogenetic and functional diversity of endophytic degrading bacteria. *Molecular Ecology* 23:1392–1404.

Pal, A., Wauters, G., Paul, A.K. 2007. Nickel tolerance and accumulation by bacteria from rhizosphere of nickel hyperaccumulators in serpentine soil ecosystem of Andaman, India. *Plant and Soil* 293:37–48.

Peng, A., Liu, J., Gao, Y., Chen, Z. 2013. Distribution of endophytic bacteria in *Alopecurus aequalis* Sobol and *Oxalis corniculata* L. from soils contaminated by polycyclic aromatic hydrocarbons. *PLoS ONE* 8(12): e83054. doi:10.1371/journal.pone.0083054.

Phillips, L.A., Germida, J.J., Farrell, R.E., Greer, C.W. 2008. Hydrocarbon degradation potential and activity of endophytic bacteria associated with prairie plants. *Soil Biology and Biochemistry* 40:3054–3064.

Płociniczak, T., Sinkkonen, A., Romantschuk, M., Sułowicz, S., Piotrowska-Seget, Z. 2016. Rhizosphere bacterial strain *Brevibacterium casei* MH8a colonizes host plant tissues and enhances Cd, Zn, Cu phytoextraction by white mustard. *Frontier in Plant Science* 7:101. doi: 10.3389/fpls.2016.00101.

Prapagdee, B., Khonsue, N. 2015. Bacterial-assisted cadmium phytoremediation by *Ocimum gratissimum* L. in polluted agricultural soil: A field trial experiment. *International Journal of Environmental Science and Technology* 12:3843–3852.

Rajkumar, M., Freitas, H. 2008. Influence of metal resistant-plant growth-promoting bacteria on the growth of *Ricinus communis* in soil contaminated with heavy metals. *Chemosphere* 71:834–842.

Rosenblueth, M., Martinez Romero, E. 2004. *Rhizobium etli* maize populations and their competitiveness for root colonization. *Archievs of Microbiology* 181:337–344.

Rosenblueth, M., Martínez-Romero, E. 2006. Bacterial endophytes and their interactions with hosts. *Molecular Plant-Microbe Interaction* 19:827–837.

Sandermann Jr, H. 1992. Plant metabolism of xenobiotics. *Trends in Biochemical Sciences* 17:82–84.

Schell, M.A. 1985. Transcriptional control of the nah and sal hydrocarbon-degradation operons by the nahR gene product. *Gene* 36:301–309.

Seghers, D., Wittebolle, L., Top, E.M., Verstraete, W., and Siciliano, S.D. 2004. Impact of agricultural practices on the *Zea mays* L. endophytic community. *Applied and Environmental Microbiology* 70:1475–1482.

Sessitsch, A., Kuffner, M., Kidd, P., Vangronsveld, J., Wenzel, W.W., Fallmann, K., Puschenreiter, M. 2013. The role of plant-associated bacteria in the mobilization and phytoextraction of trace elements in contaminated soils. *Soil Biology Biochemistry* 60:182–194.

Shehzadi, M., Fatima, K., Imran, A., Mirza, M.S., Khan, Afzal, Q.M.M. 2015. Ecology of bacterial endophytes associated with wetland plants growing in textile effluent for pollutant-degradation and plant growth-promotion potentials. *Plant Biosystems* 150(6):1261–1270.

Sheng, X., He, L., Wang, Q., Ye, H., Jiang, C. 2008b. Effects of inoculation of biosurfactant-producing *Bacillus* sp. J119 on plant growth and cadmium uptake in a cadmium-amended soil. *Journal of Hazardous Materials* 155:17–22.

Sheng, X.F., Chen, X.B., He, L.Y. 2008a. Characteristics of an endophytic pyrene-degrading bacterium of *Enterobacter* sp. 12J1 from *Allium macrostemon* Bunge. *International Biodeterioration & Biodegradation* 62(2):88–95.

Shrivastava, S., Egamberdieva, D., Varma, A. 2015. Plant growth-promoting rhizobacteria (PGPR) and medicinal plants: The state of the art, in *Plant-Growth Promoting Rhizobacteria (PGPR) and Medicinal Plants,* eds. D. Egamberdieva, S. Shrivastava, A. Varma, pp. 1–18.

Siciliano, S.D., Fortin, N., Mihoc, A., Wisse, G., Labelle, S., Beaumier, D., Ouellette, D., Roy, R., Whyte, L.G., Banks, M.K., Schwab, P., Lee, K., and Greer, C.W. 2001. Selection of specific endophytic bacterial genotypes by plants in response to soil contamination. *Applied and Environmental Microbiology* 67:2469–2475.

Siciliano, S.D., Germida, J.J. 1998. Mechanisms of phytoremediation: Biochemical and ecological interactions between plants and bacteria. *Environmental Reviews* 6:65–79.

Siciliano, S.D., Goldie, H., Germida, J.J. 1998. Enzymatic activity in root exudates of dahurian wild rye (*Elymus dauricus*) that degrades 2- chlorobenzoic acid. *Journal of Agricultural and Food Chemistry* 46:5–7.

Soleimani, M., Hajabbasi, M.A., Afyuni, M., Mirlohi, A., Borggaard, O.K., Holm, P.E. 2010. Effect of endophytic fungi on cadmium tolerance and bioaccumulation by *Festuca arundinacea* and *Festuca pratensis*. *International Journal of Phytoremediation* 12:535–549.

Sylvia, D., Fuhrmann, J., Hartel, P., Zuberer, D. 2005. *Principles and Applications of Soil Microbiology*. New Jersey: Pearson Education Inc.

Taghavi, S., Weyens, N., Vangronsveld, J., van der Lelie, D. 2011. Improved phytoremediation of organic contaminants through engineering of bacterial endophytes of trees, in *Endophytes of Forest Trees*, eds. A.M. Pirttilä, A.C. Frank, pp 205–216. Netherlands: Springer.

Tang, J., Wang, R., Niu, X., Zhou, Q. 2010. Enhancement of soil petroleum remediation by using a combination of ryegrass (*Lolium perenne*) and different microorganisms. *Soil & Tillage Research* 110:87–93.

Thijs, S., Van Dillewijn, P., Sillen, W., Truyens, S., Holtappels, M., D'Haen, J., Carleer, R., Weyens, N., Ameloot, M., Ramos, J.-L., Vangronsveld, J. 2014. Exploring the rhizospheric and endophytic bacterial communities of *Acer pseudoplatanus* growing on a TNT-contaminated soil: Towards the development of a rhizocompetent TNT-detoxifying plant growth promoting consortium. *Plant and Soil* 385:15–36.

Trapp, S., Zambrano, K.C., Kusk, K.C., Karlson, U. 2000. A phytotoxicity test using transpiration of willows. *Archives of Environmental Contamination and Toxicology* 39:154–160.

Turner, T.R., James, E.K., Poole, P.S. 2013. The plant microbiome. *Genome Biology* 14(6):209.

United States Environmental Protection Agency (USEPA). A citizen guide to bioremediation. EPA 542-F-12-003.

Van Aken, B., Tehrani, R., Schnoor, J.L. 2011. Endophyte-assisted phytoremediation of explosives in poplar trees by *Methylobacterium populi* BJ001T, in *Endophytes of Forest Trees: Biology and Applications*, eds. A.M. Pirttila, A.C. Frank, pp. 217–236. London: Springer.

Van Aken, B. et al. 2004. Biodegradation of nitro-substituted explosives 2,4,6-Trinitrotoluene, Hexahydro-1,3,5-Trinitro-1,3,5- Triazine, and Octahydro-1,3,5,7-Tetranitro-1,3,5-Tetrazocine by a *Phytosymbiotic Methylobacterium* sp. associated with poplar tissues (*Populus deltoidesxnigra* DN34). *Applied and Environmental Microbiology* 70:508–517.

Van Loon, L.C. 2007. Plant responses to plant growth-promoting rhizobacteria. *European Journal of Plant Pathology* 119(3):243–254.

Villacieros, M., Whelan, C., Mackova, M., Molgaard, J., Sánchez-Contreras, M., Lloret, J., Aguirre de Cárcer, D., Oruezábal, R.I., Bolaños, L., Macek, T., Karlson, U., Dowling, D.N., Martín, M., Rivilla, R. 2005. Polychlorinated biphenyl rhizoremediation by *Pseudomonas fluorescens* F113 derivatives, using a *Sinorhizobium meliloti* nod system to drive bph gene expression. *Applied and Environmental Microbiology* 71:2687–2694.

Wani, P.A., Khan, M.S., Zaidi, A. 2007. Chromium reduction, plant growth-promoting potentials, and metal solubilizatrion by *Bacillus* sp. isolated from alluvial soil. *Current Microbiology* 54(3):237–243.

Weston, L.A., Ryan, P.R., Watt, M. 2012. Mechanisms for cellular transport and release of allelo-chemicals from plant roots into the rhizosphere. *Journal of Experimental Botany* 63:3445–3454.

Weyens, N., Croes, S., Dupae, J., Newman, L., van der Lelie, D., Carleer, R., Vangronsveld, J. 2010. Endophytic bacteria improve phytoremediation of Ni and TCE co-contamination. *Environmental Pollution* 158:2422–2427.

Weyens, N., Schellingen, K., Beckers, B., Janssen, J., Ceulemans, R., Lelie, D., Taghavi, S., Carleer, R., Vangronsveld, J. 2013. Potential of willow and its genetically engineered associated bacteria to remediate mixed Cd and toluene contamination. *Journal of Soils and Sediments* 13:176–188.

Weyens, N., Taghavi, S., Barac, T., van der Lelie, D., Boulet J., Artois, T., Carleer, R., Vangronsveld, J. 2009c. Bacteria associated with oak and ash on a TCE-contaminated site: Characterization of isolates with potential to avoid evapotranspiration of TCE. *Environmental Science and Pollution Research International* 16:830–843.

Weyens, N., van Der Lelie, D., Taghavi, S., Newman, L., Vangronsveld, J. 2009a. Exploiting plant-microbe partnerships to improve biomass production and remediation. *Trends in Biotechnology* 27:591–598.

Weyens, N., van der Lelie, D., Taghavi, S., Vangronsveld, J. 2009b. Phytoremediation: Plant–endophyte partnerships take the challenge. *Current Opinion in Biotechnology* 20:248–254.

Whiting, S.N., De Souza, M., Terry, N. 2001. Rhizosphere bacteria mobilize Zn for hyperaccumulator by *Thlaspi caerulescens*. *Environmental Science and Technology* 35:3144–3150.

Yee, D.C., Maynard, J.A., Wood, T.K. 1998. Rhizoremediation of trichloroethylene by a recombinant, root-colonising *Pseudomonas fluorescens* stain expressing toluene *ortho*-monooxygenase consti-tuitively. *Applied and Environmental Microbiol*ogy 64:112–118.

Yousaf, S., Afzal, M., Reichenauer, T.G., Brady, C.L., Sessitsch, A. 2011. Hydrocarbon degradation, plant colonization and gene expression of alkane degradation genes by endophytic *Enterobacter ludwigii* strains. *Environmental Pollution* 159:2675–2683.

Yousaf, S., Andria, V., Reichenauer, T.G., Smalla, K., Sessitsch, A. 2010a. Phylogenetic and functional diversity of alkane degrading bacteria associated with Italian ryegrass (*Lolium multiflorum*) and birdsfoot trefoil (*Lotus corniculatus*) in a petroleum oil-contaminated environment. *Journal of Hazardous Materials* 184:523–532.

Zamani, N., Sabzalian, M.R., Khoshgoftarmanesh, A., Afyuni, M. 2015. Neotyphodium endophyte changes phytoextraction of zinc in *Festuca arundinacea* and *Lolium perenne*. *International Journal of Phytoremediation* 17(1-6):456–463.

Zhang, X., Wang, Z., Liu, X., Hu, X., Liang, X., Hu, Y. 2013. Degradation of diesel pollutants in Huangpu-Yangtze River estuary wetland using plant–microbe systems. *International Biodeterioration & Biodegradation* 76:71–75.

Zhang, Y., He, L., Chen, Z., Zhang, W., Wang, Q., Qian, M., Sheng, X. 2011. Characterization of lead-resistant and ACC deaminase-producing endophytic bacteria and their potential in promoting lead accumulation of rape. *Journal of Hazardous Materials* 186:1720–1725.

8

Nutrient Availability and Plant–Microbe Interactions in Phytoremediation of Metalliferous Soils

Dipanwita Saha, Shibu Das, Prosenjit Chakraborty, and Aniruddha Saha

CONTENTS

8.1 Introduction

Rapid increases in certain industries and intensive mining of metals have resulted in heavily contaminated sites throughout the world. These metal contaminants migrate easily to uncontaminated areas as dust or leachates through the soil and water, and this migration leads to contamination of the surrounding environment and causes disturbances in biogeochemical cycles. Agricultural activities, such as use of agrochemicals, contaminated sewage sludge, and wastewater, pose serious problems as they can directly affect human health (Ali et al. 2013, Ullah et al. 2015). Metal contamination also poses health risks to other living organisms that are part of our food chain (Gall et al. 2015). Heavy metals can be toxic to plants, leading to oxidative stress and cell injury (Schutzendubel and Polle 2002, Tak et al. 2013). Conventional cleanup of metal contaminants is often very costly and insufficient. Moreover, it can lead to further air and groundwater pollution (Lombi et al. 2001). It is therefore essential to find less expensive and environmentally friendly alternative methods to restore metal-contaminated soils.

Use of plants for removing toxic metals from the environment holds enormous promise as a green alternative that is cost-effective and also has high public acceptance (Ullah et al. 2015). Despite the promise of this option, phytoremediation as a process has not been very successful, because plants in general do not mineralize contaminants (Segura et al. 2009). In order to enhance the process of phytoremediation, chemical chelators, such as EDTA, have been used as soil amendments (Zheng et al. 2011), but some of these chemicals have been found to be toxic to both plants and plant-associated beneficial microorganisms (Saifullah et al. 2009, Banaaraghi et al. 2010). The use of microbe-associated processes for improving the performance of plants for bioremediation (instead of chemical amendments) is more desirable because of their low toxicity and biocompatibility. Plant growth-promoting rhizobacteria (PGPR) and arbuscular mycorrhizal fungi (AMF) are soil microbes which can act to improve plant growth and health and facilitate phytoremediation processes, including bioaccumulation, phytoextraction, and phytostabilization, by changing the bioavailability of nutrients and metals (Segura et al. 2009, Ma et al. 2011). Microbes resistant to heavy metals can increase plant biomass and facilitate metal translocation from soil to root or from root to shoot tissues (Rajkumar et al. 2012, Ma et al. 2016). In this chapter, we primarily discuss recent progress in understanding the beneficial association between plants and microbes and how the utilization of this association may be a useful strategy to increase plant growth under contamination stress and thereby aid in the remediation of heavy metal-contaminated soils.

8.2 Heavy Metal Pollution: Sources and Environmental Impacts

Elements with metallic properties and an atomic number greater than 20 are traditionally defined as heavy metals, although this definition varies (Tangahu et al. 2011, Ullah et al. 2015). The most common toxic heavy metals are lead (Pb), mercury (Hg), cadmium (Cd), chromium (Cr), copper (Cu), zinc (Zn), manganese (Mn), aluminium (Al), and nickel (Ni). In addition to these metals, metalloids such as antimony (Sb) and arsenic (As) are often reported as toxic contaminants in the environment (Ullah et al. 2015). All these elements are found naturally in the soil in low concentrations, and some are required in trace amounts for human health. Cu, Zn, Mn, and Ni are micronutrients that are necessary for plant metabolic activities. A chemical is considered a pollutant when it is present in the environment in excess of a certain tolerance limit. That is, beyond the tolerance level, it becomes poisonous because it can interfere with vital physiological functioning (De Vries et al. 2013). Sources of heavy metals can be classified as natural and anthropogenic. Natural sources include volcanic activities, weathering of rocks, and erosion. Prominent anthropogenic sources are mining operations, smelting of ores, industrial emissions, fertilizers, pesticides, paints, lamps, batteries, pharmaceuticals, petrochemicals, urban sewage, and sludge. Sources of selected heavy metals are summarized in Figure 8.1 (Mahar et al. 2016).

Heavy metals and metalloids are toxic to the human body when ingested in excess of admissible limits. Individually, each metal induces specific harmful effects, depending upon its concentration, oxidation state, and how it is combined with other elements. In general, heavy metals cause oxidative stress by generating high amounts of reactive oxygen species (ROS) that exceed a cell's intrinsic defense capacities and lead to extensive membrane damage (Ali et al. 2013). The metals also interfere with vital physiological processes, such as the electron transport chain, and disrupt functioning of metalloenzymes by replacing the essential metal. Some symptoms of heavy metal toxicity following inhalation of volatile vapors or fumes containing heavy metals include vomiting, tremor, gastrointestinal disorders,

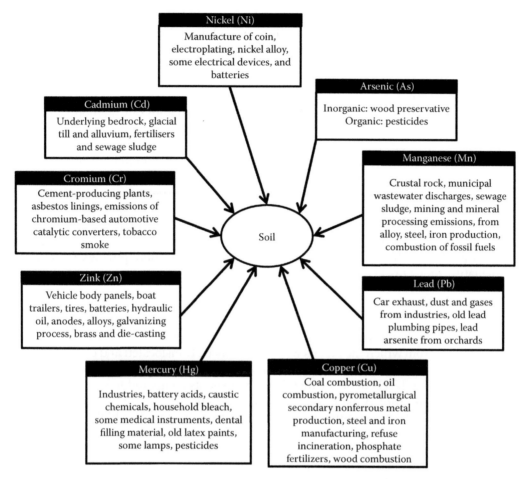

FIGURE 8.1
Sources of some heavy metals found in soil.

diarrhea, stomatitis, paralysis, convulsion, depression, and pneumonia (McCluggage 1991, Järup 2003). Toxic effects may manifest as acute, chronic, or subchronic effects, or the effects may be neurotoxic or carcinogenic (Graeme and Pollack 1998).

8.3 Phytoremediation: A Natural Method for Restoration of Metal-Contaminated Soils

Phytoremediation is defined as the use of plants and associated microbes to extract, sequester, or detoxify pollutants. This method is affordable, solar driven, and eco-friendly, and it is presently receiving considerable global attention as a method for the removal of heavy metals (Glick 2010). It is particularly cost-effective for application at very large contaminated sites and can prevent metal leaching and erosion (Prasad 2003). Phytoremediation is based on the abilities of some plant species to hyperaccumulate heavy metals in their tissues (Singh and Prasad 2011). Hyperaccumulator plants differ from other plants in their tolerance to very high external concentrations of metals and metalloids, which they are able to accumulate; the majority of these elements are translocated from roots to shoots and ultimately accumulate in the aboveground biomass (Wenzel et al. 2004). However, success with phytoremediation is limited by a plant's capabilities to accumulate and tolerate only finite amounts of metals. In addition, most metal-hyperaccumulating plants are slow growing and produce little biomass. Thus, remediation with this method takes more time than traditional treatment methods to attain full remediation (Prasad 2004). Other limiting factors include bioavailability of the metal in soil and also plant root development (Glick 2010). For effective phytoremediation, plants must be tolerant to multiple metal pollutants, be fast growing, and be able to produce substantial biomass (Glick 2010, Ma et al. 2011). Plant–microbe associations with metal tolerance properties have received attention for their potential for enhancing bioaccumulation of metals and metal mobilization and immobilization in soils, leading to improved plant biomass production (Glick 2010, Ma et al. 2011). Unlike organic pollutants, heavy metals cannot be degraded in the soil. The techniques of phytoremediation of heavy metals are phytostabilization, phytoextraction, and phytovolatilization (Tak et al. 2013, Ma et al. 2016).

8.3.1 Phytoextraction

Phytoextraction is the utilization of plants for absorption of contaminants from the soil into plant roots, from which the metals will eventually be translocated and concentrated in shoots and other aboveground biomass (Prasad 2004). Hyperaccumulator plants can accumulate much higher levels of heavy metals in their shoots compared to nonaccumulating plants (Mahar et al. 2016). A plant used for phytoextraction requires high tolerance to the heavy metal(s), a high bioaccumulation factor, a fast growth rate with a high biomass yield, a profuse root system, and a high metal-accumulating ability in the aboveground harvestable parts (Wuana and Okieimen 2011). For taking up metals from soil, the soil-bound metal must be mobilized. Plant roots can accomplish this by secreting metal-chelating molecules, substances that lower the soil pH, or enzymes such as metal reductases that alter the redox state of metals, thereby increasing their bioavailability (Raskin et al. 1994, Prasad 2004).

8.3.2 Phytostabilization

Phytostabilization refers to the use of plants to mechanically stabilize polluted soils in order to prevent bulk erosion and leaching of contaminants into other environments (Wenzel et al. 2004). The process depends on plants or on compounds secreted by the plants to immobilize or restrict the bioavailability of low levels of contaminants in the environment by adsorption onto the roots or precipitation within the root zone, thus preventing their migration to groundwater or their entry into the food chain (Tangahu et al. 2011, Ashraf et al. 2015). Mechanisms of phytostabilization include sequestering the contaminants in cell wall lignins (lignification), absorption of contaminants to soil humus (humification), binding to soil organic matter, or metal valence reduction in the rhizosphere to reduce toxicity (Prasad 2004, Ali et al. 2013). The objective of phytostabilization is to reduce the mobility, and therefore the risk, of heavy metal contaminants without necessarily removing them from the site (Bolan et al. 2011). This method is applicable when very large areas of contamination that pose probable human health impacts are to be treated. Under such situations, exposure to potential contaminants can be reduced to tolerable levels via containment (Berti and Cunningham 2000, Padmavathiamma and Li 2007). A major disadvantage of phytostabilization is that, since the contaminants remain at the site, there is the risk of future adverse environmental changes, such as the oxidation of soil carbonates leading to a rapid pH drop and remobilization of the metals. Thus, periodical monitoring of the restored site is essential.

8.3.3 Phytovolatilization

Phytovolatilization is a specific type of phytoextraction which can be used for those metals that are highly volatile. In this technique, the pollutants are taken up by plant roots, converted to volatile form(s), and subsequently released into the atmosphere (Pilon-Smits et al. 1999, Ali et al. 2013). This technique can be used for some heavy metals, like Hg and Se (Rugh 1996, Hansen et al. 1998).

8.4 Beneficial Plant–Microbe Associations in the Rhizosphere

Plant–microbe interactions in the rhizosphere are complex, and microbes can be beneficial, neutral, or unfavorable; associations with microorganisms that promote plant growth or provide protection from pathogens are favored (Bell et al. 2014, Imam et al. 2016). The microbiome in the rhizosphere is essential for plants for the production of essential vitamins, improving nutrient solubility (Figure 8.2), and increasing the overall functional potential of the host (Ma et al. 2011). Such beneficial microorganisms are able to regulate the expression of plant traits, and this can result in an improvement of the plant's physiological state

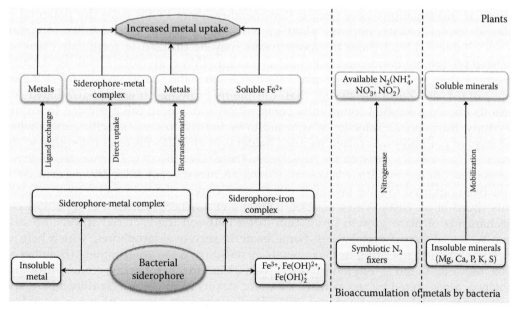

FIGURE 8.2
Roles of PGPR in mobilizing nutrients in soil and increasing their bioavailability.

(Mendes et al. 2013). In addition, recent findings have indicated that the success of phytoremediation strongly depends on plant microbiome activities (Hassan et al. 2014, Ma et al. 2016). Microorganisms such as PGPR or AMF, which form direct associations with plants, can facilitate the uptake of heavy metal pollutants by plant roots, with subsequent translocation of the metals to other components of the plant (Bell et al. 2014).

8.4.1 Association with PGPR

It is well known that rhizospheric bacteria can promote plant growth by a variety of mechanisms, such as nutrient mobilization, nitrogen fixation, production of plant growth regulators, solubilization of minerals (e.g., phosphates), production of antibiotics to provide protection against pathogens, and synthesis of the enzyme 1-aminocyclopropane-1-carboxylate (ACC) deaminase, which can reduce plant ethylene levels (Glick 2010). Plant-associated bacteria that migrate from the bulk soil to aggressively colonize the rhizosphere and impart beneficial effects on the plant are termed plant growth-promoting rhizobacter, or PGPR (Nelson 2004, Chaparro et al. 2012). The plant roots may regulate the soil microbial community through the exudation of an array of different compounds that not only provide a carbon- and energy-rich environment but also initiate communication by producing signals that modulate the process of bacterial colonization (Neal et al. 2012, Haichar et al. 2014). There is strong evidence to indicate that root exudates contain compounds which exert a selective influence on microbial community structures in the rhizosphere (Fang et al. 2013, Turner et al. 2013). However, although root exudates can act as signaling molecules, the functioning of each microorganism occurs in coordination with the overall microbiome. In addition, soil properties, such as N and P content, soil pH, soil texture, and cation content, also influence the composition of the rhizosphere microbiome (Rousk et al. 2010, Faoro et al. 2010, Chaparro et al. 2012). It is increasingly being accepted that plant hosts can select microbes out of a huge group of candidates in the bulk soil that carry genes that may serve a particular need (Siciliano et al. 2001, Thijs et al. 2016).

Several investigations have shown that metal-resistant PGPR have the potential to alleviate metal toxicity, improve plant growth, and influence the mobility of metals in soil (Table 8.1), and these microorganisms may be utilized to formulate microbe-assisted phytoremediation and restoration strategies (Ma et al. 2016). Muehe et al. (2015) observed that *Arabidopsis halleri* plants growing on natural soil accumulated higher levels of Cd and Zn in the tissues than did plants growing on gamma-irradiated soil with its altered microbial community composition and overall cell numbers. In a similar study, rhizospheric selection was found to be limited to some specific genera when the level of pollution was raised (Gomez-Balderas et al. 2014). Bacteria belonging to the genera *Achromobacter, Arthrobacter, Azotobacter, Azospirillum, Bacillus, Enterobacter, Kluyvera, Pseudomonas, Rhizobium, Streptomyces,* and *Serratia* (Carrasco et al. 2005, Zhuang et al. 2007) have been reported to have favorable effects on different plants in metal-contaminated soils (Burd et al. 2000, Wu et al. 2006, Dimkpa et al. 2008, Sessitsch et al. 2013). The mechanisms of plant growth promotion differ between the bacterial strains, because they secrete different metabolites. Some bacteria secrete siderophores, which help in chelation of nutrients (iron) or toxic metals (arsenic) from the environment (Braud et al. 2009, Mesa et al. 2017). Phytohormones, such as auxins, cytokinins, gibberellins, and ethylene, are secreted by the same or different strains of rhizobacteria (Bloemberg and Lugtenberg 2001, Nihorimbere et al. 2011). These hormones, along with other secondary metabolites, can have favorable effects on plant growth (Verma et al. 2010, Ahemad and Kibret 2014). Beneficial enzymes which help to mobilize nutrients, such as phosphorous

TABLE 8.1
Recently Reported PGPR Bacteria Associated with Phytoremediation of Metal-Contaminated Soils

Microorganisms	Habitat(s)	Feature(s) Related to Phytoremediation	Host Plant(s)	Activity Related to Plant Growth	Reference
Burkholderia sp., *Entarobacter* sp., *Achromobacter* sp., *Azospirillum* sp., *Chryseobacterium* sp.	Roots and rhizosphere	Siderophore production, Cs accumulation, N_2 fixation, P solubilization	*Helianthus annuus*	Stimulated plant growth	Ambrosini et al. 2012
Bacillus thuringiensis	Roots of *Pinus sylvestris*	ACC deaminase, IAA, siderophore production, P solubilization	*Alnus firma*	Increased biomass, chlorophyll content, nodule number, and As, Cu, Pb, Ni, and Zn accumulation	Babu et al. 2013
Massilia niastensis, *Pseudomonas* sp., *Streptomyces* sp., *Tsukamurella* sp.	Rhizospheric soil	Cd and Zn tolerant, P solubilization, biosurfactant producer, siderophore producer, IAA producer	*Betula celtiberica*, *Cytisus scoparius*, *Festuca rubra*	Improved growth, established metal-contaminated soils	Becerra-Castro et al. 2012
Bacillus licheniformis	Heavy metal-containing compost mixed with soil	Accumulation of Cr, Cu, Pb, and Zn in plants	*Brassica napus*	Helped in phytoremediation of Cr, Cu, Pb, and Zn from metal-contaminated soil	Brunetti et al. 2011
Arthrobacter nicotinovorans	Ni-rich serpentine soil, sewage sludge-affected agricultural soil	Ni accumulation	*Alyssum pintodasilvae*	Promoted plant growth, helped in phytoextraction of Ni from soil	Cabello-Conejo et al. 2014
Burkholderia sp., *Pseudomonas thivervalensis*	Cd-contaminated soils	Cd accumulation	*Brassica napus*	Increased dry weight of plants, enhanced water-extractive Cd content in rhizospheric soil	Chen et al. 2013
Bacillus amyloliquefaciens	Soil from mangroves	Pb and Al solubilization, siderophore production	Mangroves	Produced siderophores in presence of metals	Gaonkar & Bhosle 2013

(Continued)

TABLE 8.1 (CONTINUED)

Recently Reported PGPR Bacteria Associated with Phytoremediation of Metal-Contaminated Soils

Microorganisms	Habitat(s)	Feature(s) Related to Phytoremediation	Host Plant(s)	Activity Related to Plant Growth	Reference
Pseudomonas sp., *Paenibacillus jamilae*, *Arthrobacter* sp.	Zn mine soils	ACC deaminase, IAA, siderophore production	*Chenopodium ambrosioides*	Increased dry weight, Pb uptake (phytoextraction)	Zhang et al. 2012
Ralstonia sp., *Pantoea agglomerans*, *Pseudomonas thivervalensis*	Cu mine wasteland	IAA, ACC deaminase, siderophore production, P solubilization	*Brassica napus*	Plant growth, Cu uptake, metal resistance	Zhang et al. 2011
Pantoea sp., *Lysinibacillus* sp., *Bacillus* spp., *Brevibacillus* sp.	Seeds from Cd- and Ni-contaminated plots	ACC deaminase, IAA, siderophore production	*Agrostis capillaris*	Metal tolerance, plant growth, Cd uptake	Truyens et al. 2014
Bacillus methylotrophicus SMT38, *Bacillus aryabhattai* SMT48, *B. aryabhattai* SMT50, *Bacillus licheniformis* SMT51	*Spartina maritime* rhizosphere	ACC deaminase, IAA production, P solubilization, N_2 fixation, biofilm formation	*Spartina maritime*	Increased root length, root hair, lateral roots	Mesa et al. 2015
Pseudomonas brassicacearum, *Rhizobium leguminosarum*	*Brassica* and clover plant rhizosphere	Metal chelation	*Brassica juncea*	Increased plant growth in presence of Zn	Adediran et al. 2015
Burkholderia spp., *Bacillus megaterium*, *Sphingomonas* sp.	Top soil from abandoned farmland near Cu mine wasteland	Solubilized $Cu_2(OH)_2CO_3$	*Zea mays*	Promoted plant growth and Cu uptake	Sheng et al. 2012
Ralstonia eutropha, *Chrysiobacterium humi*	Metal-contaminated soil	Zn and Cd accumulation	*Helianthus annuus*	Zn accumulation in shoot and root, Cd accumulation in root	Marques et al. 2013
Bacillus pumilus, *Bacillus* spp., *Achromobacter* sp., *Stenotrophomonas* sp.	Endophytic bacteria	IAA, siderophore, and ACC deaminase production, P solubilization, pectinase and cellulase production	*Sedum plumbizincicola*	Cd, Zn, and Pb uptake, ACC deaminase production, antibiotic resistance	Ma et al. 2015
Phyllobacterium myrsinacearum	Rhizospheric soil	ACC deaminase, IAA, siderophore production, P solubilization	*Sedum plumbizincicola*	Cd, Zn, and Pb uptake, plant growth promotion	Ma et al. 2013

(Continued)

TABLE 8.1 (CONTINUED)

Recently Reported PGPR Bacteria Associated with Phytoremediation of Metal-Contaminated Soils

Microorganisms	Habitat(s)	Feature(s) Related to Phytoremediation	Host Plant(s)	Activity Related to Plant Growth	Reference
Actinobacteria, Proteobacteria, Bacteroidetes, Firmicutes	Endophytic bacteria from root, stem, and leaf	Plant growth promotion, Cd resistance	*Solanum nigrum*	Plant biomass increase, Cd uptake	Luo et al. 2011
Pseudomonas putida	Soils from noncontaminated agricultural fields	IAA, ACC deaminase, siderophore production	*Eruca sativa*	Plant growth, Ni uptake	Kamran et al. 2016
Bacillus spp., *Staphylococcus* spp., *Aerococcus* sp.	Rhizosphere soil, roots and shoots of *Prosopis juliflora*	ACC deaminase, IAA, siderophore production, P solubilization	*Prosopis juliflora*	Metal resistance, plant growth promotion	Khan et al. 2015
Acinetobacter spp., *Rhodococcus* sp., *Streptomyces* spp., *Enterobacter* spp., *Rhodococcus* sp., *Citrobacter freundii*, *Psychrobacillus psychrodurans*, *Pseudomonas mohnii*, *Lysinibacillus fusiformis*, *Paenibacillus humicus*	Multi-metal polluted substrates from uranium mine and copper mines	Siderophore production, N_2 fixation, P solubilization	*Agrostis capillaris, Deschampsia flexuosa, Festuca rubra, Helianthus annuus*	Metal bioavailability and uptake (phytostabilization)	Langella et al. 2013
Pseudomonas spp.	Heavy metal-polluted soils, wastewater	IAA, siderophore production, solubilized mineral phosphates and metals	*Zea mays, Helianthus annuus*	Plant growth, Cu uptake, total biomass	Li and Ramakrishna 2011
Sedum alfredii	Rhizosphere of *S. alfredii* growing in Pb and Zn mine	Organic acid production, metal mobilization	*Burkholderia cepacia*	Metal uptake (phytoextraction)	Li et al. 2010
Rhizobium sp. CCNWSX0481, *Rhizobium leguminosarum* bv. *viciae, Enterobacter clocae, Pseudomonas* spp.	*Vicia faba* rhizospheric soil contaminated with metals		*Vicia faba*	Cu tolerance, plant growth promotion	Fatnassi et al. 2015
Bacillus sp. MPV12, *Variovorax* sp. P4III4, *Pseudoxanthomonas* sp. P4V6	Metal-contaminated rhizospheric soil, different autochthonous plants		*Pteris vittata*	Improved As phytoextraction through hyperaccumulation	Lampis et al. 2015

and potassium, are also secreted by some bacteria, and thus the bacteria improve nutrient availability and plant growth (Sharma et al. 2013). Other beneficial compounds of rhizobacterial origin include antibiotics (Ahemad and Kibret 2014), nitric oxide (Molina-Favero et al. 2007), organic acids (Gupta et al. 2015), and biosurfactants (Sachdev and Cameotra 2013). However, the most important property that a bacterium should possess to be suitable for improving phytoremediation of metal-contaminated soils is metal resistance. Extensive research has identified several metal-resistant bacteria with multiple beneficial traits that would be effective in alleviating metal toxicity along with improving plant growth (Rajkumar and Freitas 2008). Many such genes identified in beneficial bacteria have been used to create genetically modified plants for use in specific contaminated environments, but their application remains doubtful due to regulatory restrictions (Davison 2005, Gerhardt 2009).

8.4.2 Association with AMF

Mycorrhiza are associated with more than 80% of terrestrial plants, and they are important members of the rhizosphere microbiome (Badri et al. 2009). Symbiotic interactions between host plants and mycorrhiza begin with colonization of the AMF hyphae on a compatible plant root, followed by penetration into the cortex through appresoria formation, leading to development of arbuscules, which are the principal sites for exchange of nutrients between a host plant and a colonizing fungus. Mycorrhizal colonization of roots causes an increase in the root surface area for better nutrient acquisition (Khan et al. 2000). Prior to colonization, however, a continuous exchange of signals takes place between the symbionts and also between the soil microbes and the mycorrhiza (Badri et al. 2009). Constituents of plant root exudates, such as organic acids, flavonoids, and strigolactones, are known to act as fungal stimulators (Steinkellner et al. 2007). However, availability of nutrients is a key factor in the release of signal molecules by the plant roots. Studies have shown that deficiencies of N and P enhance release of strigolactones in sorghum plants (Yonema et al. 2007). Strigolactones are now regarded as plant hormones capable of regulating adaptive responses to low phosphate levels in roots by modulating the balance between auxin and ethylene signaling, thus influencing changes in the architecture of the root system under differential P supply conditions (Koltai 2013).

As stated above, mycorrhizal colonization of roots causes an increase in the root surface area for better nutrient acquisition. Thus, plants in symbiotic association can take up heavy metals from an enlarged soil volume (Clark and Zeto 2000, Gohre and Paszkowski 2006). Metal-tolerant AMF, which are efficient in protecting plants from heavy metal-induced toxicity (Table 8.2), have been used in phytoremediation studies (Gaur and Adholeya 2004). They can change the availability of metals by changing the soil pH and can also affect metal translocation from roots to shoots (Azcon et al. 2010, Andrade et al. 2010). In addition, AMF influence plant community development, act as bioprotectants against pathogens, and improve aboveground productivity (Jeffries et al. 2003). Vogel-Mikus et al. (2006) observed that colonization by AMF caused significant improvements in nutrients and decreases in Cd and Zn uptake by the hyperaccumulator plant *Thlaspi praecox* Wulfen. AMF can also modify the pattern of host root exudation and alter the rhizosphere microflora in terms of quality, abundance, and overall rhizosphere microbial activity (Khan 2006). There is growing evidence that combined associations of AMF and rhizobacteria improve the bioremediation effects of mycorrhiza in heavy metal-contaminated soils (Miransari 2009, Azcón et al. 2010).

TABLE 8.2

Recently Reported AMF Species Associated with Phytoremediation of Metal-Contaminated Soils

Microorganisms	Habitat	Feature(s) Related to Phytoremediation	Host Plant	Activity Related to Plant Growth	Reference
Glomus mosseae, *Glomus etunicatum*	Rhizospheric soil	Phenanthrene and pyrene accumulation	*Medicago sativa*	Degraded PAHs in shoot and soil	Gao et al. 2011
Piriformospora indica, *Glomus mosseae*	Metallotoxic soil	Cd uptake	*Triticum aestivum*	Growth, chlorophyll content, performance index improvement, reduced shoot Cd content	Shahabivand et al. 2012
Glomus intraradices	Farm soil	Cd, Pb, and Ni accumulation in plant	*Ocimum basilicum*	Enhanced quality of volatile oil and yield in metal-contaminated soils	Prasad et al. 2011
Glomus sp. *Diversispora* sp., *Archaeospora* sp., *Paraglomus* sp.	Heavy metal-contaminated, semiarid soil composted with sugar beet waste	Phytostabilization	*Piptatherum miliaceum*, *Retama sphaerocarpa*, *Psoralea bituminosa*, *Coronilla juncea*, *Anthyllis cytisoides*, *Lolium perenne*	Seedling growth	Alguacil et al. 2011
Glomus etunicatum	Rhizospheric soil	Not determined	*Sorghum vulgare*, *Alphitonia neocaledonica*, *Cloezia artensis*	Increase plant biomass and P uptake, decreased Ni concentration in roots and shoots	Amir et al. 2013
Glomus clarum, *Gigaspora margarita*, *Acaulospora*	Rhizospheric soil	Malondialdehyde, proline, amino acid	*Coffea arabica*	Metal toxicity, nutrient uptake	Andrade et al. 2010
Glomus sp.	Heavy metal-contaminated soil	IAA, antioxidant enzymes, proline	*Trifolium repens*	Improved plant growth and nutrition	Azcon et al. 2010
Scutellospora sp., *Glomus* sp.	Fly ash-contaminated pond soil, *Aspergillus tubingensis* in soil	P, K, and Mg increased	*Dendrocalamus strictus*	Increased plant growth and nutrient uptake, reduced metal translocation	Babu and Reddy 2011

(Continued)

TABLE 8.2 (CONTINUED)

Recently Reported AMF Species Associated with Phytoremediation of Metal-Contaminated Soils

Microorganisms	Habitat	Feature(s) Related to Phytoremediation	Host Plant	Activity Related to Plant Growth	Reference
Glomus intraradices	Contaminated rhizospheric soil	Phytoextraction	*Salix viminalis, Populus generosa*	High biomass yields, Cd, Zn, and Cu uptake in shoots	Bissonnette et al. 2010
Rhizophagus intraradices, Funneliformis geosporum, Rhizophagus clarus, Glomus sp.	Saline soil and waste with Zn, Pb, and As	Phytoaccumulation	*Plantago lanceolata*	Zn, Pb, Cd, and As uptake from soil, plant growth promotion, P solubilization	Orlowska et al. 2012
Glomus sp., *Phytolacca americana, Glomus etunicatum, Glomus sinuosum, Glomus lamellosum, Glomus intraradices, Glomus irregular, Kuklospora colombiana, Kuklospora colombiana, Joinvillea ascendens, Ambispora leptoticha*	Heavy metal-polluted soil	Not determined	*Phytolacca americana, Rehmannia glutinosa, Perilla frutescens, Litsea cubeba, Dysphania ambrosioides*	Cu, Zn, Pb, and Cd uptake, plant growth promotion	Long et al. 2010
Glomus intraradices, Glomus constrictum, Glomus mosseae	*In vitro* culture	Increase in SOD and CAT activities, decrease in peroxidase activity	*Tagetes erecta*	Induced plant growth resistance to Cd stress	Liu et al. 2011
Glomus intraradices	*In vitro* culture	Phytostabilization	*Tagetes erecta*	Cu tolerance increased	Castillo et al. 2011
Glomus sp.	Multi-metal-polluted or nonpolluted soil	Metallothioneins, polyamines	*Populus alba*	Plant growth, enhanced stress tolerance, transcriptional upregulation of stress-related genes	Cicatelli et al. 2010
Glomus mosseae, Glomus versiforme	Rare earth element mine tailing substrates and topsoil	Increased N, P, and K uptake, decreased C:N:P stoichiometry	*Sorghum bicolor, Zea mays*	Increase dry weight of roots and shoots	Guo et al. 2013

8.5 Functions of Microbes Related to Nutrient Availability in Metal-Contaminated Soils

Heavy metals can disturb mineral nutrition in plants, causing deficiencies or imbalances in the essential elements and leading to impaired plant growth (Siedlecka 1995). Plant–microbe associations can protect plants against metal toxicity in addition to being beneficial for uptake of nutrients (Andrade et al. 2010). The uptake of essential minerals by plant roots largely depends on the mineral's bioavailability in the particular soil environment. To be taken up from soil, the soil-bound metal must be mobilized. This can be achieved by secreting either metal-chelating molecules or metal reductases from the plant roots (Raskin et al. 1994, Prasad 2003). Availability of adequate N is an important limiting factor that affects plant growth, especially in metal-contaminated soils (Mantelin and Touraine 2004). Application of fertilizer provides an immediate solution for this limiting factor, but its effects are not permanent (Bolan et al. 2003). Microorganisms, however, can provide a continuous supply of N and long-term support for plant growth. Excess heavy metal contamination can adversely affect growth of legumes and N_2 fixation (Ahmad et al. 2012). Legume–rhizobium symbiosis is an important system for phytoremediation of heavy metal-contaminated soils due to the beneficial activity of N_2 fixation. This process induces excess cation uptake by legume plants, resulting in release of more H^+ ions by the roots. Thus, the soil pH is lowered, and this plays a crucial role in enhancing metal mobilization in the soil environment (Yang et al. 2016). Carrasco et al. (2005) isolated *Rhizobium* strains resistant to As, Cu, and Pb which were symbiotically active in association with native legume plants (*Medicago sativa*, *Trifolium subterraneum*, *Vicia sativa*, *Lotus corniculatus*) in sites heavily contaminated with several toxic metals. Dary et al. (2009) observed that coinoculation of the legume plant *Lupinus luteus*, which is capable of phytostabilization of heavy metals such as Cd, Pb, and Cu, with *Bradyrhizobium* and resistant PGPR strains produced improved plant biomass in *in situ* remediation experiments. Babu et al. (2013) isolated a metal-resistant strain of *Bacillus thuringiensis* which increased growth and biomass production of *Alnus firma* plants grown in the presence of heavy metal contamination. The inoculated seedlings contained more nodules than the uninoculated controls. The authors observed that increased nodulation was likely due to better establishment of indigenous N_2-fixing bacteria, such as *Frankia* spp. in seedlings collected in the vicinity of contaminated mine sites. Deletion of the genes *copA* and *lipA*, which encode proteins involved in the copper resistance mechanism of PGPR strain *Mesorhizobium amorphae* 186, by transposon mutagenesis showed that mutants defective in *copA*, which is involved in Cu transport, did not exhibit any alterations in symbiotic potential in association with *Robinia pseudoacacia* under contaminated conditions. However, *lipA*-deficient mutants had an impaired N_2 fixation capacity and also accumulated less Cu than either the wild-type or *copA*-deficient mutants, thereby indicating that a healthy symbiotic relationship between legumes and rhizobia is more important than metal resistance for effective phytostabilization (Hao et al. 2015). Mycorrhizal associations of legume plants have been found to induce higher nutrient uptake than nonmycorrhizal legumes in contaminated soils. Hyphae of AMF were able to utilize inorganic N efficiently and transferred more N from soil to plant roots (Subramanian and Charest 1999). Yang et al. (2016) observed higher N, P, and Mg concentrations in mycorrhiza than did the nonmycorrhizal legume tree *Robinia pseudoacacia* in the presence of excess Pb concentrations. Those authors concluded that AMF play a vital role in phytoremediation by enhancing plant biomass through influencing plant photosynthesis and macronutrient acquisition.

Iron, an essential nutrient for all organisms, has a significant role as a cofactor in many enzymatic reactions. In the soil, iron occurs primarily in the ferric state (Fe^{3+}), which remains unavailable to plants and microorganisms at physiological pH (Saha et al. 2016). Most microbes produce iron-chelating siderophores that are secreted outside the cell and can bind Fe^{3+} with high affinity and allow its uptake. Additionally, siderophores can form stable complexes with other metal cations, including those of Zn, Cu, Cd, Ni, Cr, Mg, Mn, Ga, Pl, Pb, As, Co, Sn, and U (Schalk et al. 2011, Lukasz et al. 2014). Since siderophores can solubilize iron and other metals and make them available or unavailable for uptake by plant roots and microbes, siderophore-producing rhizobacteria play an important role in phytoextraction of metals (Rajkumar et al. 2012). A siderophore-overproducing mutant *Kluyvera ascorbata* strain exhibited a pronounced effect on plant growth promotion, as it protected tomato, canola, and Indian mustard plants against the inhibitory effects of high concentrations of nickel, lead, and zinc (Burd et al. 2000). The data suggested that this ability is related to the bacteria providing sufficient iron to the plants. Dimkpa et al. (2008) reported that hydroxamate siderophores in culture filtrates of *Streptomyces acidiscabies* E13 promoted growth of cowpea (*Vigna unguiculata*) plants under nickel contamination by binding to both iron and nickel, thus making iron available to plants and also protecting the plants against nickel toxicity by reducing the mobility of nickel. It was reported that siderophore-producing bacteria selectively support the uptake of iron from the pool of trace element cations competing for import and causes increased chlorophyll content and growth of various crop plants in contaminated soils, along with reducing metal-induced free radical generation and oxidative stress (Dimkpa et al. 2009, Sessitsch et al. 2013). In addition to siderophores, other metal binding chelators, such as molybdate binding tetradentate catecholates, the pigment melanin, antibiotics, and proteins such as phytochelatins and metallotheionins, also act as metal binding ligands in contaminated soil environments (Sessitsch et al. 2013).

P is a major macronutrient that limits plant growth, despite being abundant in soils in both organic and inorganic forms. Soils are deficient in P because P compounds occur either as insoluble salts (when bound to Ca, Fe, and Al) or as organic P in the form of phytate (Gyaneshwar et al. 2002). Use of microbes with a phosphate-solubilizing capability has been regarded as an economically sound alternative to P fertilizers under diverse soil conditions with greater agronomic utility (Khan et al. 2009). Mechanisms of phosphate solubilization include proton extrusion, chelation, gaseous exchange (O_2–CO_2) reactions, release of organic acids, and release of extracellular phosphatases (Mohammadi 2012). Apart from solubilizing phosphates in soil, phosphate-solubilizing microbes can enhance plant growth by increasing the efficiency of biological nitrogen fixation or increasing the availability of other trace elements, such as Fe and Zn (Sharma et al. 2013). High concentrations of heavy metals can interfere with P uptake by plants, resulting in retardation of plant growth (Zaidi et al. 2006). Phosphate-solubilizing rhizobacteria and AMF can provide P to plants under conditions of heavy metal stress, and the increased P may increase plant biomass and probably helps to alleviate the toxicity of metals via precipitation or dilution, or via adsorption of metals onto polyphosphate granules (Entry et al. 2002, Amir et al. 2013). Generally, uptake of P in nonmycorrhizal plants occurs by the direct pathway via epidermal P_i transporters, whereas mycorrhizal plants take up P via root epidermal cells and P transporters present in the hyphae of AMF over longer distances that may be 10 cm or more from the root surface. Studies have shown that essential elements such as K, P, and Mg are enriched in roots of *Glomus*-colonized maize plants grown in heavy metal-contaminated soils; however, the mycorrhiza-colonized plants contained lower heavy metal levels than noncolonized plants (Kaldorf et al. 1999, Hildebrandt et al. 2007). Chen et al. (2003) observed enhanced uptake of Zn by red clover plants inoculated with the AMF *Glomus mosseae* under Zn-deficient conditions.

8.6 Functions of Microbes in Cleanup of Metalliferous Soils

Use of plants with the ability to absorb high amounts of heavy metals in association with bacteria and/or in symbiosis with AMF can lead to successful reduction of heavy metal concentrations in soils. Microbes in the soil can mobilize heavy metals by release of chelating compounds, reducing soil pH, or changing the redox status and consequently altering the availability, solubility, and transport of heavy metals (Ma et al. 2011). Such abilities are very important for phytoextraction of heavy metals from the soil, which can contribute to establishment of a healthy and productive soil (Christie et al. 2004). Microbial siderophores play a significant role as metal chelators in phytoextraction of heavy metals. Braud et al. (2009) observed that the siderophore-producing bacteria *Pseudomonas aeruginosa*, *Pseudomonas fluorescens*, and *Ralstonia metallidurans* showed greater mobilization of Pb and Cr along with increased phytoextraction. Karimzadeh et al. (2012) reported that addition of desferrioxamine-B, a microbial siderophore naturally produced in the rhizosphere, considerably enhanced Cd sorption by roots and translocation to the aerial parts of plants. Jeong et al. (2014) investigated the effects of siderophores as microbial iron chelators on As uptake by the hyperaccumulating Cretan brake fern *Pteris cretica* L. during phytoextraction. Those authors found enhanced As uptake in siderophore-amended soils. In addition, translocation of As from roots to shoots was also mediated by siderophores. Colonization by AMF also improves As uptake by hyperaccumulating plants. For example, *Glomus mosseae*, which forms a stable association with Chinese brake fern (*Pteris vittata* L.), is resistant to As toxicity and increases plant biomass, leading to an increase in the quantity of As removed from the soil by the plant (Liu et al. 2005). Mechanisms other than siderophore production may be involved in phytoextraction of heavy metals. For example, the metal tolerant bacterium *Brevibacterium casei* MH8a, which was isolated from a heavy metal-contaminated rhizosphere soil of *Sinapis alba* L., was able to promote plant growth and enhance the Cd, Zn, and Cu uptake by white mustard (Płociniczak et al. 2016). This strain did not produce siderophores but was able to produce ACC deaminase and indole 3-acetic acid (IAA), which have beneficial effects on plant growth and may contribute to increased biomass production.

ACC deaminase-containing PGPR can improve plant performance under various stressful conditions (Glick 2014). ACC is the immediate precursor of ethylene, an important plant hormone whose levels increase under various environmental stresses, including heavy metal toxicity. Elevated ethylene levels can inhibit root elongation, lateral root growth, and root hair formation (Ma et al. 2003). However, ACC, which is exuded by plants, is cleaved by bacterial ACC deaminase to α-ketobutyrate and ammonia before it is converted to ethylene. The bacteria use the ammonia as a sole nitrogen source, thereby helping to decrease the ethylene levels and reduce the growth-inhibitory action (Gamalero et al. 2009, Ma et al. 2011). Burd et al. (1998) reported that a metal-resistant and ACC deaminase-producing *Kluyvera ascorbata* strain lowered Ni toxicity in canola seedlings. Safronova et al. (2006) observed that the inoculation of pea plants with ACC deaminase-containing PGPR enabled the plants to counteract Cd-induced inhibition of nutrient uptake in plants by stimulation of root growth and enhancement of nutrient active processes. Cheng et al. (2009) studied the mechanisms specific for heavy metal resistance of an ACC deaminase-producing *Pseudomonas putida* strain in comparison to nonproducing mutants by using a proteomics approach. Assessment of protein functions involved in bacterial heavy metal detoxification and with differential expression levels revealed several mechanisms, including general stress adaptation, antioxidative stress, and heavy metal efflux. Han et al. (2015) studied the role of ACC deaminase in a nitrogen-fixing *Pseudomonas stutzeri* strain capable of plant growth promotion under salt and heavy metal stress and observed that the enzyme was essential not only for enhancing the salt or heavy metal tolerance of bacteria but also for improving the growth of rice plants. Many phytoremediation-assisting microbes are capable of producing the plant hormone IAA, thereby influencing the proliferation of lateral and adventitious roots, stimulation of root exudation, and induction of synthesis of ACC synthase (Ma et al. 2013, 2016). Rajkumar and Freitas (2008) observed that *Ricinus communis* plants produced greater biomass when inoculated with the metal-resistant strain *Pseudomonas jessenii* (PjM15) in comparison to plants inoculated by *Pseudomonas* sp. (PsM6) in pot experiments in both contaminated and noncontaminated soils. The effect was attributed to production of higher amounts of IAA and a higher rate of phosphate solubilization by strain PjM15. However, activation of plant growth-promoting traits depends on environmental conditions, and these processes may be suppressed or delayed by high concentrations of heavy metals in soils. Marques et al. (2013) observed that the PGPR strains *Ralstonia eutropha* (B1) and *Chrysiobacterium humi* (B2) exhibited protective effects on metal-exposed *Helianthus annuus* plants via phytostabilization of Zn and Cd, but the bacteria did not promote growth or increase the biomass of the plants, regardless of whether the strains possessed PGPR traits (Marques et al. 2010). Therefore, growth promotion is not always related to PGP traits in bacteria with potentials for phytostabilization or phytoextraction of heavy metals (Becerra-Castro et al. 2012). Additionally, organic acids released by plant-associated microbes have been proposed to play important roles in heavy metal solubility, mobilization of mineral nutrients, and heavy metal detoxification in the soil. Organic acids, such as citrate, malate, gluconate, and oxalate, each with one or more carboxylic groups, can bind metal ions in soil solutions by complexation reactions and thus make the metal ions available for plants (Rajkumar et al. 2012, Ma et al. 2016).

AMF are actively involved in establishing plants in metal-contaminated soils and they also aid in nutrient uptake and in protecting plants from heavy metal toxicity. The capacity for metal sorption was found to be higher in AMF than plants colonized by other microorganisms (Joner et al. 2000). Metal binding studies conducted by Joner and his colleagues with different *Glomus* spp. revealed that adsorption of Cd was 3-fold higher in a metal-tolerant *Glomus mosseae* isolate than in other nontolerant fungi (Joner et al. 2000). Other studies showed that mycorrhiza provided protection to red clover plants by lowering Zn uptake or translocation of Zn to shoots when exposed to high Zn concentrations (Li and Christie 2001, Chen et al. 2003). These effects may be linked to changes in Zn solubility resulting from mycorrhiza-mediated alterations in soil pH, or the effects may be a consequence of binding and immobilization of Zn in extraradical mycelia (Li and Christie 2001). Yang et al. (2016) observed that the legume tree *Robinia pseudoacacia* inoculated with AMF, *Rhizophagus intraradices*, improved the phytoremediation efficiency of Pb-polluted soil in pot experiments. Those authors further reported that a higher efficiency of Pb removal was achieved by intercropping *R. pseudoacacia* with other selected legume herbs.

8.7 Conclusion and Future Perspectives

Increasing evidence suggests that rhizosphere microorganisms, such as PGPR and AMF, can significantly improve the phytoremediation potentials of plants in heavy metal-polluted soils. However, bioaugmentation with microbes in phytoremediation may not always produce desirable results. Plant growth-promoting characteristics of metal-resistant rhizobacteria may not be evident under conditions of excess metal contamination. Selection of plants for phytoremediation is also important when using an AMF–plant system, because excess heavy metals may influence the relationship between AMF and the host plant. Notwithstanding the abundance of promising results on microbe-assisted phytoremediation obtained under *in vitro* and greenhouse conditions, field studies that are focused on this topic are scarce. Since environmental factors play a significant role in phytoextraction or phytostabilization, more investigations should focus on applications of microbial bioinoculants in phytoremediation under field conditions.

Considerable progress has been made in characterization of microbes, host plants, and host–microbe associations. However, underlying mechanisms of host–microbe interactions, particularly the exchange of signals and the role of a polluted environment in influencing such signals between the host and microbe, are not clearly understood. More research should therefore focus on regulation and the influence of functional genes in host–microbe–metal interactions and also on the utilization of such knowledge to design specific bioinoculants which may be applicable in the field for cleanup of metalliferous soils.

Acknowledgment

Financial support from the University Grants Commission, India, to Shibu Das in the form of a fellowship (RGNF-SRF) is greatly appreciated.

References

Adediran, G.A., Ngwenya, B.T., Mosselmans, J.F.W., Heal, K.V., Harvie, B.A. 2015. Mechanisms behind bacteria induced plant growth promotion and Zn accumulation in *Brassica juncea*. *Journal of Hazardous Materials* 283:490–499.

Ahemad, M., Kibret, M. 2014. Mechanisms and applications of plant growth promoting rhizobacteria: Current perspective. *Journal of King Saud University Science* 26:1–20.

Ahmad, E., Zaidi, A., Khan, M., Oves, M. 2012. Heavy metal toxicity to symbiotic nitrogen-fixing microorganism and host legumes, pp. 29–44. *Toxicity of Heavy Metals to Legumes and Bioremediation*, Zaidi, A., Wani, P.A., Khan, M.S. (eds.). Vienna: Springer.

Alguacil, M.M., Torrecillas, E., Caravaca, F., Fernández, D.A., Azcón, R., Roldán, A. 2011. The application of an organic amendment modifies the arbuscular mycorrhizal fungal communities colonizing native seedlings grown in a heavy-metal-polluted soil. *Soil Biology and Biochemistry* 43:1498–1508.

Ali, H., Khan, E., Sajad, M.A. 2013. Phytoremediation of heavy metals: Concepts and applications. *Chemosphere* 91:869–881.

Ambrosini, A., Beneduzi, A., Stefanski, T., Pinheiro, F.G., Vargas, L.K., Passaglia, L.M.P. 2012. Screening of plant growth promoting rhizobacteria isolated from sunflower (*Helianthus annuus* L.). *Plant and Soil* 356:245–264.

Amir, H., Lagrange, A., Hassaïne, N., Cavaloc, Y. 2013. Arbuscular mycorrhizal fungi from New Caledonian ultramafic soils improve tolerance to nickel of endemic plant species. *Mycorrhiza* 23:585–595.

Andrade, S.A.L., Silveira, A.P.D., Mazzafera, P. 2010. Arbuscular mycorrhiza alters metal uptake and the physiological response of *Coffea arabica* seedlings to increasing Zn and Cu concentrations in soil. *Science of the Total Environment* 408:5381–5391.

Ashraf, U., Kanu, A.S., Mo, Z., Hussain, S., Anjum, S.A., Khan, I., Abbas, R.N., Tang, X. 2015. Lead toxicity in rice: Effects, mechanisms, and mitigation strategies. A minireview. *Environmental Science and Pollution Research International* 22:18318–18332.

Aslantas, R., Cakmakci, R., Sahin, F. 2007. Effect of plant growth promoting rhizobacteria on young apple tree growth and fruit yield under orchard conditions. *Scientia Horticulturae* 11:371–377.

Azcón, R., Perálvarez, M.D., Roldán, A., Barea, J.M. 2010. Arbuscular mycorrhizal fungi, *Bacillus cereus*, and *Candida parapsilosis* from a multicontaminated soil alleviate metal toxicity in plants. *Microbial Ecology* 59:668–677.

Babu, A.G., Reddy, M.S. 2011. Dual inoculation of arbuscular mycorrhizal and phosphate solubilizing fungi contributes in sustainable maintenance of plant health in fly ash ponds. *Water Air and Soil Pollution* 219:3–10.

Babu, A.G., Jong-Dae, K., Byung-Taek, O. 2013. Enhancement of heavy metal phytoremediation by *Alnus firma* with endophytic *Bacillus thuringiensis* GDB-1. *Journal of Hazardous Materials* 250–251:477–483.

Badri, D.V., Weir, T.L., van der Lelie, D., Vivanco, J.M. 2009. Rhizosphere chemical dialogues: Plant microbe interactions. *Current Opinion in Biotechnology* 20:642–650.

Banaaraghi, N., Hoodaji, M., Afyuni, M. 2010. Use of EDTA and EDDS for enhanced *Zea mays* phytoextraction of heavy metals from a contaminated soil. *Journal of Residuals Science and Technology* 7:139–145.

Barac, T., Taghavi, S., Borremans, B., Provoost, A., Oeyen, L., Colpaert, J.V., Vangronsveld, J., van der Lelie, D. 2004. Engineered endophytic bacteria improve phytoremediation of water-soluble, volatile, organic pollutants. *Nature Biotechnology* 22:583–588.

Becerra-Castro, C., Monterroso, C., Prieto-Fernández, A., Rodríguez-Lamas, L., Loureiro-Viñas, M., Acea, M.J., Kidd, P.S. 2012. Pseudometallophytes colonising Pb/Zn mine tailings: A description of the plant–microorganism–rhizosphere soil system and isolation of metal-tolerant bacteria. *Journal of Hazardous Materials* 217–218:350–359.

Bell, T.H., El-Din, H.S., Lauron-Moreau, A., Al-Otaibi, F., Hijri, M., Yergeau, E., St-Arnaud, M. 2014. Linkage between bacterial and fungal rhizosphere communities in hydrocarbon-contaminated soils is related to plant phylogeny. *ISME Journal* 8:331–343.

Berti, W.R., Cunningham, S.D. 2000. Phytostabilization of metals, pp. 71–88. *Phytoremediation of Toxic Metals: Using Plants to Clean-Up the Environment*, Raskin, I., Ensley, B.D. (eds.). New York: Wiley.

Bissonnette, L., St-Arnaud, M., Labrecque, M. 2010. Phytoextraction of heavy metals by two Salicaceae clones in symbiosis with arbuscular mycorrhizal fungi during the second year of a field trial. *Plant and Soil* 332:55–67.

Bloemberg, G.V., Lugtenberg, B.J. 2001. Molecular basis of plant growth promotion and biocontrol by rhizobacteria. *Current Opinion in Plant Biology* 4:343–350.

Bolan, N.S., Adriano, D.C., Curtin, D. 2003. Role of carbon, nitrogen, and sulfur cycles in soil acidification, pp. 29–56. *Handbook of Soil Acidity*, Z. Rengel (ed.). New York: Marcel Dekker.

Bolan, N.S., Park, J.H., Robinson, B., Naidu, R., Huh, K.Y. 2011. Phytostabilization: A green approach to contaminant containment. *Advances in Agronomy* 112:145–204.

Braud, A., Jézéquel, K., Bazot, S., Lebeau, T. 2009. Enhanced phytoextraction of an agricultural Cr-, Hg- and Pb-contaminated soil by bioaugmentation with siderophore producing bacteria. *Chemosphere* 74:280–286.

Brunetti, G., Farrag, K., Rovira, P.S. Nigro, F., Senesi, N. 2011. Greenhouse and field studies on Cr, Cu, Pb and Zn phytoextraction by *Brassica napus* from contaminated soils in the Apulia region, southern Italy. *Geoderma* 160:517–523.

Burd, G.I., Dixon, D.G., Glick, B.R. 1998. A plant growth promoting bacterium that decreases nickel toxicity in plant seedlings. *Applied and Environmental Microbiology* 64:3663–3668.

Burd, G.I., Dixon, D.G., Glick, B.R. 2000. Plant growth-promoting bacteria that decrease heavy metal toxicity in plants. *Canadian Journal of Microbiology* 46:237–245.

Cabello-Conejo, M.I., Becerra-Castro, C., Prieto-Fernández, A., Monterroso, C., Saavedra-Ferro, A., Mench, M., Kidd, P.S. 2014. Rhizobacterial inoculants can improve nickel phytoextraction by the hyperaccumulator *Alyssum pintodasilvae*. *Plant and Soil* 379:35–50.

Carrasco, J.A., Armario, P., Pajuelo, E., Burgos, A., Caviedes, M.A., Lopez, R., Chamber, M.A., Palomares, A.J. 2005. Isolation and characterization of symbiotically effective *Rhizobium* resistant to arsenic and heavy metals after the toxic spill at the Aznalcollar pyrite mine. *Soil Biology and Biochemistry* 37:1131–1140.

Castillo, O.S., Dasgupta-Schubert, N., Alvarado, C.J., Zaragoza, E.M., Villegas, H.J. 2011. The effect of the symbiosis between *Tagetes erecta* L. (marigold) and *Glomus intraradices* in the uptake of copper(II) and its implications for phytoremediation. *New Biotechnology* 29:156–164.

Chaparro, J.M., Sheflin, A.M., Manter, D.K., Vivanco, J.M. 2012. Manipulating the soil microbiome to increase soil health and plant fertility. *Biology and Fertility of Soils* 48:489–499.

Chen, B.D., Li, X.L., Tao, H.Q., Christie, P., Wong, M.H. 2003. The role of arbuscular mycorrhiza in zinc uptake by red clover growing in a calcareous soil spiked with various quantities of zinc. *Chemosphere* 50:839–846.

Chen, Z., Sheng, X., He, L., Huang, Z., Zhang, W. 2013. Effects of root inoculation with bacteria on the growth, Cd uptake and bacterial communities associated with rape grown in Cd-contaminated soil. *Journal of Hazardous Materials* 244–245:709–717.

Cheng, Z., Wei, Y.Y., Sung, W.W., Glick, B.R., McConkey, B.J. 2009. Proteomic analysis of the response of the plant growth-promoting bacterium *Pseudomonas putida* UW4 to nickel stress. *Proteome Science* 7:18–25.

Christie, P., Li, X., Chen, B. 2004. Arbuscular mycorrhiza can depress translocation of zinc to shoots of host plants in soils moderately polluted with zinc. *Plant and Soil* 261:209–217.

Cicatelli, A., Lingua, G., Todeschini, V., Biondi, S., Torrigiani, P., Castiglione, S. 2010. Arbuscular mycorrhizal fungi restore normal growth in a white poplar clone grown on heavy metal-contaminated soil, and this is associated with upregulation of foliar metallothionein and polyamine biosynthetic gene expression. *Annals of Botany* 106:791–802.

Clark, R.B., Zeto, S.K. 2000. Mineral acquisition by arbuscular mycorrhizal plants. *Journal of Plant Nutrition* 23:867–902.

Dary, M., Chamber, M.A., Palomares, A.J., Pajuelo, E. 2010. "*In situ*" phytostabilisation of heavy metal polluted soils using *Lupinus luteus* inoculated with metal resistant plant-growth promoting rhizobacteria. *Journal of Hazardous Materials* 177:323–330.

Davison, J. 2005. Risk mitigation of genetically modified bacteria and plants designed for bioremediation. *Journal of Industrial Microbiology and Biotechnology* 32:639–650.

De Vries, W., Kros, J., Kroeze, C., Seitzinger, S.P. 2013. Assessing planetary and regional nitrogen boundaries related to food security and adverse environmental impacts. *Current Opinion in Environmental Sustainability* 5:392–402.

Dimkpa, C.O., Merten, D., Svatoš, A., Büchel, G., Kothe, E. 2008. Hydroxamate siderophores produced by *Streptomyces acidiscabies* E13 bind nickel and promote growth in cowpea (*Vigna unguiculata* L.) under nickel stress. *Canadian Journal of Microbiology* 54:163–172.

Dimkpa, C.O., Merten, D., Svatoš, A., Büchel, G., Kothe, E. 2009. Metal-induced oxidative stress impacting plant growth in contaminated soil is alleviated by microbial siderophores. *Soil Biology and Biochemistry* 41:154–162.

Entry, J.A., Rygiewicz, P.T., Watrud, L.S., Donnelly, P.K. 2002. Influence of adverse soil conditions on the formation and function of arbuscular mycorrhizas. *Advances in Environmental Research* 7:123–138.

Fang, S., Liu, D., Tian, Y., Deng, S., Shang, X. 2013. Tree species composition influences enzyme activities and microbial biomass in the rhizosphere: A rhizobox approach. *PLoS One* 8:e61461.

Faoro, H., Alves, A.C., Souza, E.M., Rigo, E.U., Cruz, L.M.,Al-Janabi, S.M., Monteiro, R.A., Baura, V.A., Pedrosa, F.O. 2010. Influence of soil characteristics on the diversity of bacteria in the southern Brazilian Atlantic forest. *Applied and Environmental Microbiology* 76:4744–4749.

Fatnassi, I.C., Chiboub, M., Saadani, O., Jebara, M., Jebara, S.H. 2015. Phytostabilization of moderate copper contaminated soils using co-inoculation of *Vicia faba* with plant growth promoting bacteria. *Journal of Basic Microbiology* 55:303–311.

Forchetti, G., Masciarelli, O., Alemano, S., Alvarez, D., Abdala, G. 2007. Endophytic bacteria in sunflower (*Helianthus annuus* L.): Isolation, characterization, and production of jasmonates and abscisic acid in culture medium. *Applied Microbiology and Biotechnology* 76:1145–1152.

Gamalero, E., Lingua, G., Berta, G., Glick, B.R. 2009. Beneficial role of plant growth promoting bacteria and arbuscular mycorrhizal fungi on plant responses to heavy metal stress. *Canadian Journal of Microbiology* 55:501–504.

Gao, Y., Li, Q., Ling, W., Zhu, X. 2011. Arbuscular mycorrhizal phytoremediation of soils contaminated with phenanthrene and pyrene. *Journal of Hazardous Materials* 185:703–709.

Gaonkar, T., Bhosle, S. 2013. Effect of metals on a siderophore producing bacterial isolate and its implications on microbial assisted bioremediation of metal contaminated soils. *Chemosphere* 93:1835–1843.

Gaur, A., Adholeya, A. 2004. Prospects of arbuscular mycorrhizal fungi in phytoremediation of heavy metal contaminated soils. *Current Science* 86:528–534.

Gerhardt, K.E., Huang, X.-D., Glick, B.R., Greenberg, B.M. 2009. Phytoremediation and rhizoremediation of organic soil contaminants: Potential and challenges. *Plant Science* 176:20–30.

Glick, B.R. 2005. Modulation of plant ethylene levels by the bacterial enzyme ACC deaminase. *FEMS Microbiology Letters* 251:1–7.

Glick, B.R. 2010. Using soil bacteria to facilitate phytoremediation. *Biotechnology Advances* 28:367–374.

Glick, B.R. 2014. Bacteria with ACC deaminase can promote plant growth and help to feed the world. *Microbiological Research* 169:30–39.

Gohre, V., Paszkowski, U. 2006. Contribution of arbuscular mycorrhizal symbiosis to heavy metal phytoremediation. *Planta* 223:1115–1122.

Gomez-Balderas, C.D.C., Cochet, N., Bert, V., Tarnaud, E., Sarde, C.O. 2014. 16S rDNA analysis of bacterial communities associated with the hyperaccumulator *Arabidopsis halleri* grown on a Zn and Cd polluted soil. *European Journal of Soil Biology* 60:16–23.

Graeme, K.A., Pollack, C.V. 1998. Heavy metal toxicity. Part I: Arsenic and mercury. *Journal of Emergency Medicine* 16:45–56.

Gray, E.J., Smith, D.L. 2005. Intracellular and extracellular PGPR: Commonalities and distinctions in the plant–bacterium signaling processes. *Soil Biology and Biochemistry* 37:395–412.

Guo, W., Zhao, R., Zhao, W., Fu, R., Guo, J., Bi, N., Zhang, J. 2013. Effects of arbuscular mycorrhizal fungi on maize (*Zea mays* L.) and sorghum (*Sorghum bicolor* L. Moench) grown in rare earth elements of mine tailings. *Applied Soil Ecology* 72:85–92.

Gupta, G., Parihar, S.S., Ahirwar, N.K., Snehi, S.K., Singh, V. 2015 Plant growth promoting rhizobacteria (PGPR): Current and future prospects for development of sustainable agriculture. *Journal of Microbial and Biochemical Technology* 7:96–102.

Gyaneshwar, P., Kumar, G.N., Parekh, L.J., Poole, P.S. 2002. Role of soil microorganisms in improving P nutrition of plants. *Plant and Soil* 245:83–93.

Haichar, F.Z., Santaella, C., Heulin, T., Achouak, W. 2014. Root exudates mediated interactions belowground. *Soil Biology and Biochemistry* 77:69–80.

Han, Y., Wang, R., Yang, Z., Zhan, Y., Ma, Y., Ping, S., Zhang, L., Lin, M., Yan, Y. 2015. 1-Aminocyclopropane-1-carboxylate deaminase from *Pseudomonas stutzeri* A1501 facilitates the growth of rice in the presence of salt or heavy metals. *J. Microbiology and Biotechnology* 25:1119–1128.

Hansen, D., Duda, P.J., Zayed, A., Terry, N. 1998. Selenium removal by constructed wetlands: Role of biological volatilization. *Environmental Science and Technology* 32:591–597.

Hao, X., Xie, P., Zhu, Y.G., Taghavi, S., Wei, G., Rensing, C. 2015. Copper tolerance mechanisms of *Mesorhizobium amorphae* and its role in aiding phytostabilization by *Robinia pseudoacacia* in copper contaminated soil. *Environmental Science and Technology* 49:2328–2340.

Hassan, S.D., Bell, T.H., Stefani, F.O., Denis, D., Hijri, M., St-Arnaud, M. 2014. Contrasting the community structure of arbuscular mycorrhizal fungi from hydrocarbon-contaminated and uncontaminated soils following willow (*Salix* spp. L.) planting. *PLoS One* 9:e102838.

Hildebrandt, U., Regvar, M., Bothe, H. 2007. Arbuscular mycorrhiza and heavy metal tolerance. *Phytochemistry* 68:139–146.

Imam, J., Singh, P.K., Shukla, P. 2016. Plant microbe interactions in post genomic era: Perspectives and applications. *Frontiers in Microbiology* 7:1488.

Järup, L. 2003. Hazards of heavy metal contamination. *British Medical Bulletin* 68:167–182.

Jeffries, P., Gianinazzi, S., Perotto, S., Turnau, K., Barea, H.M. 2003. The contribution of arbuscular mycorrhizal fungi in sustainable maintenance of plant health and soil fertility. *Biology and Fertility of Soils* 37:1–16.

Jeong, S., Moon, H.S., Nam, K. 2014. Enhanced uptake and translocation of arsenic in cretan brake fern (*Pteris cretica* L.) through siderophore-arsenic complex formation with an aid of rhizospheric bacterial activity. *Journal of Hazardous Materials* 280:536–543.

Joner, E.J., Briones, R., Leyval, C. 2000. Metal-binding capacity of arbuscular mycorrhizal mycelium. *Plant and Soil* 226:227–234.

Kaldorf, M., Kuhn, A.J., Schröder, W.H., Hildebrandt, U., Bothe, H. 1999. Selective element deposits in maize colonized by a heavy metal tolerance conferring arbuscular mycorrhizal fungus. *Journal of Plant Physiology* 154:718–728.

Kamran, M.A., Eqani, S.A., Bibi, S., Xu, R.K., Amna, Monis, M.F, Katsoyiannis, A., Bokhari, H., Chaudhary, H.J. 2016. Bioaccumulation of nickel by *E. sativa* and role of plant growth promoting rhizobacteria (PGPRs) under nickel stress. *Ecotoxicology and Environmental Safety* 126:256–263.

Karimzadeh, L., Heilmeier, H., Merkel, B.J. 2012. Effect of microbial siderophore DFO-B on Cd accumulation by *Thlaspi caerulescens* hyperaccumulator in the presence of zeolite. *Chemosphere* 88:683–687.

Khan, A.A., Jilani, G., Akhtar, M.S., Naqvi, S.M.S., Rasheed, M. 2009. Phosphorus solubilizing bacteria: Occurrence, mechanisms and their role in crop production. *Journal of Agricultural and Biological Science* 1:48–58.

Khan, A.G. 2006. Mycorrhizoremediation: An enhanced form of phytoremediation. *Journal of Zhejiang University Science B* 7:503–514.

Khan, A.G., Kuek, C., Chaudhry, T.M., Khoo, C.S., Hayes, W.J. 2000. Role of plants, mycorrhizae and phytochelators in heavy metal contaminated land remediation. *Chemosphere* 41:197–207.

Khan, M.U., Sessitsch, A., Harris, M., Fatima, K., Imran, A., Arslan, M., Shabir, G., Khan, Q.M., Afzal, M. 2015. Cr-resistant rhizo- and endophytic bacteria associated with *Prosopis juliflora* and their potential as phytoremediation enhancing agents in metal-degraded soils. *Frontiers in Plant Science* 5:1–10.

Koltai, H. 2013. Strigolactones activate different hormonal pathways for regulation of root development in response to phosphate growth conditions. *Annals of Botany* 112:409–415.

Lampis, S., Santi, C., Ciurli, A., Andreolli, M., Vallini, G. 2015. Promotion of arsenic phytoextraction efficiency in the fern *Pteris vittata* by the inoculation of As-resistant bacteria: As oil bioremediation perspective. *Frontiers in Plant Science* 6:1–12.

Langella, F., Grawunder, A., Stark, R., Weist, A., Merten, D., Haferburg, G., Büchel, G., Kothe, E. 2013. Microbially assisted phytoremediation approaches for two multi-element contaminated sites. *Environmental Science and Pollution Research International* 21:6845–6858.

Li, K., Ramakrishna, W. 2011. Effect of multiple metal resistant bacteria from contaminated lake sediments on metal accumulation and plant growth. *Journal of Hazardous Materials* 189:531–539.

Li, W.C., Ye, Z.H., Wong, M.H. 2010. Metal mobilization and production of short-chain organic acids by rhizosphere bacteria associated with a Cd/Zn hyperaccumulating plant, *Sedum alfredii*. *Plant and Soil* 326:453–467.

Li, X., Christie, P. 2001. Changes in soil solution Zn and pH and uptake of Zn by arbuscular mycorrhizal red clover in Zn-contaminated soil. *Chemosphere* 42:201–207.

Liu, L.Z., Gong, Z.Q., Zhang, Y.L., Li, P.J. 2011. Growth, cadmium accumulation and physiology of marigold (*Tagetes erecta* L.) as affected by arbuscular mycorrhizal fungi. *Pedosphere* 21:319–327.

Liu, X.M., Wu, Q.T., Banks, M.K. 2005 Effect of simultaneous establishment of *Sedum alfridii* and *Zea mays* on heavy metal accumulation in plants. *International Journal of Phytoremediation* 7:43–53.

Lombi, E., Zhao, F.J., Dunham, S.J., McGrath, S.P. 2001. Phytoremediation of heavy metal-contaminated soils: Natural hyperaccumulation versus chemically enhanced phytoextraction. *Journal of Environmental Quality* 30:1919–1926.

Long, L.K., Yao, Q., Guo, J., Yang, R.H., Huang, Y.H., Zhu, H.H. 2010. Molecular community analysis of arbuscular mycorrhizal fungi associated with five selected plant species from heavy metal polluted soils. *European Journal of Soil Biology* 46:288–294.

Lukasz, D., Liwia, R., Aleksandra, M., Aleksandra, S. 2014. Dissolution of arsenic minerals mediated by dissimilatory arsenate reducing bacteria: Estimation of the physiological potential for arsenic mobilization. *BioMed Research International* 2014:841892.

Luo, S., Chen, L., Chen, J., Xiao, X., Xu, T., Wan, Y., Rao, C., Liu, C., Liu, Y., Lai, C., Zeng, G. 2011. Analysis and characterization of cultivable heavy metal-resistant bacterial endophytes isolated from Cd-hyperaccumulator *Solanum nigrum* L. and their potential use for phytoremediation. *Chemosphere* 85:1130–1138.

Ma, Y., Prasad, M.N.V., Rajkumar, M., Freitas, H. 2011. Plant growth promoting rhizobacteria and endophytes accelerate phytoremediation of metalliferous soils. *Biotechnology Advances* 29:248–258.

Ma, Y., Rajkumar, M., Luo, Y., Freitas, H. 2013. Phytoextraction of heavy metal polluted soils using *Sedum plumbizincicola* inoculated with metal mobilizing *Phyllobacterium myrsinacearum* RC6b. *Chemosphere* 93:1386–1392.

Ma, Y., Oliveira, R.S., Nai, F., Rajkumar, M., Luo, Y., Rocha, I., Freitas, H. 2015. The hyperaccumulator *Sedum plumbizincicola* harbors metal-resistant endophytic bacteria that improve its phytoextraction capacity in multi-metal contaminated soil. *Journal of Environmental Management* 156:62–69.

Ma, Y., Rui, S.O., Helena, F., Chang, Z. 2016. Biochemical and molecular mechanisms of plant-microbe-metal interactions: Relevance for phytoremediation. *Frontiers in Plant Science* 7:918.

Ma, Z., Baskin, T.I., Brown, K.M., Lynch, J.P. 2003. Regulation of root elongation under phosphorus stress involves changes in ethylene responsiveness. *Plant Physiology* 131:1381–1390.

Mahar, A., Wang, P., Ali, A., Awasthi, M.K., Lahori, A.H., Wang, Q., Li, R., Zhang, Z. 2016. Challenges and opportunities in the phytoremediation of heavy metals contaminated soils: A review. *Ecotoxicology and Environmental Safety* 126:111–121.

Mantelin, S., Touraine, B. 2004. Plant growth-promoting bacteria and nitrate availability: Impacts on root development and nitrate uptake. *Journal of Experimental Botany* 55:27–34.

Marques, A.P.G.C., Pires, C., Moreira, H., Rangel, A.O.S.S., Castro, P.M.L. 2010. Assessment of the plant growth promotion abilities of six bacterial species using *Zea mays* as indicator plant. *Soil Biology and Biochemistry* 42:1229–1235.

Marques, A.P., Moreira, H., Franco, A.R., Rangel, A.O., Castro, P.M. 2013. Inoculating *Helianthus annuus* (sunflower) grown in zinc and cadmium contaminated soils with plant growth promoting bacteria: Effects on phytoremediation strategies. *Chemosphere* 92:74–83.

McCluggage, D. 1991. Heavy metal poisoning. *NCS Magazine*. Columbus, OH: Bird Hospital.

Mendes, R., Garbeva, P., Raaijmakers, J.M. 2013. The rhizosphere microbiome: Significance of plant beneficial, plant pathogenic, and human pathogenic microorganisms. *FEMS Microbiology Review* 37:634–663.

Mesa, J., Mateos-Naranjo, E., Caviedes, M.A., Redondo-Gómez, S., Pajuelo, E., Rodríguez-Llorente, I.D. 2015. Scouting contaminated estuaries: Heavy metal resistant and plant growth promoting rhizobacteria in the native metal rhizoaccumulator *Spartina maritime*. *Marine Pollution Bulletin* 90:150–159.

Mesa, V., Navazas, A., González-Gil, R., González, A., Weyens, N., Lauga, B., Gallego, J.L.R., Sánchez, J., Peláez, A.I. 2017. Use of endophytic and rhizosphere bacteria to improve phytoremediation of arsenic-contaminated industrial soils by autochthonous *Betula celtiberica*. *Applied and Environmental Microbiology* 83:e03411-16.

Miransari, M. 2010. Contribution of arbuscular mycorrhizal symbiosis to plant growth under different types of soil stress. *Plant Biology* 12:563–569.

Mohammadi, K. 2012. Phosphorus solubilizing bacteria: Occurrence, mechanisms and their role in crop production. *Resources and Environment* 2:80–85.

Molina-Favero, C., Creus, C.M., Simontacchi, M., Puntarulo, S., Lamattina, L. 2008. Aerobic nitric oxide production by *Azospirillum brasilense* Sp245 and its influence on root architecture in tomato. *Molecular Plant-Microbe Interactions* 21:1001–1009.

Muehe, E.M., Weigold, P., Adaktylou, I.J., Planer-Friedrich, B., Kraemer, U., Kappler, A., Behrens, S. 2015. Rhizosphere microbial community composition affects cadmium and zinc uptake by the metal-hyperaccumulating plant *Arabidopsis halleri*. *Applied and Environmental Microbiology* 81:2173–2181.

Neal, A.L., Ahmad, S., Gordon-Weeks, R., Ton, J. 2012. Benzoxazinoids in root exudates of maize attract *Pseudomonas putida* to the rhizosphere. *PLoS One* 7:e35498.

Nelson, L.M. 2004. Plant growth promoting rhizobacteria (PGPR): Prospects for new inoculants. *Crop Management* 3. doi:10.1094/CM-2004-0301-05-RV.

Nihorimbere, V., Ongena, M., Smargiassi, M., Thonart, P. 2011. Beneficial effect of the rhizosphere microbial community for plant growth and health. *Biotechnology Agronomy Society and Environment* 15:327–337.

Orlowska, E., Godzik, B., Turnau, K. 2012. Effect of different arbuscular mycorrhizal fungal isolates on growth and arsenic accumulation in *Plantago lanceolata* L. *Environmental Pollution* 168:121–130.

Padmavathiamma, P.K., Li, L.Y. 2007. Phytoremediation technology: Hyperaccumulation metals in plants. *Water Air and Soil Pollution* 184:105–126.

Perrig, D., Boiero, M.L., Masciarelli, O.A., Penna, C., Ruiz, O.A., Cassán, F.D., Luna, M.V. 2007. Plant-growth promoting compounds produced by two agronomically important strains of *Azospirillum brasilense*, and implications for inoculant formulation. *Applied Microbiology and Biotechnology* 75:1143–1150.

Pilon-Smits, E.A.H., Hwang, S., Lyle, C., Zhu, Y., Tai, J.C., Bravo, R.C., Chen, Y., Leustek, T., Terry, N. 1999. Overexpression of ATP sulfurylase in Indian mustard to increased selenate uptake, reduction, and tolerance. *Plant Physiology* 119:123–132.

Płociniczak, T., Sinkkonen, A., Romantschuk, M., Sułowicz, S., Piotrowska-Seget, Z. 2016. Rhizospheric bacterial strain *Brevibacterium casei* MH8a colonizes plant tissues and enhances Cd, Zn, Cu phytoextraction by white mustard. *Frontiers in Plant Science* 7:101.

Prasad, A., Kumar, S., Khaliq, A., Pandey, A. 2011. Heavy metals and arbuscular mycorrhizal (AM) fungi can alter the yield and chemical composition of volatile oil of sweet basil (*Ocimum basilicum* L.). *Biology and Fertility of Soils* 47:853–861.

Prasad, M.N.V. 2003. Metal hyperaccumulators in plants: Biodiversity prospecting for phytoremediation technology. *Electronic Journal of Biotechnology* 6:276–372.

Prasad, M.N.V. 2004. Phytoremediation of metals in the environment for sustainable development. *Proceedings of the Indian National Science Academy* 70:71–98.

Rajkumar, M., Freitas, H. 2008. Influence of metal resistant-plant growth-promoting bacteria on the growth of *Ricinus communis* in soil contaminated with heavy metals. *Chemosphere* 71:834–842.

Rajkumar, M., Sandhya, S., Prasad, M.N.V., Freitas, H. 2012. Perspectives of plant-associated microbes in heavy metal phytoremediation. *Biotechnology Advances* 30:1562–1574.

Raskin, I., Kumar, P.B.A.N., Dushenkov, S., Salt, D. 1994. Bioconcentration of heavy metals by plants. *Current Opinion in Biotechnology* 5:285–290.

Rousk, J., Baath, E., Brookes, P.C., Lauber, C.L., Lozupone, C., Caporaso, J.G., Knight, R., Fierer, N. 2010. Soil bacterial and fungal communities across a pH gradient in an arable soil. *ISME Journal* 4:1340–1351.

Rugh, C.L., Wilde, H.D., Stack, N.M., Thompson, D.M., Summers, A.O., Meagher, R.B. 1996. Mercuric ion reduction and resistance in transgenic *Arabidopsis thaliana* plants expressing a modified bacterial *merA* gene. *Proceedings of the National Academy of Sciences USA* 93:3182–3187.

Ryu, C.M., Hu, C.H., Locy, R.D., Kloepper, J.W. 2005. Study of mechanisms for plant growth promotion elicited by rhizobacteria in *Arabidopsis thaliana*. *Plant and Soil* 268:285–292.

Sachdev, D.P., Cameotra, S.S. 2013. Biosurfactants in agriculture. *Applied Microbiology and Biotechnology* 97:1005–1016.

Safronova, V.I., Stepanok, V.V., Engqvist, G.L., Alekseyev, Y.V., Belimov, A.A. 2006. Root-associated bacteria containing 1-aminocyclopropane-1-carboxylate deaminase improve growth and nutrient uptake by pea genotypes cultivated in cadmium supplemented soil. *Biology and Fertility of Soils* 42:267–272.

Saha, M., Sarkar, S., Sarkar, B., Sharma, B.K., Bhattacharjee, S., Tribedi, P. 2016. Microbial siderophores and their potential applications: A review. *Environmental Science and Pollution Research International* 23:3984–3999.

Saifullah, Meers, E., Qadir, M., de Caritat, P., Tack, F.M., Du Laing, G., Zia, M.H. 2009. EDTA-assisted Pb phytoextraction. *Chemosphere* 74:1279–1291.

Schalk, I.J., Hannauer, M., Braud, A. 2011. New roles for bacterial siderophores in metal transport and tolerance. *Environmental Microbiology* 13:2844–2854.

Schützendübel, A., Polle, A. 2002. Plant responses to abiotic stresses: Heavy metal-induced oxidative stress and protection by mycorrhization. *Journal of Experimental Botany* 53:1351–1365.

Segura, A., Rodríguez-Conde, S., Ramos, C., Ramos, J.L. 2009. Bacterial responses and interactions with plants during rhizoremediation. *Microbial Biotechnology* 2:452–464.

Sessitsch, A., Kuffner, M., Kidd, P., Vangronsveld, J., Wenzel, W.W., Fallmann, K., Puschenreiter, M. 2013. The role of plant-associated bacteria in the mobilization and phytoextraction of trace elements in contaminated soils. *Soil Biology and Biochemistry* 60:182–194.

Shahabivand, S., Maivan, H.Z., Goltapeh, E.M., Sharifi, M., Aliloo, A.A. 2012. The effects of root endophyte and arbuscular mycorrhizal fungi on growth and cadmium accumulation in wheat under cadmium toxicity. *Plant Physiology and Biochemistry* 60:53–58.

Sharma, S.B., Sayyed, R.Z., Trivedi, M.H., Gobi, T.A. 2013. Phosphate solubilizing microbes: Sustainable approach for managing phosphorus deficiency in agricultural soils. *Springerplus* 2:587–600.

Sheng, X., Sun, L., Huang, Z.., He, L., Zhang, W., Chen, Z. 2012. Promotion of growth and Cu accumulation of bio-energy crop (Zea mays) by bacteria: Implications for energy plant biomass production and phytoremediation. *Journal of Environmental Management* 103:58–64.

Siciliano, S.D., Fortin, N., Mihoc, A., Wisse, G., Labelle, S., Beaumier, D., Ouellette, D., Roy, R., Whyte, L.G., Banks, M.K., Schwab, P., Lee, K., Greer, C.W. 2001. Selection of specific endophytic bacterial genotypes by plants in response to soil contamination. *Applied and Environmental Microbiology* 67:2469–2475.

Siedlecka, A. 1995. Some aspects of interactions between heavy metals and plant mineral nutrients. *Acta Societatis Botanicorum Poloniae* 64:265–272.

Singh, A., Prasad, S.M. 2011. Reduction of heavy metal load in food chain: Technology assessment. *Reviews in Environmental Science and Bio/Technology* 10:199–214.

Steinkellner, S., Lendzemo, V., Langer, I., Schweiger, P., Khaosaad, T., Toussaint, J.P., Vierheilig, H. 2007. Flavonoids and strigolactones in root exudates as signals in symbiotic and pathogenic plant-fungus interactions. *Molecules* 12:1290–1306.

Subramanian, K.S., Charest, C. 1999. Acquisition of N by external hyphae of an arbuscular mycorrhizal fungus and its impact on physiological responses in maize under drought-stressed and well-watered conditions. *Mycorrhiza* 9:69–75.

Tak, H. I., Ahmad, F., Babalola, O.O. 2013. Advances in the application of plant growth-promoting rhizobacteria in phytoremediation of heavy metals. *Reviews of Environmental Contamination and Toxicology* 223:33–52.

Tangahu, B.V., Abdullah, S.R.S., Basri, H., Idris, M., Anuar, N., Mukhlisin, M. 2011. A review on heavy metals (As, Pb, and Hg) uptake by plants through phytoremediation. *International Journal of Chemical Engineering* 2011:939161.

Thijs, S., Sillen, W., Rineau, F., Weyens, N., Vangronsveld, J. 2016. Towards an enhanced understanding of plant–microbiome interactions to improve phytoremediation: Engineering the metaorganism. *Frontiers in Microbiology* 7:341.

Tokala, R.K., Strap, J.L., Jung, C.M., Crawford, D.L., Salove, M.H., Deobald, L.A., Bailey, J.F., Morra, M.J. 2002. Novel plant-microbe rhizosphere interaction involving *Streptomyces lydicus* WYEC108 and the pea plant (*Pisum sativum*). *Applied and Environmental Microbiology* 68:2161–2171.

Truyens, S., Jambon, I., Croes, S., Janssen, J., Weyens, N., Mench, M., Carleer, R., Cuypers, A., Vangronsveld, J. 2014. The effect of long-term Cd and Ni exposure on seed endophytes of *Agrostis capillaris* and their potential application in phytoremediation of metal-contaminated soils. *International Journal of Phytoremediation* 16:643–659.

Turner, T.R., Ramakrishnan, K., Walshaw, J., Heavens, D., Alston, M., Swarbreck, D., Osbourn, A., Grant, A., Poole, P.S. 2013. Comparative metatranscriptomics reveals kingdom level changes in the rhizosphere microbiome of plants. *ISME Journal* 7:2248–2258.

Ullah, A., Heng, S., Munis, M.F.H., Fahad, S., Yang, X. 2015. Phytoremediation of heavy metals assisted by plant growth promoting (PGP) bacteria: A review. *Environmental and Experimental Botany* 117:28–40.

Ullah, H., Khan, I., Ullah, I. 2012. Impact of sewage contaminated water on soil, vegetables, and underground water of peri-urban Peshawar, Pakistan. *Environmental Monitoring and Assessment* 184:6411–6421.

Verma, J.P., Yadav, J., Tiwari, K.N., Lavakush, Singh, V. 2010. Impact of plant growth promoting rhizobacteria on crop production. *International Journal of Agricultural Research* 5:954–983.

Vogel-Mikus, K., Pongrac, P., Kump, P., Necemer, M., Regvar, M. 2006. Colonisation of a Zn, Cd and Pb hyperaccumulator *Thlaspi praecox* Wulfen with indigenous arbuscular mycorrhizal fungal mixture induces changes in heavy metal and nutrient uptake. *Environmental Pollution* 139:362–371.

Wenzel, W. W., E. Lombi, E., Adriano, D.C. 2004. Root and rhizosphere processes in metal hyperaccumulation and phytoremediation technology, pp. 313–344. *Heavy Metal Stress in Plants: From Biomolecules to Ecosystems*, Prassad, M.N.V. (ed.). Berlin: Springer-Verlag.

Wu, S.C., Cheung, K.C., Luo, Y.M., Wong, M.H. 2006. Effects of inoculation of plant growth-promoting rhizobacteria on metal uptake by *Brassica juncea*. *Environmental Pollution* 140:124–135.

Wuana, R.A., Okieimen, F.E. 2011. Heavy metals in contaminated soils: A review of sources, chemistry, risks and best available strategies for remediation. *Communications in Soil Science and Plant Analysis* 42:111–122.

Yang, Y., Liang, Y., Han, X., Chiu, T.-Y., Ghosh, A., Chen, H., Tang, M. 2016. The roles of arbuscular mycorrhizal fungi (AMF) in phytoremediation and tree-herb interactions in Pb contaminated soil. *Scientific Reports* 6:20469.

Yoneyama, K., Yoneyama, K., Takeuchi, Y., Sekimoto, H. 2007. Phosphorus deficiency in red clover promotes exudation of orobanchol, the signal for mycorrhizal symbionts and germination stimulant for root parasites. *Planta* 225:1031–1038.

Zaidi, S., Usmani, S., Singh, B.R., Musarrat, J. 2006. Significance of *Bacillus subtilis* strain SJ-101 as a bioinoculant for concurrent plant growth promotion and nickel accumulation in *Brassica juncea*. *Chemosphere* 64:991–997.

Zhang, W., Huang, Z., He, L., Sheng, X. 2012. Assessment of bacterial communities and characterization of lead-resistant bacteria in the rhizosphere soils of metal-tolerant *Chenopodium ambrosioides* grown on lead–zinc mine tailings. *Chemosphere* 87:1171–1178.

Zhang, Y., He, L., Chen, Z., Wang, Q., Qian, M., Sheng, X. 2011. Characterization of ACC deaminase-producing endophytic bacteria isolated from copper-tolerant plants and their potential in promoting the growth and copper accumulation of *Brassica napus*. *Chemosphere* 83:57–62.

Zheng, L.-J., Liu, X.-M., Lütz-Meindl, U., Peer, T., Zheng, L.-J. 2011. Effects of lead and EDTA-assisted lead on biomass, lead uptake and mineral nutrients in *Lespedeza chinensis* and *Lespedeza davidii*. *Water Air and Soil Pollution* 220:57–68.

Zhuang, X., Chen, J., Shim, H., Bai, Z. 2007. New advances in plant growth-promoting rhizobacteria for bioremediation. *Environment International* 33:406–413.

9

Phosphate-Solubilizing Bacteria as Plant Growth Promoters and Accelerators of Phytoremediation

Munees Ahemad and Jawed Iqbal

CONTENTS

9.1 Introduction

Phosphorous (P) is one of the most vital macronutrients, as it is used to sustain growth and development of both plants and soil microorganisms and plays diverse regulatory, structural, and energy transfer roles, from photosynthesis, respiration, and signal transduction to nucleic acids biochemistry, etc. (Yadav et al. 2010). Regardless of its abundance in soils, it is scarcely available for plants because of the insolubility of bulk fractions of both inorganic forms (insoluble mineral complexes, e.g., hydroxyapatite, rock phosphate) and organic forms (which represent 20–80% of soil P, e.g., phytate, phosphoesters) in soil reservoirs. Because plants absorb P only in soluble forms, such as HPO_4^{2-} and $H_2PO_4^{-}$, treatment with phosphatic fertilizers of P-depleted soils is very common in agriculture for the purpose of accelerating plant growth and productivity (Ahemad et al. 2009). However, more than 90% of applied fertilizer P accumulates and is fixed in soils in insoluble forms which cannot be utilized by plants and gradually percolates into deep soil to further contaminate ground water (Lara et al. 2013). In addition, indiscriminate use of these agrochemicals not only affects the soil microflora but also decreases their beneficial activities, as well as soil fertility (Cheng-Hsiung and Shang-Shyng 2009, Ahemad et al. 2012). In this regard, exploitation of phosphate-solubilizing bacteria (PSB), in order to solubilize a large reservoir of P and to improve the chemical, physical, and biological properties of soils in an eco-friendly and sustainable manner, is a better choice among the different possible approaches to overcome P deficiency in soils (Vyas and Gulati 2009, Sahay and Patra 2013). PSB as bioinoculants facilitate the hydrolysis of several kinds of P compounds of inorganic and organic natures and lead to greater plant productivity with a concomitant reduction of chemical hazards to agro-ecosystems (J.F. Li et al. 2013, Zhao et al. 2014).

Recent research findings regarding PSB have revealed new physiological traits implicated in plant growth promotion under various agronomical and environmental stresses, such as heavy metal contamination (Ahemad et al. 2009, Yuan et al. 2013), pesticides (Ahemad et al. 2009, Rajasankar et al. 2013, Kryuchkova et al. 2014), osmotic stress (Chakraborty et al. 2013), phytopathogens (Zhao et al. 2014), temperature (Cheng et al. 2007, Cheng-Hsiung and Shang-Shyng 2009), and salinity (Sahay and Patra 2013). Thus, the roles of PSB are not limited to phosphate mobilization in soils and plants. Their functionally versatile nature that allows them to flourish under diverse soil conditions, vis-à-vis possession of plant- and soil-beneficial physiological traits, makes them a better alternative than existing approaches of plant production in P-deficient and stressed soils. This review presents recent updates regarding multifunctional PSB of environmental and agronomical significance and presents key aspects of their implications for growth promotion and acceleration of phytoremediation in the different plant species.

9.2 Sources of Phosphorus in the Environment

Phosphorus is present on the Earth in various compound forms, such as organic and mineral P. It is present in water, soil, and sediments. The amounts of phosphorus in the soil are usually low, and these low levels confine plant growth. Phosphorus is an essential nutrient for both animals and plants, and it plays an important role in cell development. It is a vital component of molecules such as ATP, DNA, and phospholipids (Richardson 1994, Holford 1997, Schachtman et al. 1998).

Phosphorus circulates in a cycle through rocks, water, soil and sediments, and organisms. Over periods of time, rain and weathering cause rocks to release phosphate ions and other minerals. Then, inorganic phosphate is distributed in soils and water and is available for uptake by plants. Furthermore, plants are consumed by animals. Once the phosphate is incorporated into organic molecules, such as DNA in plants or animals, and later when the plant and animal dies and decays, organic phosphate is returned into the soil. Organic forms of phosphate in the soil can be available to plants with the help of bacteria that break down organic matter to inorganic forms of phosphorus. Phosphorus in soil can end up in waterways and ultimately oceans; over time, it can be incorporated into sediments (Richardson 1994, Holford 1997, Schachtman et al. 1998).

9.3 Mechanisms of Bacterial Phosphate Solubilization

A number of bacterial genera from diverse ecological niches have been reported to exhibit phosphate-solubilizing activities that mobilize P in soils (Ahemad et al. 2009, Sridevi and Mallaiah 2009, Kumar et al. 2013, Zhao et al. 2014, Kryuchkova et al. 2014, Iqbal and Ahemad 2015). Various mechanisms of bacteria-mediated phosphate solubilization have been proposed, but the most accepted and evident mechanism of phosphate solubilization is the lowering of soil pH due to secretion of organic acids (e.g., gluconic acid, oxalic acid, 2-ketogluconic acid, lactic acid, succinic acid, formic acid, citric acid, malic acid) (Vyas and Gulati 2009, Sashidhar and Podile 2010, Roca et al. 2013). Interestingly, the amount of solubilized P that results when soil pH declines is highly dependent upon pH changes for specific phosphate compounds as well as P substrates for PSB (Walpola et al. 2012).

9.3.1 Effects of Secreted Organic Acids

The evidence shows that addition of organic acids to soils increases plant P uptake and in turn plant growth (Ahemad et al. 2009). In many studies, gluconic acid is one of the most prevalent organic acids secreted by PSB and is produced by direct oxidation of glucose via a membrane-bound quinoprotein, glucose dehydrogenase enzyme, which exhibits broad substrate specificity in some organisms (Park et al. 2009, Sashidhar and Podile 2010, Castagno et al. 2011, Shen et al. 2012, Zhao et al. 2014). General mechanisms for bacterial phosphate solubilization are summarized in Figure 9.1.

9.4 Mechanisms of PSB-Assisted Plant Growth Promotion and Phytoremediation

Phytoremediation is the means by which plants with inherent physiological characteristics remove soil pollutants or render them nontoxic in the soil. Plant-based remediation can occur via several mechanisms, such as phytoextraction, phytostabilization, phytovolatilization, and rhizodegradation. The process deployed depends upon the chemical or physical features of the contaminants available in the soils. PSB with various plant growth-promoting traits and simultaneous metal-detoxifying capabilities can increase the phytoremediation efficacy of plants by promoting their growth and condition. Many studies have shown substantial enhancement in plant biomass, yields, and other growth characteristics of different plants inoculated with PSB exhibiting the sole property of phosphate solubilization (Lara et al. 2013, Ahemad et al. 2009). In contrast, several researchers have found in their PSB-inoculated plant studies that the overall plant growth promotion is the result of the interplay of other plant-beneficial traits in addition to the phosphate-mobilizing feature of PSB, such as production of phytohormones, 1-aminocyclopropane-1-carboxylate (ACC) deaminase, siderophores, and antimicrobial substances (Ahemad et al. 2009, Rajkumar et al. 2013, Zhao et al. 2014, Iqbal and Ahemad 2015) (Figure 9.2). How these PSB properties contribute to acceleration and promotion of plant growth and development can be analyzed by focusing on each major trait separately.

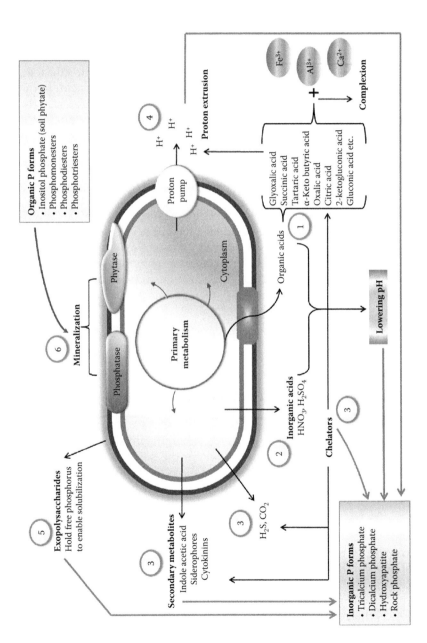

FIGURE 9.1

Mechanisms of bacterial phosphate solubilization in soils. (1 and 2) The inorganic phosphate-solubilizing property is attributable to a drop in pH, which is due to secretion of low-molecular-weight organic acids (1) or due to inorganic acids (2) produced by nitrifying bacteria and thiobacilli during oxidation of nitrogenous or inorganic sulfur-containing compounds. (3) Chelating substances like H_2S, CO_2, organic acid anions, siderophores, IAAs, cytokinins, and gibberllines also facilitate phosphate solubilization. (4) Insoluble metal forms may be solubilized by protons exported into the external medium by membrane-associated pumps or liberated by dissociation of organic acids. The released organic acid anions form complexes with metal cations present in soil. (5) Exopolysaccharides also help to accelerate phosphate solubilization by holding free phosphorus synergistically with the solubilizing property of organic acids. (6) Solubilization of organic phosphates, ie., mineralization, which occurs through the synthesis of a variety of different phytases and phosphatises that catalyze the hydrolysis of phosphoric esters. Light gray arrows indicate solubilization of inorganic forms, while the dark gray arrow denotes organic phosphate solubilization.

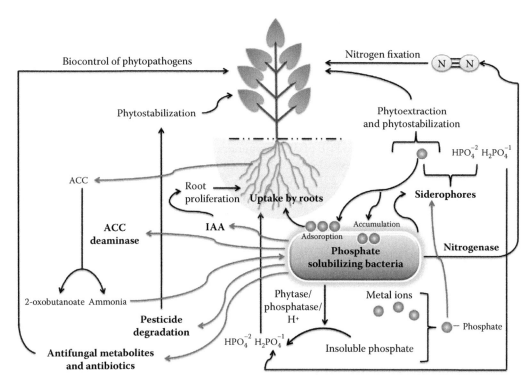

FIGURE 9.2
Mechanisms of plant growth promotion by PSB.

9.4.1 Siderophores

Iron occurs abundantly in the Earth's crust and is required for various important cellular processes, like the electron transport chain, respiration, photosynthesis, DNA synthesis, legume–*Rhizobium* symbiosis, and as a cofactor for many metabolic enzymes (Hider and Kong 2010, Chu et al. 2010, Zuo and Zhang 2011, Traxler et al. 2012, Vigani et al. 2013, Iqbal and Ahemad 2015). Despite its presence in soils in high amoutns amount, it is not available to organisms in the Fe^{3+} form due to formation of insoluble hydroxides and oxides in soil and its limited solubility ($\sim 10^{-24}$ mol/L) (Ahemad et al. 2009, Traxler et al. 2012). Under iron limitation conditions in soils, bacteria secrete low-molecular-weight and structurally diverse iron-scavenging molecules, known as siderophores, which have very strong affinity to Fe^{3+} (Ahemad et al. 2009, Hider and Kong 2010). Siderophores are categorized into four groups on the basis of their chemical structure, the functional group that coordinates iron, and the types of ligands, as follows: pyoverdines, hydroxamates, carboxylates, and phenol catecholates (Beneduzi et al. 2012).

Fe^{3+}–siderophore complexes are recognized by membrane receptor proteins, and they are unable to cross channels (porins) in bacterial membranes due to their large size. In Gram-negative bacteria, ABC transporters and the TonB protein are involved in siderophore transport, while this process is less understood for Gram-positive bacteria, which lack TonB (Sessitsch et al. 2013). In addition to Fe^{3+}, bacterial siderophores also chelate a number of other metals, including Zn, Cu, Ni, As, Cr, and Pb (Ahemad et al. 2009). Hence, this bacterial strategy has been implicated in several plant protection protocols, e.g., biocontrol of phytopathogens and bioremediation of heavy metals (Ahemad et al. 2009,

Wensing et al. 2010, Beneduzi et al. 2012). Moreover, bacteria that synthesize a specific siderophore may be proficient in utilizing an iron-loaded siderophore complexes secreted by other bacterial genera (heterologous siderophores) (Iqbal and Ahemad 2015). Although plants acquire iron through reduction of iron ions under low-pH conditions or by secreting phyto-siderophores, these approaches have limitations in alkaline soils, and binding affinities of phyto-siderophores are considerably lower than those of other bacterial siderophores (Brear et al. 2013; Radzki et al. 2013). So, it is of prime significance to provide iron to plants through inoculation with siderophore-producing organisms. Siderophore production has been reported in PSB of diverse genera. Commonly, the PSB species belong to the genera *Pseudomonas* (Ahemad et al. 2009, Roca et al. 2013, Duan et al. 2013), *Bacillus* (Ahemad et al. 2009, Kumar et al. 2013, Chakraborty et al. 2013), *Enterobacter* (Ahemad et al. 2009), *Klebsiella* (Ahemad et al. 2009, Ahemad and Khan 2012a), and *Rahnella* (Ahemad et al. 2009, Yuan et al. 2013).

Various findings in studies of siderophore-producing PSB inoculation, including metal remediation studies, have shown significant improvements in plant growth, owing to bacterial siderophores and their actions to indirectly alleviate metal toxicity by providing iron to plants to overcome iron deficiency. However, inoculation with siderophore-producing bacteria has been observed to both promote and decrease heavy metal uptake, depending upon the plant, bacterium, and metal species (Sessitsch et al. 2013). For instance, Kumar et al. (2008) observed that the phosphate-solubilizing *Enterobacter* sp. strain NBRI K28 and its siderophore-overproducing mutant, NBRI K28 SD1, significantly stimulated plant biomass and also enhanced phytoextraction of Ni, Zn, and Cr in *Brassica juncea* (Indian mustard). In contrast, the mutant NBRI K28 strain facilitated higher metal accumulation and plant growth than the wild type. In another study, siderophore-producing *Pseudomonas* PSB inoculated in chickpea plants, which were grown in nickel-stressed soils in pots, had increased growth and biomass compared to uninoculated plants (Ahemad et al. 2009).

9.4.2 Indole-3-Acetic Acid

Phytohormones, such as auxins, abscisic acid, gibberellins, cytokinins, brassinosteroids, ethylene, salicylic acid, jasmonates, and strigolactones, play vital roles in modulating and regulating developmental processes (e.g., cell division and differentiation, shoot and root development, and cell death) and signaling pathways are involved in plant responses to various biotic and abiotic stresses (e.g., responses against pests and diseases) (Bari and Jones 2009). Auxins, the most-studied growth regulators, are considered indispensable for plant growth, since no plant is reported to be devoid of their synthesis. Indole-3-acetic acid (IAA) is the most common auxin in plants; other compounds with auxin-like activity, for instance indole-3-butyric acid, phenyl acetic acid, and 4-chloro-IAA, also occur in plants, but little is known about their physiological roles (Ahmed et al. 2010).

Moreover, IAA that is also synthesized by most of the soil bacteria through different pathways using tryptophan as a precursor influences levels of endogenous poold of plant–IAA acceleration of root proliferation and increasing root surface area, in this manner allowing plants to access more water and nutrients from soils (Malfanova 2013). In addition, IAA triggers plant cell metabolism to adapt to pollutants like metals or pesticides under stressed environments (Idris et al. 2004, Ahemad et al. 2009).

Bacillus, Pseudomonas, Acinetobacter, Microbacterium, Alcaligenes, Comamonas, Azomonas, Ochrobactrum, Klebsiella, Serratia, Enterobacter, Phyllobacterium, Rahnella, Burkholderia, Rhizobium, Ensifer, Azotobacter, and *Tetrathiobacter* are some of the reported IAA-producing genera that also exhibit phosphate-solubilizing potential (Ahemad et al. 2009, Ahemad and Khan 2012, Chakraborty et al. 2013, Kumar et al. 2013, Prabha et al. 2013, Zhao et al. 2014, Kryuchkova et al. 2014, Iqbal and Ahemad 2015). Many studies have evaluated the potential role of IAA, when inoculated into plants with PSB, in boosting plant growth and development. For example, phosphate-solubilizing *Sinorhizobium meliloti* strain RD64 (a mutant strain), which overproduces IAA compared to the wild-type strain, improved nitrogen fixation and growth in *Medicago truncatula* plants under P starvation in a study by Bianco and Defez (2010). They linked the better plant growth to overproduction of bacterial IAA. Similarly, Quecine et al. (2012) studied the growth promotion of sugarcane by the phosphate-solubilizing *Pantoea agglomerans* strain 33.1 under a controlled environment and observed that *P. agglomerans* 33.1-inoculated sugarcane plants had higher dry biomass. They corroborated the correlation of biomass augmentation in inoculated sugarcane plants with the capability of *P. agglomerans* 33.1 to synthesize IAA and solubilize phosphate.

9.4.3 1-Aminocyclopropane-1-Carboxylate Deaminase

The endogenously produced gaseous plant hormone ethylene, which requires S-adenosyl-L-methionine (SAM) and ACC for biosynthesis, regulates many physiological processes in plants, including root initiation and root hair formation, flowering, fruit ripening and abscission, sex determination, germination of seeds, senescence, and responses to biotic (e.g., pathogenic) and abiotic (e.g., wounding, salinity, water logging, soil compaction, heavy metal exposure, hypoxia, drought, chilling) stresses in plants. It is synthesized in three steps: (i) SAM synthetase converts methionine to SAM, (ii) SAM is converted to ACC by ACC synthase, and (iii) ACC is degraded via ACC oxidase to release ethylene (Ahemad et al. 2009, Lin and Zhong 2009, Agrawal et al. 2012, Tatsuki et al. 2013, Iqbal and Ahemad 2015). Although ethylene as a stress hormone is involved in plant responses to various stresses, high concentrations of this phytohormone are inhibitory to plant growth and many other agriculturally desirable and important physiological processes in plants (Ahemad et al. 2009, Shaharoona et al. 2011, M. Li et al. 2013). In this context, ACC deaminase-producing organisms can alleviate ethylene-induced stress responses in plants and, in turn, accelerate plant growth (Ahemad et al. 2009). Diverse genera of PSB with the ACC deaminase-producing trait have been reported and have been successfully implicated in plant growth promotion under different environmental stresses (Ahemad et al. 2009, Yuan et al. 2013, Iqbal and Ahemad 2015).

9.5 Nitrogen Fixation and Plant Growth

Nitrogen (N) is the most important nutrient for plant growth and development, and it is abundantly present in the atmosphere in its gaseous form. Plants cannot utilize it in this form, but they do utilize ammonium (NH_4^+) and nitrate (NO_3^-) ions (Malfanova et al. 2013). To acquire N, legume plants form beneficial symbiotic partnerships with the rhizobia that reside in N_2-fixing root nodules. The symbiosis commences when flavonoids (present in root exudates of host plants) activate *nod* genes in rhizobia (Figure 9.3). Thereafter, N_2 is fixed in nodules by a complex enzymatic process, centered on nitrogenase, in a series of energy-dependent biochemical reactions (Dixon and Kahn 2004, Ahemad et al. 2009, Sohm et al. 2011, Santi et al. 2013, Ahemad and Khan 2013).

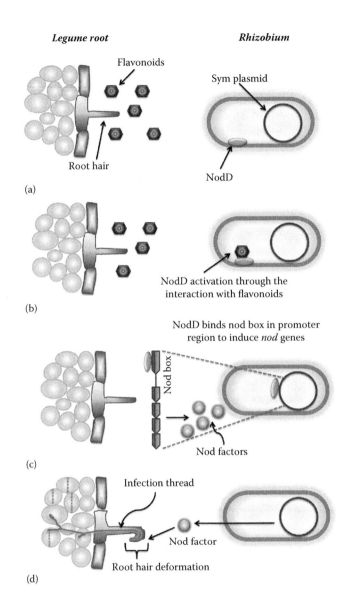

FIGURE 9.3
Schematic of the interaction between legume roots and rhizobia. (a) The legume root secretes flavonoids to *Rhizobium*, (b) Plant-secreted flavonoids passively enter the rhizobial cytoplasm through the cell membrane and activate the rhizobial NodD proteins by conformational changes, (c) Flavonoid-activated NodD binds the Nod box in the promoter region of *nod* genes to induce expression of enzymes needed for Nod factor synthesis. (d) Nod factors secreted by *Rhizobium* attach to Nod factor receptors on plant root hairs and induce nodule formation via root hair deformation, division in cortex cells, and promotion of some steps of the infection process. (*Source:* Ahemad, M., Khan, M.S., *Biochemistry and Molecular Biology* 1:63–75, 2013.)

9.5.1 Role of Nitrogenase in N$_2$ Fixation

Nitrogenase genes are required for nitrogen fixation; these genes include structural genes and genes involved in Fe protein activation, biosynthesis of an iron–molybdenum cofactor, as well as electron donation. The regulatory genes are necessary for the synthesis and function of the enzyme. In nitrogen-fixing bacteria, nitrogenase genes are typically found in a cluster of different proteins. Although opinions vary, some scientists believe that if nitrogenase genes were isolated and characterized, it would be possible to genetically engineer plants in order to improve their capability for nitrogen fixation. However, some scientists have reasoned that it could be possible for the genetically engineered plants to fix their own nitrogen. There are numerous free-living bacteria, e.g., *Azospirillum* spp., that are able to fix nitrogen and provide it to plants (Glick 2012).

9.5.2 Effects of Nitrogen-Fixing Rhizobia

Exploitation of nitrogen-fixing rhizobia with phosphate-solubilizing properties in agronomy should take precedence over other PSB, as they provide both macronutrients, N, and P to plants. Alternatively, PSB devoid of a nitrogen-fixing capability may be genetically modified to fix nitrogen using the genes that encode nitrogenase, and thus they may be used on a wide range of crop plants extending beyond *Rhizobium*–legume host specificity (Setten et al. 2013). A number of species of rhizobia, e.g., *Rhizobium* spp. (Sridevi and Mallaiah 2009), *Sinorhizobium meliloti* (Bianco and Defez 2010), *Rhizobium leguminosarum* (Mikanova and Novakova 2002, Alikhani et al. 2006, Zarei et al. 2006, Xie et al. 2009, Prabha et al. 2013), *Bradyrhizobium* spp. (Farnandez et al. 2005), *Rhizobium trifolii* (Rivas et al. 2006), *Mesorhizobium mediterraneum* (J.F. Li et al. 2013), *Mesorhizobium ciceri* (Zarei et al. 2006, Rivas et al. 2006), and *Rhizobium meliloti* (J.F. Li et al. 2013), have been explored for the phosphate-solubilizing trait. Many authors have shown intriguing increments in growth parameters of different plants following inoculation of phosphate-solubilizing rhizobia (Zarei et al. 2006, Rivas et al. 2006, J.F. Li et al. 2013, Prabha et al. 2013).

9.5.3 Role of PGPR in N$_2$ Fixation

Rhizospheric bacteria have been reported to be beneficial in phytoremediation as they enhance plant growth via phytohormones, siderophore production, and mobilization of phosphate by immobilizing heavy metals to less toxic forms. An inoculum should have a specific bacterial strain which might have various properties such as heavy metal tolerance, N$_2$ fixation, and phosphate solubilization (Iqbal and Ahemad 2015). Plant growth-promoting bacteria can fix atmospheric N$_2$ and facilitate its supply to plants. They can synthesize siderophores which can sequester iron from soil and supply iron to plant cells. Eventually, they can help phytoremediation by enhancing the different stages of plant growth via phytohormone, auxin, cytokinin, and gibberellin, and also by solubilizing minerals such as phosphorus (Amora-Lazcano et al. 2010).

TABLE 9.1

Plant Growth-Promoting Substances Released by Phosphate-Solubilizing Bacteria

PGPR	Plant-Growth Promoting Trait(s)	References
Burkholderia cepacia SCAUK0330	Antifungal ability	Zhao et al. 2014
Alcaligenes sp. K1, *Comamonas* sp. K4, *Azomonas* sp. K5, *Pseudomonas* sp. K3, *Enterobacter cloacae* K7	Nitrogen fixation, IAA	Kryuchkova et al. 2014
Pseudomonas putida BIRD-1	IAA, siderophore	Roca et al. 2013
Bacillus safensis W10, *Ochrobactrum pseudogregnonense* IP8	IAA, siderophore, ACC deaminase	Chakraborty et al. 2013
Rahnella sp. JN6	IAA, siderophore, ACC deaminase	Ahemad and Kibret 2014
Acromobacter sp. PB-01, *Tetrathiobacter* sp. PB-03 and *Bacillus* sp. PB-13	IAA, siderophore, antifungal activity	Zhao et al. 2014
Serratia marcescens KiSII, *Enterobacter* sp. RNF 267	IAA, ACC deaminase, ammonia	Iqbal and Ahemad 2015
Pseudomonas sp. UW4	IAA, siderophore	Duan et al. 2013
Phyllobacterium myrsinacearum RC6b	IAA, siderophore, ACC deaminase	Iqbal and Ahemad 2015
Pseudomonas aeruginosa A11	siderophores, HCN, ammonia	Ahemad and Kibret 2014
Bacillus megaterium SR28C	IAA	Rajkumar et al. 2013
Ochrobactrum sp. MSP9, *Enterobacter* sp. MSP10	IAA	Tariq et al. 2013
Rahnella sp. JN27	IAA, siderophore, ACC deaminase	Ahemad and Kibret 2014
Rhizobium leguminosarum PCC2, *Ensifer meliloti* PCC7	IAA, siderophore, ACC deaminase, nitrogen fixation	Prabha et al. 2013
Pseudomonas aeruginosa strain OSG41	IAA, siderophores	Ahemad and Kibret 2014
Pseudomonas sp. strain SGRAJ09	IAA	Rajasankar et al. 2013
Pseudomonas spp.	IAA, HCN	Ahemad and Kibret 2014
Acinetobacter haemolyticus RP19	IAA	Ahemad and Kibret 2014
Pseudomonas putida	IAA, siderophores, HCN, ammonia	Ahemad and Kibret 2014
Pseudomonas fluorescens strain Psd	IAA, siderophores, HCN, antibiotics and biocontrol activity	Ahemad and Kibret 2014
Agrobacterium tumefaciens CCNWGS0286	IAA	Ahemad and Kibret 2014
Bacillus thuringiensis	IAA	Ahemad and Kibret 2014
Pseudomonas aeruginosa PS1	IAA, siderophores, HCN, ammonia	Ahemad and Kibret 2014
Pseudomonas sp. TLC 6-6.5-4	IAA, siderophore	Ahemad and Kibret 2014
Bacillus spp.	IAA, HCN	Karuppiah & Rajaram 2011
Klebsiella spp.	IAA, siderophores, HCN, ammonia	Ahemad & Khan 2012
Sinorhizobium meliloti strain RD64	IAA	Bianco & Defez 2010
Enterobacter asburiae	IAA, siderophores, HCN, ammonia	Ahemad and Kibret 2014
Bacillus species PSB10	IAA, siderophores, HCN, ammonia	Ahemad and Kibret 2014
Arthrobacter sp. MT16, *Microbacterium* sp. JYC17, *Pseudomonas chlororaphis* SZY6, *Azotobacter vinelandii* GZC24, *Microbacterium lactium* YJ7	IAA, siderophore, ACC deaminase	Ahemad and Kibret 2014

(Continued)

TABLE 9.1 (CONTINUED)

Plant Growth-Promoting Substances Released by Phosphate-Solubilizing Bacteria

PGPR	Plant-Growth Promoting Trait(s)	References
Pseudomonas sp.	IAA, siderophore, HCN, biocontrol potentials	Ahemad and Kibret 2014
Enterobacter aerogenes NBRI K24, *Rahnella aquatilis* NBRI K3	ACC deaminase, IAA, siderophore	Ahemad and Kibret 2014
Enterobacter sp.	ACC deaminase, IAA, siderophore	Ahemad and Kibret 2014
Burkholderia	ACC deaminase, IAA, siderophore	Ahemad and Kibret 2014
Pseudomonas aeruginosa	ACC deaminase, IAA, siderophore	Ahemad and Kibret 2014

9.6 Implications of PSB in Plant Growth Promotion and Phytoremediation

Multifunctional PSB of different genera have been exploited in plant growth promotion studies conducted under different environmental conditions and soil types, using a wide range of plant species. Owing to production of various plant growth-promoting substances (Table 9.1), PSB have not only facilitated accelerated growth and development but also have heightened resistance of plants against diseases and stresses (Table 9.2).

9.6.1 Phosphate-Solubilizing Rhizobia

From Table 9.2, it is obvious that very few studies have been dedicated to assess the performance of PSB among rhizobial species. Employment of phosphate-solubilizing rhizobia as biofertilizers or growth boosters would be beneficial in terms of supplying major nutrients (N and P) to plants. By this approach, inefficiency in colonization, encountered in cases where non-rhizobial PSB are used as inoculants, would also be minimized to a considerable extent due to the prevalence of host specificity in rhizobial organisms.

9.6.2 Phytoremediation of Heavy Metals

Most of the studies summarized Table 9.2 have been directed towards the phytoremediation of heavy metals, and a small number of those studies were devoted to overcoming pesticide-induced phytotoxicity and osmotic stress. As pesticides form the bulk of agrochemicals used in plant cultivation, exploration of pesticide-degrading PSB and investigations regarding the assessment of their application as bioinoculants would fortify sustainability of, in particular, pesticide-polluted and P-deficient soils.

9.6.3 Efficacy of PSB Strains

Almost all studies (Table 9.2) concerning the evaluation of efficacy of PSB strains that exhibit specific phytoremediation-related traits in plant models have been conducted in pots or greenhouses or under gnotobiotic conditions. Field trials are rarely conducted and reported. Field experiments would provide more realistic information, especially as new practical impediments to expected results could be revealed and show the presence of constraints which are otherwise lacking in controlled environments.

TABLE 9.2

Plant Growth Promotion by PSB in the Absence or Presence of Stress

PSB(s)	Plant(s)	Growth Condition	Improved Plant Growth (Parameter, Activity, or Stress Tolerance)	Reference
Enterobacter cloacae K7	Common sunflower (*Helianthus annuus* L.)	–	Growth and development of seedlings	Kryuchkova et al. 2014
Burkholderia cepacia SCAUK0330	Maize	Pots	Growth of both healthy and fungal pathogen-infected plants	Zhao et al. 2014
Rhizobium leguminosarum PCC2, *Ensifer meliloti* PCC7	*Psoralea corylifolia* L.	Fields	Early vegetative and late reproduction parameters	Prabha et al. 2013
Enterobacter cloacae Z$_{32}$, *Klebsiella oxytoca* Z$_{42}$	Radish (*Raphanus sativus* L.)	Pots	Dry weight and height, leaf area and number	Lara et al. 2013
Bacillus spp.	Rice	Pots	Plant biomass and root growth	Ahemad and Kibret 2014
Ralmella sp. JN27	*Amaranthus hypochondriacus, Amaranthus mangostanus, Solanum nigrum, Zea mays*	Pots	Plant biomass, phytoextraction of metals	Yuan et al. 2013
Klebsiella sp. RSN219, *Rhizobium meliloti* LW135	Alfalfa (*Medicago sativa* L.)	Pots	Survival rate of seedlings, shoot height, root length, root volume, leaf area, number of leaves per plant, biomass, and phosphorus uptake	Li et al. 2013
Bacillus megaterium SR28C	*Brassica juncea, Luffa cylindrical, Sorghum halepense*	Pots	Plant biomass, alleviation of nickel phytotoxicity	Rajkumar et al. 2013
Serratia marcescens KiSII, *Enterobacter* sp. RNF 267	Coconut (*Cocos nucifera* L.)	Greenhouse	Seedling growth and nutrient uptake	Iqbal and Ahemad 2015
Pseudomonas aeruginosa strain OSG41	Chickpea (*Cicer arietinum*)	Pots	Dry matter, symbiosis, grain yield and grain protein, phytostabilization	Ahemad and Kibret 2014
Tetrathiobacter sp. PB-03, *Bacillus* sp. PB-13	Indian mustard (*Brassica juncea*)	Greenhouse	Plant biomass and P content	Kumar et al. 2013
Phyllobacterium myrsinacearum RC6b	*Sedum plumbizincicola*	Pots	Plant growth, phytoextraction of metals	Iqbal and Ahemad 2015
Ralmella sp. JN6	Rape (*Brassica napus*)	Pots	Plant growth, metal accumulation	Ahemad et al. 2009
Bacillus safensis W10, *Ochrobactrum pseudogregnonense* IP8	Wheat (*Triticum aestivum*)		Root and shoot biomass, plant height, yield, chlorophyll content, osmotic stress tolerance	Chakraborty et al. 2013

(Continued)

TABLE 9.2 (CONTINUED)

Plant Growth Promotion by PSB in the Absence or Presence of Stress

PSB(s)	Plant(s)	Growth Condition	Improved Plant Growth (Parameter, Activity, or Stress Tolerance)	Reference
Agrobacterium tumefaciens CCNWGS0286	*Robinia pseudoacacia*	Greenhouse	Stem dry weight	Ahemad and Kibret 2014
Acinetobacter haemolyticus RP19	Pearl millet (*Pennisetum glaucum*)	Pots	Root length, shoot length, fresh weight and root biomass, phytostabilization	Ahemad and Kibret 2014
Pseudomonas aeruginosa strain PS1	Greengram [*Vigna radiata* (L.) Wilczek]	Pots	Plant biomass, nodulation, nutrient uptake, seed yield and seed protein, phytostabilization	Ahemad and Kibret 2014
Pseudomonas sp. A3R3	*Alyssum serpyllifolium, Brassica juncea*	Pots	Plant biomass, phytoextraction of metal	Ahemad and Kibret 2014
Pseudomonas sp. TLC 6-6.5-4	*Zea mays, Helianthus annuus*	Pots	Plant biomass, phytoextraction of metal	Ahemad and Kibret 2014
Pseudomonas aeruginosa strain PS1	Greengram [*Vigna radiata* (L.) Wilczek]	Pots	Plant dry weight, nodule number, total chlorophyll, leghemoglobin, nitrogen and phosphorus content, seed yield and seed protein, phytostabilization	Ahemad and Kibret 2014
Psychrobacter sp. SRS8	*Ricinus communis, Helianthus annuus*	Pots	Plant biomass, chlorophyll, and protein content	Ahemad and Kibret 2014
Sinorhizobium meliloti strain RD64	*Medicago truncatula* cv. Jemalong 2HA	Hydroponic culture	Plant fresh and dry weight	Bianco et al. 2010
Arthrobacter sp. MT16, *Microbacterium* sp. JYC17, *Pseudomonas chlororaphis* SZY6, *Azotobacter vinelandii* GZC24, *Microbacterium lactium* YJ7	*Brassica napus*	Gnotobiotic	Root length	Ahemad and Kibret 2014
Pseudomonas aeruginosa strain PS1	Greengram [*Vigna radiata* (L.) Wilczek]	Pots	Plant biomass, symbiosis, chlorophyll, nitrogen, phosphorus, seed yield and grain protein, phytostabilization	Ahemad and Kibret 2014
Bacillus sp. PSB10	Chickpea (*Cicer arietinum*)	Pots	Plant growth, nodulation, chlorophyll, leghaemoglobin, seed yield and grain protein, phytostabilization	Ahemad and Kibret 2014
Pseudomonas sp. SRI2, *Psychrobacter* sp. SRS8, *Bacillus* sp. SN9	*Brassica juncea, Brassica oxyrrhina*	Pots	Plant biomass, phytoextraction of metal	Ahemad and Kibret 2014

(Continued)

TABLE 9.2 (CONTINUED)

Plant Growth Promotion by PSB in the Absence or Presence of Stress

PSB(s)	Plant(s)	Growth Condition	Improved Plant Growth (Parameter, Activity, or Stress Tolerance)	Reference
Psychrobacter sp. SRA1, *Bacillus cereus* SRA10	*Brassica juncea, Brassica oxyrrhina*	Pots	Metal extraction	Ahemad and Kibret 2014
Achromobacter xylosoxidans strain Ax10	*Brassica juncea*	Pots	Root length, shoot length, fresh weight and dry weight, phytoextraction of metal	Ahemad and Kibret 2014
Pseudomonas spp.	Chickpea	Pots	Fresh and dry weight of plants, phytostabilization	Ahemad and Kibret 2014
Pseudomonas trivialis BIHB 745, *Pseudomonas trivialis* BIHB 747, *Pseudomonas* sp. BIHB 756, *Pseudomonas poae* BIHB 808	Maize	Pots	Plant growth and total N, P, and K content	Vyas and Gulati 2009
Enterobacter aerogenes NBRI K24, *Rahnella aquatilis* NBRI K3	*Brassica juncea*	Pots	Root length, dry weight, leaf protein and chlorophyll content, phytoextraction of metal	Ahemad and Kibret 2014
Bacillus weihenstephanensis strain SM3	*Helianthus annuus*	Pots	Plant biomass, phytoextraction of metal	Ahemad and Kibret 2014
Pseudomonas aeruginosa strain MKRh3	Black gram	Pots	Plant growth and rooting, phytostabilization	Ahemad and Kibret 2014
Enterobacter sp. NBRI K28, mutant strain NBRI K28 SD1	*Brassica juncea*	Pots	Plant biomass, chlorophyll and protein content, phytoextraction of metal	Ahemad and Kibret 2014
Burkholderia sp. J62	*Lycopersicon, Esculentum*	Pots	Root and shoot dry weight, phytoextraction of metal	Ahemad and Kibret 2014
Rhizobium leguminosarum bv. *viciae*, *Mesorhizobium ciceri*	Lentils (*Lens culinaris* cv. Ziba)	Greenhouse	Plant growth and nutrient uptake	Zarei et al. 2006
Bacillus subtilis SJ-101	*Brassica juncea*	Growth chamber	Plant growth, phytoextraction of metal	Ahemad and Kibret 2014
Pseudomonas spp., *Bacillus* spp.	Mustard	Pots	Plant growth, phytostabilization	Ahemad and Kibret 2014
Mesorhizobium ciceri, Mesorhizobium mediterraneum	*Cicer arietinum*	Pots	Mobilize phosphorus for plants	Rivas et al. 2006

9.6.4 Performance of PSB

Performances of PSB are not consistent in different soils and environments, as bacterial genes are extremely inducible and several factors, like ecological fitness and inconsistency in inoculant delivery systems, affect study outcomes (Rainey 1999, Farnandez et al. 2005).

9.7 Conclusions and Future Perspectives

PSB that possess versatile plant growth-promoting traits along with a phosphate-solubilizing potential are far superior for remediation purposes compared to other bacterial inoculants that lack phosphate-solubilizing activity. Organic acids produced by PSB solubilize many metal salts in soils, and some metals are utilized as essential nutrients while others can be mobilized in plant tissues for phytoextraction. In addition, they increase the soil quality through mineralization. Although several genes for phosphate solubilization have been isolated, manipulation of these genes to achieve PSB exhibiting greater phosphate-solubilizing activities would further open new avenues to acquire potential bioinoculants to furnish P to soils and plants. Several activities, such as production of IAA, ACC deaminase, and siderophores, by PSB would be a better option to assist phytoremediation of contaminated soils. Recent developments concerning the functional diversity, rhizosphere-colonizing ability, and modes of action could facilitate the use of PSB as reliable components in agricultural systems. A sufficient number of studies related to phosphate-solubilizing microbes and their importance in agriculture have been reported over the last few years. Cutting edge research must focus on the interactions of plants and microbes that benefit the plants. Furthermore, scientists need to address how nutritional and root exudation aspects could be better managed in order to provide maximum benefits from phosphate-solubilizing microbe applications. Molecular and genetic engineering technologies could lead to development of increased understanding about phosphate-solubilizing microbes' modes of action, which could lead to more successful uses for plant–microbe interactions.

References

Agarwal, G., Choudhary, D., Singh, V.P., Arora, A. 2012. Role of ethylene receptors during senescence and ripening in horticultural crops. *Plant Signaling and Behavior* 7:827–846.

Ahemad, M., Khan, M.S. 2012. Biotoxic impact of fungicides on plant growth promoting activities of phosphate-solubilizing *Klebsiella* sp. isolated from mustard (*Brassica compestris*) rhizosphere. *Journal of Pest Science* 85:29–36.

Ahemad, M., Kibret, M. 2014. Mechanisms and applications of plant growth promoting rhizobacteria: Current perspective. *Journal of King Saud University Science* 26:1–20.

Ahemad, M., Khan, M.S. 2013. Pesticides as antagonists of rhizobia and the legume-*Rhizobium* symbiosis: A paradigmatic and mechanistic outlook. *Biochemistry and Molecular Biology* 1:63–75.

Ahemad, M., Zaidi, A., Khan, M.S., Oves, M. 2009. Biological importance of phosphorus and phosphate solubilizing microbes, pp. 1–14. *Phosphate-Solubilizing Microbes for Crop Improvement*, Khan, M.S., Zaidi, A. (eds.). New York: Nova Science Publishers, Inc.

Ahmed, M., Stal, L.J., Hasnain, S. 2010. Production of indole-3-acetic acid by the cyanobacterium *Arthrospira platensis* strain MMG-9. *Journal of Microbiology and Biotechnology* 20:1259–1265.

Alikhani, H.A., Saleh-Rastin, N., Antoun, H. 2006. Phosphate solubilization activity of rhizobia native to Iranian soils. *Plant and Soil* 287:35–41.

Amora-Lazcano, E., Guerrero-Zuniga, L.A., Rodriguez-Tovar, A., Rodriguez-Dorantes, A., Vasquez-Murrieta, M.S. 2010. Rhizospheric plant-microbe interactions that enhance the remediation of contaminated soils, pp. 251–256. *Current Research Technology and Education Topics in Applied Microbiology and Microbial Biotechnology*, Mendez, A. (ed.). Badajoz, Spain: Formatex.

Bari, R., Jones, J.D. 2009. Role of plant hormones in plant defence responses. *Plant Molecular Biology* 69:473–488.

Beneduzi, A., Ambrosini, A., Passaglia, L.M.P. 2012. Plant growth-promoting rhizobacteria (PGPR): Their potential as antagonists and biocontrol agents. *Genetics and Molecular Biology* 35:1044–1051.

Bianco, C., Defez, R. 2010. Improvement of phosphate solubilization and *Medicago* plant yield by an indole-3-acetic acid-overproducing strain of *Sinorhizobium meliloti*. *Applied and Environmental Microbiology* 76:4626–4632.

Brear, E.M., Day, D.A., Collina Smith, P.M. 2013. Iron: An essential micronutrient for the legume-rhizobium symbiosis. *Frontiers in Plant Science* 4:359.

Castagno, L.N., Estrella, M.J., Sannazzaro, A.I., Grassano, A.E., Ruiz, O.A. 2011. Phosphate-solubilization mechanism and *in vitro* plant growth promotion activity mediated by *Pantoea eucalypti* isolated from *Lotus tenuis* rhizosphere in the Salado River Basin (Argentina). *Journal of Applied Microbiology* 110:1151–1165.

Chakraborty, U., Chakraborty, B.N., Chakraborty, A.P., Dey, P.L. 2013. Water stress amelioration and plant growth promotion in wheat plants by osmotic stress tolerant bacteria. *World Journal of Microbiology and Biotechnology* 29:789–803.

Cheng, Z., Park, E., Glick, B.R. 2007. 1-Aminocyclopropane-1-carboxylate deaminase from *Pseudomonas putida* UW4 facilitates the growth of canola in the presence of salt. *Canadian Journal of Microbiology* 53:912–918.

Cheng-Hsiung, C., Shang-Shyng, Y. 2009. Thermo-tolerant phosphate-solubilizing microbes for multi-functional biofertilizer preparation. *Bioresource Technology* 100:1648–1658.

Chu, B.C., Garcia-Herrero, A., Johanson, T.H. et al. 2010. Siderophore uptake in bacteria and the battle for iron with the host: A bird's eye view. *Biometals* 23:601–611.

Dixon, R., Kahn, D. 2004. Genetic regulation of biological nitrogen fixation. *Nature Reviews in Microbiology* 2:621–631.

Duan, J., Jiang, W., Cheng, Z., Heikkila, J.J., Glick, B.R. 2013. The complete genome sequence of the plant growth-promoting bacterium *Pseudomonas* sp. UW4. *PLoS One* 8:e58640.

Fernández, L.A., Zalba, P., Gómez, M.A., Sagardoy, M.A. 2005. Bacterias solubilizadoras de fosfato inorgánico aisladas de suelos de la región sojera. *Cienc Suelo* 23:31–37.

Glick, B.R. 2012. Plant growth-promoting bacteria: Mechanism and applications. *Scientifica* 2012:963401.

Hider, R.C., Kong, X. 2010. Chemistry and biology of siderophores. *Natural Product Reports* 27:637–657.

Holford, I.C.R. 1997. Soil phosphorus: Its measurement, and its uptake by plants. *Australian Journal of Soil Research* 35:227–239.

Idris, R., Trifonova, R., Puschenreiter, M., Wenzel, W.W., Sessitsch, A. 2004. Bacterial communities associated with flowering plants of the Ni hyperaccumulator *Thlaspi goesingense*. *Applied and Environmental Microbiology* 70:2667–2677.

Iqbal, J., Ahemad, M. 2015. Recent advances in bacteria-assisted phytoremediation of heavy metals from contaminated soil. *Advances in Biodegradation and Bioremediation of Industrial Waste*, Chandra, R. (ed.). Boca Raton: CRC Press.

Karuppiah, P., Rajaram, S. 2011. Exploring the potential of chromium reducing *Bacillus* sp. and their plant growth promoting activities. *Journal of Microbiology Research* 1:17–23.

Kryuchkova, Y.V., Burygin G.L., Gogoleva, N.E. et al. 2014. Turkovskaya: Isolation and characterization of a glyphosate-degrading rhizosphere strain, *Enterobacter cloacae* K7. *Microbiological Research* 169:99–105.

Kumar, K.V., Singh, N., Behl, H.M., Srivastava, S. 2008. Influence of plant growth promoting bacteria and its mutant on heavy metal toxicity in *Brassica juncea* grown in fly ash amended soil. *Chemosphere* 72:678–683.

Kumar, V., Singh, P., Jorquera, M.A. et al. 2013. Isolation of phytase-producing bacteria from Himalayan soils and their effect on growth and phosphorus uptake of Indian mustard (*Brassica juncea*). *World Journal of Microbiology and Biotechnology* 29:1361–1369.

Lara, C., Sanes, S.C., Oviedo, L.E. 2013. Impact of native phosphate solubilizing bacteria on the growth and development of radish (*Raphanus sativus* L.) plants. *Biotecnología Aplicada* 30:276–279.

Li, J.F., Zhang, S.Q., Huo, P.H., Shi, S.L., Miao, Y.Y. 2013. Effect of phosphate solubilizing *Rhizobium* and nitrogen fixing bacteria on growth of alfalfa seedlings under P and N deficient conditions. *Pakistan Journal of Botany* 45:1557–1562.

Li, M., Zhang, Y., Zhang, Z. et al. 2013. Hypersensitive ethylene signaling and *ZMdPG1* expression lead to fruit softening and dehiscence. *PLoS One* 8:e58745.

Lin, Z., Zhong, S., Grierson, D. 2009. Recent advances in ethylene research. *Journal of Experimental Botany* 60:3311–3336.

Malfanova, N.V. 2013. Endophytic bacteria with plant growth promoting and biocontrol abilities. Ph.D. thesis, Leiden University, Leiden, Netherlands.

Mikanova, O., Novakova, J. 2002. Evaluation of the P-solubilizing activity of soil microorganisms and its sensitivity to soluble phosphate. *Rostlinna Vyroba* 48:397–400.

Park, K.H., Lee, C.Y., Son, H.J. 2009. Mechanism of insoluble phosphate solubilization by *Pseudomonas fluorescens* RAF15 isolated from ginseng rhizosphere and its plant growth-promoting activities. *Letters in Applied Microbiology* 49:222–228.

Prabha, C., Maheshwari, D.K., Bajpai, V.K. 2013. Diverse role of fast growing rhizobia in growth promotion and enhancement of psoralen content in *Psoralea corylifolia* L. *Pharmacognosy Magazine* 9:57–65.

Quecine, M.C., Araujo, W.L., Rossetto, P.B. et al. 2012. Sugarcane growth promotion by the endophytic bacterium *Pantoea agglomerans* 33.1. *Applied and Environmental Microbiology* 78:7511–7518.

Radzki, W., Gutierrez Mañero, F.J., Algar, E. 2013. Bacterial siderophores efficiently provide iron to iron-starved tomato plants in hydroponics culture. *Antonie Van Leeuwenhoek* 104:321–330.

Rainey, P.B. 1999. Adaptation of *Pseudomonas fluorescens* to the plant rhizosphere. *Environmental Microbiology* 1:243–257.

Rajasankar, R., Manju Gayathry, G., Sathiavelu, A., Ramalingam, C., Saravanan, V.S. 2013. Pesticide tolerant and phosphorus solubilizing *Pseudomonas* sp. strain SGRAJ09 isolated from pesticides treated *Achillea clavennae* rhizosphere soil. *Ecotoxicology* 22:707–717.

Rajkumar, M., Ma, Y., Freitas, H. 2013. Improvement of Ni phytostabilization by inoculation of Ni resistant *Bacillus megaterium* SR28C. *Journal of Environmental Management* 128:973–980.

Richardson, A.E. 1994. Soil microorganisms and phosphorus availability, pp. 50–62. *Soil Biota Management in Sustainable Farming Systems*, Pankhurst, C.E., Gupta, V.V., Grace, P.R. (eds.). Canberra, Australia: CSIRO.

Rivas, R., Peix, A., Mateos, P.F. et al. 2006. Biodiversity of populations of phosphate solubilizing rhizobia that nodulates chickpea in different Spanish soils. *Plant and Soil* 287:23–33.

Roca, A., Pizarro-Tobías, P., Udaondo, Z. et al. 2013. Analysis of the plant growth-promoting properties encoded by the genome of the rhizobacterium *Pseudomonas putida* BIRD-1. *Environmental Microbiology* 15:780–794.

Sahay, R., Patra, D.D. 2013. Identification and performance of stress-tolerant phosphate-solubilizing bacterial isolates on *Tagetes minuta* grown in sodic soil. *Soil Use and Management* 29:494–500.

Santi, C., Bogusz, D., Franche, C. 2013. Biological nitrogen fixation in non-legume plants. *Annals of Botany* 111:743–767.

Sashidhar, B., Podile, A.R. 2010. Mineral phosphate solubilization by rhizosphere bacteria and scope for manipulation of the direct oxidation pathway involving glucose dehydrogenase. *Journal of Applied Microbiology* 109:1–12.

Schachtman, D.P., Reid, R.J., Ayling, S.M. 1998. Phosphorus uptake by plants: From soil to cell. *Plant Physiology* 116:447–453.

Sessitsch, A., Kuffner, M., Kidd, P. et al. 2013. The role of plant-associated bacteria in the mobilization and phytoextraction of trace elements in contaminated soils. *Soil Biology and Biochemistry* 60:182–194.

Setten, L., Soto, G., Mozzicafreddo, M. et al. 2013. Engineering *Pseudomonas protegens* Pf-5 for nitrogen fixation and its application to improve plant growth under nitrogen-deficient conditions. *PLoS One* 8:e63666.

Shaharoona, B., Imran, M., Arshad, M., Khalid, A. 2011. Manipulation of ethylene synthesis in roots through bacterial ACC deaminase for improving nodulation in legumes. *Critical Reviews in Plant Science* 30:279–291.

Shen, Y.Q., Bonnot, F., Imsand, E.M. et al. 2012. Distribution and properties of the genes encoding the biosynthesis of the bacterial cofactor, pyrroloquinoline quinine. *Biochemistry* 51:2265–2275.

Sohm, J.A., Webb, E.A., Capone, D.G. 2011. Emerging patterns of marine nitrogen fixation. *Nature Reviews in Microbiology* 9:499–508.

Sridevi, M., Mallaiah, K.V. 2009. Phosphate solubilization by *Rhizobium* strains. *Indian Journal of Microbiology* 49:98–102.

Tariq, M., Hameed, S., Yasmeen, T., Zahid, M., Zafar, M. 2013. Molecular characterization and identification of plant growth promoting endophytic bacteria isolated from the root nodules of pea (*Pisum sativum* L.). *Word Journal of Microbiology and Biotechnology* 30:719–725.

Tatsuki, M., Nakajima, N., Fujii, H. et al. 2013. Increased levels of IAA are required for system 2 ethylene synthesis causing fruit softening in peach (*Prunus persica* L. Batsch). *Journal of Experimental Botany* 64:1049–1059.

Traxler, M.F., Seyedsayamdost, M.R., Clardy, J., Kolter, R. 2012. Interspecies modulation of bacterial development through iron competition and siderophore piracy. *Molecular Microbiology* 86:628–644.

Vigani, G., Zocchi, G., Bashir, K., Philippar, K., Briat, J.F. 2013. Cellular iron homeostasis and metabolism in plant. *Frontiers in Plant Science* 4:490.

Vyas, P., Gulati, A. 2009. Organic acid production in vitro and plant growth promotion in maize under controlled environment by phosphate-solubilizing fluorescent *Pseudomonas*. *BMC Microbiology* 9:174.

Walpola, B.C., Keum, M.J., Yoon, M.H. 2012. Influence of different pH conditions and phosphate sources on phosphate solubilization by *Pantoea agglomerans* DSM3493. *Korean Journal of Soil Science and Fertilizer* 45:998–1003.

Wensing, A., Braun, S.D., Büttner, P. et al. 2010. Impact of siderophore production by *Pseudomonas syringae* pv. syringae 22d/93 on epiphytic fitness and biocontrol activity against *Pseudomonas syringae* pv. glycinea 1a/96. *Applied and Environmental Microbiology* 76:2704–2711.

Xie, J., Knight, J.D., Leggett, M.E. 2009. Comparison of media used to evaluate *Rhizobium leguminosarum* bivar *viciae* for phosphate-solubilizing ability. *Canadian Journal of Microbiology* 55:910–915.

Yadav, V., Kumar, M., Deep, D.K. et al. 2010. A phosphate transporter from the root endophytic fungus *Piriformospora indica* plays a role in phosphate transport to the host plant. *Journal of Biological Chemistry* 285:26532–26544.

Yuan, M., He, H., Xiao, L. et al. 2013. Enhancement of Cd phytoextraction by two *Amaranthus* species with endophytic *Rahnella* sp. JN27. *Chemosphere* 103:99–104.

Zarei, M., Saleh-Rastin, N., Alikhani, H.A., Aliasgharzadeh, N. 2006. Responses of lentil to coinoculation with phosphate-solubilizing rhizobial strains and arbuscular mycorrhizal fungi. *Journal of Plant Nutrition* 29:1509–1522.

Zhao, K., Penttinen, P., Zhang, X. et al. 2014. Maize rhizosphere in Sichuan, China, hosts plant growth promoting *Burkholderia cepacia* with phosphate solubilizing and antifungal abilities. *Microbiological Research* 169:76–82.

Zuo, Y., Zhang, F. 2011. Soil and crop management strategies to prevent iron deficiency in crops. *Plant and Soil* 339:83–95.

10

Quorum Sensing and Siderophore Formation Mechanism of Rhizospheric Bacteria during Phytoremediation of Environmental Pollutants

Sangeeta Yadav and Ram Chandra

CONTENTS

10.1 Introduction

Cell-to-cell signaling, known as quorum sensing (QS), plays an important role in biofilm formation. Bacterial gene expression in some bacterial species may be regulated by quorum sensing, a cell density–dependent signaling system mediated by chemical autoinducer molecules produced by bacteria. The autoinducer molecules bind to the appropriate transcription regulator(s) when the bacterial population reaches the quorum level (that is, the signal concentration reaches a threshold concentration sufficient to facilitate binding to the receptor). Binding of the autoinducers is followed by activation or repression of target genes. Thus, quorum sensing allows bacteria to display a unified response that benefits the population. Bacterial quorum sensing systems enhance access to nutrients and more favorable environmental niches and they enhance action against competing bacteria and environmental stresses. QS-regulated processes deal with multicellular behaviors, that is, bioluminescence, cell competency, horizontal gene transfer, virulence, motility, biofilm formation, siderophore formation, production of antibiotics transfer of conjugative plasmids, sporulation, antimicrobial peptide synthesis, regulation of virulence, and other secondary metabolites as shown in Figure 10.1. Quorum sensing can be divided into at least four steps: (1) production of small biochemical signal molecules by the bacterial cell; (2) release of the signal molecules, either actively or passively, into the surrounding environment; (3) recognition of the signal molecules by specific receptors once they exceed a threshold concentration; (4) activation of specific gene accordingly. A number of quorum sensing systems have now been well characterized and extensively documented, such as those that regulate the production of bioluminescence in the marine bacteria *Vibrio harveyi*

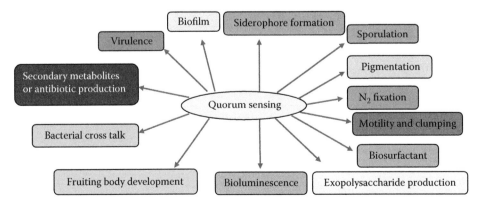

FIGURE 10.1
Phenotyping changes due to quorum sensing.

and *Allivibrio fischeri*. There still remains a wide range of organisms and environments in which quorum sensing has yet to be identified or characterized.

Microorganisms attach themselves on either abiotic or biotic surfaces (providing a suitable and optimal environment for the growth, activity, and interaction of different bacterial species) and forming a complex matrix of biopolymers known as *biofilm* that protect them from environmental stresses. Biofilm may be composed of a single bacterial species, but more frequently they are formed by a complex and diverse community of microorganisms (bacteria, algae, fungi, and protozoa) embedded in an extracellular matrix of polysaccharides, exudates, and detritus. Biofilms also provide a shelter for coping with transient or permanent stress conditions, also favoring metabolic interactions and genetic interchange between different bacterial species struggling for survival in a changing environment. The biofilm development on surfaces is a stepwise process involving adhesion, growth, motility, and extracellular polysaccharide production. Once an initial biofilm is established, cell-to-cell communication (i.e., quorum sensing) via extracellular signaling molecules regulates further modification and development of the biofilm.

Siderophores are relatively low molecular weight complexes, which are specific ferric chelating agents. They are mostly produced by bacteria and fungi growing under low iron stress. The origin of the word *siderophore* is from a Greek word meaning "iron bearer." The range of siderophore association constant for F^{3+} lies between 10^{12}–10^{52}. Iron is an essential element for both plants and microbes. It plays important roles in numerous metabolic activities: electron transport system, formation of heme, as cofactors for enzymes, and in the synthesis of chlorophyll. In the presence of oxygen, iron converts to oxyhydroxide, which is relatively a less soluble complex. Hence, there is requirement of specific molecules in the environment that can release Fe from the immobilized complex. Siderophore is the molecule that can release iron and get it in an available form. Plants and microbes both have the ability to synthesis siderophores. Siderophores produced by plants are known as *phytosiderophore*. Moreover, actinomycetes and certain algae growing under low ionic stress are also able to produce siderophores. The yeast *Saccharomyces cerevisiae*, on the other hand, does not produce siderophores but has the siderophore–iron uptake system and can utilize siderophores produced by other microorganisms. Siderophores are useful for phytoremediation, which is an efficient and environmentally friendly method for heavy metal and xenobiotic compound remediation, an aspect in which many research studies have been focused during the past two decades. In addition, rhizosphere microbes play an important role in many plant physiological activities which are useful for the remediation of environmental pollution. These microbes may form biofilm with the help of quorum-sensing molecules.

So many microorganisms are associated with the rhizospheric zone of plants as plant growth promoting rhizobacteria (PGPR). This group of microorganisms directly or indirectly enhances the growth of plants and gives tolerance against stress conditions. Initially these were used in agriculture and forestry to increase plant yield, growth, and tolerance to diseases. Now they have recently been used in environmental pollution remediation (Lucy et al., 2004). PGPR can be divided into two main groups based on their relationships: namely, free-living (ePGPR), which live outside the plant cells; and symbiotic (iPGPR), which live inside the plant cells and produce nodules (Gray and Smith, 2005). PGPR can promote the growth of plants using direct and indirect mechanisms. Direct mechanisms include lowering the production levels of ethylene through synthesis of 1-aminocylopropane-1-carboxylate (ACC) deaminase in plants, providing bioavailable phosphorus for plant uptake and atmospheric nitrogen fixation for plant use, siderophores solubilized the phosphorous, potassium, zinc, and production of plant hormones like gibberellins, cytokinins,

TABLE 10.1

Different Processes Involved in Phytoremediation

S. No.	Process of Phytoremediation	Descriptions
1	Phytoextraction or phytoaccumulation	Uptake of metal contaminants in the soil by plant roots into the above ground portions of the plants
2	Phytostabilization	Reduce the mobility and bioavailability of pollutants in the soil by plant roots
3	Phytodegradation	Degradation of organic xenobiotics by plant enzymes within plant tissues
4	Rhizodegradation	Degradation of organic xenobiotics in the rhizosphere by rhizospheric microorganisms
5	Phytovolatilization	Conversion of pollutants to volatile form and their subsequent release to the atmosphere
6	Phytofiltration	Sequestration of pollutants from contaminated waters by plants

and auxins (Glick et al., 1995). In plants, ethylene inhibits cell division, DNA synthesis, and growth in the meristems of roots, shoots, and axillary buds without influencing RNA synthesis. Indirect impact of PGPR is usually achieved by increasing the plant tolerance to diseases (Guo et al., 2004). PGPR species belong to many different genera, such as *Azoarcus, Azospirillum, Rhizobium, Azotobacter, Arthrobacter, Bacillus, Clostridium, Enterobacter, Gluconoacetobacter, Pseudomonas,* and *Serratia* (Cheng, 2009; Arora, 2015). PGPR play a major role in the phytoremediation of different pollutants and heavy metals. Phytoremediation is the power of plants to clean the environment. Phytoremediation is categorized into the six groups: phytoextraction, phytostabilization, phytodegradation, rhizodegradation, phytovolatilization, and phytofiltration, as shown in Table 10.1.

10.1.1 Phytoextraction (Phytoaccumulation, Phytoabsorption, or Phytosequestration)

Uptake of pollutants or metal contaminants from the soil by plant roots into the aboveground portions of the plants is called *phytoextraction*. It is also known as phytoaccumulation, phytoabsorption, or phytosequestration. Phytoextraction is primarily used for the treatment of contaminated soils. In this approach plants absorb, concentrate, and precipitate the toxic metals from contaminated soils into the shoots and leaves. This technique preferentially uses hyperaccumulator plants, which have ability to store high concentrations of specific metals in their aerial parts (0.01 percent to 1 percent dry weight, depending on the metal). *Elsholtzia splendens, Alyssum bertolonii, Thlaspi caerulescens, Pteris vittata, Phragmites cummunis, Typha angustata,* and *Cyperus esculentus* are known examples of hyperaccumulator plants for copper (Cu), nickel (Ni), zinc (Zn)/cadmium (Cd), arsenic (As), iron (Fe), and lead (Pb). There are several advantages of phytoextraction. The cost of phytoextraction is fairly low when compared to conventional methods for the removal of pollutants. Another benefit of phytoextraction is that the contaminant is permanently removed from the soil.

10.1.2 Phytostabilization (Phytoimmobilization)

This is also referred to as in-place inactivation or immobilization. It is used for the remediation of soil, sediment, and sludge. In this process some specific plant species immobilize contaminants in the soil and groundwater through absorption and accumulation by roots, adsorption onto roots, and precipitation within rhizospheric zone of plants. Organic or inorganic contaminants are incorporated into the lignin of the cell wall of root cells or into humus. Metals are precipitated as insoluble forms by direct action of root exudates and subsequently trapped in the soil matrix. This process decreases the mobility of the contaminants, prevents migration to the groundwater, and reduces bioavailability of metals in the food chain. It is useful for the treatment of lead Pb, As, Cd, Cr, Cu, and Zn. Some of the advantages associated with this technology are that the disposal of hazardous material/ biomass is not required and it is very effective when rapid immobilization is needed to preserve ground and surface waters. The presence of plants also reduces soil erosion and decreases the amount of water available in the system. However, this clean-up technology has several major disadvantages, including contaminant remaining in soil.

10.1.3 Phytodegradation (Phytotransformation)

Phytodegradation is also known as *phytotransformation*. In the phytodegradation process, the organic contaminants are metabolized or mineralized inside plant cells when they have been taken by the plants. During this process many specific enzymes that include nitroreductases (degradation of nitroaromatic compounds), dehalogenases (degradation of chlorinated solvents and pesticides), and laccases (degradation of anilines) are involved. *Populus* species and *Myriophyllium spicatum* are examples of plants that have these enzymatic systems. Phytodegradation has been showed to remediate some organic contaminants, such as chlorinated solvents, herbicides, and munitions.

10.1.4 Rhizodegradation (Phytostimulation)

This is also referred to as *phytostimulation*. Microbial diversity exists in to the rhizospheric zone of plants breakdown the contaminants within rhizosphere. It is believed to be carried out by bacteria or other microorganisms. It is documented in several studies that 100 times more microorganisms are present in rhizosphere soil as compare to the soil outside the rhizosphere. Microbial diversity may be so prevalent in the rhizosphere because the plant exudes sugars, amino acids, enzymes, and other compounds that can stimulate bacterial growth. The roots also provide additional surface area for microbes to grow. The localized nature of rhizodegradation means that it is primarily useful in contaminated soil, and it has been investigated and found to have at least some successes in treating a wide variety of mostly organic chemicals, including petroleum hydrocarbons, polycyclic aromatic hydrocarbons (PAHs), chlorinated solvents, pesticides, polychlorinated biphenyls (PCBs), benzene, toluene, ethylbenzene, and xylenes.

10.1.5 Phytovolatilization

In this process plants take up contaminants from the soil, transforming them into nontoxic volatile forms and transpiring them into the atmosphere. Some element ions of the groups IIB, VA, and VIA of the periodic table (specifically Hg, Se, and As) are absorbed by the roots, converted into nontoxic forms, and then released into the atmosphere. Phytovolatilization may also refer to the diffusion of contaminants from the stems or other plant parts that the contaminant travels through before reaching the leaves. Phytovolatilization can occur with contaminants present in soil, sediment, or water. Mercury is the primary metal contaminant that this process has been used for. It has also been found to occur with volatile organic compounds, including trichloroethene, also as well as inorganic chemicals that have volatile forms, such as selenium and arsenic. The advantage of this method is that the contaminant, mercuric ion, may be transformed into a less toxic substance. The disadvantage of this is that the mercury released into the atmosphere is likely to be recycled by precipitation and then redeposited back into lakes and oceans, repeating the production of methylmercury by anaerobic bacteria.

10.1.6 Phytofiltration

In this process plants are absorbed, concentrated, or precipitated contaminants, particularly heavy metals or radioactive elements, from an aqueous medium through their root or other submerged organs. Plants with high root biomass or high absorption surface with more accumulation capacity (aquatic hyperaccumulators) and tolerance to contaminants achieve the best results. Promising examples include *Helianthus annus, Brassica juncea, Phragmites australis, Fontinalis antipyretica* and several species of *Salix, Populus, Lemna* and *Callitriche*. Many plant species have been successfully applied in the removal of organic and inorganic pollutants from water and soil. For example, poplars remove and metabolize polychlorinated biphenyls (PCBs), 2,4,6-trinitrotoluene (TNT), hexahydro-1,3,5-trinitro-1,3,5-triazine (RDX), and trichloroethylene (TCE); the Brassica family (Indian mustard and broccoli), some wetland plants, that is, *Phragmites cummunis, Typha angustifolia,* and *Cyperus esculentus* take up heavy metals such as cadmium, manganese, copper, iron, chromium, nickel, and zinc (Chandra and Sangeeta, 2011; Sangeeta and Chandra, 2011). However, not all the plants can play such a big role in phytoremediation. First, they should survive and grow fast in a contaminated environment and tolerate high concentrations of pollutants. Second, at best they should be plants native to the desired site because every plant has its own habitat. Third, they should have high capacities to remove one or more pollutants. Therefore, it is a very challenging to find the right plants for each and every pollutant.

10.2 Siderophore and Its Types

Bacteria, fungi, and plants that secrete some small molecules that have capacity to chelate iron are known as *siderophores*. Siderophores are usually classified by the ligands used to chelate the ferric iron. Citric acid can also act as a siderophore. The wide variety of siderophores may be produced due to evolutionary pressures placed on microbes to produce structurally different siderophores which cannot be transported by other microbes specific active transport systems or in the case of pathogens deactivated by the host organism. Molecular structures of siderophores can be characterized by mass spectrometry, NMR spectroscopy, and X-ray diffraction. The major groups of siderophores include the hydroxamates, catecholates (phenolates), carboxylates (e.g., derivatives of citric acid), and mixed types as shown in Figure 10.2.

10.2.1 Hydroxamate Siderophores

10.2.1.1 Desferrioxamine (DFO)

Desferrioxamine is a biological product derived from ferrioxamine B. Ferrioxamine B is an iron-bearing metabolite produced by actinomycetes, *Streptomyces pilosus*, and *Streptomyces coelicolor* belongs to the group of the sideramines. Desferrioxamine B also known as

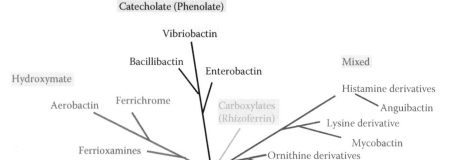

FIGURE 10.2
Different types of siderophores.

Deferoxamine, desferoxamine B (DFO-B), DFO-A (brand name, Desferal; Figure 10.3a). Desferrioxamine is a biological product with a chainlike molecule with three hydroxamic acid groups. In a crab-like manner, the chain-like molecule of desferrioxamine entraps and wraps itself around the iron or aluminum ion, binds it, and chelates it, resulting in stable octahedral complexes, namely, ferrioxamine B. This complex is readily water soluble and easily removes from the kidney. Hence, it has medical applications as a chelating agent

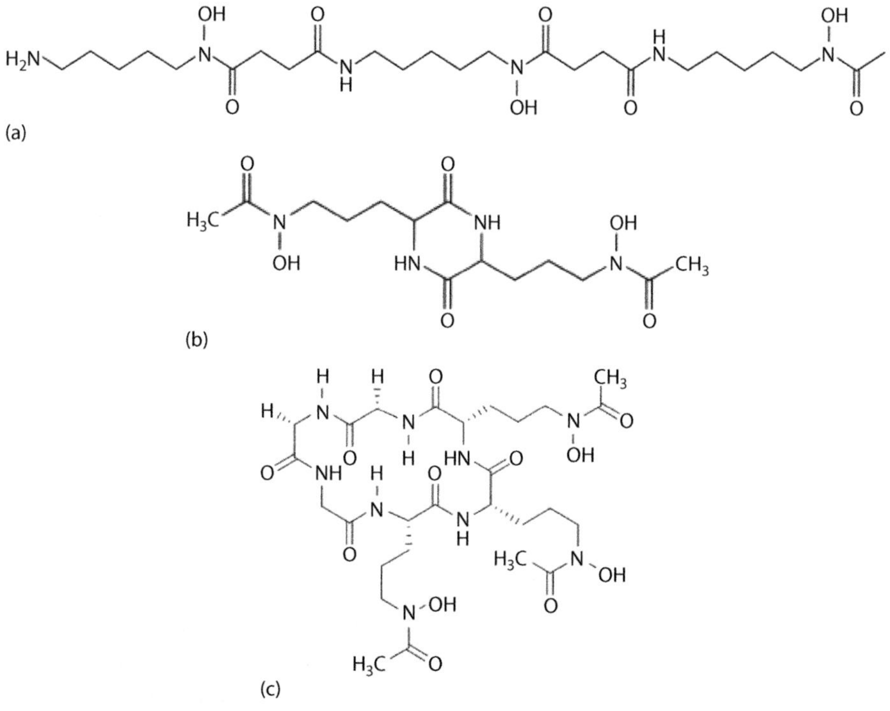

FIGURE 10.3
Structures of different types of hydroxymate siderophores: (a) Deferoxamine, (b) ferrichrome, and (c) rhodo-torulic acid.

used to remove excess iron from the body. It is specifically used in iron overdoses; hemo-chromatosis, either due to multiple blood transfusions or an underlying genetic condition; and aluminum toxicity in people on dialysis. It is used by injection into a muscle, vein, or under the skin. The mesylate salt of DFO-B is commercially available. Deferoxamine E is also secreted by *Streptomyces coelicolor*.

10.2.1.2 Ferrichrome

Ferrichrome was first isolated in 1952. It has been found to be produced by fungi of the genera *Aspergillus*, *Ustilago*, and *Penicillium*. Ferrichrome is a cyclic hexapeptide that forms a complex with iron atoms as shown in Figure 10.3b. It is a siderophore composed of three glycine and three modified ornithine residues with hydroxamate groups [-N(OH)C(=O) C-]. The six oxygen atoms from the three hydroxamate groups bind Fe(III) in near-perfect octahedral coordination.

10.2.1.3 Rhodotorulic Acid

It is the smallest of the 2,5-diketopiperazine family of hydroxamate siderophores which are high-affinity chelating agents for ferric iron, produced by bacterial and fungal phyto-pathogens for scavenging iron from the environment. 2,5-Diketopiperazine, also known as piperazine-2,5-dione and as the cyclodipeptide cyclo (Gly-Gly), is an organic compound and the smallest cyclic dipeptide that consists of a six-membered ring containing two amide linkages where the two nitrogen atoms and the two carbonyls are at opposite posi-tions in the ring.

Rhodotorulic acid is a tetradentate ligand; it binds one iron atom in four locations (two hydroxamate and two ketone moieties), and forms $Fe_2(siderophore)_3$ complexes to fulfil an octahedral coordination for iron (Figure 10.3c). Rhodotorulic acid occurs in basidio-mycetous yeasts and was found to retard the spore germination of the fungus *Botrytis cinerea*. In combination with yeast *R. glutinis* it was found to be effective in the biocontrol of iprodione-resistant *B. cinerea* of apple wounds caused by the disease.

10.2.1.4 Aerobactin

Aerobactin is a bacterial iron chelating agent (siderophore) found in *E. coli*. It is a virulence factor, enabling *E. coli* to sequester iron in iron-poor environments.

10.2.2 Catecholate Siderophores

10.2.2.1 Enterobactin (Enterochelin)

It is a high-affinity siderophore that acquires iron for microbial systems. It is primarily found in Gram-negative bacteria, such as *Escherichia coli* and *Salmonella typhimurium*. The structure of enterobactin is shown in Figure 10.4a. Enterobactin is the strongest known siderophore, binding to the ferric ion (Fe^{3+}) with the affinity (K = 1052 M^{-1}). This value is substantially larger than even some synthetic metal chelators, such as EDTA. Due to its high affinity, enterobactin is capable of chelating even in environments where the con-centration of ferric ion is held very low, such as within living organisms. Enterobactin can extract iron even from the air. Pathogenic bacteria can steal iron from other living organ-isms using this mechanism, even though the concentration of iron is kept extremely low due to the toxicity of free iron.

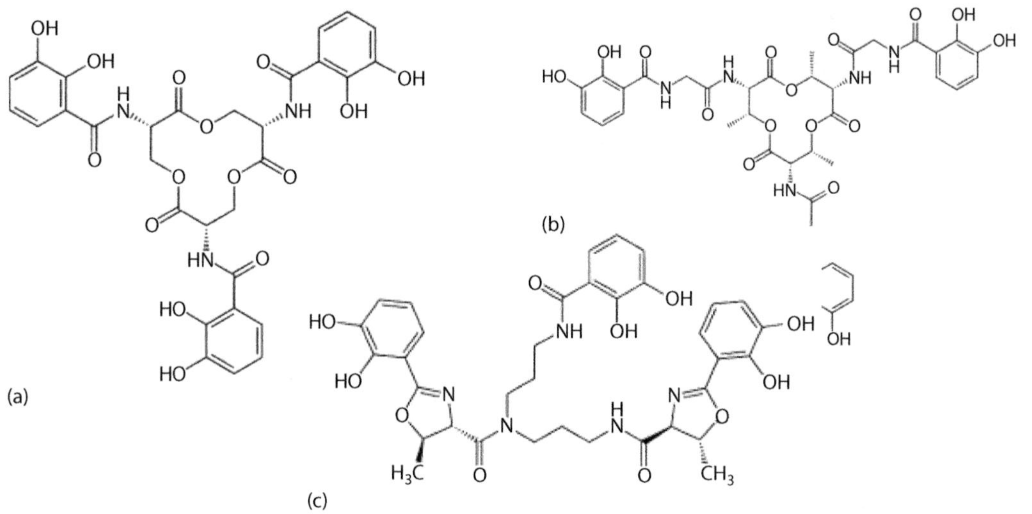

FIGURE 10.4
Structures of different types of catecholate siderophores: (a) Enterobactin, (b) bacillibactin, and (c) vibriobactin.

10.2.2.2 Bacillibactin

Bacillibactin is a catechol-based siderophore secreted by members of the genus *Bacillus*, including *Bacillus anthracis* and *Bacillus subtilis*. The structure of bacillibactin is shown in Figure 10.4b. It is involved in the chelation of ferric iron (Fe^{3+}) from the surrounding environment and is subsequently transferred into the bacterial cytoplasm via the use of ABC transporters.

10.2.2.3 Vibriobactin

A novel siderophore has been isolated from low iron cultures of *Vibrio cholera*. The structure of vibriobactin is shown in Figure 10.4c. Vibriobactin belongs to the catecholamide family of chelators, and has been shown to contain three residues of 2,3-dihydroxybenzoic acid and two residues of threonine. Both threonine moieties are present in the form of oxazoline rings.

10.2.3 Carboxylate-Type Siderophore

This is a novel class of siderophores whose members possess neither hydroxamate nor phenolate ligands; rather, iron binding is achieved by hydroxyl carboxylate and carboxylates. Rhizoferrin is a member of a new class of siderophores (microbial iron transport compounds) based on carboxylate and hydroxy donor groups rather than the commonly encountered hydroxamates and catecholates. Rhizoferrin is composed of diaminopropane symmetrically acylated with citric acid via amine bonds to the terminal carboxylate of citric acid. These siderophores are found in the kingdom of bacteria as well in fungi. Rhizoferrin is the only known carboxylate siderophore produced by fungi, specifically synthesized by members of the zygomycetes. Carboxylate siderophore include rhizobacteria, staphyloferrin, and rhizopherrin. Interestingly, both fungi and bacteria produce rhizoferrin, fungi produce only R, R-rhizoferrin, while a few bacteria produce enantio-rhizoferrin S, S-rhizoferrin.

10.2.4 Siderophores with Mixed Ligands

10.2.4.1 Pyoverdine

Pyoverdine, or pyoverdin (PVD), are fluorescent siderophores produced by certain *Pseudomonas* bacteria species as shown in Figure 10.5a. PVD was discovered in 1892 and was originally described using different names (fluorescein; pseudobactin in soil isolates). All are composed of three parts: (1) A conserved fluorescent dihydroxyquinoline chromophore; (2) an acyl side chain (either dicarboxylic acid or amide) bound to the amino group of the chromophore; and (3) a variable peptide chain linked by an amide group bound to the C1 (rarely C3) carboxyl group of the chromophore. PVD represents the primary iron uptake system of fluorescent *Pseudomonas*, although some species can also synthesize additional siderophores such as pyochelin or quinolobactin, or can acquire iron bound to a variety of exogenous chelators, including many heterologous siderophores. The structure of PVD is highly variable among species and even between strains. In *Pseudomonas aeruginosa* PAO1 there are 14 PVD genes involved in the biosynthesis of pyoverdine. Unlike enterobactin, pyoverdine/pseudobactins are iron-binding non-ribosomal peptides containing a dihydroxyquinoline derivative. The structure of the peptide differs between *Pseudomonas* and more than 40 structures have been described, while the chromophore, (1S)-5-amino-2,3-dihydro-8,9-dihydroxy-1H-pyrimido[1,2-a]quinoline-1-carboxylic acid, is the same with the exception of azobactin from *Azotobacter vinelandii*, which possesses an extra urea ring.

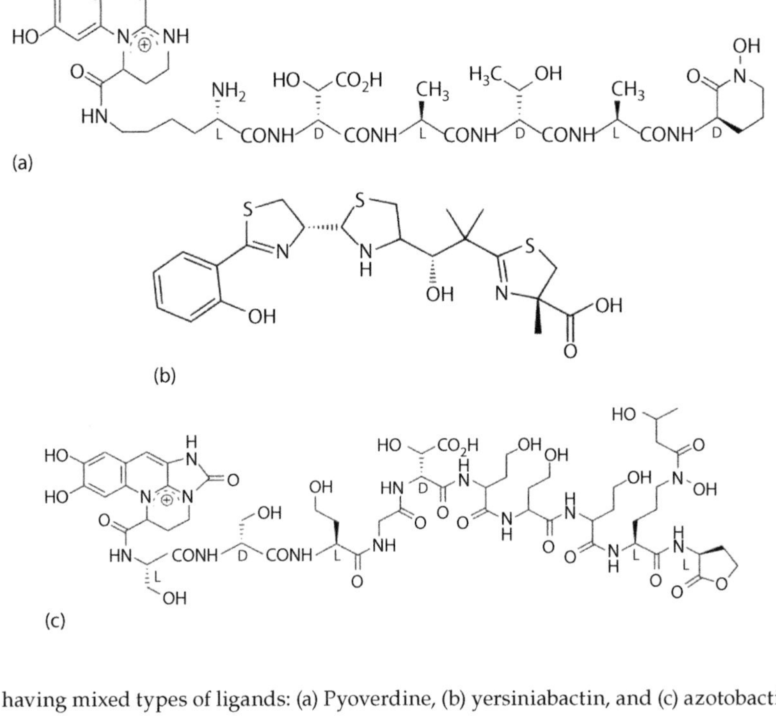

FIGURE 10.5
Siderophores having mixed types of ligands: (a) Pyoverdine, (b) yersiniabactin, and (c) azotobactin.

10.2.4.2 Mycobactin

Mycobactin is an intracellular lipophilic siderophore whereas carboxymycobactin and exochelin are extracellular siderophores. The main siderophores in *Mycobacterium tuberculosis* are mycobactin and carboxymycobactin. Carboxymycobactin is a large family of siderophores bearing a short alkyl side chain of variable length and unsaturation. They incorporate a terminal methyl ester motif, which enhances polarity and solubility and is essential for iron chelation. Both mycobactins and carboxymycobactins are salicylate derivatives with one modified serine/threonine and two lysine molecules. Most *Mycobacterium* siderophores are mycobactin derivatives.

10.2.4.3 Anguibactin

This class of siderophores utilizes N-hydroxy amino side chains with an oxygen atom as one of the ligands for Fe^{3+}. *Vibrio anguillarum* have at least two different siderophore-mediated systems, namely anguibactin and vanchrobactin. *V. anguillarum* is a marine pathogen that causes serious hemorrhagic septicemia in wild and cultured fish. The non-ribosomal peptide anguibactin represents a unique structural class of siderophores with both a catechol and hydroxamate ligand and a thiazole core. The biosynthetic genes encoding this compound are found on a 65-kb virulence plasmid in some *V. anguillarum* strains. Knockout of genes involved in anguibactin production attenuated virulence, confirming that anguibactin is a prerequisite for successful host invasion of this bacterium. In contrast, the catechol vanchrobactin is chromosome-encoded, and interestingly, the coding genes are silenced in anguibactin-producing strains. Vanchrobactin activity was not detected from the strain that produces anguibactin due to the competition for iron between two siderophores as anguibactin has higher iron affinity than vanchrobactin.

10.2.4.4 Yersiniabactin

Yersiniabactin is a siderophore found in the pathogenic bacteria *Yersinia pestis*, *Yersinia pseudotuberculosis*, and *Yersinia enterocolitica*, as well as several strains of enterobacteria, including enteropathogenic *Escherichia coli*. The structure of yersiniabactin is shown in Figure 10.5b. Siderophores, compounds of low molecular mass with high affinities for ferric iron are important virulence factors in pathogenic bacteria. Iron an essential element for life utilized by such cellular processes as respiration and DNA replication is extensively chelated by host proteins like lactoferrin and ferritin; thus, the pathogen produces molecules with an even higher affinity for Fe^{3+} than these proteins in order to acquire sufficient iron for growth. As a part of such an iron uptake system, yersiniabactin plays an important role in pathogenicity of *Y. pestis*, *Y. pseudotuberculosis*, and *Y. entercolitica*.

10.2.4.5 Azotobactin

The free-living N_2-fixing bacterium *Azotobacter vinelandii* uses catechol siderophores for the uptake of molybdenum (Mo) and vanadium (V), in addition to Fe. Under diazotrophic growth (an organism that is able to grow without external sources of fixed nitrogen), *A. vinelandii* releases compounds such as the tris-catechol protochelin and the bis-catechol azotochelin, which bind MoO_4^{2-} and VO_4^{3-} in strong complexes. These complexes are then taken up through regulated transport systems. The complexation of molybdate and

vanadate by the catechol compounds releases them from organic matter and oxide surfaces to which they are bound in soils and makes them available for uptake. At high concentrations, where the free concentrations of MoO_4^{2-} and VO_4^{3-}, along with that of WO_4^{2-},
are toxic, catechol complexation and down regulation of the uptake systems serve as an
effective detoxification mechanism. Catecholate siderophores, which may more appropriately be called *metallophores*, thus serve multiple functions and help control the acquisition of several metals by *A. vinelandii*. In addition to catechol siderophores, *A. Vinelandii*
releases the pyoverdine-like siderophore azotobactin, which is also produced by other
Azotobacter isolates. Catechol siderophores, particularly protochelin, are released by
A. vinelandii (wildtype) under a variety of conditions, but azotobactin seems to be released
only under severe Fe limitation. Therefore, azotobactin is sometimes considered the "true"
siderophore of this bacterium. Azotobactin possesses three different types of coordinating
moieties, a hydroxamate, an R-hydroxy acid, and a catechol, making it a representative of
several classes of siderophores (Figure 10.5c).

10.2.5 Phytosiderophores (PS)

Phytosiderophores are organic substances (such as nicotinamine, mugineic acids, avenic
acid, etc.) produced by plants under Fe-deficient conditions, which can form organic complexes or chelates with Fe^{3+} and increase the movement of iron in soil. It is nonproteinaceous,
with low molecular weight acids released by the graminaceous species under Fe and Zn
deficiency stress. The PS mobilizes micronutrients Fe, Zn, Mn, and Cu from the soils to the
plant in deficient condition. Fe is an essential micronutrient for plant growth. Under conditions of Fe deficiency, graminaceous plants (e.g., barley and wheat) have developed an
efficient strategy for acquiring Fe from insoluble sources and possibly other metals from
the environment. These plants secrete Fe^{3+}-chelating compounds called *phytosiderophores*
that form specific strong complexes with Fe^{3+}. Phytosiderophores have been shown to
enhance soil mobility of Fe, Mn, and Zn as much as microbial siderophores and more than
some synthetic chelators. Some researchers have shown that phytosiderophores are up to
100 times more efficient than anthropogenic or bacterial iron chelators for iron uptake into
graminaceous plants. They are notably similar to ethylenediaminetetraacetic acid (EDTA)
in structure and metal-binding chemistry. Both have multi-carboxylate chelating groups
and form extremely strong complexes with iron and other transition metal ions. Fe^{3+} is relatively insoluble in neutral and alkaline soils which make it unavailable for plants. Hence,
two alternative mechanisms for Fe acquisition have evolved in plants. Nongraminaceous
monocots and dicots, the "strategy I plants," acidify the rhizosphere (presumably via an
H^+-ATPase) to increase Fe solubility and use a ferric-reductase to reduce Fe^{3+} to Fe^{2+}, which
is transported into the roots via an Fe^{2+} transporter. Graminaceous species utilize strategy II, whereby a metal-binding ligand, mugineic acid (MA), is synthesized enzymatically
from three molecules of S-adenosyl methionine and is secreted from roots to bind Fe^{3+} in
the rhizosphere (Figure 10.6). The phytosiderophores are hexadentate ligands that coordi-
nate Fe^{3+} with their amino and carboxyl groups. When the phytosiderophore is released to
the rhizosphere, it chelates Fe from the soil by forming Fe^{3+}-phytosiderophore complexes
that can be subsequently transported across the root plasma membrane.

In comparison with the molecular mass of microbial siderophores, which range between
200 and 2000 Da, phytosiderophores range between 500 and 1000 Da (Neilands, 1981).
Mugineic acid is the most common and the first identified phytosiderophore. The stability constant of the MA-Fe^{3+} complex is K = 1020, which is low compared with the stability constant of microbial siderophores, such as ferrichrome (K = 1029), ferrioxamine B

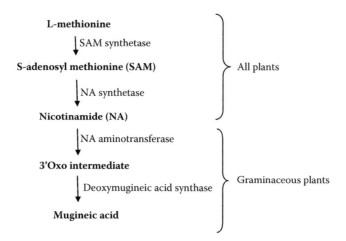

FIGURE 10.6
Biosynthesis of phytosiderophore nicotinamide and mugineic acid.

(K = 1031), and enterobactin (K = 1052). MA is closely related to its biochemical precursor, nicotinamine, and to a number of other compounds that have also been identified as phytosiderophores in graminaceous plants: 3-hydroxymugineic acid, 2'-deoxymugineic acid, avenic acid, and distichonic acid. The binding coordinating groups are two amine-N, two carboxylate-O, and one a-hydroxycarboxylate site (two O)—a total of two Ns and four Os, forming a tight octahedron in which the central Fe(III) atom resides. Other types of phytosiderophores have also been isolated from graminaceous plants, such as avenic acid A from oats (*Avena sativa*) and distichonic acid from beer barley (*Hordeum vulgate*). In general, several studies showed that plant species such as barley, rye, and wheat, which produce a high concentration of phytosiderophores are more resistant to Fe deficiency than other species such as maize, sorghum, and rice, which produce a lower concentration of phytosiderophores.

10.3 Synthesis of Siderophore and Mechanism of Siderophore Export

During intracellular iron deficiency conditions, microorganism secrete siderophores into the environment to scavenge the iron. Although iron deficiency is the key factor regulating the synthesis of siderophores, other external factors like pH, temperature, carbon source, metals, and other pollutants play an important role in siderophore production. There are two pathways involved in the synthesis of siderophores: (1) Dependent on nonribosomal peptide synthatases (NRPS) and (2) independent of NRPS. The synthesis of aryl-capped siderophores such as pyochelin and pyoverdine in *P. aeruginosa*, enterobactin in *E. coli enterica*, *Klebsiella* spp. and mycobactin in *M. tuberculosis* are primarily dependent on NRPS. NRPS are large multi-enzyme complexes responsible for the synthesis of several biologically important peptidic products without an RNA template. In general, NRPS consists of three domains: (1) adenylation domain; (2) peptidyl carrier protein domain (PCP or thiolation); and (3) condensation domain, responsible for the assembly of a wide array of amino, carboxy, and hydroxy acids in various combinations to produce polypeptides with high structural variability. The adenylation domain first activates and recognizes the amino acid which is then bound to a cofactor in the thiolation domain and then is incorporated in the growing polypeptide chain through the formation of the peptide chain by the condensation domain. Finally, the polypeptide chain is released from the synthetase by a cyclization event catalyzed by the C terminal thioesterase domain. Aryl-capped siderophores are synthesized by the NRPS. The genes encoding the enzymes responsible for the synthesis of aryl acids (2,3-dihydrobenzoic acid [DHB] and salicylate) and NRPSs are regulated by the Fur repressor. In *E. coli* enterobactin biosynthesis, the product of genes entB, entC, and entA are responsible for the synthesis of DHB. Once the aryl acid (DHB) is synthesized, it, together with amino acids (L-serine), leads to the assembly of enterobactin by the NRPSs. The enterobactin NRPS system consists of three enzymes EntE, EntB (C terminal), and EntF responsible for enterobactin assembly. The hydroxymate and carboxylate siderophores, for example, petrobactin in *B. anthrasis*, alcaligin in *Bordetella pertussis* staphyloferrin A, and staphyloferrin B *in S. aureus*, are assembled by the NRPS-independent mechanism. The regulation of iron balance and siderophore utilization in Gram-negative and low GC-content Gram-positive bacteria (e.g., *B. subtilis*) is carried out by Fur protein. Whereas in high GC-content Gram-positive bacteria, such as *Streptomyces* and *Mycobacteria*, the diphtheria toxin regulator (DtxR) performs the function. In addition to the global repressor Fur, there are several transcriptional regulators that control siderophore biosynthesis and utilization. These mainly act as activators by sensing the iron–siderophore complex, either intracellularly or extracellularly, and can be classified in various classes such as (1) the two-component sensory transduction system; (2) alternative sigma factors, for example, the FecA–FecR–FecI regulatory proteins in *E. coli* and the FpvI/Pvd-FpvRFpvA system in *P. aeruginosa*; (3) other transcriptional factors, for example, LysR family regulators IrgB in *V. cholerae* and FetR in *V. anguillarum* 775(pJMI); (4) AraC-type regulators, for example, the PchR in *P. aeruginosa*, PdtC in *P. stutzeri*, MpeR in pathogenic *Neisseria*, YbtA in *Yersinia pestis*, and AlcR in *Bordetella*. The extra cytoplasmic function (ECF) sigma factor and AraC systems are in turn regulated by Fur.

The mechanism involved in the secretion or export of siderophores outside the cell is carried out by a transport protein or a pump. There are three major types of proteins identified as involved in this process: The major facilitator superfamily (MFS); the resistance, nodulation, and cell division (RND) superfamily; and the ABC superfamily. In *E. coli*, enterobactin export is carried out by a MFS protein named EntS encoded by the gene ybda. Recently, bacillibactin secretion in *B. subtilis* was found to be carried out by a similar MFS-type transporter YmfE. In *P. aeruginosa*, the secretion of pyoverdine was thought to be carried out by the efflux system MexA–MexB–OprM, which is a typical RND superfamily transport protein. Representatives of the ABC type of transporters involved in siderophore export are found in *S. aureus*, *M. tuberculosis*, and *M. smegmatis*.

10.4 Transport of Iron-Siderophore Complex

Once the Fe^{3+}-siderophore complex is accessible for cellular uptake, it is internalized in either of these two general ways: (1) the iron is released from the complex and enters the cell as a single ion (in filamentous algae and fungi) or (2) the whole Fe^{3+}-siderophore complex is entered into the cell, for example, in most bacterial systems (Figure 10.7).

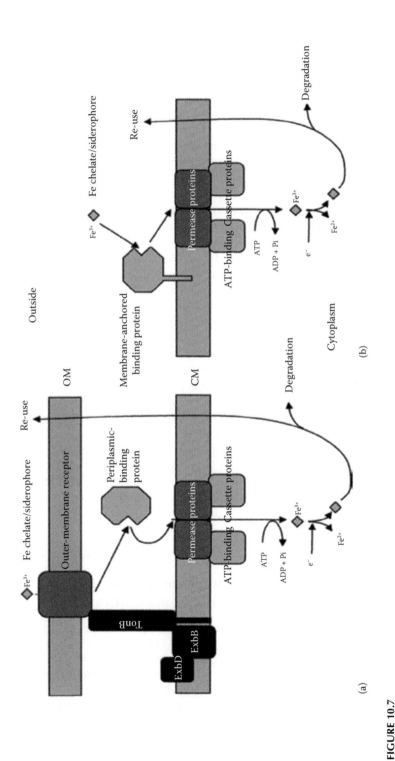

FIGURE 10.7

Schematic representation of siderophore-mediated iron uptake in Gram-negative (a) and Gram-positive (b) bacteria. (Reprinted from Andrews, S.C., Robinson, A.K., Rodriguez-Quinones, F., *FEMS Microbiology Reviews*, 27, 3, 215–237, 2003.)

Once inside the periplasm, the Fe^{3+}-siderophore complex is transported across the inner membrane either through the ABC transporters (ferrichrome and ferrienterobactin in *E. coli*) or permeases (ferripyoverdine in *P. aeruginosa*). The release of iron from the Fe^{3+}-siderophore complex is also different between microbes and different siderophore types. For example, in *E. coli*, ferrichrome and ferrienterobactin pathway iron is released in the cytoplasm, whereas in *P. aeruginosa* ferric-pyoverdine pathway iron is released in the periplasm. The uptake of Fe^{3+}-siderophore complex is mediated through the outer membrane receptor proteins, such as FepA (for enterobactin), FhuA (for ferrichrome), and FecA (for ferric citrate) in *E. coli* and FpvA and FptA in *P. aeruginosa*. The structure and mechanism of the action of these receptors have recently been well studied, which suggests that a conformational change of particular domains of these receptors led to the passage of substrate. The energy required for the transport is supplied by the TonB protein complex, which consists of two other inner-membrane proteins, ExbD and ExbB. The role of the TonB–ExbD–ExbB complex in ferric–siderophore transfer has been extensively studied in *P. aeruginosa* and *E. coli*. The ABC-type transporter proteins are also involved in the delivery of iron–siderophore complexes into the cytosol. In Gram-negative bacteria these are extracytoplasmic substrate-binding proteins located in the periplasm, whereas in Gram-positive bacteria, these are present as lipoproteins attached to the external surface of the cell membrane. Upon interaction with the extracytoplasmic substrate binding unit of the ABC transporter, the Fe-siderophore complex is channeled through the membrane. The energy required for this process is provided by the cytoplasmic subunits of the ABC receptors that undergo a dimerization or conformational change due to NTP binding or hydrolysis.

Four subtypes of ABC transporters have been described in Gram-negative bacteria which are associated with Fe-siderophore uptake: (1) The *E. coli* FepBDGC system for Fe-enterobactin uptake is the most common subtype, with one extra cytoplasmic protein for substrate binding, two membrane proteins acting as a transmembrane channel, and one cytoplasmic subunit; (2) the *E. coli* FhuDCB system where the transmembrane component comprises only a single polypeptide chain of FhuB; (3) the *Vibrio cholera* ViuP system for Fe-vibriobactin uptake, which comprises lipoproteins for substrate uptake, a feature typical for Gram-positive bacteria; and (4) in the *Y. pestis* YbtPQ system for Fe-yersiniabactin uptake, which consists of a transmembrane segment and a nucleotide binding fold. In yeast and filamentous fungi, the iron–siderophore complex is imported by MFS transporters.

10.5 Role of Quorum-Sensing in Siderophore and Biofilm Formation in Rhizospheric Bacterial Communities

Siderophore production is most common among plant growth–promoting rhizosphere bacteria, which exhibit their optimum growth and siderophore production activity at extreme environmental conditions, including the scarcity of nutrients or the presence of elevated concentrations of heavy metals and other pollutants in the environment. It has been also reported that siderophore biosynthesis in rhizospheric bacteria and pathogenic bacterium, that is, *Pseudomonas aeruginosa*, is controlled by a quorum-sensing system. Quorum sensing is cell density dependent on regulations of specific gene expressions in response to extracellular chemical signals produced by the bacteria themselves, called *autoinducers* (AIs). Quorum sensing can be divided into at least four steps: (1) production of small biochemical signal molecules by the bacterial cell; (2) release of the signal molecules, either actively or passively, into the surrounding environment; and (3) recognition of the signal molecules by specific receptors once they exceed a threshold concentration, leading to (4) changes in gene regulation.

The first example of quorum sensing was observed in the Gram-positive, pathogenic bacteria *Streptococcus pneumoniae*. *S. Pneumoniae* cells produce and release a small protein into the environment. As population size increases, the amount of peptide increases. The concentration of the peptide is high enough to convert some cells in the population from a noncompetent state to a competent state. In the competent state, cells are able to take up DNA, a process called *transformation*. In addition, the competent cells release a chemical called *bacteriocin* that lyses the cells in the population that did not become competent. When the noncompetent cells lyse, they release DNA and virulence factors that allow the competent cells to invade tissue in a host organism, causing serious diseases such as pneumonia and meningitis. After the quorum-sensing system of *S. pneumoniae* was discovered, the term was coined years later when the quorum-sensing system of a very different bacterium was discovered. The quorum sensing reported in the bioluminescent marine bacterium *Vibrio fischeri* was considered the paradigm for quorum sensing in most Gram-negative bacteria. *V. fischeri* colonizes the light organ of the Hawaiian squid *Euprymna scolopes*. Luminescent bacteria are found in free-living, symbiotic, saprophytic, or parasitic relationships. The symbiotic relationship between *E. scolopes* and *V. fischeri* provides an example of specific cooperativity during the development and growth of both. For instance, once *V. fischeri* cells utilize type IV pili (some pili, called type IV pili, generate motile forces) to interact with the squid host, the bacteria grow to high cell density and induce the expression of genes required for bioluminescence. *V. fischeri* are helpful to the squid, a nocturnal forager, by erasing the shadow that would normally be seen as the moon's rays strike the squid, therefore protecting the squid from its predators. The squid, in turn, provides the bacteria with shelter and a stable source of nutrients.

The QS system involves the following elements:

1. The autoinducers
2. The signal synthase
3. The signal receptor
4. The signal response regulator
5. The regulated genes

Gram-positive and Gram-negative bacteria use different types of QS systems. Gram-positive bacteria regulate a variety of processes in response to an increasing cell population density. Gram-positive bacteria use peptides, called *autoinducer peptides* (AIPs) as signaling molecules. AIPs are processed and actively exported outside the cell. When concentration of the AIP is high to the threshold concentration, they interact to a cognate membrane-bound two component histidine kinase receptor. Usually binding activates the receptor's kinase activity, it autophosphorylates, and passes phosphate to a cognate cytoplasmic response regulator. These regulators influence the transcription of specific genes so that each sensor protein is highly selective for a given peptide signal (Figure 10.8a). In some cases of Gram-positive bacterial QS, AIPs are transported back into the cell cytoplasm

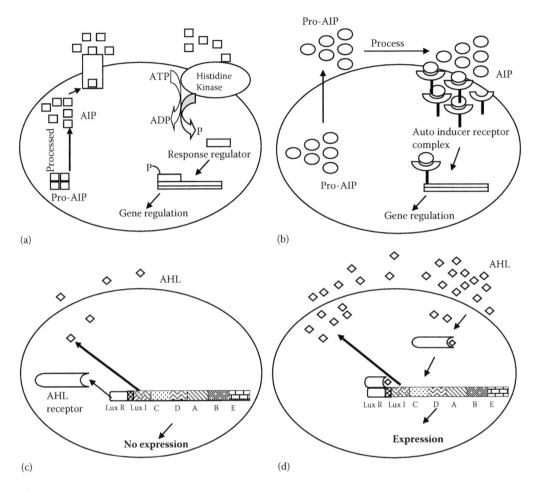

FIGURE 10.8
Bacterial quorum sensing circuits. Autoinducing peptide (AIP) in Gram-positive bacteria by (a) two-component signaling, (b) an AIP-binding transcription factor, (c) QS in Gram-negative bacteria at low cell density and acyl-homoserine lactone (AHL) results in no expression, and (d) at high AHL concentration leading to expression of gene.

where they interact with transcription factors to modulate the transcription factor's activity and in turn modulate gene expression changes (Figure 10.8b).

While Gram-negative bacteria communicate with each other using small molecules, such as AIs. Generally they produce acylated homoserine lactone (AHL). AHL are composed of a lactonized homoserine moiety with an acyl chain. The length of an acyl chain varies from four to eighteen carbons (Marketon et al., 2002), usually by increments of two carbon units (C4, C6, C8, etc.; Fuqua and Greenberg, 2002). It can be modified by a 3-oxo substituent or a 3-hydroxy substituent a terminal methyl branch and can contain varied degrees of unsaturation. Siderophores or other molecules, such as autoinducers, which are synthesized by a *LuxI*-type enzyme (signal synthase), are encoded by the first gene of the lux operon, as shown in Figure 10.8c,d. At a low bacterial cell density, the low level of transcription of the lux operon is insufficient for the activation of *LuxR*. When the cell density increases and signal levels reach a specified threshold level, *LuxR* activation can take place. The *LuxR*/3-oxo-C6-HSL complex thereby activates transcription via the *lux* operon promoter, giving rise to the expression of other genes, including (in *Vibrio fischeri*) *lux* AB genes encoding luciferase and *lux* CDE, which encodes the enzymes that produce the substrate for luciferase as bioluminescence. Importantly, the *LuxR*-AHL complex also induces expression of *luxI* because it is encoded in the luciferase operon. Two proteins, *LuxI* and *LuxR*, control expression of the luciferase operon (luxICDABE) required for light production. Structural studies of *LuxI*-type proteins indicate that each possesses an acyl-binding pocket that precisely fits a particular side-chain moiety. This structural feature apparently confers specificity in signal production. Thus, each *LuxI* protein produces the correct signal molecule with high fidelity. There are some *LuxI*-type proteins that produce multiple AHLs, although it is not clear if all are biologically relevant. The structures of *LuxR* proteins suggest that *LuxR* proteins also possess specific acyl-binding pockets that allow each *LuxR* to bind and be activated only by its cognate signal. In a wide range of Gram-negative bacteria, quorum sensing was identified to be based on one or more AHL. AHL is a monocarboxylic acid amide that is the N-acyl derivative of homoserine lactone. All AHLs thus far reported are composed of an acyl chain with an even number of carbon atoms ranging from 4 to 14 in length, ligated to the homoserine lactone moiety. Although quorum-sensing–related siderophore biosynthesis has also been reported in the pathogenic bacterium *Burkholderia cepacia*, siderophores produced by this strain and *P. aeruginosa* were thought to be the virulence factors related to their pathogenesis.

To survive under iron-limited conditions or high pollution load conditions, microorganisms are known to biosynthesize siderophores. Recently some marine bacteria have been found to acquire iron through the siderophore desferroixamine.

Also some marine bacteria were found to have stimulated growth under iron-limited nutrient-poor conditions similar to marine environments in the presence of an exogenous siderophore or an exogenous siderophore plus AHL. AHLs are known to be chemical signals for the quorum-sensing activity in many Gram-negative bacteria. The entire well-known quorum sensing systems is based on symbiotic or pathogenic conditions for bacteria in "closed" or "semi-closed" environments. Both siderophores and AHLs have been suggested to play roles as chemical signals for interspecies communication among bacteria resulted into the biofilm formation. Van Leeuwenhoek, using his simple microscopes, first observed microorganisms on tooth surfaces and can be credited with the discovery of microbial biofilms. Heukelekian and Heller observed the "bottle effect" for marine microorganisms, that is, that bacterial growth and activity were substantially enhanced by the incorporation of a surface to which these organisms could attach. However, a detailed examination of biofilms would await the electron microscope, which allowed high-resolution photomicroscopy at much higher magnifications than did the light microscope. In 1969, Jones, Roth, and Saunders used scanning and transmission electron microscopy to examine biofilms on trickling filters in a wastewater treatment plant and showed them to be composed of a variety of organisms (based on cell morphology). By using a specific polysaccharide-stain called ruthenium red and coupling this with osmium tetroxide fixative, these researchers were also able to show that the matrix material surrounding and enclosing cells in these biofilms was polysaccharide. A microbial biofilm is a complex community of microorganisms growing on a biotic or abiotic surface in an aqueous environment. It can be composed of multiple species of organisms including Gram-positive or Gram-negative bacteria along with yeast and protozoa. These adherent cells are frequently embedded within a self-produced matrix of extracellular polymeric substance (EPS).

The EPS is composed with polysaccharide, protein, nucleic acid, and other substances, which help to protect biofilm organisms from various environmental stress factors, such as UV radiation, extreme pH condition, osmotic shock, dehydration, antimicrobial substances, and predators. The development of a biofilm depends on the availability of nutrients and surface for attachment. Biofilm development comprises of four different steps: Attachment of planktonic cells to a surface, formation of microcolonies, maturation of microcolonies, and finally detachment of the mature biofilm structure. Microorganisms constituting biofilm communities have extremely complex and heterogeneous physiology and are very different from their planktonic stage. Biofilm formation can be beneficial or harmful; it varies from situation to situation. One of the best examples of a successful, beneficial application of biofilms to solve a huge problem is in the treatment of wastewater and remediation of contaminated soil and groundwater. In another side, biofilm-forming bacteria become less susceptible to antimicrobial agents and are thus difficult to control, causing biofouling and pathogenicity.

In contrast to QS, quorum quenching (QQ) also works. QQ blocks QS systems and inhibits gene expression for bacterial behaviors, that is, biofilm formation. These are two antagonistic processes coexisting in various bacterial communities during biofilm formation. Many enzymes work as QQ molecules. The first reports of an enzymatic degradation of AHL molecule were reported in soil bacterial isolates of *Variovorax* and *Bacillus genera* (Dong et al., 2000; Leadbetter and Greenberg, 2000). The numerous enzymes involved in degradation or modification of AHL have been reported. They represent four catalytic classes (Figure 10.9): Class I, the lactonases that open the homoserine lactone ring (Zhang et al., 2002; Uroz et al., 2008); Class II, the acylases (amidohydrolases or amidases) that cleave AHLs at the amide bond and release fatty acid and homoserine lactone (Lin et al., 2003); Class III, the reductases that convert 3-oxo-substituted AHL to their cognate 3-hydroxyl-substituted AHL (Bijtenhoorn et al., 2011); and Class IV, cytochrome oxidases that catalyze oxidation of the acyl chain (Chowdhary et al., 2007). They occur in bacteria, archaea, and eukaryotes. But deep research is needed in the areas of QS and QQ during bioremediation of pollutants in situ and ex situ conditions.

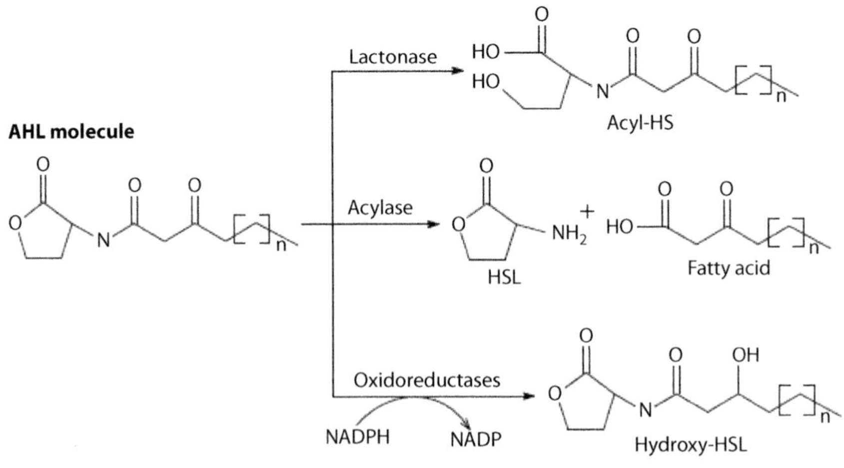

FIGURE 10.9
Possible linkage degraded by quorum-quenching enzymes in quorum-sensing molecule N-acyl homoserine lactone.

10.6 Role of Siderophore

10.6.1 Siderophores Promote Plant Growth

Although iron is a micronutrient, it is required for chlorophyll biosynthesis, redox reactions, and some important physiological activities in plants. Therefore, iron starvation significantly reduces the quantity and quality of crop production. This reduction in crop production also alters the natural food web of the ecosystem. The level of available iron required by plants at neutral pH is around 10–17 mol/L while the level of available iron required by microorganism is 10^{-6} mol/L under the similar condition. For several decades, it has been known that different *Pseudomonas* species can enhance plant growth by producing pyoverdine siderophores. These types of bacteria are therefore considered as plant growth–promoting bacteria. To investigate the role of soil microbial activity in Fe uptake by plant, an experiment was carried out by Masalha et al. (2000) where plants were grown under both sterile and nonsterile conditions on a loess loam soil. After the incubation, it was observed that plants cultivated under nonsterile conditions grew well, exhibiting higher Fe concentrations in the roots.

In contrast, plants grown in the sterile condition showed very little growth and suffered from severe iron deficiency. Through this experiment, Masalha et al. (2000) showed that the production of microbial siderophores was totally suppressed when the plants were grown under sterile conditions.

Thus they concluded that microbial siderophores might be considered an efficient iron source for plants. In agreement with this observation, Crowley (2006) also showed that microbial siderophores are used as the major source of iron in plants. *Escherichia coli* from endo-rhizosphere of sugarcane (*Saccharum sp.*) and rye grass (*Lolium perenne*) is associated with maximum siderophore production and thus enhances plant growth considerably. Siderophores produced by an endophytic *Streptomyces* sp. isolated from the roots of a Thai jasmine rice plant induced plant growth and markedly elevated root and shoot biomass and lengths. Recently, *Trichoderma asperellum* was found to produce siderophore which had a potential role in enhancing cucumber growth by ameliorating salt stress. An investigation conducted on the plant growth–promoting activities of fungi revealed that the siderophores produced by *Aspergillus niger*, *Penicillium citrinum*, and *Trichoderma harzianum* increase the shoot and root lengths of chickpeas (*Cicer arietinum*). Ectomycorrhizal is a type of symbiotic relationship that occurs between a fungal symbiont and the roots of various plant species. In this symbiotic relationship, it was reported that a fungal symbiont depends on fungal siderophores in order to supply iron to the host roots of plants. Sometimes the plant also modifies the structure of root soil microbial community and favors the growth of more siderophore-secreting microbes by secreting phenolic exudates from their roots. This improves the solubility of insoluble iron and enhances plant uptake of iron via microbial siderophores. Besides microbial siderophores, plants can also synthesize phytosiderophore which can chelate the iron directly. In some plants, the sign of iron shortage decreased completely with the rapid consumption of phytosiderophore. Thus, the siderophores originated either from microbes or from plants and are recognized as the potential source of iron for their survival and growth.

10.6.2 Siderophores as Potential Biocontrol Agents

Siderophores play a significant role in the biological control mechanism against certain phytopathogens (Figure 10.10). Siderophores bind with the iron tightly and reduce the bioavailable iron for the plant pathogens, thus facilitating the killing of phytopathogens.

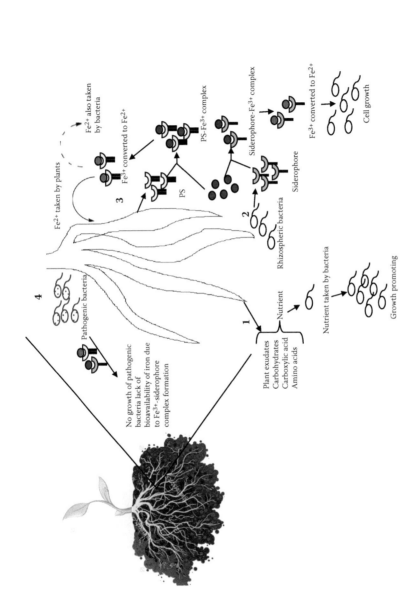

FIGURE 10.10

Diagrammatic presentation of role of phytosiderophore and bacterial siderophore. (1) During plant growth, roots either actively or passively release a range of organic compounds. Among them are exudates, mainly carbohydrates, carboxylic acids, and amino acids, which are passively released along concentration gradients that serve as nutrients for microbes in the rhizosphere. Microbes colonize the rhizosphere of many plants and often confer beneficial effects. (2) In iron-limited conditions, bacterial cells release siderophores. Siderophore forms complex with the insoluble ferric iron and bind to the surface of the bacterial cell. The Fe^{3+}-siderophore complex gets transported inside the cell and the insoluble ferric iron (Fe^{3+}) is converted into the soluble ferrous form (Fe^{2+}). The siderophore either gets degraded inside the cell or is released in free form outside the cell. Bacterial cell utilizes this ferrous form of iron for their growth and thereby increase in number. (3) Plants can also release phytosiderophore (PS) for enhancing the metal and other pollutant accumulation by forming PS-Fe^{2+} complex, it can also be used by rhizospheric bacteria. (4) Suppressing the growth of pathogenic bacteria due to formation of phytosiderophore and bacterial siderophore leads to lack of Fe^{3+} availability which is essential for pathogenic bacterial growth.

10.6.3 Siderophores Enhance Bioremediation of Environmental Pollutants

10.6.3.1 Heavy Metals

Soils may become contaminated by the rapid accumulation of heavy metals and metal-loids coming from rapidly growing industry, mine tilling, improper disposal of metal wastes, indiscriminate application of chemical fertilizers, pesticides, wastewater irrigation, abnormal spillage of petrochemicals, and atmospheric deposition. Although the principal role of siderophores is to chelate ferric iron, they can also play significant roles in detoxifying heavy metal–contaminated samples by binding to a wide array of toxic metals, for example, Cr^{3+}, Al^{3+}, Cu^{2+}, and Pb^{2+}. Therefore, siderophores can become a useful eco-friendly agent for heavy metal remediation. Most of the metals at low concentrations promote the growth of the bacteria while at higher concentrations they are toxic to bacteria. For example, low concentrations of copper promotes bacterial growth by participating in electron transport chain and enzymatic functionality, but at high concentrations, it generates oxidative stress that can cause damage in DNA. Siderophores bound to other heavy metals do not enter the cell efficiently, whereas siderophore bound to iron moves into the cell adequately. Siderophores produced by *Pseudomonas azotoformans* were associated with the removal of arsenic from contaminated soil. It has been reported that certain rhizobacteria can improve the plant growth by reducing the severity associated with nickel toxicity. Pyochelin, a siderophore produced by *Pseudomonas aeruginosa*, can chelate a variety of metals like Ag^+, Al^{3+}, Cd^{2+}, Co^{2+}, Cr^{2+}, Cu^{2+}, Hg^{2+}, Mn^{2+}, Ni^{2+}, Pb^{2+}, and Zn^{2+} and prevents the entry of these metals into the bacteria. Azotochelin and azotobactin, two siderophores produced from *Azotobacter vinelandii*, have the ability to pursue molybdenum (Mo) and vanadium (V) acquisition. It was observed that siderophores play a crucial role in mobilizing metals from metal-contaminated soils. In a separate study, it was reported that the siderophores synthesized by *Agrobacterium radiobacter* removed approximately 54 percent of the arsenic from metal-contaminated soil. Siderophores also played vital roles in mobilizing metals from mine waste. Several metals (Fe, Ni, and Co) were mobilized from waste material (acid-leached ore) of a uranium mine with the aid of siderophores produced by *Pseudomonas fluorescens*. Considering all the above information, siderophores can be used as an efficient bioremediating agent for metals. Siderophores enhance plant growth by killing pathogenic bacteria through iron sequestration (Figure 10.10). Phytopathogens associate with the root of the plant and cause pathogenesis. Siderophores bind with the irons that result in iron depletion of the phytopathogens, causing the death of the pathogens.

10.6.3.2 Petroleum Hydrocarbons

Petroleum hydrocarbons in marine ecosystems are a major environmental problem. Microorganisms could play an important role in the remediation of petroleum hydrocarbons from the marine environment. Microbial siderophores participate in the biodegradation of petroleum hydrocarbons through an indirect mechanism, by facilitating the Fe acquisition for the degraded microorganisms under Fe-limiting conditions. Petrobactin was the first structurally characterized siderophore produced by the oil-degrading marine bacterium *Marinobacter hydrocarbonoclasticus* (Barbeau et al., 2002). Hickford and colleagues (2004) identified another sulfonated siderophore called *Petrobactin sulfonate* isolated from the same oil-degrading marine bacterium. Few studies suggested that the use of siderophores may be a good strategy for oil spill clean-up. Gauglitz and colleagues (2012) showed recently that marine *Vibrio* spp. isolated from the Gulf of Mexico after the 2010 *Deepwater Horizon* oil spill produce amphiphilic sidero-phores called *ochrobactins* that were suggested to efficiently contribute to the degrada-tion of petroleum hydrocarbons.

10.6.3.3 Bio-Bleaching of Pulps

The pulp and paper industry is a primary source of many environmental problems, includ-ing global warming, human toxicity, ecotoxicity, photochemical oxidation, acidification, nitrification, and solid wastes. The main problem of pulp and paper manufacturing results from the bleaching process. Some pollutants are emitted into the air, while others are dis-charged in wastewater. Siderophores are considered effective agents in pulp treatment, where they can reduce 70 percent of the chemicals needed to bleach Kraft pulp, making siderophores environment-friendly alternatives. Brown rot fungi are considered to be one of the most important groups of wood-decaying microorganisms. Some studies reported the production of catecholate and hydroxamate siderophores by wood-decaying fungi. For instance, hydroxamate siderophores isolated from brown rot fungus (*Gloephyllum trabeum*) had the ability to mediate the reduction of Fe in redox-cycling processes. The reduced Fe can then react with hydrogen peroxide to generate oxygen radical species that depoly-merize cellulose, hemicellulose, and lignocellulose. This depolymerization process was considered the main role of siderophores in the bio-bleaching of pulps. It has also been observed that siderophores produced by *Coriolus versicolor* could alter the lignin structure to make it more susceptible to degradation and accordingly contribute to the bio-bleaching of pulps (Wang et al. 2008).

10.6.4 Siderophores as Medicine

The importance of siderophores extends their application in biotechnology and medi-cine. Siderophores produced by *Pseudomonas* spp. have been employed efficiently as bio-control agents against certain soilborne plant pathogens. Also, siderophores are used in medicine for iron and aluminum overload therapy and antibiotic for better target-ing. One potentially powerful application of siderophores is to use the iron transport abilities of siderophores to carry drugs into cells by preparation of conjugates between siderophores and antimicrobial agents. Because microbes recognize and utilize only cer-tain siderophores, such conjugates are anticipated to have selective antimicrobial activ-ity. In the treatment of thalassemia (iron overload diseases) and certain other anemias, periodic whole blood transfusions are required. Since there is no specific physiological mechanism for the excretion of iron in humans, continued transfusion therapy leads to a steady build-up of iron. These iron excesses, as well as the primary iron overload diseases such as hemochromatosis and hemosiderosis and accidental iron poisoning, require the removal of iron from the body, especially from the liver. Such diseases can be efficiently treated with siderophore-based drugs and siderophores acting as the prin-cipal model. Desferrioxamine B has also found therapeutic applications for various pathological conditions due to aluminum overload. Accumulation of this toxic metal is frequently observed in chronically dialyzed patients who have lost the ability to clear via renal excretion. Desferrioxamine B has also been recommended for the diagnosis of such an overload state.

10.7 Perspectives, Questions, and Conclusions

On the basis of overall assessment, it has become clear that siderophores work as central organic compounds for iron uptake in many plants and microorganisms. Siderophores might work as quorum-sensing molecules, which leads to the formation of biofilm. But it is not clear whether siderophores work as quorum-sensing molecules, because not much research has been reported in this area. The chemical structures of different siderophores and the membrane receptors involved in Fe uptake has opened new areas for research. The importance of siderophores is obvious and they play a significant role in environmental applications, even if there are many questions remaining. What is the specific role of plants and microorganisms in the selectivity of metal uptake by siderophores? Why do microorganisms secrete more than one type of siderophores to meet their mineral nutritional needs? What is the relative importance of the different siderophore structures involved in environmental applications? Can modified genetic methods such as labeled DNA be useful tools for the direct detection of siderophore functional genes in the environment? More research is focusing on finding effective ways to use siderophores in bioremediation and biocontrol, which should enhance their application in the environment. Siderophore variability and their structural and functional characteristics in relation to microbial communities must be vigorously investigated to improve the role of siderophores in environmental applications.

References

Arora, N.K. 2015. *Plant Microbes Symbiosis: Applied Facets*. Dordrecht, Netherlands: Springer.

Barbeau, K., Zhang, G.P., Live, D.H., Butler, A. 2002. Petrobactin, a photoreactive siderophore produced by the oil-degrading marine bacterium *Marinobacter hydrocarbonclasticus*. *Journal of the American Chemical Society* 124:378–379.

Bijtenhoorn, P., Schipper, C., Hornung, C., Quitschau, M., Grond, S., Weiland, N., Streit, W.R. 2011. BpiB05, a novel metagenome-derived hydrolase acting on N-acylhomoserine lactones. *Journal of Biotechnology* 155:86–94.

Chandra, R., Sangeeta, Y. 2011. Phytoremediation of Cd, Cr, Cu, Mn, Fe, Ni, Pb, and Zn from Aqueous solution using *Phragmites cummunis*, *Typha angustifolia*, and *Cyperus esculentus*. *International Journal of Phytoremediation* 13:580–591.

Cheng, W., 2009. Rhizosphere priming effect: Its functional relationships with microbial turnover, evapotranspiration and C–N budgets. *Soil Biology and Biochemistry* 41:1795–1801.

Chowdhary, P.K., Keshavan, N., Nguyen, H.Q., Peterson, J.A., González, J.E., Haines, D.C. 2007. *Bacillus megaterium* CYP102A1 oxidation of acyl homoserine lactones and acyl homoserines. *Biochemistry* 46:14429–14437.

Crowley, D.A. 2006. Microbial siderophores in the plant rhizosphere, in *Iron Nutrition in Plants and Rhizospheric Microorganisms*, eds. L.L Barton and J. Abadia, pp. 169–189. Netherlands: Springer.

Dong, Y.H., Xu, J.L., Li, X.Z., Zhang, L.H. 2000. AiiA, an enzyme that inactivates the acylhomoserine lactone quorum-sensing signal and attenuates the virulence of *Erwinia carotovora*. *Proceedings of the National Academy of Sciences* 97:3526–3531.

Fuqua, C., Greenberg, E.P. 2002. Listening in on bacteria: Acyl-homoserine lactone signalling. *Nature Reviews Molecular Cell Biology* 3:685–695.

Gauglitz, J.M., Zhou, H., Butler, A. 2012. A suite of citrate-derived siderophores from a marine Vibrio species isolated following the *Deepwater Horizon* oil spill. *Journal of Inorganic Biochemistry* 107:90–95.

Glick, B.R. 1995. The enhancement of plant growth by free-living bacteria. *Canadian Journal of Microbiology* 41:109–117.

Gray, E.J., Smith, D.L. 2005. Intracellular and extracellular PGPR: Commonalities and distinctions in the plant–bacterium signaling processes. *Soil Biology and Biochemistry* 37:395–412.

Guo, J., Chi, J. 2014. Effect of Cd-tolerant plant growth promoting Rhizobium on plant growth and Cd uptake by Lolium multiflorum Lam. and Glycine max (L.) Merr. in Cd-contaminated soil. *Plant Soil* 375:205–214.

Hickford, S.J.H., Kupper, F.C., Zhang, G., Carrano, C.J., Blunt, J.W., Butler, A. 2004. Petrobactin sulfonate, a new siderophore produced by the marine bacterium *Marinobacter hydrocarbonoclasticus*. *Journal of Natural Products* 67:1897–1899.

Leadbetter, J.R., Greenberg, E.P. 2000. Metabolism of acyl-homoserine lactone quorum-sensing signals by *Variovorax paradoxus*. *Journal of Bacteriology* 182:6921–6926.

Lin, Y.H., Xu, J.L., Hu, J., Wang, L.H., Ong, S.L., Leadbetter, J.R., Zhang, L.H. 2003. Acyl-homoserine lactone acylase from *Ralstonia* strain XJ12B represents a novel and potent class of quorum-quenching enzymes. *Molecular Microbiology* 47:849–860.

Lucy, M., Reed, E., Glick, B.R. 2004. Applications of free living plant growth promoting rhizobacteria. *Antonie Van Leeuwenhoek* 86:1–25.

Marketon, M.M., Gronquist, M.R., Eberhard, A., Gonzalez, J.E. 2002. Characterization of the Sinorhizobium meliloti sinR/sinI locus and the production of novel N-acyl homoserine lactones. *Journal of Bacteriology* 184.

Masalha, J., Kosegarten, H., Elmaci, O., Mengel, K. 2000. The central role of microbial activity for iron acquisition in maize and sunflower. *Biology and Fertility of Soils* 30:433–439.

Neilands, J.B. 1981. Iron absorption and transport in microorganisms. *Annual Review of Nutrient* 27:46.

Sangeeta, Y., Chandra, R. 2011. Heavy metals accumulation and ecophysiological effect on *Typha angustifolia* L. and *Cyperus esculentus* L. growing in distillery and tannery effluent polluted natural wetland site, Unnao, India. *Environmental Earth Science* 62:1235–1243.

Uroz, S., Oger, P.M., Chapelle, E., Adeline, M.T., Faure, D., Dessaux, Y. 2008. A Rhodococcus qsdA-encoded enzyme defines a novel class of large-spectrum quorum-quenching lactonases. *Applied Environmental Microbiology* 74:1357–1366.

Wang, L., Yan, W., Chen, J., Huang, F., Gao, P. 2008. Function of the iron-binding chelator produced by *Coriolus versicolor* in lignin biodegradation. *Science in China Series C Life Sciences* 51:214–221.

Zhang, H.B., Wang, L.H., Zhang, L.H. 2002. Genetic control of quorum sensing signal turnover in *Agrobacterium tumefaciens*. *Proceedings of the National Academy of Sciences* 99:4638–4643.

11

Common Weeds as Potential Tools for In Situ Phytoremediation and Eco-Restoration of Industrially Polluted Sites

Dhananjay Kumar, Sanjeev Kumar, and Narendra Kumar

CONTENTS

11.1 Introduction

Soil and water are the two basic rudiments of our ecosystem in general, and they are essential for survival and development of human beings. To attain better economic growth and employment opportunities, most nations have encouraged the setup of various industries. Besides contributing to economic growth and the emergence of various luxurious facilities and commodities, these industries and several other anthropogenic activities, such as municipal waste disposal, mining, and application of chemical fertilizers and pesticides, have generated a large amount of organic and inorganic toxic wastes, and these wastes have caused the degradation of precious natural resources, like soil and water (Kumar et al. 2013, Van and Maggio 2015). Organic contaminants mainly comprise pesticides, chlorinated solvents, hydrocarbons, explosives such as trinitrotoluene, and polyaromatic hydrocarbons (PAHs), while inorganic contaminants include radioactive elements, such as uranium, and heavy metals and metalloids, like zinc, copper, arsenic, cadmium, and lead. Degradation of the qualities of soils and waters by organic and inorganic pollutants poses a serious threat to food safety, ecological balance, and agricultural sustainability. For the past few decades, considerable attention has been given to controlling environmental pollution for the sake of sustainability and human survival (Sud et al. 2008, Wichelns et al. 2015). However, in order to achieve the desired goals, advances beyond the conventional environmental pollution prevention and control techniques are needed. Further, the use of eco-friendly techniques, such as phytoremediation, should be harnessed for the restoration of degraded environments.

Several conventional chemical- and engineering-based techniques have been utilized for the treatment and restoration of contaminated environments, e.g., via soil washing, soil vapor extraction, electrokinetics, solidification, and encapsulation (Kulkarni et al. 2008, Dadrasnia et al. 2013). However, these techniques are neither eco-friendly nor economically viable; furthermore, they also produce by-products which are not only hazardous but also produce problems in terms of safe disposal. Therefore, there is a prerequisite global demand to develop more effective and economically viable remediation techniques. To cope with the situation, phytoremediation has emerged as an eco-friendly, cost-efficient, and aesthetically accepted alternative technique.

Phytoremediation is a solar-driven green technology that utilizes the potential of plants and aquatic macrophytes to degrade, accumulate, extract, immobilize, and reduce the potential risks posed by toxic metals, organic xenobiotics, and radionuclides in contaminated environments (Khan et al. 2004, Rajkumar et al. 2010). Summaries of some conventional treatment techniques are presented in the following sections.

11.2 Conventional Treatment Techniques

We provide here some glimpses of conventional techniques used for treatment of environmental pollutants.

11.2.1 Soil Washing

Soil washing is an *ex situ* mode of remediation that is applicable for removal of metal contaminants from soil via chemical or physical treatment approaches with the soil in an aqueous suspension. With physical approaches for soil washing, the soil particles (sand and gravel), which contain the majority of metal contaminants, are separated from the bulk of the soil fraction (silt and clay). With chemical approaches, metal contaminants are removed from the soil via application of aqueous chemicals and then recovered on a solid substrate (Khan et al. 2004).

11.2.2 Soil Vapor Extraction

Soil vapor extraction (also known as soil venting) is a simple, cheap, and *in situ* technique that is applied to the unsaturated (vadose) zone of soil. This method is applicable for removal of volatile organic compounds (Fischer et al. 1996). During soil vapor extraction, a vacuum is applied through extraction wells to create pressure and a concentration gradient in the contaminated soil in order to induce a controlled airflow for the removal of volatile and semivolatile organic compounds and contaminants (U.S. EPA 2006).

11.2.3 Solidification and Stabilization

Solidification and stabilization refer to closely related techniques that reduce the potential risks posed by hazardous wastes by transmuting the pollutants into less-soluble, immobile, and nonhazardous or less toxic forms. The physical characteristics and handling aspects for the toxicants are not fundamentally changed through this mode of remediation. The solidification technique involves encapsulation of wastes in a highly structural integrated monolithic solid. This encapsulation may entail fine waste particles (microencapsulation) or use of large blocks or containers (macroencapsulation) (U.S. EPA 2006). Solidification does not include any chemical interactions between toxicants and the solidifying reagents, but it may mechanically bind the waste into the monolith. The mobility of contaminants is restricted by decreasing the surface area exposed for potential leaching and by isolating the wastes within an impervious capsule. Further, the migration of contaminants can be reduced by various process, including precipitation, complexation, and adsorption reactions (U.S. EPA 2006).

11.2.4 Encapsulation

Encapsulation does not filter contaminants from the soil; instead, it involves the mixing of contaminated soil with other materials, such as cement, lime, concrete, or asphalt, leading to prevention of spread of the contaminants to surrounding clean soil strata (Christensen et al. 2005). Lime or concrete encapsulation is applicable and effective for heavy metals, while encapsulation by asphalt is applicable for hydrocarbon-contaminated soils. In particular, silica encapsulation is effective and applicable for remediation of heavy metals from soil or removal of hydrocarbon-contaminated soil (Christensen et al. 2005; U.S. EPA 2006).

11.2.5 Electrokinetics

Electrokinetics is an *in situ* remediation technique applied for the removal of heavy metals, radionuclides, and selected organic pollutants from contaminated, low-permeability soil, mud, sludge, and marine dredging spoils. It involves the application of low-intensity, direct current through the soil for desorption and removal of metals, radionuclides, and polar organics from contaminated strata (Gomes et al. 2012). Additionally, the electric current also induces an electro-osmotic hydraulic flow that provides a driving force for movement of neutrally charged soluble contaminants (Saichek and Reddy 2005, U.S. EPA 2006).

11.3 Weeds as Accumulators and Extractors of Heavy Metals and Other Contaminants

Contamination of land and water by toxic pollutants, e.g., heavy metals, has occurred in developing and developed countries. In recent years, phytoremediation has been a leading technology for removal of toxic metals from contaminated land and water. This chapter focuses on different native weed plant species that have a high accumulation potential for heavy metals, e.g., arsenic, cadmium, mercury, lead, zinc, iron, and copper, from various contaminated soil types (Table 11.1). Kumar et al. (2013) suggested that weeds may be applied for phytoremediation because such species grow naturally in contaminated environments (water or soil) and have adapted to higher concentrations of metals, even as these metals accumulate in plant parts. Kumar et al. (2013) revealed that heavy metal accumulation varied within plant tissue types, i.e., roots and shoots. They reported the concentrations (in micrograms of metal per gram of plant root) of metals in roots in the following ranges: Cr (3.97–67.72), Cu (3.44–45.47), Ni (2.52–14.19), Pb (6.13–58.28), and Cd (0.32–3.96). Concentrations in shoots were in the following ranges: Cr (5.45–125.67), Cu (2.87–19.60), Ni (1.81–15.39), Pb (5.18–44.28), and Cd (0.23–3.17) in various weeds, including *Amaranthus cruentus*, *Parthenium hysterophorus*, and *Solanum nigrum* (Table 11.2).

Metal transfer from the roots to the shoots is an important feature of phytoremediator weed plant species, and the extent of transfer can be calculated based on the translocation factor (TF) (Kumar et al. 2013). Singh et al. (2010) reported that TFs higher than unity and lower than 1.0 are features of metal accumulators and metal excluder plant species, respectively. Usually, species which accumulate 100 mg Cd/kg, 1000 mg Ni, Cu, Co, Cr, or Pb per kg, or 10,000 mg Zn or Mn per kg are called hyperaccumulator species (Baker and Brooks 1989, Brooks 1998, Kumar et al. 2013, Yuan et al. 2016). Varun et al. (2012) concluded that native weed species have adapted themselves to soil contamination with multiple metals (i.e., Zn, Mn, Co, Cd, Pb, Cr, Ni, Cu, and As), and the following species are recommended

TABLE 11.1

Types of Phytoremediation Techniques

Method	Contaminated Medium	Description	Accumulation and Reduction Zone	References
Phytoextraction/ Phytoaccumulation	Soil or water	The plant absorbs a pollutant through its root system and translocates it to aerial parts, where they accumulate	Shoot	Sharma 2011, Sharma et al. 2015, Van and Maggio 2015
Rhizofiltration/ Phytofiltration	Water	Absorption or adsorption of pollutant by plant roots (rhizhofiltration) of seedlings (blastofiltration); applicable mainly for metals from water by aquatic plants	Roots and shoots	Prasad and Freitas 2003, Thakur et al. 2016
Phytostabilization	Soil or water	Contaminants are stabilized or immobilized with the help of plant roots and rhizobacterial interactions	Rhizosphere	Thakur et al. 2016, Sharma 2011, Ali et al. 2013
Phytovolatilization	Soil or water	Extraction of a specific pollutant from soil and transformation of it into a volatile form	Released into atmosphere	Van and Maggio 2015, Thakur et al. 2016
Phytodegradation	Soil or water	Degradation of pollutant by enzymes	Plant tissues	Van and Maggio 2015, Thakur et al. 2016

TABLE 11.2

Heavy Metal Concentrations in Weed Species Growing at Contaminated Sites

Weed Name	Common Name	Family	Contamination at Site	Metal(s)	Reference(s)
Cyperus rotundus	Nut grass	Cyperaceae	Tannery waste	Cr	Jaison and Muthukumar 2016
Cyperus rotundus	Nut grass	Cyperaceae	Tanning sludge	Cr, Cu, Zn, Cd, Pb	Yuan et al. 2016
Parthenium hysterophorus	Feverfew	Asteraceae	Paper mill and distillery waste	Mg, Fe, Pb, Zn, Mn, Ni, Cd, Pb	Mazumdar and Das 2015, Chandra and Kumar, 2016
Amaranthus cruentus	Red amaranth	Amaranthaceae	Paper mill waste	Mg, Fe, Pb, Zn	Mazumdar and Das 2015
Solanum americanum	American nightshade	Solanaceae	Paper mill waste	Mg, Fe, Pb	Mazumdar and Das 2015
Croton bonplandianum	Ban Tulsi	Euphorbiaceae	Flash light manufacturer waste	Cr, Cu, Ni, Pb	Kumar et al. 2013
Cyperusrotundus	Nut grass	Cyperaceae	Flash light manufacturer waste	Cr, Cu, Ni, Pb, Cd	Kumar et al. 2013
Solanum nigrum	Black nightshade	Solanaceae	Glass industry waste	Zn, Mn, Co, Cd, Zn, Ni, Cu, Pb	Varun et al. 2012, Chandra and Kumar 2016
Cyperus rotundus	Nut grass	Cyperaceae	Ex tin mining	Pb, Cu, Zn, As, Sn	Ashraf et al. 2011
Solanum nigrum	Black nightshade	Solanaceae	Industrial area	Pb, Cu, Zn, Ni, Co, Cr	Malik et al. 2010
Cannabis sativa	Hemp	Cannabaceae	Pulp and paper	Pb, Cu, Zn, Ni, Co, Cr, Ni, Mn, Fe, Cd, Pb	Malik et al. 2010, Chandra et al. 2017
Cyperus rotundus	Nut grass	Cyperaceae	Tanning sludge	Cr, Cu, Zn, Cd, Pb	Yuan et al. 2016

as phytoextractors: *Datura stramonium* for Mn, Cr, Cu, and As; *Chenopodium murale* for Zn, Cd, and Cu; *Lycoersicon esculentum* for Cd, Cr, and As; *Poa annua* for Pb and As.

11.4 Phytoremediation: An Eco-Friendly and Cost-Effective Technique

Some plants have a special genetic setup to survive in metal-contaminated environments. They can not only survive but can also significantly extract and accumulate the metals in their tissues, even at concentrations several orders of magnitude higher than in the substrate. Identification and application of such plants that have this kind of adaptive mechanism can help to a great extent to remediate a metal-contaminated matrix (Lasat 2002). Phytoremediation involves the application of green plants and their associations with microbes for the remediation of contaminants, like heavy metals, radionuclides, and organic xenobiotics (Mathur and Bohra 2007, Malik et al. 2010). It is a solar-driven process,

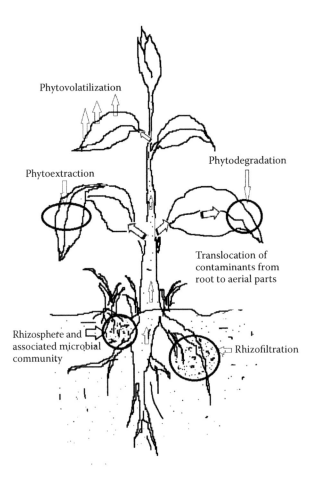

FIGURE 11.1
Different forms of phytoremediation techniques.

and it is also known as botanical bioremediation (Chaney et al. 1997). Unlike other remediation techniques, phytoremediation has been proven to be sustainable and the most aesthetically acceptable, cost-effective, and eco-friendly approach (Kumar et al. 2013, Bauddh et al. 2015, Wang et al. 2016). Further, plant-based remediation systems take advantage of the roots for unique and selective uptake potential along with translocation, bioaccumulation, and stabilization of potentially toxic elements (Figure 11.1).

11.4.1 Categorization of Plants on the Basis of Their Responses to Metal Stress

On the basis of avoidance, accumulation, and tolerance potentials, weeds can be classified as indicators, excluders, accumulators, and hyperaccumulators (Prasad 2004).

11.4.1.1 Indicators

Indicators are those plants in which metal uptake and translocation imitate metal concentrations in the soil strata, and the plants display toxic symptoms. Growth and development of such plant is reduced as the metal concentration(s) in the soil increases.

11.4.1.2 Excluders

Excluders are those plants that hamper the uptake and accumulation of toxic metals into the aerial biomass. These plants accumulate high levels of metals in their roots, and hence the TF is always less than unity. Generally, such plants are used for stabilization and avoidance of further soil contamination.

11.4.1.3 Accumulators

Accumulators are those plants in which the uptake of heavy metals and their translocation reflect background metal concentrations without displaying toxic indication. These plants accumulate a high level of metals in their shoots and hence the translocation factor is generally greater than unity.

11.4.1.4 Hyperaccumulators

Hyperaccumulators are those plants which have the potential of accumulating metals at concentrations up to 100 times greater than those naturally measured in nonaccumulator plants. The translocation factor and extraction coefficient for a hyperaccumulator are always greater than unity. The metal accumulation potential of a plant depends on the uptake capability and activity of intracellular binding sites. The steps involved in metal accumulation are uptake of metal from soil, xylem loading, mobilization and translocation, compartmentalization, sequestration, distribution in the aerial corpus, and storage in leaves. Further, the amounts and affinities of metal chelators and selectivities of transport channels can also alter the extent of metal accumulation (Clemens et al. 2002, Dalvi and Bhalerao 2013).

11.4.1.4.1 Bioavailability and Uptake of Metals

Usually, metal exists in complexed and insoluble forms, which are generally not available for plant uptake. Thus, it is a prerequisite step that a metal be made bioavailable for plants. Root exudates naturally solubilize the insoluble metal complexes upon their secretion (e.g., low-molecular-weight compounds, such as amino acids, organic acids, sugars, and phenolics, and several other secondary metabolites), whereas high-molecular-weight exudates mainly include mucilage (high-M_r polysaccharides) and proteins. Exudation of mugenic and aveic acid lead to solubilization of metals and cause acidification of rhizospheric soil and chelation of metal (Salt et al. 1995, Walker et al. 2003, Mahmood 2010). Metals which are bioavailable for plants enter through root systems, either via the apoplastic pathway (intercellular spaces) or by the symplastic pathway (crossing plasma membranes). Further, the positively charged metal ions are adsorbed, at negatively charged extracellular root cell wall sites (COO-) which act as preliminary blockades for translocation. Moreover, the impermeable cellular suberin layers also curtail metal translocation from root apoplast to xylem (Taiz et al. 2002, Peer et al. 2006). Consequently, to cross this apoplastic barrier, metals must cross the plasma membrane (Ghosh and Singh 2005).

11.4.1.4.2 Translocation and Accumulation of Metals

Subsequently, the movement of metals from the root to xylem occurs by three routes, i.e., metal sequestration in root cells, symplastic transport into stele, and final release into the xylem. Normally, the high cation exchange capacity of xylem cell walls restricts the further transportation of metals. However, in case of hyperaccumulators, metal complexation with chelators permits easy transportation to leaves (Clemen et al. 2002, Mahmood 2010). Mahmood (2010) reported that transportation and distribution of metals in leaves occurs via the apoplast or symplast pathways. In case of such plants having a high accumulation potential, complexation of metals with organic ligands offers high metal tolerance (Peer et al. 2006). Furthermore, cell types where metal deposition takes place vary with the metal and plant species.

11.4.1.4.3 Sequestration

The final step of accumulation is the sequestration of metal in a vacuole, which prevents free metal ions in cytosol. Further, metals may persist in cell walls due to interactions of multivalent cations with negatively charged sites (Peer et al. 2006). Chandra and Kumar (2016) reported that transmission electron microscope observations of tested hyperaccumulator plants exhibit the formation of multinucleolus, multivacuoles, and metal granule depositions in their cellular components, and this is considered an adaptation approach against metal stress (Chandra and Kumar, 2016).

11.5 Weeds: Potential Tools for Remediation of Contaminated Soils

Weeds are undesirable plants that are considered vexations for cultivated cereals and other economically important crops, as competitors for nutrients, light, water, and other vital requirements. Weeds are more aggressive in terms of their growth and development in soil and water. Aggressive behavior, high tolerance, a built-in competitive nature, and adaptability towards abiotic and biotic stresses make weeds promising candidates for use in remediation of contaminated environments (Gardea-Torresdey et al. 2004, Wei et al. 2005, Wei and Zhou 2006). Further, along with their remediation potential, weeds also offer opportunities for eco-restoration of industrial polluted sites. To date, the use of over 500 plant species belonging to more than 45 plant families has been documented, based on the species abilities for tolerance and accumulation of potentially toxic elements (Krämer 2010, Hemen 2011, Sharma 2011). Some of these weeds having wide applicability and are discussed below.

11.5.1 *Solanum nigrum*

Solanum nigrum L., commonly known as black nightshade or poison berry, is an annual herb found in almost every part of the world. It can grow up to 0.3–1 m in height. The biomass of *S. nigrum* is greater than other plants which have been used for remediation, e.g., *Rhamnus globosa*, *A. haller*, and *Thlaspi caerulescens* (Baker et al. 1994, Dahmani-Muller et al. 2000, Wei and Zhou 2006). Having a high biomass and being easily harvestable, along with a good phytoremediation potential, *S. nigrum* has been given much attention in recent years. *Solanum* has been well documented as a feasible plant for phytoremediation of heavy metals, particularly Cd, Cu, Zn, Pb, and As (Ji et al. 2011, Sun et al. 2008, Yu et al. 2015). Ji et al. (2011) reported the remediation efficiency of *S. nigrum* for a Cd-contaminated environment. They also reported that *S. nigrum* has a good potential to tolerate and grow on Cd-contaminated strata, with high accumulation and translocation capacities. *S. nigrum* translocates and accumulates Cd in its aerial part, making it easy to harvest (Ji et al. 2011). Further, Sun et al. (2008) reported that there is no reduction in the height or the shoot biomass of *S. nigrum* grown on Cd-concentrated medium (≤25 mg/kg) and meets the basic features of a hyperaccumulator plant. Yang et al. (2011) also investigated *S. nigrum*'s effectiveness as a phytoextractor of Cd when it is grown in a Cd- and PAH-contaminated environment. *S. nigrum* possesses a considerable degradation potential for PAHs in the presence of chelators such as EDTA, cysteine, salicylic acid, and Tween 80 (Yang et al. 2011). Further, Yu et al. (2015) reported that the phytoremediation potential of *S. nigrum* was not affected by multiple-heavy metal contamination in growing strata. Moreover, the bioaccumulation coefficient (BC) factors for multiple heavy metals were higher than in Cd-contaminated soil (Yu et al. 2015).

11.5.2 *Parthenium hysterophorus*

P. hysterophorus L. is an aggressive, noxious, and invasive annual herbaceous weed species found in America, Asia, Africa, and Australia. This ubiquitous weed is commonly known as carrot grass, bitter weed, white top, ragweed, *Parthenium*, congress grass, cofitillo. This invasive weed was introduced India as a contaminant in PL 480 wheat (Public Law 480) imported from the United States in the 1950s and is now widely prevalent in India (Patel 2011, Singh et al. 2008). *P. hysterophorus* has been well reported for its phytoremediation potential (Hadi and Bano 2009, Sanghamitra 2011, Hadi et al. 2014, Ahmad and Al-Othman 2014, Ali and Hadi 2015). Ahmad and Al-Othman (2014) investigated how *P. hysterophorus* can be used as a potential tool for the remediation of soil contaminated with Pb, Cd, Cu, Ni, Zn, and Fe. They also reported that the accumulation potentials were in the order Fe > Zn > Cu > Pb > Cd > Ni. Further, the plants more efficiently translocated Pb, Ni, and Cd in aerial tissues than Fe, Zn, and Cu (Ahmad and Al-Othman 2014). Hadi and Bano (2009) investigated the role of plant growth regulators, chelating agents, and plant growth-promoting bacteria in the remediation of Pd-contaminated soil. Accumulation and trans-location of Pb in aboveground tissues were significantly higher in plants grown in soil amended with plant growth regulators and chelating agents than the microbe-amended soil. It was observed that *P. hysterophorus* has has a high phytoextraction potential for Cd at its reproductive stage. Application of gibberellic acid (GA$_3$) and ethylenediaminetetraaceticacid (EDTA) increased the bioconcentration and accumulation of Cd. Further, those authors suggested that the concentration of free proline and total phenolics in plant tissue were increased with the increase in Cd concentration, especially in the GA$_3$ plant, and these findings advocates that GA$_3$ plays some role in the biosynthesis of free proline and total phenolics in plants during Cd stress (Ali and Hadi 2015).

11.5.3 *Amaranthus* Species

Amaranthus cruentus is a flowering plant species. The plant can grow up to 6 feet tall. *Amaranthus* has a great potential to remediate toxic substances and heavy metals from contaminated soil (Iori et al. 2013, Li et al. 2013, Carvalho et al. 2014, Jin et al. 2016). Li et al. (2013) showed that *Amaranthus hypochondriacus* has a great phytoremediation potential in Cd-contaminated soil and recommended it to maximize the phytoextraction efficiency of *Amaranthus hypochondriacus* by repeated harvested techniques with NPK fertilizers at the base application. Iori et al. (2013) studied the accumulation of Ni in 3week-old *Amaranthus paniculatus* L. plants that were subjected to different concentrations of nickel chloride solution, i.e., 25, 50, 100, and 150 μM, and revealed a progressive decrease in plant organ dry mass with the enhancement of Ni concentration in the solution, suggesting a good metal tolerance at 25 μM Ni and a marked sensitivity at 150 μM Ni. Jin et al. (2016) found higher accumulation of cadmium in the roots of *Amaranthus hybridus* than in the other organs. A total of 28 differentially expressed proteins were identified in response to Cd stress in *Amaranthus* plants. Chinmayee et al. (2012) observed that accumulation of Cu, Pb, and Cd was highest in the roots, followed by levels in stems and leaves, and that of Zn and Cr remained high in aerial parts of *Amaranthus spinosus*. A steady increase was noticed in the bioaccumulation of Cu, Zn, and Cd, enhancing the concentration of the corresponding metal in the contaminated soil.

11.5.4 *Ricinus* sp.

Ricinus sp. is an oilseed crop that belongs to the Euphorbiaceae family. It is a fast-growing C_3 plant and is used as an industrial crop because of the quality of its oil. *Ricinus* sp. is a bioenergy plant with high biomass and with the potential to accumulate heavy metals. *R. communis* has attracted attention because of its ability to grow on contaminated and polluted sites (Babita et al. 2010, Bauddh and Singh 2012, Bauddh et al. 2015, Wang et al. 2016). Huang et al. (2011) showed that total uptake of DDT and Cd varied from 83.1 to 267.8 µg per pot and 66.0–155.1 µg per pot, indicating that *R. communis* has great potential for DDT and Cd removal from contaminated soils. Costa et al. (2012) evaluated Cd and Pb accumulation in castor beans grown in Hoagland and Arnon's nutrient solution and with increasing concentrations of Cd and Pb. They observed that the increase in Cd concentration in the nutrient solution decreased root and shoot growth. Wang et al. (2016) investigated the accumulation of Cd, Cu, and Zn and their interaction in *R. communis* L. after exposure to these metals for 4 months. They found that *R. communis* grew well, produced a high biomass, and showed a high phytoremediation potential, with tolerance to Cd, Cu, and Zn. Bauddh et al. (2016) studied the effectiveness of *R. communis* for remediation of Cd from contaminated soil. They found that 8-month-old plants stabilized approximately 51% of the Cd in its roots, and the rest of the metal was transferred to the stem and leaves. There were no significant differences in growth, biomass, or yield between control and Cd-treated plants, with the exception of shoot fresh weight.

11.6 Application of Weed Plants in Different Fields

In addition to phytoremediation, weeds have several other important applications. These other applications include use as medicinal products, as well as contributions to societal development, employment generation for local people, carbon sequestration, reduction in greenhouse gases, and biofuel production. They are also used to increase the fertility of the soil and reduce soil erosion.

11.7 Eco-Restoration of Industrial Polluted Sites

Eco-restoration is the process of renewing and restoring degraded, damaged, and destroyed ecosystems. Ecosystem destruction is usually the consequence of heavy metals and other waste materials being discharged into or on soil and water ecosystems.

In the last five decades, industrial contaminated sites have been a major issue in developed and developing countries. Several physical, chemical, and biological techniques are being used to remediate polluted soils, but these methods do not result in stabilization and they leave residues in the contaminated sites. Soils contaminated with heavy metals are generally excavated and landfilled. However, some sites are treated with the help of acid leaching. Physical separation and electrochemical processes are also used for remediation of contaminated lands. Biological soil treatment systems include *in-situ* land farming and an *ex situ* bio-pills and bioreactors. The costs of these techniques for soil remediation are highly variable and depend on the type and quantity of contaminant, soil properties, site conditions, and the volume of material to be remediated. On average, remediation procedures are not cost-effective or eco-friendly, whereas phytoremediation is a cost-effective and eco-friendly technique for eco-restoration of industrial polluted sites. Most plants growing on contaminated soils effectively exclude heavy metals. Nonedible and native weed species have also been studied for metal remediation potentials and they have been found capable of eco-restoration of polluted and contaminated sites.

11.8 Conclusions

Native weed species have the ability to accumulate heavy metals in their body parts (roots, shoots, and leaves) without any ostensible alterations. Further, they neither leave any residue nor cause any secondary waste disposal problems. Therefore, they can be considered for phytostabilization and revegetation of polluted sites. The nonedible character and fast growth of native weed species at the contaminated site are added advantages and makes them more appropriate for the purpose of eco-friendly and economical phytoremediation and restoration of waste land in developed and developing countries.

References

Ahmad, A., Al-Othman, A.A.S. 2014. Remediation rates and translocation of heavy metals from contaminated soil through *Parthenium hysterophorus*. *Chemistry and Ecology* 30(4):317–327.

Ali, H., Khan, E., Sajad, M.A. 2013. Phytoremediation of heavy metals: Concepts and applications. *Chemosphere* 91:869–881.

Ali, N., Hadi, F. 2015. Phytoremediation of cadmium improved with the high production of endogenous phenolics and free proline contents in *Parthenium hysterophorus* plant treated exogenously with plant growth regulator and chelating agent. *Environmental Science and Pollution Research* 22:13305–13318.

Ashraf, M.A., Maah, M.J., Yusof, I. 2011. Heavy metals accumulation in plants growing in ex tin mining catchment. *International Journal of Environmental Science and Technology* 8(2):401–416.

Babita, M., Maheswari, M., Rao, L.M., Shanker, A.K., Rao, D.G. 2010. Osmotic adjustment, drought tolerance and yield in castor (*Ricinus communis* L.) hybrids. *Environmental and Experimental Botany* 69:243–249.

Baker, A.J.M., Brooks, R.R. 1989. Terrestrial higher plants which hyperaccumulate metallic elements. A review of their distribution, ecology and phytochemistry. *Biorecovery* 1(2):81–126.

Baker, A.J.M., Reeves, R.D., Hajar, A.S.M. 1994. Heavy metal accumulation and tolerance in British populations of the metallophyte *Thlaspi caerulescens* J.&C. Presl (*Brassicaceae*). *New Phytology* 127:61–68.

Bauddh, K., Singh, R.P. 2012. Growth, tolerance efficiency and phytoremediation potential of *Ricinus communis* (L.) and *Brassica juncea* (L.) in salinity and drought-affected cadmium contaminated soil. *Ecotoxicology and Environmental Safety* 85:13–22.

Bauddh, K., Singh, K., Singh, B., Singh, R. 2015. *Ricinus communis*: A robust plant for bio-energy and phytoremediation of toxic metals from contaminated soil. *Ecological Engineering* 84:640–652.

Bauddh, K., Singh, K., Singh, R.P. 2016. *Ricinus communis* L.: A value added crop for remediation of cadmium contaminated soil. *Bulletin of Environmental Contamination and Toxicology* 96:265–269.

Brooks, R.R. 1998. Phytochemistry of hyperaccumulators, pp. 15–54. *Plants That Hyperaccumulate Heavy Metals: Their Role in Phytoremediation, Microbiology, Archaeology, Mineral Exploration and Phytomining*, Brooks, R.R. (ed.). Wallingford: CAB International.

Carvalho, P.N., Basto, M.C.P., Almeida, C.M.R., Brix, H. 2014. A review of plant–pharmaceutical interactions: From uptake and effects in crop plants to phytoremediation in constructed wetlands. *Environmental Science and Pollution Research* 21:11729–11763.

Chandra, R., Kumar, V. 2016. Phytoextraction of heavy metals by potential native plants and their microscopic observation of root growing on stabilised distillery sludge as a prospective tool for in situ phytoremediation of industrial waste. *Environmental Science and Pollution Research* 24(3):2605–2609.

Chandra, R., Yadav, S., Yadav, S. 2017. Phytoextraction potential of heavy metals by native wetland plants growing on chlorolignin containing sludge of pulp and paper industry. *Ecological Engineering* 98:134–145.

Chaney, R.L., Malik, M., Li, M.Y., Brown, S.L., Angle, J.S., Baker, A.J. 1997. Phytoremediation of soil metals. *Current Opinion in Biotechnology* 8:279–284.

Chinmayee, M.D., Mahesh, B., Pradesh, S., Mini, I., Swapna, T.S. 2012. The assessment of phytoremediation potential of invasive weed *Amaranthus spinosus* L. *Applied Biochemistry and Biotechnology* 167:1550–1559.

Clemens, S., Palmgren, M.G., Kramer, U. 2002. A long way ahead: Understanding and engineering plant metal accumulation. *Trends in Plant Science* 7(7):309–315.

Costa, E.N.S., Guilherme, L.R.G., Melo, E.E.C., Ribeiro, B.T., Inacio, E.S., Severiano, E.C., Faquin, V., Hale, B.A. 2012. Assessing the tolerance of castor bean to Cd and Pb for phytoremediation purposes. *Biological Trace Element Research* 145:93–100.

Dadrasnia, A., Shahsavari, N., Emenike, C.U. 2013. Remediation of contaminated sites. *INTECH* 2013:51591.

Dahmani-Muller, H., Van-Oort, F., Gélie, B., Balabane, M. 2000. Strategies of heavy metal uptake by three plant species growing near a metal smelter. *Environmental Pollution* 109:231–238.

Dalvi, A.A., Bhalerao, S.A. 2013. Response of plants towards heavy metal toxicity: An overview of avoidance, tolerance and uptake mechanism. *Annals of Plant Sciences* 2(9):362–368.

Fischer, U., Rainer, S., Martin, K. 1996. Experimental and numerical investigation of soil vapor extraction. *Water Resources Research* 32(12):3413–3427.

Gardea-Torresdey, J.L., Peralta-Videa, J.R., Montes, M., de la-Rosa, G., Corral-Diaz, B. 2004. Bioaccumulation of cadmium, chromium and copper by *Convolvulus arvensis* L.: Impact on plant growth and uptake of nutritional elements. *Bioresource Technology* 92:229–235.

Ghosh, M., Singh, S.P. 2005. A comparative study of cadmium phytoextraction by accumulator and weed species. *Environmental Pollution* 133(2):365–371.

Gomes, H.I., Dias-Ferreira, C., Ribeiro, A.B. 2012. Electrokinetic remediation of organochlorines in soil: Enhancement techniques and integration with other remediation technologies. *Chemosphere* 87:1077–1090.

Hadi, F., Ali. N., Ahmad, A. 2014. Enhanced phytoremediation of cadmium-contaminated soil by *Parthenium hysterophorus* plant: Effect of gibberellic acid (GA3) and synthetic chelator, alone and in combinations. *Bioremediation Journal* 18(1):46–55.

Hadi, F., Bano, A. 2009. Utilization of *Parthenium hysterophorus* for the remediation of lead-contaminated soil. *Weed Biology Management* 9(4):307–314.

Hemen, S. 2011. Metal hyperaccumulation in plants: A review focusing on phytoremediation technology. *Journal of Environmental Science and Technology* 4(2):118–138.

Huang, H., Yu, N., Wang, L., Gupta, D.K., He, Z., Wang, K., Zhu, Z., Yan, X., Li, T., Yang, X. 2011. The phytoremediation potential of bioenergy crop *Ricinus communis* for DDTs and cadmium co-contaminated soil. *Bioresource Technology* 102:11034–11038.

Iori, V., Pietrini, F., Cheremisina, A., Shevyakova, N.I., Radyukina, N., Kuznestov, V.V., Zacchini, M. 2013. Growth responses, metal accumulation and phytoremoval capability in Amaranthus plants exposed to nickel under hydroponics. *Water Air Soil Pollution* 224:1450.

Jaison, S., Muthukmar, T. 2016. Chromium accumulation in medicinal plants growing naturally on tannery contaminated and non-contaminated soils. *Biological Trace Element Research* 175(1):223–235.

Ji, P., Song, Y., Sun, T., Liu, Y., Cao, X., Xu, D., Yang, X., Mc-Rae, T. 2011. In-situ cadmium phytoremediation using *Solanum nigrum*: The bio-accumulation characteristics trail. *International Journal of Phytoremediation* 13:1014–1023.

Jin, H., Xu, M., Chen, H., Zhang, S., Han, X., Tang, Z., Sun, R. 2016. Comparative proteomic analysis of differentially expressed proteins in *Amaranthus hybridus* L. roots under cadmium stress. *Water Air Soil Pollution* 227:220.

Khan, F.I., Husain, T., Hejazi, R. 2004. An overview and analysis of site remediation technologies. *Journal of Environmental Management* 71:95–112.

Krämer, U. 2010. Metal hyperaccumulation in plants. *Annual Review of Plant Biology* 61:517–534.

Kulkarni, P.S., Crespo, J.G., Cam, A. 2008. Dioxins sources and current remediation technologies: A review. *Environment International* 34:139–153.

Kumar, N., Bauddha, K., Kumar, S., Dwivedi, N., Singh, D.C., Barman, S.C. 2013. Accumulation of metals in weed species grown on the soil contaminated with industrial waste and their phytoremediation potential. *Ecological Engineering* 61:491–495.

Lasat, M.M. 2002. Phytoextraction of toxic metals: A review of biological mechanisms. *Journal of Environmental Quality* 31:109–120.

Li, N., Li, Z., Fu, Q., Zhuang, P., Guo, B., Li, H. 2013. Agricultural technologies for enhancing the phytoremediation of cadmium-contaminated soil by *Amaranthus hypochondriacus* L. *Water Air Soil Pollution* 224:1673.

Mahmood, T. 2010. Review phytoextraction of heavy metals: The process and scope for remediation of contaminated soils. *Soil and Environment* 29(2):91–109.

Malik, R.N., Husain, S.Z., Nazir, I. 2010. Heavy metal contamination and accumulation in soil and wild plant species from industrial area of Islamabad. *Pakistan Journal of Botany* 42(1):291–301.

Mazumdar, K., Das, S. 2015. Phytoremediation of Pb, Zn, Fe, and Mg with 25 wetland plant species from a paper mill contaminated site in North East India. *Environmental Science and Pollution Research* 22:701–710.

Patel, S. 2011. Harmful and beneficial aspects of *Parthenium hysterophorus*: An update. *3 Biotech* 1:1–9.

Peer, W.A., Baxter, I.R., Richards, E.L., Freeman, J.L., Murphy, A.S. 2006. Phytoremediation and hyperaccumulator plants: Molecular biology of metal homeostasis and detoxification. *Topics in Current Genetics* 14(84):299–340.

Prasad, M.N.V. 2004. *Heavy Metal Stress in Plants from Biomolecules to Ecosystems*, 2nd ed. Berlin: Springer Verlag.

Prasad, M.N.V., Freitas, H. 2003. Metal hyperaccumulation in plants: Biodiversity prospecting for phytoremediation technology. *Electronic Journal of Biotechnology* 6(3):285–321.

Rajkumar, M., Ae, N., Prasad, M.N.V., Freitas, H. 2010. Potential of siderophore-producing bacteria for improving heavy metal phytoextraction. *Trends in Biotechnology* 28:142–149.

Saichek, R., Reddy, K. 2005. Electrokinetically enhanced remediation of hydrophobic organic compounds in soils: A review. *Critical Reviews in Environmental Science and Technology* 35:115–192.

Salt, D.E., Blaylock, M., Kumar, P.B.A.N., Dushenkov, V., Ensley, B.D., Chet, I., Raskin, I. 1995. Phytoremediation: A novel strategy for the removal of toxic metals from the environment using plants. *Biotechnology* 13:468–475.

Sanghamitra, K., Prasada, R.P.V.V., Naidu, G.R.K. 2011. Heavy metal tolerance of weed species and their accumulations by phytoextraction. *Indian Journal of Science and Technology* 4(3):285–290.

Sharma, H. 2011. Metal hyperaccumulation in plants: A review focusing on phytoremediation technology. *Journal of Environmental Science and Technology* 4(2):118–138.

Sharma, S., Singh, B., Manchanda, V.K. 2015. Phytoremediation: Role of terrestrial plants and aquatic macrophytes in the remediation of radionuclides and heavy metal contaminated soil and water. *Environmental Science and Pollution Research* 22:946–962.

Singh, R., Singh, D.P., Kumar, N., Bhargava, S.K., Barman, S.C. 2010. Accumulation and translocation of heavy metals in soil and plants from fly ash contaminated area. *Journal of Environmental Biology* 31:421–430.

Singh, R.K., Kumar, S., Kumar, S., Kumar, A. 2008. Development of *Parthenium* based activated carbon and its utilization for adsorptive removal of p-cresol from aqueous solution. *Journal of Hazardous Materials* 155:523–535.

Sud, D., Mahajan, G., Kaur, M.P. 2008. Agricultural waste material as potential adsorbent for sequestering heavy metal ions from aqueous solutions: A review. *Bioresource Technology* 99:6017–6027.

Sun, Y.B., Zhou, Q.X., Diao, C.Y. 2008. Effects of cadmium and arsenic on growth and metal accumulation of Cd-hyperaccumulator *Solanum nigrum* L. *Bioresource Technology* 99:1103–1110.

Taiz L., Zeiger E. 2002. *Plant Physiology*, 3rd ed. Sunderland, MA: Sinauer Associates.

Thakur, S., Singh, L., Wahid, Z.A., Siddiqui, M.F., Atnaw, S.M., Dinv, M.F.M. 2016. Plant-driven removal of heavy metals from soil: Uptake, translocation, tolerance mechanism, challenges, and future perspectives. *Environmental Monitoring and Assessment* 188:1–11.

U.S. Environmental Protection Agency. 2006. *Solid Waste and Emergency Response*. EPA report 542/F-06/013. U.S. EPA, Washington, DC.

Van Oosten, M.J., Maggio, A. 2015. Functional biology of halophytes in the phytoremediation of heavy metal contaminated soils. *Environmental and Experimental Botany* 111:135–146.

Varun, M., D'Souza, R., Pratas, J., Paul, M.S. 2012. Metal contamination of soils and plants associated with the glass industry in North Central India: Prospects of phytoremediation. *Environmental Science and Pollution Research* 19:269–281.

Walker, T.S., Bais, H.P., Grotewold, G., Vivanco, J.M. 2003. Root exudation and rhizosphere biology. *Plant Physiology* 132:44–51.

Wang, S., Zhao, Y., Guo, J., Zhou, L. 2016. Effects of Cd, Cu and Zn on *Ricinus communis* L. growth in single element or co-contaminated soils: Pot experiments. *Ecological Engineering* 90:347–351.

Wei, S.H., Zhou, Q.X. 2006. Phytoremediation of cadmium-contaminated soils by *Rorippa globosa* using two-phase planting. *Environmental Science and Pollution Research* 13:151–155.

Wei, S.H., Zhou, Q.X., Wang, X. 2005. Cadmium-hyperaccumulator *Solanum nigrum* L. and its accumulating characteristics. *Environmental Sciences* 26:167–171.

Wichelns, D., Drechsel, P., Qadir, M. 2015. Wastewater: Economic asset in an urbanizing world, pp. 3–14. *Wastewater*. Dordrecht, Netherlands: Springer.

Yang, C., Zhou, O., Wei, S., Hu, Y., Bao, Y. 2011. Chemical-assisted phytoremediation of Cd-PAHs contaminated soils using *Solanum nigrum*. *International Journal of Phytoremediation* 13:818–833.

Yu, C., Peng, X., Yan, H., Li, X., Zhou, Z., Yan, Z. 2015. Phytoremediation ability of *Solanum nigrum* L. to Cd-contaminated soils with high levels of Cu, Zn, and Pb. *Water Air Soil Pollution* 226:157.

Yuan, Y., Yu, S., Banuelos, G.S., He, Y. 2016. Accumulation of Cr, Cd, Pb, Cu, and Zn by plants in tanning sludge storage sites: Opportunities for contamination bioindication and phytoremediation. *Environmental Science and Pollution Research* 23:22477–22487.

12

Endophytic Bacterial Diversity in Roots of Wetland Plants and Their Potential for Enhancing Phytoremediation of Environmental Pollutants

Ram Chandra and Kshitij Singh

CONTENTS

12.1 Introduction

Endophytic bacteria are microbes which colonize healthy plant tissues and establish harmonious relations with plants. They are essential components in a wetland ecosystem, as they play important roles in wetland water purification, maintenance of ecosystem balance, and stabilization of plant growth in their symbiotic relationships with wetland plants. Endophytic bacteria are widely distributed in different plant tissues, including roots, stems, leaves, flowers, and fruits of host plants. The number and species of bacteria are differentially distributed, depending on the different host plants that grow in different climates and under different nutrient conditions. The microbes play a key role in host plant adaptation to polluted environments, and they can enhance phytoremediation by mobilizing and degrading or immobilizing contaminants in the aquatic region, promoting plant growth, decreasing phytotoxicity, improving plant metal tolerance, controlling plant diseases, insects, and pests, and strengthening the purification ability of a wetland, as well as maintaining the wetland's ecological balance. In addition, these microbes can supply nutrients to plants by fixing atmospheric nitrogen and solubilizing iron, and they can protect their host plants from infection by plant pathogens through competition for space and nutrients, the production of hydrolytic enzymes, antibiosis, and by inducing plant defense mechanisms.

Endophytic bacteria in wetland plants are rich in species diversity because of their habitat, which is called a transitional zone. This habitat consists of the microbial community in both dry and wet regions. However, knowledge regarding endophytic bacteria for phytoremediation of environmental pollutants in wetland ecosystems is limited. Current research on endophytic bacteria is the latest topic, so there are many questions still to be answered about their reason for residing inside the plants. Until now, such bacteria have been known as symbionts with respect to the plant with which they associate and for their role in phytoremediation. In wetland ecosystems, only a small amount of contaminants can be absorbed by the plants, while endophytic bacteria are the major decomposers in wetland ecosystems. Another emerging function for these bacteria is their role in carrying exogenous genes. Genetically engineered endophytic bacteria carrying additional genes for improved colonization, interactions with plants, or degradation are a promising strategy for improving phytoremediation of soil and water polluted with organic compounds. However, in the case of genetic engineering, an assessment of potential side effects should be performed. Moreover, to maximize the outcomes of plant–endophyte interactions, high-throughput strategies are needed to evaluate all or most of the genes involved in organic pollutant remediation. In addition, recent studies have shown that endophytic bacteria have the potential to enhance the removal of soil contaminants through phytoremediation. For example, a genetically modified endophytic strain of *Burkholderia cepacia*, together with yellow lupine *Lupinus luteus*, could improve degradation of toluene. Some endophytic bacteria from poplar trees exhibit the ability to degrade benzene, toluene, ethylbenzene, and xylene (BTEX). Some strains have been shown to degrade naphthalene and pesticides, and other strains have shown potential to dissolve insoluble phosphate.

The reason that endophytic bacteria have become a hot research topic at home and abroad is their development and extensive application value. However, little research has been undertaken to understand more about endophytic bacteria in aquatic plants, in particular, their diversity, distribution within plants, bioremediation abilities, and their associations with the plants. Community structure, dynamics, and biological mechanisms are popular topics currently. Endophytic bacteria could also have broader applications in the study of ecological functions of wetlands. However, concentrated efforts are needed to explore the beneficial traits of pollutant-degrading endophytic bacteria to get maximum outcome from their combined use with plants. Therefore, this chapter describes the general mechanism of plant and endophytic bacteria associations in wetland ecosystems. Furthermore, the chapter also describes the diveristy of dominant endophytic bacteria in their use for phytoremediation of various environmental pollutants in wetland ecosystems.

12.2 Microbiology of Wetlands

Wetlands provide a transition zone between dry land and water bodies, and they therefore also have been described as ecotones, based on the microbial diversity present in wetlands and the sophisticated hierarchy of metabolic systems, each bringing a unique enzymatic capability to the bioconversion process. Wetland microbes, plants, and wildlife are part of global cycles for water, nitrogen, and sulfur. Interactions of different microbial communities in wetland ecosystems result in a richness of diversity of microbes. Due to the transition zone, they are the main link of interaction with living biota and various environmental pollutants of aquatic ecosystems, which have direct or indirect interactions with the environmental food chain. Bacterial communities in wetland ecosystems choose their habitat according to their needs, as some microorganisms reside in wetland soils while some enter the plant tissues, i.e., endophytic bacteria. The close proximity of oxic–anoxic conditions, often created by wetland plant roots, facilitates the simultaneous activity of aerobic as well as anaerobic bacterial communities. Input of nutrients and fast recycling due to active aerobes and anaerobes make these systems highly productive and therefore attractive for humans as well as many other organisms. Roots may also increase microbial activity through the production of organic carbon and release of substances such as sugars and amino acid exudates. Wetland ecosystems are characterized by their abundant microflora, which includes different heterotrophic and autotrophic microorganisms. To maintain the wetland ecosystem, various bacterial communities involved, such as methanotrophs, sulfate reducers, acetate producers (acetogenic bacteria), nitrogen fixers, denitrifiers, and iron oxidizers, which are highly abundant in wetlands.

Wetland soil or sediment environment has a number of physical, chemical, and biological characteristics due to the periodic or permanent flooding of soils and sediments. The most important ramification of the water layer on top of flooded soils and sediments is the restricted entry of atmospheric oxygen. The diffusion of oxygen in water is 10,000 times slower than in air. Oxygen is rapidly depleted in the upper layers of inundated soil due to chemical and biological oxidation processes, resulting in a soil profile where the presence of oxygen is limited to the upper millimeters. In most cases, microorganisms are the cause of anoxic conditions. They take in oxygen for cellular respiration and release CO_2 as part of the chemical processes that make energy for cells. Wetland plants obviously have developed mechanisms to equip themselves for colonization and growth in anoxic flooded soils and sediments. Changes in anatomy, morphology, and metabolism are of paramount importance for surviving in anoxic root environments. So-called aerenchymatous tissue in shoots and roots facilitates the transport of oxygen from the atmosphere to the roots, thereby supplying the roots with oxygen for respiration. However, roots also lose substantial amounts of oxygen to the surrounding anoxic soil, as demonstrated in numerous studies (Armstrong et al. 2000). By means of this process of radial oxygen loss, wetland plants create oxic–anoxic interfaces and thereby provide habitats for both aerobic and anaerobic microbes, facilitating nutrient recycling (Figure 12.1). Along with the oxygen,

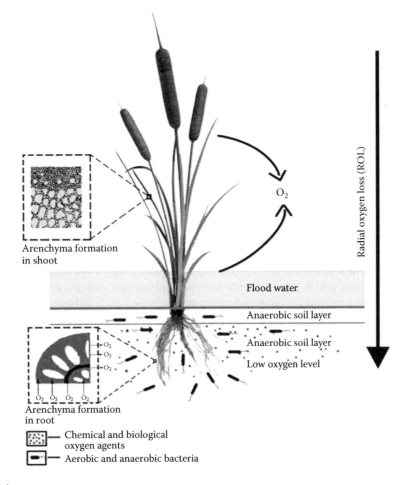

FIGURE 12.1
Aerobic and anaerobic conditions in a wetland. Wetland plants have aerenchymatous tissue in the shoots and roots and this facilitates transport of oxygen from the atmosphere to the roots. In contrast, radial oxygen loss helps in creating both oxic and anoxic conditions and thus provides habitats for both aerobic and anaerobic microbes.

plants also excrete some organic carbon compounds that contribute to rhizospheric microbial diversity as they process the compounds for their necessary energy to perform their various biogeochemical reactions. The presence of both oxic and anoxic interfaces in wetland soils implies that aerobic as well anaerobic bacteria have to adapt to this environment. Anaerobic acetate-producing (acetogenic) bacteria have also been reported to be closely associated with wetland plant roots. It is hardly surprising that a wide variety of microbial taxa has been identified in the field of contemporary microbial ecology, ranging from obligate anaerobes to obligate aerobes, and they are intimately associated with higher plant roots in flooded soils.

The oxygen concentration within and around roots varies from nearly atmospheric values to strict anoxia, with steep gradients in oxygen concentration in tissues and rhizospheres. Concentrations vary not only spatially but also temporally, with a constantly shifting mosaic of anoxic and oxic microsites as roots grow, explore soil, age, and die. Plants growing in wetlands significantly oxidize the sediment, where new young tissue colonizes previously unvegetated soil, but plant growth also makes the sediment more reducing as older biomass is incorporated into the soil organic matter. A mature root is compartmentalized internally with anoxic and oxic habitats, while its external surface can form both anoxic surfaces and oxidized rhizospheres, depending on age and branching patterns.

12.3 Important Wetland Plant Species and Their Ecosystems

The primary factor that distinguishes wetlands from other land forms or water bodies is the characteristic vegetation of aquatic plants. Many wetlands have distinctive plants which are adapted to the wetting and drying cycles of wetlands. The types of plants found in a wetland depend on (i) whether a wetland has mostly fresh, saline, or brackish water, (ii) surface and underground drainage, (iii) the frequency of inundation, and (iv) other factors, such as soil characteristics, temperature, rainfall, and topography. The larger aquatic plants growing in wetlands are usually called macrophytes. This term includes aquatic vascular plants (angiosperms and ferns), aquatic mosses, and some larger algae that have tissues that are easily visible. Although fern genera like *Salvinia* and *Azolla* and large algae like *Cladophora* are widespread in wetlands, it is usually the flowering plants (i.e., angiosperms) that dominate. The presence or absence of aquatic macrophytes is one of the characteristics used to define wetlands, and as such macrophytes are an indispensable component of these ecosystems. Some examples of wetland plants are shown in Figure 12.2. Types of wetlands and its associated plants are as follows:

1. Permanent wetlands, which are always or nearly always flooded and are dominated by aquatic plants such as ribbon weed (*Vallisneria* species) and wavy marshwort (*Nymphoides crenata*)

2. Semipermanent wetlands, which are usually flooded every year and are characterized by sedges (e.g., cumbungi and *Cyperus* species), rushes (e.g., *Juncus* species, marsh club rush), spike rushes (*Eleocharis* species), water couch (*Paspalum distichum*), common reed (*Phragmites australis*), and herbs and forbs, such as isotomes (*Isotoma* species), primrose (*Ludwigia*), nardoo (*Marsilea*), and *Ranunculus* species

FIGURE 12.2
Some common aquatic wetland plants, showing their histological structures of root cross-sections. (a) *Typha angustifolia*; (b) *Phragmites australis*; (c) *Eichornia crassipes*.

3. Ephemeral wetlands, which experience irregular flooding and long dry periods and contain lignum (*Muehlenbeckia cunninghamiana*), river red gum, black box, coolabah, and other dry land species.

With recent advances in research, it has been determined that plants play a major role in removing environmental pollutants from wetland ecosystems via phytoremediation. Wetland plant species with potential for phytoremediation should possess the following properties: (i) an extensive root system, (ii) release root exudates (because they are more important than other sources of rhizodeposits in structuring rhizosphere bacterial communities), (iii) accumulate, extract, transform, degrade, or volatilize contaminants at levels that are toxic to ordinary plants, and (iv) exhibit fast growth and high yields and have the ability to remediate multiple pollutants simultaneously.

12.4 Endophytic Bacteria, Their Distribution, and Potential Roles in Phytoremediation of Pollutants in Aquatic Ecosystems

12.4.1 Endophytic Bacteria

The kinds of associated bacteria which reside in plant tissues beneath the epidermal cell layers, where they can colonize the internal tissues and follow a range of different lifestyles with their host, including symbiotic, mutualistic, commensalistic, and trophobiotic, are known as endophytic bacteria. Endophytic bacteria can be classified as obligate or facultative. Obligate endophytes are strictly dependent on the host plant for their growth and survival. Transmission of obligate endophytes to other plants occurs vertically or via vectors. Facultative endophytes are characterized as biphasic, alternating between living in plants and living in the environment. Facultative endophytes have a stage in their life cycle in which they exist outside host plants. They show chemotactic movement towards plant roots in response to various root exudates that help bacteria colonize and enter plant roots via wounds, or they penetrate by self-enzymatic activity (Figure 12.3). Endophytic bacteria mainly reside in the apoplasm or symplasm of plants. Recent studies have shown that endophytic bacteria are not host specific. Endophytic bacteria can also invade a wide host range of plants. Therefore, endophytic bacteria have been isolated from various parts of

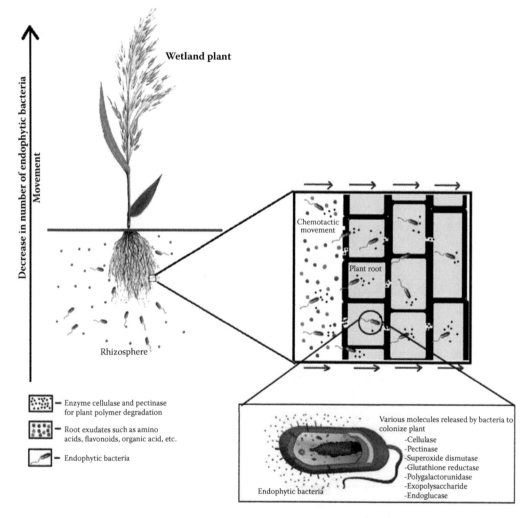

FIGURE 12.3
Chemotactic movement of bacteria towards plant roots. Various root exudates released by plant roots act as chemical attractants which help bacteria colonize and enter into plant roots via wounds, or they penetrate by self-enzymatic activity. At the site of entry, they are present in large numbers, which decline as they toward the phyllosphere.

different plants that belong to different families and classes and grow under different eco-logical and geographical conditions. Endophytic bacteria are often capable of triggering physiological changes that promote both pollutant-degrading and plant growth-promoting activities and as such can be superior in improving plant growth and phytoremediation activity compared to microbes with only one of these activities. These bacteria are known to change root morphology and increase their biomass, thus enabling the roots to take up more soil nutrients. They span a wide range of bacterial phyla, and some are known to play an important role in plant yield and growth promotion, plant health, and protection. Some endophytic bacteria show resistance to heavy metals and degrade organic compounds in a plant, soil, or water niche and thus also play an important role in pollution control and phytoremediation. Consequently, opportunities for newly identified, beneficial endo-phytic microorganisms growing among a diversity of plants in different ecosystems are considerable.

It has not been resolved whether plants benefit more from endophytic bacteria than from a rhizospheric bacterium, or if it is more advantageous for bacteria to become endophytic than rhizospheric. Nevertheless, benefits conferred by endophytic bacteria are well rec-ognized. There are several common properties of endophytic bacteria (Figure 12.4): (i) they are omnipresent in a large diversity of plant species and colonize roots at a high density, with fewer colonizing stems and leaves. (ii) They colonize inside a plant without causing any pathogenicity to the plant and have potential to survive in both biotic and abiotic stress conditions. (iii) Most of the endophytic bacteria have been found to originate from epiphytic bacterial communities in the rhizosphere or phyllosphere or from other plant parts. Their transmission takes place through damaged tissues or the cracks formed in lateral root junctions, and then they quickly spread to the intercellular spaces in the root (Bacon and Hinton 2006). (iv) They respond to plant root exudates by showing chemo-tactic movement towards them. The responses may differ in different endophytic bac-teria. (v) They have potentials for both plant growth-promoting and pollutant-degrading activities and are therefore highly promising for certain phytoremediation methods

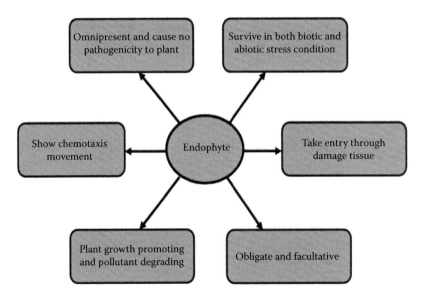

FIGURE 12.4
Various properties of endophytic bacteria.

(Khan et al. 2013a). However, the intimate association of endophytic bacteria with plants offers a unique opportunity for their potential application in plant protection and biological control. Bacteria living inside plant tissues form associations ranging from pathogenic to symbiotic. Multitudes of bacteria reside in leaves, stems, roots, seeds, and fruits of plants that are seemingly neutral in terms of plant health (Surette et al. 2003). However, several studies have also suggested that many endophytic associations are not neutral at all, but are beneficial to plants.

12.4.2 Diversity and Distribution of Endophytic Bacteria in Plants

Plant–bacteria associations can improve the degradation of organic pollutants in soil. However, little is known about the diversity and distribution of endophytic bacteria associated with wetland plants and their potential to enhance phytoremediation of wetland environments. To date, the diversity of endophytic bacteria in wetlands has been studied in the following aquatic plants: *Phragmites communis, Potamogeton crispus, Nymphaea tetragona, Najas marina, Typha* spp., *Eichornia* spp., *Pistia* spp., *Najas marina, Juncus acutus, Oryza sativa,* and *Scirpus triqueter* (Table 12.1). The isolated bacteria were classified into 12 genera within the classes *Gammaproteobacteria, Bacilli, Alphaproteobacteria, Flavobacteria,* and *Actinobacteria* (Wu et al. 2012) However, reports about the biodegradation capabilities of endophytic bacteria have been mostly focused on rhizospheric bacteria in terrestrial plants. Apart from Toyama et al. (2009), who showed that aquatic plant–bacteria collaboration can accelerate contaminant degradation in aquatic environments, there has been little research on endophytic bacteria associated with aquatic plants. Environmental water pollution is a serious problem in China and other developing countries because of the discharge of organic pollutants such as pesticides and nutrients, including phosphorus (Wang et al. 2007). Organic pollutants in aquatic environments are generally present in low concentrations; therefore, bioremediation or phytoremediation may be the most economic and reliable approach to address the problem. In addition, phosphorus, which can stimulate eutrophication, usually precipitates in sediments. This makes phosphorus largely unavailable to plants. Furthermore, the phosphorus precipitate can persist in a lake for a considerable length of time and continue to periodically cause algal blooms. The release of insoluble phosphorus and its removal from the environment via aquatic plants is a so-called eco-engineering approach in which released phosphorus is taken up by plants and then removed from the aquatic environment by harvesting the plants. However, little research has been undertaken to understand more about endophytic bacteria in aquatic plants, in particular, their diversity, distribution within the plant, bioremediation abilities, and their associations with the plants.

Wu et al. (2012) isolated endophytic bacteria from four different plants: *Phragmites communis, Potamogeton crispus, Najas marina,* and *Nymphaea tetragona.* After a comprehensive comparison among the 16S rRNA genotypes of the different parent plants and plant parts, a total of 39 isolates were selected from the 25 groups, sequenced, and phylogenetically analyzed. The 100 isolates were then assigned to 12 bacterial genera (Figure 12.5). *Pseudomonas,* which contained 27 strains of the 100 isolates (27% of the total isolates), *Enterobacter* (16%), *Aeromonas* (12%), *Klebsiella* (4%), and *Pantoea* (5%) in *Gammaproteobacteria* (64% in total); *Bacillus* (8%), *Paenibacillus* (11%), *Lactococcus* (2%), and *Staphylococcus* (5%) in the *Bacilli* (26% in total); *Delftia* (1%) in the *Alphaproteobacteria; Flavobacterium* (1%) in the *Flavobacteria;* and *Microbacterium* (8%) in the *Actinobacteria* (Figures 12.5 and 12.6). The phylogenetic analyses also revealed that the endophytic isolates shared high sequence similarities (greater than 98%) with their recognized relative species. For example, *Pseudomonas* isolates were

TABLE 12.1

Diversity of Bacterial Endophytes in Various Parts of Wetland Plants

Host Plant	Plant Part	Bacterial Endophyte Species	Reference
Typha sp.	Root	*Rhizobium* sp., *Cloacibacterium normanense*, *Bacillus* sp., *Halomonas hamiltonii*, *Sphingobium* sp., *Pannonibacter phragmitetus*, *Microbacterium oleivorans*, *Bacillus safensis*, *Bacillus pumilus*, *Psychrobacter alimentarius*, *Bacillus subtilis*	Shehzadi et al. 2015
Typha sp.	Shoot	*Microbacterium arborescens*, *Halomonas stevensii*, *Microbacterium* sp., *Microbacterium schleiferi*, *Bacillus safensis*, *Bacillus* spp., *Janibacter melonis*, *Pseudomonas fluorescens*, *Bacillus aerophilus*	Shehzadi et al. 2015
Typha domingenesis	Root	*Microbacterium arborescens*, *Bacillus pumilus*	Shehzadi et al. 2014
Typha angustifolia L.	Root	*Rhodoferax*, *Pelomonas*, *Uliginosi bacterium*, *Pseudomonas*, *Aeromonas*, *Rhizobium*, *Sulfurospirillum*, *Ilyobacter*, *Bacteroides*	Li et al. 2011
Phragmites australis	Root	*Pleomorphomonasoryzae*, *Pleomorphomonas koreensis*, *Azospirillum picis*, *Agrobacterium vitis*, *Dechloromonas hortensi*, *Aeromonas bivalvium*, *Acetobacterium malicum*, *Desulfomicrobiumnorvegicum*	Zhang et al. 2010
Phragmites communis	Root	*Lactococcus Pantoea*, *Bacillus*, *Pseudomonas*, *Paenibacillus*, *Microbacterium*, *Enterobacter*, *Aeromonas*, *Klebsiella*	Wu et al. 2012
Phragmites communis	Stem	*Flavobacterium*, *Bacillus*, *Pseudomonas*, *Paenibacillus*, *Microbacterium*, *Enterobacter*, *Aeromonas*	Wu et al. 2012
Phragmites communis	Leaf	*Pseudomonas*, *Paenibacillus*, *Microbacterium*, *Enterobacter*, *Aeromonas*, *Klebsiella*	Wu et al. 2012
Potamogeton crispus	Root	*Delftia*, *Staphylococcus*, *Enterobacter*, *Pseudomonas*	Wu et al. 2012
Potamogeton crispus	Stem	*Paenibacillus*, *Klebsiella*, *Pantoea*, *Enterobacter*, *Pseudomonas*	Wu et al. 2012
Potamogeton crispus	Leaf	*Pseudomonas*	Wu et al. 2012
Nymphaea tetragonal	Root	*Pseudomonas*, *Bacillus*	Wu et al. 2012
Nymphaea tetragonal	Stem	*Paenibacillus*, *Enterobacter*, *Aeromonas*, *Klebsiella*, *Lactococcus*, *Pantoea*, *Bacillus*, *Pseudomonas*	Wu et al. 2012
Eichornia sp.	Shoot	*Escherichia hermannii*, *Jonesia* sp., *Halomonasvenusta*, *Alishewanella* sp.	Shehzadi et al. 2015
Eichornia sp.	Root	*Kocuriarosea*, *Bacillus marisflavi*, *Bacillus marisflavi*	Shehzadi et al. 2015
Pistia sp.	Root	*Paracoccussp.*, *Chryseobacterium* sp., *Bacillus pumilus*, *Bacillus licheniformis*	Shehzadi et al. 2015
Pistia sp.	Shoot	*Planococcus rifietoensis*, *Rhodobacter* sp., *Bacillus endophyticus*, *Bacillus pumilus*	Shehzadi et al. 2015
Najas marina	Root	*Pseudomonas*, *Bacillus*, *Paenibacillus*	Wu et al. 2012
Najas marina	Stem	*Pseudomonas*	Wu et al. 2012
Juncus acutus	Root	*Sphingomonas* sp., *Bacillus* sp., *Ochrobactrum* sp.	Syranidou et al. 2016
Oryza sativa	Root	*Herbaspirillum seropedicae*	Ladha et al. 1997
Scirpus triqueter	Root and shoot	*Pseudomonas* sp.	Zhang et al. 2014

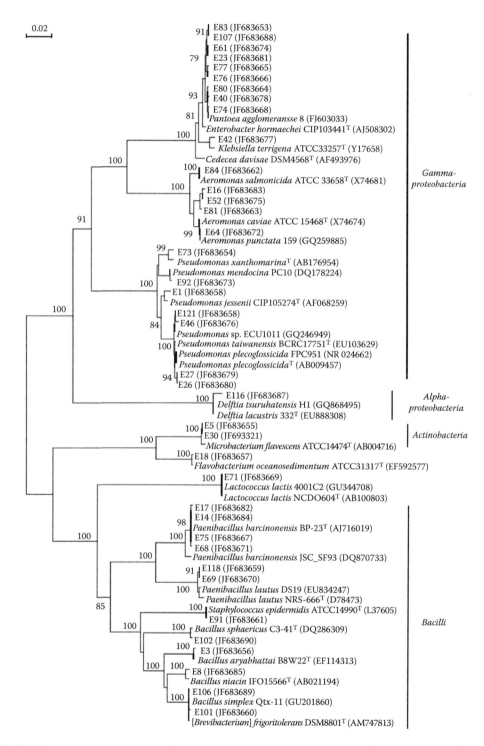

FIGURE 12.5
Neighbor-joining tree of the 16S rRNA gene sequences of endophytic bacterial strains, showing the relationships of strains isolated from wetland plants *Phragmites communis, Potamogeton crispus, Nymphaea tetragona,* and *Najas marina* and also reference species. (*Source:* Chen, W.-M. et al., *Aquatic Biology,* 15:99–110, 2012.)

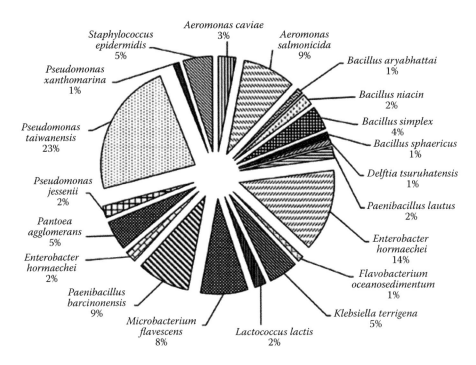

FIGURE 12.6
Relative frequencies of endophytic bacteria isolated from four wetland plants classified to the species level on the basis of 16S rRNA gene sequence analysis. Data show the percentage contributions of each group of isolates among the total 100 isolates. (*Source:* Chen, W.-M. et al., *Aquatic Biology*, 15:99–110, 2012.)

phylogenetically closely related to *Pseudomonas taiwanensis* (with 16S rRNA gene sequence similarity of 99.4%), *P. xanthomarina* (99.3%), and *P. jessenii* (98%). The isolates belonging to *Enterobacter* and *Pantoea*, most of which were isolated from roots, were closely related to *Enterobacter hormaechei* (99.3%) and *Pantoea agglomerans* (99.6%). The *Aeromonas* isolates were closely related to *Aeromonas salmonicida* (99.9%) and *A. caviae* (98.6%). Bacilli isolates were closely related to *Paenibacillus barcinonensi*s (99%), *P. lautus* (99.2%), *Staphylococcus epidermidis* (99.9%), *Bacillus simplex* (99.9%), *B. sphaericus* (99.9%), *B. aryabhattai* (99.8%), and *Lactococcus lactis* (100%). The remaining endophytic bacteria were related to *Flavobacterium oceano sedimentum* (99.7%), *Microbacterium flavescens* (99%), and *Klebsiella terrigena* (99.2%).

Although phytoremediation has been widely used to restore various contaminated sites, it is still unclear how soil microbial communities respond microecologically to plants and pollutants during the process. Endophytic bacteria involved in phytoremediation of a petroleum-contaminated wetland, including *Scirpus triqueter*, were set up to monitor the influence of plant rhizospheres on soil microbes. Palmitic acid, one of the main root exudates of *S. triqueter*, was added to strengthen the rhizosphere effect. Increasing the diesel concentration led to higher abundances of fungi and Gram-positive and Gram-negative bacteria. The addition of palmitic acid amplified the rhizosphere effect on soil microbial populations and diesel removal. Principal component analysis revealed that the plant rhizosphere effect was the dominant factor affecting microbial structure. These results provided new insights into plant–microbe–pollutant coactions responsible for diesel degradation, and they were valuable to facilitate phytoremediation of diesel contamination in wetland habitats.

12.4.3 Interactions of Plants and Endophytic Bacteria in Wetlands

The roots of wetland plants interact with a large number of different microorganisms, with these interactions being major determinants of the extent of phytoremediation. It has been established that plants communicate to attract specific microorganisms for their own ecological and evolutionary adaptation. First step in endophytic bacteria-plant interaction is chemotaxis. Chemotaxis is the directional motility of endophytic bacteria species from the chemotactic response to root exudates. Bacterial root colonization often starts with the recognition of specific compounds in the root exudates by the bacteria (Figure 12.7). A competent endophytic bacteria, moves towards active zones in the rhizosphere known as exosphere that will gear its metabolism towards a physiological state that enables optimal nutrient acquisition, competition and growth as shown in Figure 12.8.

When attached to roots, bacteria often increase in numbers by several cell divisions, resulting in the establishment of microcolonies. Invasion of root tissue might take place from such established microcolonies. In this invasive process, enzymatic activity is crucial for degradation of plant cell envelopes. It has been proposed that levels of cell wall-degrading enzymes produced by root-colonizing bacteria differentiate endophytic bacteria (low enzyme levels) from bacterial phytopathogens (deleteriously high enzyme levels). Alternatively, endophytic bacteria might enter roots or other plant tissues without the aid of cell-wall-degrading enzymes, possibly via spontaneous cracks formed between displaced epidermal cells or wounds caused by phytopathogens. Once inside the roots, competent endophytic bacteria must pass the casparian strips in the endoderm to systemically spread to the aboveground parts of the plant. Endophytic bacterial species enter

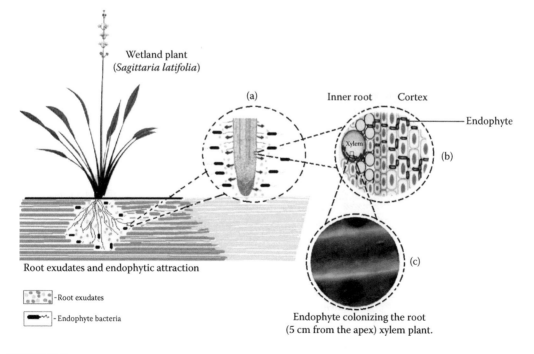

FIGURE 12.7
Interaction of bacterial endophytes with the roots of wetland plants. (a) Root exudate secretion attracts bacteria to enter the root tissue; (b) endophytic bacteria colonization inside the root; (c) endophyte colonization inside xylem tissue.

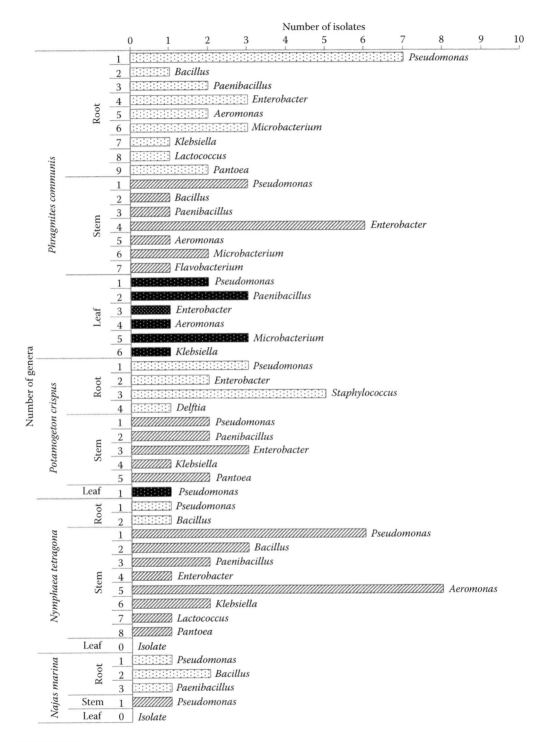

FIGURE 12.8
Number of endophytic bacteria isolated from *Phragmites communis*, *Potamogeton crispus*, *Nymphaea tetragona*, and *Najas marina* in relation to the location within the plant. (Source: Chen, W.-M. et al., *Aquatic Biology*, 15:99–110, 2012)

plant tissues due to the production of enzymes, such as endoglucanases and endopolyga-lacturonidases. The interaction of plants with bacteria can have a major impact on plant growth and health. Finally, there is enhancement of bacterial colonization. Endophytic bacterial population sizes are dependent, and positively correlated with, the plant developmental stage.

Root colonization by any plant–endophytic bacteria interaction involves several stages:

- First stage: A bacterium senses various root exudates, e.g., flavonoids, amino acids, carbohydrates, organic acids, and moves to the plant roots via chemotactic movements, either passively or actively, by specific induction of flagellar activity by the plant-released compounds.

- Second stage: Entry of endophytic bacteria in plant roots is known to occur (a) through wounds, particularly where lateral or adventitious roots occur; (b) through root hairs; and (c) between undamaged epidermal cells. Nonspecific adsorption of bacteria to the roots takes place, followed by anchoring and finally the firm attachment of bacteria to the root surface, by the help of its enzymatic activities, e.g., cellulase, pectinase, etc. It degrades the polysaccharide wall made of cellulose, hemicellulose, and pectin. Specific or complex interactions between the bacterium and the host plant occur, including the secretion of root exudates, whichi can result in the induction of bacterial gene expression.

- Third stage: Endophytic bacteria can subsequently enter their host plant at sites of tissue damage, which arise due to enzymatic activity of bacteria or sometimes naturally arise as the result of plant growth (lateral root formation) or through root hairs and at epidermal conjunctions. In addition, plant exudates leaking through these wounds provide a nutrient source for the colonizing bacteria.

Endophytic bacteria have more advantages over rhizospheric bacteria, because they get the opportunity to stay in direct contact with the plant tissues. Also, they offer more beneficial effects for the plants compared to bacteria residing outside plants. However, rhizospheric colonizers have been found to be the major source of endophytic colonization. The rhizosphere offers an environment of high competition for microorganisms to occupy spaces and obtain nutrients. It has been observed that bacterial root colonization often starts with the recognition of specific compounds in the root exudates by the bacteria. These compounds probably also have major roles in belowground community interactions. Plants produce exudates from roots to interact with microorganisms for their own ecological and evolutionary benefits. Flavonoids are considered important compounds in plant–microbe communications. They are possibly also important for competent endophytic bacteria to occupy a suitable and permanent niche in the rhizosphere and on roots. Bacteria respond to these exudates by showing chemotactic movement towards them. The response may differ in different endophytes. Mutational analyses suggest that surface characteristics of bacteria play an important role in the establishment of endophytic bacteria in roots. Root exudates act as messengers for communication to the neighboring plant and adjoining microorganisms present in the rhizosphere of the root. The chemical ingredients of the root exudates are specific to a particular plant species and also depend on the nearby biotic and abiotic environment. Root exudates mediate various positive and negative interactions, like plant–plant and plant–microbe interactions. Through the exudation of a wide variety of compounds, roots may regulate the aquatic microbial community in their immediate vicinity, encourage beneficial symbioses, change the chemical and physical properties of the aquatic environment, and inhibit the growth of competing plant species. Microbes within the rhizosphere can in turn modify root exudate composition. A survey of the literature exposes an extensive range of compounds exuded from intact and healthy roots; these include sugars, amino acids, peptides, enzymes, vitamins, organic acids, nucleotides, inhibitors, attractants, and many miscellaneous compounds (Table 12.2). Organic acids, sugars, amino acids, lipids, coumarins, flavonoids, proteins, enzymes, aliphatics, and aromatics are examples of the primary substances found within rhizosphere of the root. Among them, the organic acids have received considerable attention due to their role in providing substrates for microbial metabolism and for serving as intermediates for biogeochemical reactions in soil. The rhizosphere is defined as a zone of most intense bacterial activity around the roots of plants. However, for the sake of practical investigation, the rhizosphere is most often defined as the soil adhering to plant roots when they are rigorously shaken, throughout which the rhizosphere effect must be observed to some extent.

In a study conducted on rice plant, it was found that exopolysaccharide production is necessary for plant colonization by the nitrogen-fixing microbes. In another study, it was observed that bacterial superoxide dismutase and glutathione reductase were crucial for endophytic colonization. For endophytic colonization, not only attachment is instrumental, but also motility: type IV pili are required for twitching motility on solid surfaces, which depends on PilT-mediated retraction of pili. This suggests that endophytic spreading of bacteria inside the host is an active process involving motility on plant surfaces. It not currently known whether there is any specific form of communication (other than chemotaxis) between endophytic bacteria and plants that is involved in the colonization

TABLE 12.2

Chemical Natures of Wetland Plant Root Exudates Responsible for Bacterial Associations with Roots through Chemotaxis

Substance Group	Enzymes
Enzymes	Amylase, invertase, peroxidase, phenolase, phosphatases, polygalacturonase, lectins, proteases, hydrolases, lipase
Organic acids	Acetic, succinic, laspartic, L-glutamic, salicylic, shikimic, isocitric, chorismic, sinapic, p-hydroxybenzoic, gallic, tartaric, protocatacheuic, p-coumaric, mugineic, oxalic, citric, piscidicaconitic, ascorbic, benzoic, butyric, caffeic, citric, p-coumaric, ferulic, fumaric, glutaric, glycolic, glyoxilic, malic, malonic, oxalacetic, oxalic, p-hydroxybenzoic, propionic, succinin, syringic, tartaric, valeric, vanillic
Amino acids	α-Alanine, L-hydroxyproline, homoserine, mugineic acid, β-alanine, γ-aminobutyric, arginine, aspartate, citrulline, cystathionine, cysteine, cystine, deoxymugineic, 3-epihydroxymugineic, aminobutyricacid, glutamine, glutamic, glycine, homoserine, isoleucine, leucine, lysine, methionine, mugineic, ornithine, phenylalanine, praline, serine, threonine, tryptophan, tyrosine, valine
Growth factors	p-Amino benzoic acid, biotin, choline, N-methyl nicotinic acid, niacin, pantothenic, vitamins B_1 (thiamine), B_2 (riboflavin), and B_6 (pyridoxine)
Miscellaneous	Auxins, scopoletin, hydrocyanic acid, glucosides, unidentified ninhydrin-positive compounds, unidentified soluble proteins, reducing compounds, ethanol, glycine betaine, inositol and myo-inositol-like compounds, Al-induced polypeptides, dihydroquinone, sorgoleone
Polysaccharides and sugars	Sucrose, pentose, arabinose, fructose, galactose, glucose, maltose, mucilages of various compositions, oligosaccharides, raffinose, rhamnose, ribose, sucrose, xylose, mannitol
Fatty acids	Linoleic, palmitic, stearic, linolenic, oleic
Sterols	Sitosterol, stigmasterol, campesterol, cholesterol
Flavonones	Adenine, flavonone, guanine, uridine/cytidine

of the internal tissues of plants, comparable to the mechanisms used by various *Rhizobia* spp. As soon as their cells are inside the plant, competent endophytic bacteria respond to plant cues to enable further induction of cellular processes necessary for entering the endophytic life stage and spreading to other (intercellular) tissues of the root cortex and beyond. Production of enzymes, such as endoglucanases and endopolygalacturonidases, seems to be indispensable in this process. At this point, competent endophytic bacteria can quickly multiply inside the plant, often reaching high cell numbers. In endophytic and rhizobacteria, the processes of host–microbe signaling and colonization and the mechanisms leading to mutual benefits are less well characterized. The host–endophyte relationship may be variable from host to host and endophyte to endophyte.

Some research has shown that the host plant–endophyte relationship is able to balance pathogen–host antagonism and is not truly a symbiotic one. Root colonization is the first and the critical step in establishment of plant–microbe association, which is greatly influenced by the chemotactic responses of endophytes towards root exudates. This is probably among the strongest determinants for successful endophytic colonization. Colonization of bacteria in the rhizosphere or on plant surfaces is a complex process which involves interplay between several bacterial traits and genes. The colonization is a multistep process and includes migration of microbes towards root surface, attachment, distribution along the root, and growth and survival of the population. For endophytic bacteria, one additional step is required: entry into root and formation of microcolonies inter- or intracellularly. Each trait may vary for different associative and endophytic bacteria. In addition to providing nutritional substances, plants start cross-talk to microorganisms by secreting some signals which cause colonization by some bacteria while inhibiting the other.

The variations in the endophytic bacterial communities can be attributed to oxic and anoxic conditions of wetland regions, plant age, plant source, tissue type, plant genotype, time of sampling, and environmental conditions. It has been demonstrated that age of the plant can largely influence the variations in the endophytic community. Also, endophytic diversity might be a function of the different maturation stages specific to each plant, which might influence the different types and amounts of root exudates. Competition experiments with endophytes have shown that some endophytic bacteria are more aggressive colonizers and displace others. This was observed with *Pantoea* sp. outcompeting *Ochrobactrum* sp. in rice and with different *Rhizobium etli* strains. However, when the host ranges of a large diversity of endophytic bacteria were analyzed, a seeming lack of strict specificity was observed (Zinniel et al. 2002).The population density of endophytic bacteria is highly variable, depending mainly on the bacterial species and host genotypes but also on the host developmental stage, inoculum density, and environmental conditions. In plant tissue in general, endophytic bacterial populations have been reported at densities between 102 and 104 viable bacteria/gm (Kobayashi and Palumbo 2000).

12.5 Factors That Affect the Abundance and Expression of Catabolic Genes in the Rhizosphere and Endosphere

As with the application of endophytic bacteria for the cleanup of contaminated soil and water, there are two particular aspects that should be considered in phytoremediation: (i) pollutant-degrading endophytic bacteria have to be maintained in the rhizosphere and endosphere, and (ii) the expression of pollutant-degrading genes. It has been shown that hydrocarbon concentrations in soil can affect endophytic bacterial colonization, i.e., the abundance and expression of pollutant-degrading genes in the rhizosphere and endosphere of inoculated plant, which were highest in the presence of the highest (2%) diesel concentration in soil. The soil type (sandy, loamy sand, and loam) may additionally affect endophytic bacterial colonization and expression of alkane-degrading genes. The endophyte strain *Pseudomonas* sp. ITRI53 exhibited following inoculation the highest levels of abundance and expression of alkane-degrading *alkB* genes in the roots and shoots of ryegrass growing in loamy soil.

12.6 Biofilm Formation by Endophytic Bacteria in Wetland Ecosystems

Wetland biofilms are of interest for both ecological and applied disciplines. The term biofilm was coined and described by Costerton et al (1995). Biofilm functions are influenced by the complex interactions of bacteria inside these biofilms and as a result efficiency of the wetland system. In other words, biofilms affect wetland effluent quality. A biofilm is defined as an assemblage of microbial cells that is irreversibly associated with a surface and usually enclosed in a matrix of polysaccharide material. Close proximity of different microbial colonies can bring about complex and rapid microbial degradation requiring combined metabolic capabilities. A biofilm can be formed by a single bacterial species or by a group of microorganisms (bacteria, archaea, protozoa, fungi, and algae), with each group performing specialized metabolic functions. Cells in the biofilm are held together by extracellular polymeric substances including exopolysaccharide. Exopolysaccharide provides an optimal environment for the exchange of genetic material between cells. Cells may also communicate via quorum sensing, which may in turn affect biofilm processes such as detachment.

In terms of advantages of biofilms, there are several advantages of endophytic bacteria growing on polluted environments:

1. The dense extracellular matrix of the outer layer of cells provides increased resistance to detergents and antibiotics as well as protection against antimicrobial agents.

2. Biofilms improve the uptake of dissolved organic matter from solutions or suspensions. Organisms can concentrate scarce nutrients extracellularly and utilize these compounds through the activity of exoenzymes.

3. Biofilms that contain water channels help distribute nutrients and signaling molecules.

4. Exchange of nutrients, metabolites, or genetic material from close proximity to other microorganisms is facilitated in biofilms

5. Many sewage treatment plants have a treatment stage in which water passes over biofilms grown on a filter that can extract or digest organic compounds. Bacteria can remove organic matter and protozoa can remove suspended solids, including pathogens.

6. Biofilms can also help eliminate petroleum oil from contaminated oceans or marine systems. The oil is eliminated by the hydrocarbon-degrading activities of microbial communities, in particular by hydrocarbonoclastic bacteria.

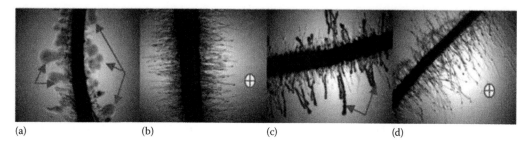

(a) (b) (c) (d)

FIGURE 12.9

Microscopic image of root colonization and biofilm formation by *Bacillus subitilis*. The arrows in panels a and c indicate biofilm formation on root system, and the circled crosses in panels b and d indicate absence of biofilm formation on the root system. *Source*: Rafique, M. et al., *The Battle Against Microbial Pathogens*, 2015.

Biofilm structures are strongly influenced by the surface properties of the plant tissue, nutrient and water availability, and the proactivities of the colonizing bacteria (Figure 12.9). Currently, the consensus model in biofilm research proposes that microbial community development on surfaces is a stepwise process involving adhesion, growth, motility, and extracellular polysaccharide production. Once an initial biofilm is established, cell-to-cell communication (i.e., quorum sensing) via extracellular signaling molecules regulates further modification and development of the biofilm. Recent reports on quorum sensing and its modification due to the presence of plant root exudates and metabolites have further compounded the complexity of microbe–plant–root interaction mechanisms. Inter- and intraspecies signal molecules (*N*-acyl homoserine lactones for Gram-negative bacteria and proteins or polypeptides for Gram-positive bacteria) are synthesized as a bacterial population reaches sufficiently high densitiy, and they play a part in regulating sets of genes involved in the production of exoenzymes, such as pectin lyase, pectatelyase, polygalactouranase, cellulase, protease, and antibiotics. Thus, root colonization by rhizobacteria is usually found to be correlated with a high inoculum density. An increasing number of other bacterial species, including pathogenic, symbiotic, and free-living bacteria, have been demonstrated to form biofilms on biotic and abiotic surfaces. It has been found that plant growth promotion by free-living diazotrophic bacteria is associated with colonization on root elongation zones and root hairs followed by the formation of biofilms. *Bacillus cereus*, a Gram-positive bacterium, develops dense surface-associated populations, and one recent study linked biocontrol with the ability of this species to form biofilms (Bais et al. 2004a). Several functions known to influence biocontrol activity are also likely to play a role in biofilm formation. A number of microbial cell structures, such as flagella and type IV pili, lipopolysaccharide, and outer membrane proteins, including adhesins, are important in colonization and biofilm formation. Similarly, bacterial products, such as exopolysaccharides, are well-associated with biofilm development in many bacteria, including *Pseudomonas aeruginosa* and *A. tumefaciens* (Ramey et al. 2004). Various factors, including surface chemistry, nutrient availability, and the intrinsic abilities of a bacterium determine the nature and type of biofilm formed.

12.6.1 Bioremediation of Pollutants by Endophytic Bacteria in Biofilms

Endophytic bioremediation is the application of biological processes for the removal of hazardous pollutants present in the environment. Endophytic biofilms can be applied for the bioremediation of wetland wastewater. There have been successful examples of the

TABLE 12.3

Reported Information on Remediation of Various Pollutants via Bacterial Biofilms

Pollutant Category	Pollutant	Biofilm-Forming Organism or Culture	Reference
Toxic compounds	Dodecylbenzenesolfonate sodium	*Stenotrophomonas maltophilia*	Farzaneh et al. 2010
	4-Aminophenol	*Pseudomonas* sp. strain ST-4	Khan et al. 2006
	Tolune	*Burkholderia vietnamiensis* G4	Amit et al.2009
	Dinitrotolune	Sludge from wastewater treatment plant, various aerobic and anaerobic bacteria	Lendermann & Smith 1998
	Polychlorinated biphenyls	Anaerobic bacteria isolated from secondary sludge wastewater	Josephine et al. 2006
	4-Chlorophenol	Bacterial consortium from rhizophere of *Phragmites australis*	Caldeira et al. 1999, Carvalho et al. 2001
	2-Chlorophenol	Anaerobic sludge from a swine wastewater treatment plant	Chang et al. 2004
	2,4,6-Trichlorophenol 2,3,4,6-tetrachlorophenol, pentachlorophenol	*Pseudomonas* sp., *Rhodococcus* sp.	Puhakka et al. 1995
	2,4-Dichlorophenol	Pseudomonas putida	Kargi & Eker 2005
	Hydrocarbon compounds	Cyanobacteria *Dermocarpella* sp. and *Acinetobacter calcoaceticus*	Radwan & Al- Hassan 2001
	Hydrocarbon compounds	Filamentous cyanobacteria, picoplankton, and diatoms	Al-Awadhi et al. 2003
	Diesel	Candida tropicalis	Das & Chandran 2011
Synthetic dyes	Acid orange 10, 14	*Methylosinus trichosporium*	Zhang et al. 1995
Heavy metals	Cr^{3+}	*Bacillus subtilis* and *Bacillus cereus*	Sundar et al. 2011
	Cr^{6+}, Cd^{2+}, Fe(III), Ni(II)	*Escherichia coli*	Quintelas et al. 2009
	Cu^{2+}, Zn^{2+}, Cd^{2+}	Activated sludge from a sewage plant	Costley & Wallis 2001
	Cr^{6+}	Escherichia coli	Gabr et al. 2009
	Zn^{2+}	*Pseudomonas putida*	Toner et al. 2005
	Cd^{2+}, Cu^{2+}, Zn^{2+}	*Pseudomonas* sp. NCIMB	Scott et al. 1995

positive use of biofilms, i.e., beneficial biofilms, which offer their member cells several benefits for the remediation of various pollutants (Table 12.3), among which protection of the environment from the hazardous effects of toxic pollutants stands first. Biofilm-based reactors are commonly used for the treatment of large volumes of industrial and municipal wastewaters. During the last few decades, biofilm reactors have become a focus of interest for the remediation of xenobiotic compounds. The main biofilm reactors are categorized according to the methods used: biofilm fluidized bed, upflow sludge blanket, expanded granular sludge blanket, and biofilm airlift suspension.

12.6.2 Bioremediation of Heavy Metals by Endophytic Bacteria in Biofilms

The potential environmental health hazards generated by the presence of toxic heavy metal contaminants in industrial effluents are well known. Heavy metal remediation can be achieved by immobilization, concentration, and partitioning into an environmental compartment, thereby minimizing the anticipated hazards. There are reports on the application of biofilms for the removal of heavy metals. Recently, bioremoval of Cr(III) using bacterial biofilms in a continuous flow reactor was reported. The ability of a biofilm of *Escherichia coli* supported on NaY zeolite for the removal of Cr(VI), Cd(II), Fe(III), and Ni(II) from wastewater was reported. The results of the study showed that the biofilm tested was very promising for the removal of metal ions from effluents. The same group of workers reported the removal of Cd(II), Cr(VI), Fe(III), and Ni(II) from aqueous solutions by *E. coli* biofilms supported on kaolin. The biosorption performance in terms of uptake followed the sequence Fe(III) > Cd(II) > Ni(II) > Cr(VI). The use of biosorption of hexavalent chromium using biofilms of *E. coli* supported on granulated activated carbon has been reported. The results demonstrated that the biofilm supported on granular activated carbon, prepared by an impregnation method, could be used as a promising biosorbent for the removal of Cr(VI) ions from aqueous environments.

12.6.3 Bioremediation of Toxic Compounds by Endophytic Bacteria in Biofilms

Over the last few decades, enormous quantities of synthetic chemicals have been released into the environment. These chemicals pose serious problems, as they are xenobiotic and highly hazardous, and they include hydrocarbons, phenols, polychlorinated biphenyl (PCB) compounds, and other recalcitrant and persistent aromatic compounds. The U.S. Environmental Protection Agency (EPA) listed some organic primary pollutants injected into the environment by human activities, including chlorinated phenols, dichlorobenzene, hexachloroethane, naphthalene, polynucleated aromatic hydrocarbons (e.g., toluene), benzopyrine, PCBs, hexachlorocyclohexane, and pesticides like aldrin, DDT, and endrin. Rafida et al. reported the removal of hydrocarbon compounds by using a reactor of biofilm in an anaerobic medium. The technique involved the use of biofilms developed on a supporting material in a reactor, which used a vertical flow biofilter under anaerobic conditions. The reactor was a cylinder made of polyvinyl chlorine and was filled with layers of supporting materials. The lower part of the unit was filled with gravel, whereas the uppermost part was filled with small bottles of polyethylene terephthalate. The biofilms were then allowed to develop and featured a diversity of microorganisms, mainly bacteria. The hydraulic retention time of the reactor was found to be 6.25 days, which meant that the reactor should be sampled every 6 days after water treatment. The total organic carbon consumption in the water after treatment was 40%, with the unit showing a hydrocarbon removal efficiency of 90%. Al-Awadhi et al. described a method for artificially establishing biofilms rich in hydrocarbon-degrading bacteria on gravel particles and glass plates. The microbial consortia in the biofilms included filamentous cyanobacteria, picoplankton, and diatoms. Phototrophic microorganisms were the pioneer colony formers. In batch cultures, it was found that artificial biofilms showed a remediation effect on crude oil-contaminated sea water samples. The potential use of these biofilms was suggested for preparing trickling filters (gravel particles) and in bioreactors (glass plates) for removing hydrocarbons from oily liquid wastes before disposal in the environment. Hydrocarbon-degrading microbial consortia (bacteria and microalgae) immobilized in biofilms on gravel particles in the intertidal zone of the Arabian Gulf coast has been described. Each gravel particle was found to be coated with about 100 mg of blue green biomass. The predominant prototroph was the cyanobacterium *Dermocarpella* sp., and the most dominant hydrocarbon-degrading bacterium in the consortium was *Acinetobacter calcoaceticus*. The biofilm-coated gravel particles were used in five successive cycles of purification of oily sea water. Chlorinated aromatic compounds are the most widespread contaminants of soil and groundwater and are present in many chemical industry effluents. They are carcinogenic at very low concentrations. To remove 2,4-dichlorophenol (DCP) from synthetic wastewater, a rotating perforated tube biofilm reactor was used which contained a mixed microbial biomass of activated sludge supplemented with DCP-degrading *Pseudomonas putida*. Nearly 100% of the DCP was degraded. Overall degradation efficiency was found to be in the range of 70–100%. Complete degradation of 2,4,6-trichlorophenol, 2,3,4,6-tetrachlorophenol, and pentachlorophenol was reported after use of a fluidized bed biofilm reactor that included *Pseudomonas* sp. and *Rhodococcus* sp. Chang et al. reported the degradation of 2-chlorophenol using a hydrogenotrophic biofilm under different reductive conditions. The biofilm was developed on cement balls and acclimatized to PCBs for 2 months by feeding the reactor alternately with PCB and biphenyl. The rate of PCB degradation was influenced by the long exposure of the biofilm to PCBs and the presence of mixed culture in the biofilm. Nitroaromatic compounds are another group of xenobiotic compounds which are resistant to biodegradation, and harmful metabolites are produced after microbial degradation. The tendency to concentrate the pollutants in a biofilm in the natural environment can serve as a negative function of providing the mechanism of entry of the pollutants into the aquatic organism food chain. This biofilm system may represent a viable means of reducing these pollutants in wastewater. Biodegradation of steroidal hormones and alkylphenols by stream biofilms have been reported. Water pollution caused by branched alkyl benzene sulfonates is a significant environmental problem in some countries. Biodegradation of dodecylbenzenesolfonate sodium by *Stenotrophomonas maltophilia* biofilm has been reported. A study showed the main advantages of *S. maltophilia* biofilms on silanized glass beads. Thiocyanate is another toxic compound which is widely used in many industries, like photofinishing, herbicide and insecticide production, dyeing, acrylic fiber production, thiourea production, metal separation, electroplating, soil sterilization, and corrosion inhibition. It is toxic to microorganisms at relatively low concentrations of 58–116 mg/L, and thyroid function is depressed upon chronic exposure to thiocyanate.

Synthetic dyes extensively used in various industries have been reported as carcinogenic and mutagenic for aquatic organisms. Thus, removal of dyes from wastewater is still a major environmental concern. Compared to other pollutants, a few reports are available on remediation of synthetic dyes using endophytic biofilms. Aerobic biodegradation of azo dyes in biofilms was reported by Jiang and Bishop. Among the three azo dyes studied, Acid Orange 8, Acid Orange 10, and Acid Red 14, only Acid Orange 8 was degraded aerobically. The azo bond cleavage occurred very easily for all three dyes under anaerobic conditions. Acid Orange 8 removal ranged from 20% to 90%. Statistically designed experiments were used to characterize the response of pseudo-steady-state biofilms. The presence of azo dyes, along with other factors such as COD loading, bulk-phase DO level, and shear force, showed impacts on biofilm accumulation. The biofilm-forming organism was *Methylosinus trichosporium*. Remediation of the azo dye Everzol Turquoise Blue G was reported with the biofilm formed by *Coriolus versicolor* in a laboratory-scale activated sludge unit.

12.7 Plant Uptake of Organic Pollutants and Their Degradation

Plants generally absorb organic pollutants from soil water through their root system and transport the compounds into the aboveground biomass. Nightshade plants near wetlands can absorb and accumulate organic pollutants in their different organs. The fate of an organic pollutant in the rhizosphere–root system largely depends on its physico-chemical properties. Both biotic and abiotic transformations of organic compounds occur simultaneously in wetland environments. During the phytoremediation of organic pollutants, endophytic bacteria produce different enzymes to mineralize organic pollutants and decrease both the phytotoxicity and evapotranspiration of volatile organic pollutants. Furthermore, endophytic bacteria improve a plant's adaptation and growth by virtue of their plant growth-promoting activities, and consequently they improve phytoremediation activity.

Organic pollutants enter the root xylem faster than the soil and rhizosphere microflora can degrade them, even if the latter is enriched with degrading bacteria. Once taken up, plants metabolize these contaminants, although some of them, or their metabolites, such as trichloroethene (TCE), which is transformed into trichloroacetic acid, can be toxic. In wetlands, polycyclic aromatic hydrocarbons, bisphenol A, BTEX, hydrocarbons including diesel range organics, glycol, DDT, PCBs, cyanide, benzene, chlorophenols, and formaldehyde are common organic compounds that are toxic to wetland biotic communities but are found in wetland environments (Table 12.4). Alternatively, plants preferentially release volatile pollutants, such as benzene, toluene, ethylene, and xylene compounds and TCE and their metabolites, into the environment by evaporation via the leaves. Three particular mechanisms for endophytic bacteria enhancement of phytoremediation of organic pollutants have been proposed: they may enhance (i) plant growth and biomass production, (ii) organic pollutant bioavailability, and (iii) population size and activity of indigenous bacteria to degrade organic pollutants through horizontal gene transfer. Endophytic bacteria strains are likely to be more successful or reliable due to their more intense interactions with the plant. Efficient endophytic bacterial colonization of degraders has been observed, and they proved to show high expression of degrading genes. The number of transcripts and number of genes present (as a proxy of activity per cell) was even higher inside the plant than in the rhizosphere, strongly indicating that degradation takes place inside the plants. Some studies have indicated that endophytic bacteria are more efficient in plant growth promotion. Different biotic and abiotic processes are involved in the uptake of organic pollutants by plants.

The symplastic pathway is involved in the transportation of dinitrotoluene and dinitromenzene into plants, whereas the apoplastic pathway is involved in the transportation of phenanthrene and pyrene. Plants can also absorb organic pollutants from the atmosphere. After transportation of organic pollutants in the cytosol of plants cells, they frequently make conjugates by covalent linkage to endogenous tripeptides and glutathione. These conjugates are transported and accumulate in the vacuoles of plant tissues. The possible fates of organic compounds in plants usually include accumulation and evapotranspiration to the atmosphere.

TABLE 12.4

Application of Plant–Endophytic Bacteria Partnerships for Remediation of Water Contaminated with Organic Pollutants

Plant Used	Endophytic Bacterium	Bacterial Characteristics	Reference
Poplar	*Pseudomonas putida* W619-TCE	TCE degradation	Weyens et al. (2013)
Poplar	*Burkholderia cepacia* VM1468	BTEX and TCE degradation	Taghavi et al. (2011)
Poplar	*Pseudomonas putida* W619-TCE	TCE degradation	Weyens et al. (2010a)
Yellow lupine	*Bacillus cepacia* VM1468	TCE degradation and Ni resistance	Weyens et al. (2010b)
Poplar	Indigenous	Toluene degradation	Barac et al. (2009)
Poplar	*Burkholderia cepacia* VM1468	Toluene degradation	Taghavi et al. (2005)
Yellow lupine	*Bacillus cepacia* sp. strain L.S.2.4	Toluene degradation	Barac et al. (2004)

Usually, organic pollutants are released from plants several hours after uptake. These complex pollutants and recalcitrant compounds cannot be broken down to basic molecules (water, carbon dioxide, etc.) by plant molecules and, hence, the term phytotransformation represents a change in chemical structure without complete breakdown of the compound. The term Green Liver is used to describe phytotransformation, as plants behave analogously to the human liver when dealing with xenobiotic compounds (i.e., foreign compounds or pollutants).

12.8 Processes of Endophytic Bacteria for Phytoremediation of Pollutants

Phytoremediation is actually a generic term for several ways in which plants can be used to clean up contaminated soils and water. The mechanisms and efficiency of phytoremediation depend on the type of contaminant, its bioavailability, and aquatic properties. There are several ways by which plants clean up or remediate contaminated sites (Figure 12.10). The uptake of contaminants in plants occurs primarily through the root system, in which

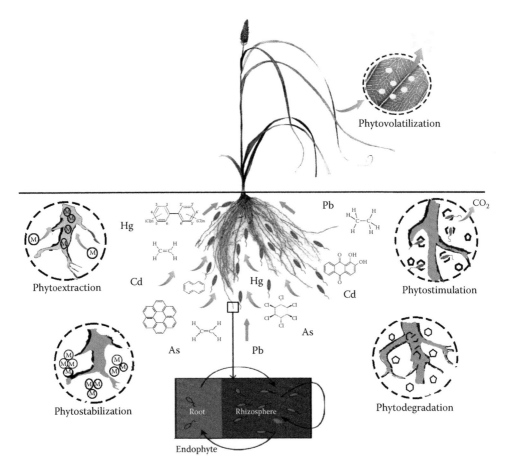

FIGURE 12.10
Various strategies involved in phytoremediation technology.

the principal mechanisms for preventing toxicity are found. The root system provides an enormous surface area that absorbs and accumulates water and nutrients essential for growth along with other nonessential contaminants.

12.8.1 Rhizofiltration

Metal pollutants in industrial process water and in groundwater are most commonly removed by precipitation or flocculation, followed by sedimentation and disposal of the resulting sludge. A promising alternative to this conventional cleanup method is rhizofiltration. It refers to the use of plant roots to absorb, concentrate, and precipitate toxic metals from contaminated groundwater. Initially, suitable plants with stable root systems are supplied with contaminated water to acclimate the plants. These plants are then transferred to the contaminated site to collect the contaminants, and once the roots are saturated, they are harvested. Rhizofiltration allows *in situ* treatment, minimizing disturbances to the environment. It removes contaminants from water and aqueous waste streams, such as agricultural runoff, industrial discharges, and nuclear material processing wastes. Absorption and adsorption by plant roots play key roles in this method, and consequently large root surface areas are usually required.

12.8.2 Phytostabilization

Phytostabilization is also known as phytorestoration, a plant-based remediation technique that stabilizes wastes and prevents exposures via wind and water erosion; provides hydraulic control, which suppresses the vertical migration of contaminants into groundwater, and it physically and chemically immobilizes contaminants by root sorption and by chemical fixation with various soil amendments. Erosion and leaching can mobilize soil contaminants, resulting in aerial or waterborne pollution of additional sites. In phytostabilization, accumulation bit by bit in plant roots or precipitation in the soil by root exudates immobilizes and reduces the availability of soil contaminants. Plants growing on polluted sites also stabilize the soil and can serve as groundcover, thereby reducing wind and water erosion and direct contact of animals with the contaminants. The goal of phytostabilization is not to remove metal contaminants from a site, but rather to stabilize the contaminants and reduce the risks to human health and the environment.

12.8.3 Phytostimulation

Phytostimulation is the process where root-released compounds enhance microbial activity in the rhizosphere. This process is critical for the applied technology of rhizoremediation, which combines phytoremediation and bioaugmentation. This type of rhizosphere phytoremediation can be used as a low-cost approach to remove organic pollutants from the soil. Phytostimulation is important for rhizoremediation for multiple reasons. First, the rhizosphere provides a specific niche for soil microorganisms to live. This niche is continuously expanding as roots grow and penetrate new soil zones. Root exudates provide many benefits for microorganism growth and activity. Root-released compounds have many roles in phytostimulation, because they often mimic human-made pollutants. The rhizosphere serves as a unique soil habitat, full of organic compounds that may serve as energy sources for microorganisms. Alternatively, these organic compounds may serve as inducers for microbial genes involved in degradation, or as cometabolites that are necessary for pollutant degradation to occur.

12.8.4 Phytoextraction

Phytoextraction is also known as phytoaccumulation, phytoabsorption, or phytoseques-tration. Phytoextraction involves the removal of toxins, especially heavy metals and metal-loids, by the roots of the plants with subsequent transport to aerial plant organs. Pollutants accumulated in stems and leaves are harvested with accumulating plants and removed from the site. Phytoextraction can be divided into two categories: continuous and induced. Continuous phytoextraction requires the use of plants that accumulate particularly high levels of the toxic contaminants throughout their lifetime. The roots of the established plants absorb metal elements from the soil and translocate them to the aboveground shoots, where they accumulate (hyperaccumulators), while induced phytoextraction take place if metal availability in the soil is not adequate for sufficient plant uptake, chelates or acidifying agents may be used to liberate them into the soil solution.

12.8.5 Phytovolatization

Phytovolatilization involves the use of plants that uptake organic pollutants and metals from soil, biologically convert them in a volatile form, and then release them into the atmo-sphere by volatilization. Furthermore, it has a limitation that it does not eliminate the pollutant completely; it only transfers it from one form (soil) to another (atmosphere) from where the pollutant can redeposit.

12.8.6 Phytodegradation

Complex organic pollutants are degraded into simpler molecules and are incorporated into the plant tissues to help the plant grow faster. Plants contain enzymes that catalyze and accelerate chemical reactions. Some enzymes break down and convert ammunition wastes, others degrade chlorinated solvents such as TCE, and others degrade herbicides.

12.9 Phytoremediation of Heavy Metals by Endophytic Bacteria in Wetland Ecosystems

Increasingly, wetlands are used for treatment of metal-contaminated water or to cover over metal-enriched mine tailings. Natural wetlands may also be contaminated with met-als from anthropogenic sources. Heavy metals are nondegradable, and without interven-tion they will persist in the environment. The persistence of heavy metals may adversely affect ecosystems, causing concerns not only for agricultural products and water quality, but also human health risks.

Heavy metals occur naturally in the environment from pedogenetic processes that weather parent materials and also through anthropogenic sources (Figure 12.11). Accumulation of heavy metals in wetlands is attributed to rapid industrialization, and it poses toxicity to plants, impairing their metabolism, biomass production, and yield. It has been well dem-onstrated that the inherent abilities of endophytic bacteria may help host plants adapt to unfavorable wetland conditions and enhance the efficiency of phytoremediation by promot-ing plant growth, alleviating metal stress, reducing metal phytotoxicity, and altering metal bioavailability in soil and metal translocation in plants by various mechanisms (Table 12.5).

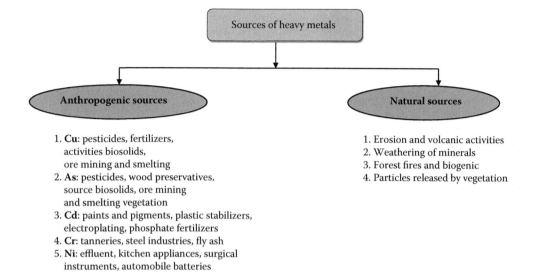

FIGURE 12.11
Sources of heavy metal contaminants in the environment.

12.9.1 Metal Uptake, Transport, and Release by Wetland Plants

The fate of metals within plant tissues is a critical issue for effectiveness of the phytoremediation process. Storage in roots is most beneficial for phytostabilization of metal contaminants, which are least available when concentrated belowground. Plants may alter the speciation of metals and may also suffer toxic effects as a result of accumulating them. The extent of uptake and how metals are distributed within plants can have important effects on the residence time of metals in plants and in wetlands, and also the potential release of metals. A mass balance of heavy metals in the soil environment can be expressed using the formula:

$$M_{total} = (M_p + M_a + M_f + M_{ag} + M_{ow} + M_{ip}) - (M_{cr} + M_l)$$

where M is the heavy metal, p is the parent material, a is atmospheric deposition, f is fertilizer sources, ag is agrochemical sources, ow is organic waste sources, ip is inorganic pollutants, cr is crop removal, and l is losses by leaching, volatilization, and other processes. It is estimated that emission of several heavy metals into the atmosphere from anthropogenic sources is 1 to 3 orders of magnitude higher than from natural sources.

Another factor than can affect the accumulation of metals in wetland plants is the presence of microbial symbionts, such as rhizospheric bacteria. The rhizosphere, an important interface of soils and plants, plays a significant role in phytoremediation of contaminated soil by heavy metals, as microbial populations are known to affect heavy metal mobility and availability to the plant through release of chelating agents, acidification, phosphate solubilization, and redox changes, and therefore rhizospheric bacteria have the potential to enhance

TABLE 12.5

Endophytic Bacteria Enhance Phytoremediation of Metal-Contaminated Wetland Ecosystems

Endophytic Species	Metal Stress	Mechanism(s)	Reference
Burkholderia sp. SaZR4, *Burkholderia* sp. SaMR10	Cd, Zn	SaZR4 only promoted Zn extraction, SaMR10 had little effect on phytoextraction	Zhang et al. 2013
Microbacterium sp. NCr-8, *Arthrobacter* sp. NCr-1, *Bacillus* sp. NCr-5, *Bacillus* sp. NCr-9, *Kocuria* sp. NCr-3	Ni	Enhanced growth and Ni translocation in plants	Visioli et al. 2014
Bacillus sp. MN3-4	Pb, Cd, Zn, Ni, Cu	Bioremoval of Pb, increased root elongation, reduced metal phytotoxicity, increased Pb accumulation in plants	Shin et al. 2012
B. pumilus E2S2, *Bacillus* sp. E1S2, *Bacillus* sp. E4S1	Cd, Pb	Bacterial inoculation increased water-extractable Cd and Zn in soil, improved plant growth and metal uptake	Ma et al. 2015a
Pseudomonas sp.	Ni	Increased biomass and Ni content in plants, high level of colonization in tissue interiors of two plant species	Ma et al. 2011a
Bacillus sp. SLS18	Cd, Mn	Improved plant biomass production and total metal uptake	Luo et al. 2012
Bacillus pumilus	Zn, Cd	Mobilized Zn in soil, thus increased soil Zn bioavailability	Long et al. 2011
Burkholderia cepacia	Cu, Cd, Co, Ni, Pb, Zn	Bioremoval of Ni, thus reducing metal toxicity, increased Ni concentration in roots	Lodewyckx et al. 2001
Pseudomonas koreensis	As, Cd, Cu, Pb, Zn	Increased plant biomass, chlorophyll, protein content, superoxide dismutase and catalase activities and metal uptake; decreased malondialdehyde content in plants	Babu et al. 2015
Bacillus thuringiensis	As, Cu, Cd, Ni, Pb, Zn	Bioremoval of Pb, Zn, As, Cd, Cu, and Ni in metal-amended and mine tailing extract medium, increased biomass, chlorophyll content, and nodule number	Babu et al. 2013

phytoremediation processes. However, microbial iron–siderophore complexes can be taken up by plants and thereby serve as an iron source for plants. It was therefore reasoned that the best way to prevent plants from becoming chlorotic in the presence of high levels of heavy metals is to provide them with an associated siderophore-producing bacterium.

12.9.2 Siderophore Formation

Siderophores are organic compounds with low molecular masses that are produced by microorganisms and plants growing under low-iron conditions. The primary function of these compounds is to chelate ferric iron [Fe(III)] from different terrestrial and aquatic habitats and make it available for microbial and plant cells. Siderophores have received much attention in recent years because of their potential roles and applications in various areas of environmental research. Their significance in these applications is because siderophores have the ability to bind a variety of metals in addition to iron, and they have a wide range of chemical structures and specific properties. In soils, the microbial communities that colonize mineral surfaces differ from the inhabitants of the surrounding soil. Microbial attachment to mineral surfaces leads to the formation of a microenvironment that protects the microorganisms against environmental stresses. In these microenvironments, mineral nutrients can be chelated directly from the soil minerals or shared among the surrounding microorganisms. Siderophores produced by soil microorganisms can promote the mineral dissolution of insoluble phases.

12.9.3 Mechanism of Amelioration of Metal Stress

Metal phytotoxicity is a critical factor in the success of phytoremediation. To overcome metal stress, a number of bacteria-mediated processes have been involved in the endophyte–host coevolution process, either by alleviating metal toxicity or by conferring plant metal tolerance. Recent studies have suggested that some endophytic bacteria can reduce metal phytotoxicity via several mechanisms that include reduced uptake, efflux, extracellular sequestration, intracellular sequestration, repair, metabolic bypass, and chemical modification (Figure 12.12). The regulation, expression, and activities of proteins involved in metal uptake (as well as efflux) are crucial for metal resistance, and different endophytic bacterial species have distinct complements of these systems. Endophytic bacteria have evolved tight regulatory mechanisms to control the activity of membrane transporters that take up metals, and some of these transporters are controlled by regulators that can bind metal ions. Some endophytic bacteria upregulate the expression of extracellular polymers or siderophores in response to metal exposure, and these molecules contain functional groups capable of coordinating metal ions. Extracellular polymers and siderophores can trap or precipitate metal ions in the extracellular environment, and soluble siderophores that are bound to toxic metals may be subject to reduced uptake or increased efflux by membrane transporters. Metals might also bind to, or precipitate at, bacterial cell surfaces through interactions involving proteins or cell-associated polysaccharides, such as lipopolysaccharide. Although the biochemical mechanism is unknown, many microorganisms precipitate metals as metal oxides, metal sulfides, metal–protein aggregates, or elemental metal crystals, and these which form particulates that are closely associated with the cytoplasmic membrane. This effectively sequesters these metals in the cytoplasm or periplasmic space. Cellular molecules with redox-sensitive functional groups, which are vulnerable to oxidation by primary reactions with metals or highly reactive, catalytic by-products of metals (such as reactive oxygen species), can be repaired by cellular chaperones, enzymes, or antioxidants. Cells can circumvent metabolic pathways containing metal-disrupted enzymes by producing alternative proteins with catalytic cores that do not bind to the toxic metal ligand or by shunting metabolites towards alternative pathways. Metals can undergo specific redox and covalent reactions in cells, and these reactions alter the chemical reactivity of the metal atoms and convert toxic metal species to less toxic or less available forms. This process can alter the redox state of the metal, create metal crystal precipitates, or generate organometallic small-molecule compounds.

Endophytic bacteria can modulate the activity of plant antioxidant enzymes (such as catalase, superoxide dismutase, glutathione peroxidase, ascorbate peroxidase) as well as lipid peroxidation (malondialdehyde formation), which confront plant defense mechanisms, especially in resisting the heavy metal-induced oxidative stress in plants. To date, most

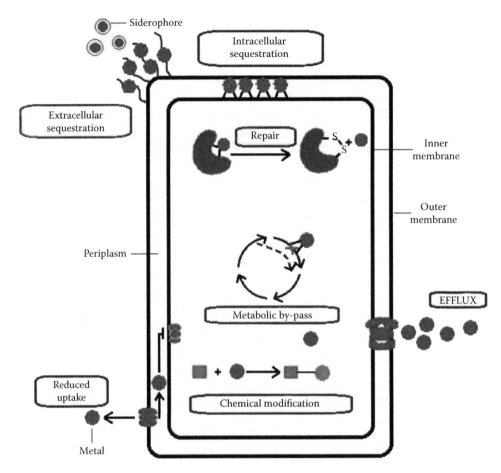

FIGURE 12.12
Mechanisms of amelioration of metal stress by endophytic bacteria, categorized by function: reduced uptake, efflux, extracellular sequestration, intracellular sequestration, repair, metabolic bypass, chemical modification.

studies have focused on the role of endophytic bacteria in heavy metal tolerance and accumulation in hyperaccumulator and nonaccumulator plants. However, it is not yet known whether plants growing in metal-polluted soils have altered colonization and survival potentials in the presence of specific metal-resistant and/or beneficial microbes. Thus, it is critical to explore the diversity, distribution, and activity of endophytic microbial communities associated with various hyperaccumulator plants in phytoremediation studies.

12.9.4 Metal Biosorption, Bioaccumulation, and Translocation

From the toxicological point of view, previous studies have demonstrated that the endophytic bacteria may contribute to reducing metal phytotoxicity through biosorption and bioaccumulation. The metal biosorption by bacteria involved two steps: (i) passive biosorption of metals by living and dead or inactive cells; this essentially takes place in the cell wall due to a number of metabolism-independent processes. In this process, metal ions are adsorbed rapidly to the cell surface by reactions between metals and functional groups on the cell surface, such as hydroxyl, carbonyl, carboxyl, sulfhydryl, thioether, sulfonate, amine, amide, and phosphonate groups. Various metal-binding mechanisms, such as ion exchange, complexation, coordination, sorption, chelation, electrostatic interaction, or microprecipitation may be synergistically or independently involved. (ii) Active biosorption (bioaccumulation), which refers to the uptake of metals (transport into cells, accumulate intracellularly, across cell membranes through the cell metabolic cycle) by living cells through a slower active metabolism-dependent transport of metal into bacterial cells. Once the metals enter living cells, they may be bound, precipitated, accumulated, and sequestered within specific intracellular organelles or translocated to specific structures, depending upon the organism and element concerned. As described above, the endophytic bacteria possessing specific and remarkable metal bioaccumulation abilities can be used in plant–endophyte symbiotic systems to facilitate detoxification of heavy metals and improve efficiency of phytoremediation. It has already been reported that specific plant-associated bacteria are responsible for metal sequestration, siderophore or organic acid production, and for the degradation pathway for the appropriate organic pollutant (Figure 12.13).

Endophytic bacteria can alter heavy metal bioavailability and their translocation in plants, hence modifying metal toxicity and plant metal uptake through secretion of a

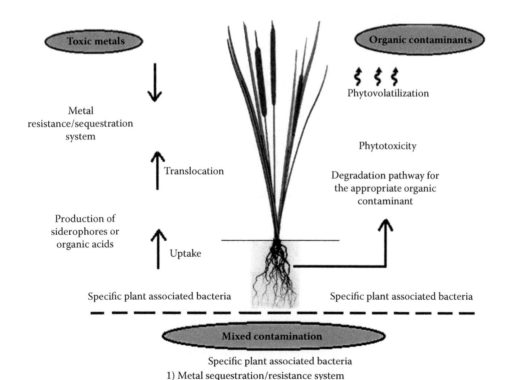

FIGURE 12.13
Phytoremediation of wetland soils and water contaminated with toxic metals, organic contaminants, and various mixed contaminants by specific plant-associated bacteria. The bacteria play roles in metal sequestration, siderophore or acid production, and degradation of organic contaminants.

variety of metabolites, including siderophores, organic acids (e.g., citric, oxalic, and acetic acids). Kuffner et al. (2010) reported inoculation with the copper-resistant endophytic bacteria with multiple plant growth-promoting characteristics, such as 1-amino-cyclopropane-1-carboxylic acid (ACC) deaminase and arginine decarboxylase activities. The plant-associated plant growth-promoting endophyte is now being considered a biotool to improve the efficiency of phytoremediation of metal-polluted wetland soils.

12.10 Ecology of Organic Pollutant-Degrading Endophytic Bacteria

Although organic pollutants are toxic to plants and microorganisms, several plants that grow in polluted wetland have different types of endophytic bacteria that are able to degrade organic pollutants. Wetlands contain high concentrations of nutrients, sugars, and amino acids. The intercellular spaces of plants support the survival and proliferation of different types of pollutant-degrading endophytic bacteria. It is likely that some endophytic bacteria might show degradation capacities to degrade plant-derived complex organic metabolites, giving them an increased competitive ability for colonizing different plants or plant tissues. It was also observed that higher numbers of hydrocarbon-degrading endophytic bacteria were present in the root interior than the aquatic plant rhizosphere and their density correlated with the occurrence and quantity of hydrocarbons in wetlands. Later on, several species of endophytic bacteria were isolated from different plants in a crude oil-polluted wetland. Endophytic bacteria, able to degrade organic pollutants, were also isolated from the internal tissues of trees planted near a wetland. For example, herbicide-degrading endophytic bacterial strains were isolated from poplar trees (Germaine et al. 2009). These strains were able to degrade 2,4-dichlorophenoxyacetic acid. Endophytic bacteria able to degrade several aromatic hydrocarbons, such as benzene, toluene, ethyl benzene, and xylene, were also isolated from roots, stems, and leaves of different cultivars of poplar tree. These strains showed marked spatial compartmentalization within the plants, suggesting species-specific and nonspecific relations between bacteria and plants. Endophytic bacteria have the potential to degrade volatile organic pollutants, such as trichloroethylene. Some organic pollutant-degrading endophytic bacteria also show resistance to heavy metals and can be applied to enhance the phytoremediation of soil and water polluted with organic pollutants and heavy metals (Weyens et al. 2010a). Endophytic bacteria, able to utilize several organic compounds as the sole carbon source, were also isolated from different wetland plants. These endophytic bacteria showed the potential to degrade different hydrocarbons and pesticides and belonged mainly to *Gammaproteobacteria, Bacilli, Alphaproteobacteria, Flavobacteria,* and *Actinobacteria.* Hydrocarbon-degrading endophytic bacteria were also isolated from different salt marsh plant species. These endophytic bacteria might be particularly important for plant colonization in urban estuarine areas exposed to petroleum hydrocarbon contamination. Many endophytic bacteria have both pollutant-degrading and plant growth-promoting activities. The population and diversity of pollutant-degrading endophytic bacteria usually depends on the presence of the pollutant concentration in the soil and water. Large numbers of toluene-degrading endophytic bacteria were observed in the roots of poplar plants that reached toluene-contaminated groundwater. In addition to pollutant concentration, the population of pollutant-degrading endophytic bacteria also depends on the plant genotype, plant development stage, and environment. Recently, it was found through metagenomic analysis that endophytic bacteria isolated from the roots of rice plants have the potential to degrade aliphatic and aromatic hydrocarbons.

This metagenomic analysis revealed that degradation genes were highly abundant, although rice plants and thus also endophytic bacteria were not exposed to a polluted environment. This may indicate that endophytic bacteria are equipped with genes to degrade complex organic substances which may be produced as plant metabolites and they may be utilized as carbon sources by the bacteria. The same enzymes might be involved in the degradation of pollutants. Pollutant-degrading endophytic bacteria may show preferential colonization for a specific plant tissue. For instance, alkane-degrading root endophytic bacteria showed better colonization in the root, whereas shoot endophytic bacteria showed better colonization in the shoot. However, higher numbers of pollutant-degrading endophytic bacteria were observed in roots of most plants compared to the shoots and leaves. Moreover, lower densities of alkane-degrading endophytic bacteria were observed at the initial stages of plant growth, indicating that first bacteria have to be established in the plant environment and later they become important for the degradation of organic pollutants.

Several hydrocarbon-degrading endophytic bacteria have also been found to be nitrogen fixers, meaning that they can make atmospheric nitrogen available to their host plant. As the bioavailability of nutrients in a contaminated wetland is one of the main constraints of successful phytoremediation, endophytic bacteria possessing both pollutant-degrading and nutrient-providing activities can be applied for phytoremediation of wetlands polluted with organic compounds with reduced fertilizer input. Many of these exhibited both alkane-degrading and plant growth-promoting ACC deaminase activities. Endophytic bacteria possessing both activities performed better than those bacteria with only one of these activities, and they showed more efficient hydrocarbon degradation, which was also evidenced by higher expression of key genes involved in degradation.

12.11　Survival and Pollutant Degradation Activity of Endophytic Bacteria in Wetlands

The understanding of interactions between plants and their associated microbiome is an exciting area of research, and it has recently raised higher interest. Although many studies have been performed to study the interactions between plants and their associated endophytic bacteria in relation to pollutant or toxin degradation, our understanding on the interaction is still limited, such as persistence and colonization of the inoculated pollutant-degrading endophytic bacteria in the rhizosphere and endosphere of plants, mechanisms involved in the colonization of endophytic bacteria, expression of pollutant-degrading genes in the plant interior, the effect of horizontal gene transfer on pollutant degradation in planta, stimulation of plant metabolites in microbial activities, and the role of endophytic bacteria in the plant's adaptation to the abiotic stress. Bacterial colonization inside plant tissues can be monitored by labeling inoculant strains with markers, such as the *gusA* and *gfp* genes. The colonization of large number of green fluorescent protein-labeled endophytic bacteria was studied in the roots and shoots of poplar and pea plants. Similarly, different endophytic bacterial strains were tagged with the *gusA* gene, and their persistence and colonization in the rhizosphere and endosphere of different plants were studied (Germaine et al. 2009, Afzal et al. 2013).

12.12 Isolation and Purification of Endophytic Bacteria

The importance of endophytic bacteria over a long period as a source of pharmaceutical bioactive compounds has been documented, as many endophytes are exposed to produce novel bioactive metabolites, such as antibacterial, antifungal, antiviral, antitumor, antioxidant, anti-inflammatory, or immunosuppressive drugs and many related compounds. Endophytic bacteria are well known for the production of various classes of natural products and have been reported to exhibit a broad range of biological activity. They are grouped into various categories, which include alkaloids, terpenoids, steroids, lactones, phenolic compounds, quinones, lignans, etc. Importantly, secondary metabolites produced by endophytic bacteria provide a variety of fitness enhancements and exert several beneficial effects on host plants, such as stimulation of plant growth, nitrogen fixation, and induction of resistance to drought, herbivorous, parasitism, etc. Endophytic bacteria are generally isolated for good reason, such as for their characterization, for studying population dynamics and diversity, to evaluate their use as microbial inoculants to improve plant growth and plant health, and as sources of novel biologically active secondary metabolites.

12.12.1 Collection of Plant Material

For the isolation of endophytic bacteria, healthy leaves, stems, and roots of plants were collected from healthy wild and cultivated plants from the plant's natural ecosystem and placed in clean plastic bags, brought to the laboratory and used for further experimental purpose.

12.12.2 Pretreatment

The leaves, stems, and roots of each plant were washed separately under tap water to remove adhering soil particles, and the majority of microbial surface epiphytes were a part of the pretreatment.

12.12.3 Surface Sterilization

Surface sterilization is the initial and mandatory step for isolation of baterial endophytes in order to kill all the surface microbes. It is usually accomplished by treatment of plant tissues with an oxidizing or general sterilizing agent for a period and then by three to five sterile rinses. The most commonly used isolation procedures combine surface sterilization of the plant tissue, plating small sterilized segments onto nutrient agar, maceration of the plant tissue and streaking onto nutrient agar, or vacuum or pressure extraction. Theoretically, the sterilizing agent should kill any microbe on the plant surface without affecting the host tissue or the endophytic microorganisms.

12.12.4 Verification of Surface Sterilization Effectiveness

Only if complete surface sterilization of the plant tissue is confirmed, then the isolated microorganisms are said to be endophytes. Validation of the surface sterilization procedure is done by imprinting the surface-sterilized plant tissue onto nutrient media.

12.12.5 Media

The choice of the growth medium is crucial, as it directly affects the number and type of endophytic microorganisms that can be isolated from the root tissue. Nutrient agar media are used for the isolation of endophytic bacteria. Since there is no component in nutrient agar which can suppress the growth of endophytic fungi, the media used for the isolation of endophytic bacteria are supplemented with an antifungal agent, nystatin, to suppress fungal growth.

12.12.6 Subculture

After proper drying of surface-sterilized plant material, using aseptic procedure the surface of the stems are removed using a sterile scalpel in the laminar air flow cabinet, and leaves were cut into pieces and each piece was placed on nutrient agar medium supplemented with antifungal agent. Plates with plant tissues are sealed using parafilm tape and incubated at 28 ± 2°C in order to recover the maximum possible colonies of endophytic bacteria. The observation can be made for 48 hours. After 24 hours, morphologically different bacterial colonies are selected and are repeatedly streaked in order to achieve bacterial isolates. All selected isolates are subcultured in nutrient agar slants and finally, all the purified endophytes are maintained at 4°C until use.

12.13 Techniques and Molecular Approaches to Detect and Study Endophytic Bacteria

Because endophytic bacteria are diverse in the natural environment, most of the bacteria are still unknown. As a vast topic of endophytic ecology, it is difficult to understand specialized species with novel enzymatic functions or new products. Therefore, bacterial identification is a growing field of interest within microbiology, and different approaches are used for the identification and classification of bacterial diversity.

12.13.1 16S rRNA Analysis

The use of 16S rRNA gene sequences to study bacterial phylogeny and taxonomy has been by far the most common housekeeping genetic marker used for a number of reasons. These reasons include (i) its presence in almost all bacteria, often existing as a multigene family, or operons; (ii) the function of the 16S rRNA gene over time has not changed, suggesting that random sequence changes are more accurate measures of time (evolution); and (iii) the 16S rRNA gene (1,500 bp) is large enough for informatics purposes.

12.13.2 DNA–DNA Hybridization

DNA–DNA hybridization values have been used by bacterial taxonomists since the 1960s to determine relatedness between strains and are still the most important criterion in the delineation of bacterial species. It is the method that provides more resolution than 16S rDNA sequencing, and the 70% criterion has been a cornerstone for describing a bacterial species. Since the extent of hybridization between a pair of strains is ultimately governed by their respective genomic sequences, we examined the quantitative relationship between DNA–DNA hybridizaiton values and genome sequence-derived parameters, such as the average nucleotide identity of common genes and the percentage of conserved DNA. The method takes advantage of the capacity provided by microarray technology. Bacterial genomes are fragmented randomly, and representative fragments are spotted on a glass slide and then hybridized to test genomes. Resulting hybridization profiles are analyzed via statistical procedures to identify test strains. Importantly, a database of hybridization profiles can be established.

12.13.3 Denaturing Gradient Gel Electrophoresis

The molecular biology approach of denaturing grandient gel electrophoresis (DGGE) is a fingerprinting methodology that has led to revolutionary changes in many of the traditional routines used in assessing microbial populations.

Steps for DGGE analysis of the microbial diversity associated with decomposing plant litter in freshwater include the following: (i) colonized plant litter in a stream is collected; (ii) DNA is extracted from colonized plant litter; (iii) PCR products are amplified with primers for the gene of interest that is present in all members of the bacterial community; (iv) an increase of the gradient of urea forming amide; (v) separation of bacterial amplicons of the same size but with different nucleotide compositions by using DGGE.

12.13.4 Detection of Bacteria by Use of DNA Probes

Within the field of microbial ecology, oligonucleotide probes are used in order to determine the presence of microbial species, genera, or microorganisms classified on a broader level, such as bacteria. These DNA probes are complementary to conserved regions of small or large ribosomal subunit RNAs and thus, for taxonomic purposes, the application of DNA probes seems to be highly useful. The hybridization probes are designed to differentiate between the phylogenetic groups of organisms.

12.13.5 Genetic Fingerprinting

Genetic fingerprinting generates a profile of microbial communities based on direct analysis of PCR products amplified from environmental DNA (Muyzer 1999). These techniques include DGGE (or temporal temperature gradient gel electrophoresis), single-strand conformation polymorphism analysis, random amplified DNA polymorphism analysis, amplified rDNA restriction analysis (ARDRA), terminal restriction fragment length polymorphism analysis, length heterogeneity PCR, and ribosomal intergenic spacer analysis, and they produce a community fingerprint based on either sequence polymorphism or length polymorphism. In general, genetic fingerprinting techniques are rapid and allow simultaneous analyses of multiple samples. Fingerprinting approaches have been devised to demonstrate an effect on microbial communities or differences between microbial communities and do not provide direct taxonomic identities.

12.13.6 Amplified rDNA Restriction Analysis

ARDRA is based on DNA sequence variations present in PCR-amplified 16S rRNA genes. Although ARDRA provides little or no information about the type of microorganisms present in the sample, the method is still useful for rapid monitoring of microbial communities over time, or to compare microbial diversity in response to changing environmental conditions.

12.13.7 DNA Microarrays

DNA microarrays have been used primarily to provide a high-throughput and comprehensive view of microbial communities in environmental samples. The PCR products amplified from total environmental DNA are directly hybridized to known molecular probes, which are attached on the microarrays (Gentry et al. 2006). After the fluorescently labeled PCR amplicons are hybridized to the probes, positive signals are scored via confocal laser scanning microscopy. The microarray technique allows samples to be rapidly evaluated with replication, which is a significant advantage in microbial community analyses.

12.14 Conclusions

Plant–endophyte partnerships can be exploited for the phytoremediation of wetland pollutants and this is a promising area of research. Urbanization and industrialization have enhanced releases of organic and inorganic pollutants in soil and water. In the case of phytoremediation of organic contaminants, endophytic bacteria possessing the appropriate degradation pathways can assist their host plant by degrading contaminants that are readily taken up by plants. Increases or decreases in contaminant levels change the composition of the root endophytic community, while the inoculated bacteria do not modify the community structure. The textile industry is one of the main contributors of wetland pollution. Textile effluents adversely affect the quality of wetlands and aquatic systems with the release of pigments and dyes, surfactants, grease and oil, metals, sulfate, and chloride into the environment. Endophytic biofilm systems in wetlands are continuously drawing attention for research. Based on published reports, it may be concluded that biofilms have the potentiality to remove xenobiotic compounds and other heavy metals from wetlands. A better understanding of endophytic bacterial diversity has led to a great variety of configurations and designs in an effort for effective removal of specific wetland pollutants. With the development and application of molecular genomic tools, the field of endophytic ecology is undergoing unprecedented changes. Postgenomic molecular approaches enable us to interrogate the structural and functional endophytic diversity. Studies in this field have revealed that on a large scale, exploitation of wetland plants and their associated endophytic diversity shows promising potential toward implementing CW systems with improved performance. However, further studies are needed to explore the metabolic activities (gene expression) of endophytic bacteria in the rhizospheres and endospheres of wetland plants.

References

Afzal, M., Yousaf, S., Reichenauer, T.G., Kuffner, M., Sessitsch, A. 2011. Soil type affects plant colonization, activity and catabolic gene expression of inoculated bacterial strains during phytoremediation of diesel. *Journal of Hazardous Materials* 186:1568–1575.

Afzal, M., Khan, S., Iqbal, S., Mirza, M.S., Khan, Q.M. 2013. Inoculation method affects colonization and activity of *Burkholderia phytofirmans* PsJN during phytoremediation of diesel-contaminated soil. *International Biodeterioration and Biodegradation* 85:331–336.

Al-Awadhi, H., Al-Hasan, R.H., Sorkhoh, N.A., Salamah, S., Radwan, S.S. 2003. Establishing oil-degrading biofilms on gravel particles and glass plates. *International Biodeterioration and Biodegradation* 51(3):181–185.

Amit, K., Dewulf, J., Wiele, T.V., Langenhove, H.V. 2009. Bacterial dynamics of biofilm development during toluene degradation by *Burkholderia vietnamiensis* G4 in a gas phase membrane bioreactor. *Journal of Microbiology and Biotechnology* 19:1028–1033.

Armstrong, W., Brandle, R., Jackson, M.B. 1994. Mechanisms of flood tolerance in plants. *Acta Botanica Neerlandica* 43:307–358.

Armstrong, W., Cousins, D., Armstrong, J., Turner, D.W., Beckett, P.M. 2000. Oxygen distribution in wetland plant roots and permeability barriers to gas-exchange with the rhizosphere: Microelectrode and modeling study with *Phragmites australis*. *Annals of Botany* 86:687–703.

Armstrong, W., Beckett, PM. 1987. Internal aeration and the development of stellar anoxia in submerged root. A multishelled mathematical model combing axial distribution of oxygen in the cortex with radial losses to the stele, the wall layers of the rhizosphere. *New Phytologist* 105:221–245.

Babu, A.G., Shea, P.J., Sudhakar, D., Jung, I., Oh, B. 2015. Potential use of *Pseudomonas koreensis* AGB-1 in association with *Miscanthus sinensis* to remediate heavy metal(loid)-contaminated mining site soil. *Journal of Environmental Management* 151:160–166.

Babu, A.G., Kim, J.D., Oh, B.T. 2013. Enhancement of heavy metal phytoremediation by *Alnus firma* with endophytic *Bacillus thuringiensis* GDB-1. *Journal of Hazardous Materials* 25:477–483.

Bacon, C.W., Hinton, D.M. 2006. Bacterial endophytes: The endophytic niche, its occupants, and its utility, pp. 155–194. *Plant-Associated Bacteria*, Gnanamanickam, S.S. (ed.). Dordrecht, Netherlands: Springer.

Bailey, B.A., Bae, H., Strem, M.D., Roberts, D.P., Thomas, S.E., Crozier, J., Samuels, G.J., Choi, I.Y., Holmes, K.A. 2006. Fungal and plant gene expression during the colonization of cacao seedlings by endophytic isolates of four *Trichoderma* species. *Planta* 224:1149– 1164.

Bais, H.P., Park, S.W., Weir, T.L., Callaway, R.M., Vivanco, J.M. 2004. How plants communicate using the underground information superhighway. *Trends in Plant Science* 9:26–32.

Barka, E.A., Gognies, S., Nowak, J., Audran, J.C., Belarbi, A. 2002. Inhibitory effect of endophytic bacteria on *Botrytis cinerea* and its influence to promote the grapevine growth. *Biological Control* 24:135–142.

Blom, C.W.P.M., Voesenek, L.A.C.J. 1996. Flooding: The survival strategies of plants. *Trends in Ecology and Evolution* 11:290–295.

Caldeira, M., Heald, S.C., Carvalho, M.F., Vasconcelos, I., Bull, A.T., Castro, P.M.L. 1999. 4-Chlorophenol degradation by a bacterial consortium: Development of a granular activated carbon biofilm reactor. *Applied Microbiology and Biotechnology* 52:722–729.

Chang, C.C., Tseng, S.K., Ho, C.M. 2004. Degradation of 2-chlorophenol via hydro-genotrophic biofilm under different reductive conditions. *Chemosphere* 56:989–997.

Chen, W.-M., Tang, Y.-Q., Mori, K., Wu, X.-L. 2012. Distribution of culturable endophytic bacteria in aquatic plants and their potential for bioremediation in polluted waters. *Aquatic Biology* 15:99–110.

Costley, S.C., Wallis, F.M. 2001. Treatment of heavy metal polluted wastewaters using the biofilms of a multistage rotating biological contactor. *World Journal of Microbiology and Biotechnology* 17:71–78.

Das, N., and Chandran, P. 2011. Microbial degradation of petroleum hydrocarbon contaminants. *Biotechnology Research International* 2011:941810.

Elbeltagy A., Nishioka, K., Suzuki, H., Sato, T., Sato, Y.I., Morisaki, H., Mitsui, H., Minamisawa, K. 2000. Isolation and characterization of endophytic bacteria from wild and traditionally cultivated rice varieties. *Soil Science and Plant Nutrition* 46:617–629.

Farzaneh, H., Fereidon, M., Noor, A., Naser, G. 2010. Biodegradation of dodecylbenzene solfonate sodium by *Stenotrophomonas maltophilia* biofilm. *African Journal of Biotechnology* 9:55–62.

Frenzel, P., Rothfuss, F., Conrad, R. 1992. Oxygen profiles and methane turnover in a flooded rice microcosm. *Biology and Fertility of Soils* 14:84–89.

Gabr, R.M., Gad-Elrab S.M., Abskharon, R.N.N., Mohammed, A.E.F. 2009. Biosorption of hexavalent chromium using biofilm of *E. coli* supported on granulated activated carbon. *World Journal of Microbiology and Biotechnology* 25:1695–1703.

Gisi, D., Hanselman, K.W., Stucki, G., Gisi, D. 1997. Biodegradation of pesticide 4,6-dinitro-ortho-cresol by microorganisms in batch cultures and in fixed bed column reactor. *Applied Microbiology and Biotechnology* 48P:441–448.

Glick, B.R. 2010. Using soil bacteria to facilitate phytoremediation. *Biotechnology Advances* 28:367–374.

Gray, E.J., Smith, D.L. 2005. Intracellular and extracellular PGPR: Commonalities and distinctions in the plant-bacterium signaling processes. *Soil Biology and Biochemistry* 37:395–412.

Hallmann, J., Quadt-Hallmann, A., Mahaffee, W.F., Kloepper, J.W. 1997. Bacterial endophytes in agricultural crops. *Canadian Journal of Microbiology* 43:895–914.

Jiang, H., Bishop, P.L. 1994. Aerobic biodegradation of azo dyes in biofilms. *Water Science and Technology* 29:525–530.

Kargi, F., Eker, S. 2005. Kinetics of 2,4-dichlorophenol degradation by *Pseudomonas putida* CP1 in batch culture. *International Biodeterioration and Biodegradation* 55:25–28.

Khan, S.A., Hamayun, M., Ahmed, S. 2006. Degradation of 4-aminophenol by newly isolated *Pseudomonas* sp. strain ST-4. *Enzyme and Microbial Technology* 38(1–2):10–13.

Khan, S., Afzal, M., Iqbal, S., Khan, Q.M. 2013. Plant-bacteria partnerships for the remediation of hydrocarbon contaminated soils. *Chemosphere* 90:1317–1332.

Kuffner, M., De Maria, S., Puschenreiter, M., Fallmann, K., Wieshammer, G., Gorfer, M., Strauss J., Rivelli, A.M., Sessitsch, A. 2010 Bacteria associated with Zn and Cd-accumulating *Salix caprea* with differential effects on plant growth and heavy metal availability. *Journal of Applied Microbiology* 108:1471–1484.

Lindemann U., Smith J.C., Smets, B.F. 1998. Simultaneous biodegradation of 2,4-dinitrotoluene and 2,6-dinitrotoluene in an aerobic fluidized bed biofilm reactor. *Environmental Science and Enginering* 32:82–87.

Lodewyckx, C., Taghavi, S., Mergeay, M., Vangronsveld, J., Clijsters, H., van der Lelie, D. 2001. The effect of recombinant heavy metal resistant endophytic bacteria in heavy metal uptake by their host plant. *International Journal of Phytoremediation* 3:173–187.

Long, X.X., Chen, X.M., Chen, Y.G., Woon-Chung, W.J., Wei, Z.B., Wu, Q.T. 2011. Isolation and characterization endophytic bacteria from hyperaccumulator *Sedum alfredii* Hance and their potential to promote phytoextraction of zinc polluted soil. *World Journal of Microbiology and Biotechnology* 27:1197–1207.

Luo, S., Xu, T., Chen, L., Chen, J., Rao, C., Xiao, X. et al. 2012. Endophyte-assisted promotion of biomass production and metal-uptake of energy crop sweet sorghum by plant-growth-promoting endophyte *Bacillus* sp. SLS18. *Applied Microbiology and Biotechnology* 93:1745–1753.

Ma, Y., Prasad, M.N.V., Rajkumar, M., Freitas, H. 2011. Plant growth promoting rhizobacteria and endophytes accelerate phytoremediation of metalliferous soils. *Biotechnology Advances* 29:248–258.

Ma, Y., Oliveira, R.S., Nai, F.J., Rajkumar, M., Luo, Y.M., Rocha, I. et al. 2015. The hyperaccumulator *Sedum plumbizincicola* harbors metal-resistant endophytic bacteria that improve its phytoextraction capacity in multi-metal contaminated soil. *Journal of Environmental Management* 156:62–69.

Marx, J. 2004. The roots of plant-microbe collaborations. *Science* 304:234–236.

Ponnamperuma, F.N. 1984. Effects on flooding of soils, pp. 9–45. *Flooding and Plant Growth*, Kozlowski, T.T. (ed.). New York: Academic Press.

Porras-Soriano, A., Soriano-Martín, M.L., Porras-Piedra, A., Azcón, R. 2009. Arbuscular mycorrhizal fungi increased growth, nutrient uptake and tolerance to salinity in olive trees under nursery conditions. *Journal of Plant Physiology* 166:1350–1359.

Puhakka, J., Melin, E., Jarvinen, K., Koro, P., Rintala, J., Hartikainen, P., Shieh, W., Ferguson, J. 1995. Fluidized-bed biofilms for chlorophenol mineralization. *Water Science and Technology* 31(1):227–235.

Quintelas, C., Rocha, Z., Silva, B., Fonseca, B., Figueiredo, H., Tavares, T. 2009. Biosorptive performance of an *Escherichia coli* biofilm supported on zeolite NaY for the removal of Cr(VI), Cd(II), Fe(III) and Ni(II). *Chemical Engineering Journal* 152:110–115.

Radwan, S.S., Al-Hassan, R.H. 2001. Potential application of coastal biofilm-coated gravel particles for treating oily waste. *Aquatic Microbiology Ecology* 23:113–117.

Rafida, A.I., Elyousfi, M.A., Al-Mabrok, H. 2011. Removal of hydrocarbon compounds by using a reactor of biofilm in an anaerobic medium. *World Academy of Science Engineering and Technology* 73:153–156.

Rafique, M., Hayat, K., Mukhtar, T. et al. 2015. Bacterial biofilm formation and its role against bacterial agricultural pathogens, pp. 373–382. *The Battle Against Microbial Pathogens: Basic Science, Technological Advances and Educational Programs*, Mendez-Vilas, A. (ed.). Badajos, Spain: Formatex.

Ramey, B.E., Matthysse, A.G., Fuqua, C. 2004. The FNR-type transcriptional regulator SinR controls maturation of *Agrobacterium tumefaciens* biofilms. *Molecular Microbiology* 52:1495–1511.

Rosenblueth, M., Martinez-Romero, E. 2004. *Rhizobium etli* maize population and their competitiveness for root colonization. *Archives of Microbiology* 181:337–344.

Scott, J.A., Karanjkar, A.M., Rowe, D.L. 1995. Biofilm covered granular activated carbon for decontamination of streams containing heavy metals and organic chemicals. *Minerals Engineering* 8:221–230.

Shehzadi, M., Afzal, M., Islam, E., Mobin, A., Anwar, S., Khan, Q.M. 2014. Enhanced degradation of textile effluent in constructed wetland system using *Typha domingensis* and textile effluent degrading endophytic bacteria. *Water Research* 58:152–159.

Shehzadi, M., Fatima, K., Imran, A., Mirza, M.S., Khan, Q.M., Afzal M. 2015. Ecology of bacterial endophytes associated with wetland plants growing in textile effluent for pollutant-degradation and plant growth-promotion potentials. *Plant Biosystems* 150(6):1261–1270.

Shukla, K.P., Singh, N.K., Sharma, S. 2010. Bioremediation: Developments, current practices and perspectives. *Genetic Engineering and Biotechnology Journal* 2010:GEBJ-3.

Sundar, I.M., Sadiq, M., Amitava, N., Chandrasekaran. 2011. Bioremoval of trivalent chromium using *Bacillus* biofilms through continuous flow reactor. *Journal of Hazardous Materials* 196:44–51.

Surette, M.A., Sturz, A.V., Lada, R.R., Nowak, J. 2003. Bacterial endophytes in processing carrots (*Daucus carota* L. var. sativus): Their localization, population density, biodiversity and their effects on plant growth. *Plant and Soil* 253:381–390.

Toner, B., Manceau, A., Marcus, M.A., Millet, D.B., Sposito, G. 2005. Zinc sorption by a bacterial biofilm. *Environmental Science and Technology* 39:8288–8294.

Toyama, T., Sato, Y., Inoue, D., Sei, K., Chang, Y.C., Kikuchi, S., Ike, M. 2009. Biodegradation of bisphenol A and bisphenol F in the rhizosphere sediment of *Phragmites australis*. *Journal of Bioscience and Bioengineering* 108:147–150.

van Loon, L.C., Bakker, P.A.H.M., Pieterse, C.M.J. 1998. Systemic resistance induced by rhizosphere bacteria. *Annual Review of Phytopathology* 36:453–483.

Visioli, G., D'Egidio, S., Vamerali, T., Mattarozzi, M., Sanangelantoni, A.M. 2014. Culturable endophytic bacteria enhance Ni translocation in the hyperaccumulator *Noccaea caerulescens*. *Chemosphere* 117:538–544.

Wang, H., He, M., Lin, C., Quan, X., Guo, W., Yang, Z. 2007. Monitoring and assessment of persistent organochlorine residues in sediments from the Daliaohe River watershed, northeast of China. *Environmental Monitoring and Assessment* 133:231–242.

Weyens, N., Croes, S., Dupae, J., Newman, L., van der Lelie, D., Carleer, R. et al. 2010. Endophytic bacteria improve phytoremediation of Ni and TCE co-contamination. *Environmental Pollution* 158:2422–2427.

Yousaf, S., Afzal, M., Reichenauer, T.G., Brady, C.L., Sessitsch, A. 2011. Hydrocarbon degradation, plant colonization and gene expression of alkane degradation genes by endophytic *Enterobacter ludwigii* strains. *Environmental Pollution* 159:2675–2683.

Zhang, T.C., Fu, Y.C., Bishop, P.L. 1995. Transport and biodegradation of toxic organics in biofilms. *Journal of Hazardous Materials* 41:267–285.

Zhang, X., Wang, Z., Liu, X., Hu, X., Liang, X., Hu, Y. 2013. Degradation of diesel pollutants in Huangpu–Yangtze River estuary wetland using plant–microbe systems. *International Biodeterioration and Biodegradation* 76:71–75.

Zhao, G., Xu, Y., Han, G., Ling, B. 2006. Biotransfer of persistent organic pollutants from a large site in China used for the disassembly of electronic and electrical waste. *Environmental and Geochemical Health* 28:341–351.

Zinniel, D.K., Lambrecht, P., Harris, N.B., Feng, Z., Kuczmarski, D., Higley, P., Ishimaru, C.A., Arunakumari, A., Barletta, R.G., Vidaver, A.K. 2002. Isolation and characterization of endophytic colonizing bacteria from agronomic crops and prairie plants. *Applied and Environmental Microbiology* 68:2198–2208.

13

Phytoremediation as a Green and Clean Tool for Textile Dye Pollution Abatement

Niraj R. Rane, Rahul V. Khandare, Anuprita D. Watharkar, and Sanjay P. Govindwar

CONTENTS

13.1 Introduction

Global industrialization has resulted in widespread contamination of the environment with persistent organic and inorganic wastes. Industrial development is often associated with costs in terms of pollution of air, water bodies, and soils by a number of toxic compounds. The rapidly developing and expanding textile industry is one of the major sources discharging toxic chemicals, in the form of dyes. Around 10 to 15% of synthetic textile dyes, many of which are associated with carcinogenic and other toxic effects, are released during the dye and finishing processes for clothing, ultimately causing threats to all life forms (Khataee et al. 2010). Attributes like high stability to light, temperature, water, and detergents and resistance to biological degradation due to the presence of antimicrobial agents make these dyes recalcitrant entities that persist in the environment by escaping conventional wastewater treatment processes (Couto 2009). Valuable water bodies needed for drinking, irrigation, and industrial purposes are being polluted by the textile dyes released with the industry's effluents. Addition of textile dyes to water reservoirs alters the vital parameters of water bodies by influencing the levels of chemical oxygen demand (COD), biological oxygen demand (BOD), total organic carbon (TOC), total suspended solids (TSS), total dissolved solids (TDS), pH, and color (Kabra et al. 2013). Therefore, addressing the issue of these environmental contaminants has become very important and needs attention from environmental scientists and environmentalists (Khandare et al. 2013a). People now realize that for the sake of progress and prosperity, there has been a failure to protect natural resources and the environment on which we all depend. Despite these concerns, industries often show a reluctance to invest in proper treatment technologies for the effluents, because they consider such measures to be a nonproductive use of money when they are trying to remain profitable in a highly competitive world.

A number of physical and chemical methods, such as adsorption, coagulation, flocculation, filtration, photodegradation, and chemical oxidation processes, are available for dealing with the pollution produced by textile dyes. These procedures are costlier and also less efficient, with a major problem of secondary pollution and an inability to treat a wide array of dyes due to their high structural diversity (Kabra et al. 2013). However, these methods are employed to remove the color; their capacity to reduce the toxicity is still a major problem. These facts certainly demand the development of an efficient, cost-effective, and solar solution for the removal of these dyes from the environment. Use of microbes for remediation has gained a lot of interest by researchers, because such microbes are ubiquitous, represent the richest source of molecular and chemical diversity, can be sustained under extreme conditions, and have high potential. The roles of microbes in purifying effluent and wastewater have been extensively studied and applied in pilot-scale studies but are still associated with some limitations as far as *in situ* treatment of pollutants is concerned (Khandare et al. 2011b). There are many obstacles with scaling up such technologies. In light of these facts, there is an urgent need to develop a cost-effective yet efficient solution to address the potential environmental and public health problems.

Phytoremediation, that is, the use of plants for environmental cleanup, is rising as a true green technology. Plants and their associated microbes enhance the prospects of remediation of pollutants via rhizodegradation, biostimulation, biostabilization, bioaccumulation, phytoextraction, and phytovolatization (Pilon-Smits 2005). Plants possess very systematic, efficient, and versatile metabolic systems that can degrade and even mineralize highly toxic persistent xenobiotic compounds. *In situ* phytoremediation is highly reasonable for public approval, as it is economical, easy to run, requires low nutrient input, and is aesthetically gratifying; however, it is still in the experimental stages and will require a great deal of attention from the scientific community (Khandare et al. 2011a). Research and improvements in phytoremediation, however, have gained increased attention in the last few years. The most important primary step in phytoremediation research is the selection of the right plants, i.e., those that have certain beneficial characteristics for the removal of textile dyes from the environment. This knowledge of the biochemical processes in plants may help to manage and reclaim contaminated sites.

Phytoremediation is a cost-effective and solar power-driven green technology provided by nature itself. Protocols of phytoremediation are relatively easy to implement, and after initial site development and planting, the maintenance costs are minimal. *In situ* or *ex situ* application is possible with effluent and soil substrates, respectively, whereas *in situ* phytoremediation processes lead to minimum site disturbances. Many times, the price of a phytotechnology has been found to be less than half the price of alternative physico-chemical and biological methods. There is no limit for sites utilizing phytoremediation, and this strategy can be employed in any geographical area that can support plant growth. Phytoremediation is visually appealing, amenable to a variety of organic and inorganic compounds, and produces less waste. Moreover, plants provide ground-cover, and their roots help to stabilize soil, which mitigates erosion from both wind and water (Govindwar and Kagalkar 2010). *In situ* phytoremediation provides organic materials, nutrients, and oxygen to soil via metabolic processes, improving the overall quality and texture of soil at remediated sites.

Extracellular and intracellular plant enzymes, like dehalogenases, nitroreductases, nitrilases, peroxidases, and laccases, have been reported to be responsible for dye degradation. Phytoremediation technologies, with their rhizosphere microorganisms, are considered more efficacious because of the synergism between plants and microbes. During phytoremediation processes, the plant root zone is the zone of interest for bacterial colonization. Plant root exudates supply nutrients and growth supplements to bacteria in the form of amino acids, organic acids, phenolics, flavonoids, and polysaccharides. These root exudates are metabolized by bacteria and form intermediates that bind with pollutant molecules, making the pollutants bioavailable while the microbes also help in degradation of pollutants, converting them to simpler forms or stabilizing them in soil, which is called the rhizospheric effect (Lefevre et al. 2013). In this way, plant root exudates enhance contaminant desorption by increasing the overall rhizospheric effect and enabling plants to take up such bioavailable pollutants by their roots and to either sequester the compounds further or cause complete mineralization. In the last several decades, some plants, like vetiver grass and tomato, have been used for the remediation of ethidium bromide (EtBr) from polluted soil environments (Uera et al. 2007). Contaminants may be transferred to shoots and leaves for accumulation. Besides, the use of aquatic plants, e.g., macrophytes, is a promising alternative to remove textile dyes, heavy metals, and other contaminants from contaminated wastewater.

Recent biological strategies, like the use of bacteria, fungi, yeast, and their combinatorial systems, have been confirmed to be successful for decolorization of dyes, but these strategies have practical hurdles when *in situ* administration of the pollutant is concerned. The interaction and bioavailability of pollutants to roots are the most important preconditions for further treatment processes, as roots are the first plant organs to come in contact with pollutants (Kagalkar et al. 2010). Plant–microbe interactions have been described as a low-input technology for the renewal of contaminated soils and waters (Abhilash et al. 2012). Soil in the root zone is generally surrounded with a 100–10,000 times-wider range of microorganisms than is bulk soil (Glick 2003). Microbe-assisted phytoremediation has been described to enhance contaminant degradation by controlling the treatment process with consistent enzymatic actions along with improvement of plant longevity (Glick 2010). Numerous microbes have stimulating effects on phytoremediation with candidate plants. Plant endophytes have the potential to fix nitrogen, solubilize phosphates and sulfates, etc., and also have the potential to synthesize phytohormones and siderophores (Bloemberg and Lugtenberg 2001). Plant–endophyte synergies have also been observed to play a key role in degradation and detoxification of many harmful pollutants, not only from rhizospheric zones but also from aerial plant parts (Stępniewska and Kuźniar 2013). Use of endophytic microorganisms, however, would be an ideal and appropriate approach for enhancement of phytoremediation efficacy. The natural interactions between an endophyte and a plant could prove to be additionally effective to minimize abiotic stress (Gerhardt et al. 2009). The plant–bacteria synergisms have also been found to achieve efficient decolorization of textile dyes and effluents compared to independent treatment systems for the two contaminant sources (Khandare et al. 2012, 2013). Genetic engineering and plant breeding techniques have been utilized to obtain effective and more efficient transgenic phytoremediators. Transgenic plants have also been obtained for complete mineralization of organic pollutants (Aken 2008).

Constructed wetland (CW) systems have been proposed for *in situ* phytoremediation of polluted matrices to certain depths beneath ground level. Constructed wetlands are large plotted reactors planned especially to imitate conditions close to real dye wastewater conditions. Treatment in a constructed wetland can ensure greater efficacies, as proper checking and monitoring of the whole bed becomes achievable by virtue of manageable hydraulics and pollutant loads. Hydroponic phyto-tunnel systems have also been utilized for the treatment of textile effluents (Khandare et al. 2013a). Laboratory-scale horizontal and vertical subsurface flow bioreactors based on a plant–bacteria synergistic approach have been developed to treat real textile effluents (Kabra et al. 2013, Khandare et al. 2013b). Some pilot-scale operational systems using macrophytes are also on record. For instance, *Phragmites australis*, *Typha domingensis*, and *Alternenthera philoxeroides* were proposed in independent constructed wetland studies for removal of textile dyes from wastewater (Rane et al. 2015). Large-scale treatment of textile dye effluents with a combinatorial system of plants has occasionally been reported. A 200-L volume of wastewater from the tie and dye industry was shown to be treated with the use of cattail and cocoyam plants in independent engineered wetland systems (Mbuligwe 2005). However, phytotechnology is recommended in combination with other mechanical processes, as it requires sufficient time to treat wastewater properly.

The increasing number of textile industries located in India have resulted in higher levels of dyes in nearby soils and water bodies, which is of great concern for the population and environment. These wastewaters are used for irrigation purposes by many farmers, and therefore their treatment is necessary. Constructed wetland plants and associated enriched dye-degrading microflora can be effective tools to degrade and detoxify dyes from textile industrial effluents, since several plants and microbes have shown the ability to degrade various textile dyes on a laboratory scale. By use of constructed wetlands, phytoremediation capabilities can be studied with potential plants from actual dye-contaminated sites and the associated microflora, which can sustain and degrade highly contaminated sites and wastewaters.

13.2 Magnitude of Textile Dye Industrial Pollution

Textile dye manufacturers and processors are considered one of the major contaminators of water and soil. Effluents from this industry largely contain high amounts of colors, acids, bases, salts, dispersants, detergents, humectants, surfactants, and oxidants, which leave the effluents aesthetically unacceptable and unusable (Figure 13.1). Additionally, textile dyes have been reported to be potent mutagenic and carcinogenic agents that are hazardous to various living beings of the ecosphere (Saratale et al. 2011). In terms of the dangers of dye pollution, especially in developing countries, it is crucial to have a solution for this issue.

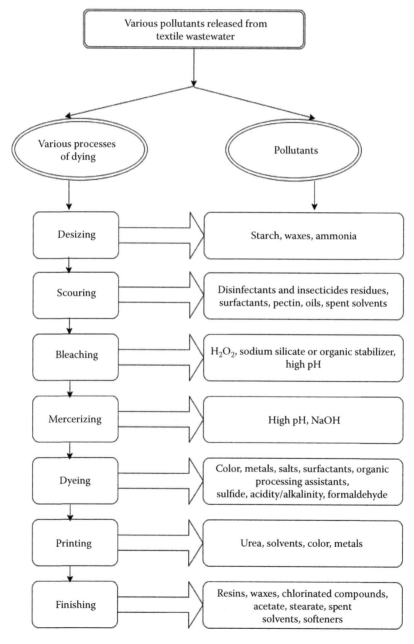

FIGURE 13.1
Various pollutants released from textile wastewater.

The World Bank has estimated that around 20% of total industrial water pollution comes from dyeing processes alone. Further, it was also reported that more than 72 different toxic compounds are purely from dye processing. Of the listed chemicals, 30 were proposed to be impossible to be removed and highly recalcitrant in nature (Chen and Burns 2006). Generally, 1.6 million L of water is utilized for production of 8000 kg of fabric. Of the consumed water, 16 and 8% is used for dying and printing, respectively. It is clearly evident that a large volume of a valuable resource, water, is directly or indirectly wasted each day. Even after these processes, enormous amounts of water are required for cleaning of the printed and dyed fabrics. For bright coloring and color fastness, a number of other chemicals, like mordants, fixing agents, surfactants, fasteners, acids, bases, defoamers, and alkalis, are utilized. All these chemical compounds of organic or inorganic origin further clearly add to the TOC, COD, TSS, TDS, and BOD of the water receiving the effluents (Kurade et al. 2011, Khandare et al. 2014). Dyes, even in minute quantities, create bright colors in water, which is aesthetically unacceptable. Considering the pollution of water on such a large scale, it is obligatory to work on treatment of water polluted by textile dyes (Govindwar and Kagalkar 2010). Simultaneously, consumption of such waters is also a key issue, as leaching and subsequent secondary pollution also requires attention. Plants may prove to be very valuable tools for treatment of polluted runoff waters and soils.

13.3 Textile Dye Effluents Present Critical Threats to All Life Forms

Textile industries emerged as the first large-scale and profit-making trade during the European industrial revolution. As a result, serious concerns over the continuous deterioration of clean water bodies began (Figure 13.2). This directly affected the water quality and useful vegetation and microflora of water reservoirs, leading to further declines in the related ecosystems (Saratale et al. 2011).

A few decades ago, Yahagi et al. (1975) reported the mutagenicity and carcinogenicity of azo dyes. The complex chemical structures (of the dyes and their derivatives) may cause mutagenesis, which can lead to malignancies. The International Agency for Research on Cancer (1982) declared that benzidine dyes are very potent carcinogens to a number of classes of mammals. A dye spray-specific disorder, called Ardystil syndrome, has been reported to affect workers, as it causes difficulty breathing and ultimately has caused the death of a few workers in parts of Spain and Algeria. Erythrosine has been reported to have xenoestrogenic, DNA-damaging, neurotoxic, allergic, and carcinogenic effects on humans (Mittal et al. 2006). Disperse dyes like orange number 1 and blue number 291 increase micronuclei formation in human hepatoma-derived HepG2 cells, indicating mutagenic and genotoxic effects of dyes (Tsuboy et al. 2007). Caco-2 and HepG2 cell lines also revealed persistence of malachite green and leuco malachite green that led to decreases in total protein content, colony-forming ability, and cell viability (Stammati et al. 2005). Toxicity of textile effluents was also shown in Swiss albino rats (Sharma et al. 2007). Malachite green was also reported to be responsible for egg mortality in *Micropterus salmoides*, a large-mouth bass species (Wright 1976). Similarly, rainbow trout showed focal liver necrosis, sinusoidal congestion, nuclear alterations, and mitochondrial damage upon exposure to malachite green. This dye also caused a severe drop in protein and calcium levels and increased blood cholesterol in *Heteropneustes fossilis*, a catfish (Srivastava

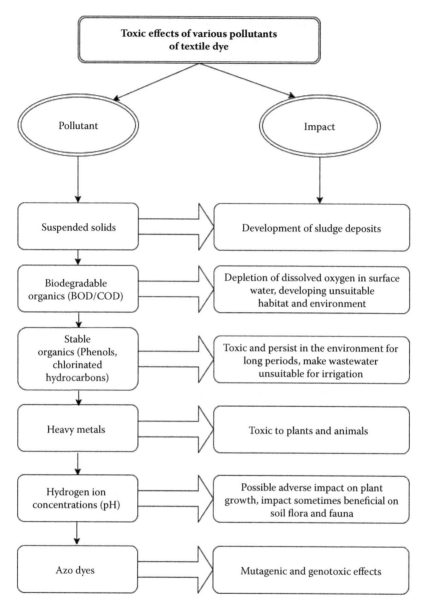

FIGURE 13.2
Toxic effects of various pollutants from textile dye.

et al. 1995). *Xenopus laevis* frogs, when exposed to dyes such as Blue G-A, Remazol turquoise, Remazol red, Cibacron red FN-3G, Cibacron blue FN-R, Astrazon blue FGRL, and Astrazon red FBL, revealed extreme toxicity in a teratogenesis assay (Birhanli and Ozmen 2005). *Daphnia magna*, a freshwater flea and a model organism for aquaculture studies, revealed the toxicity of various textile dyes (Bae and Freeman 2005). A model plant for mitosis investigations, *Allium cepa*, showed a very strong genotoxic effect upon dye exposure (Watharkar and Jadhav 2014).

A *Salmonella* assay revealed the mutagenicity of disperse dyes such as Blue 291 and Orange 1 (Chequer et al. 2009, Ferraz et al. 2011). The presence of Orange 16 azo and Congo red was shown to inhibit bacterial luminescence and algal growth in water. The alarming toxicities of these dyes make it massively imperative to detoxify and degrade textile dyes and effluents.

13.4 Commonly Adapted Treatments for Dye Removal from Effluents and the Limitations of These Treatments

Traditionally available physico-chemical methods used for treatment of textile effluents are generally followed. However, companies in the textile industry only partially follow at times do not follow all the regulations set by legislation, as these methodologies are expensive (Khandare et al. 2013a). Modern bioremediation schemes of treatment using bacteria, fungi, and even their consortial systems have also been proposed to be effective in treatment of dyes but carry practical problems and complications for their *in situ* applications.

13.4.1 Physico-Chemical Methods and Their Limitations

A number of physico-chemical approaches, such as adsorption, precipitation, coagulation, flocculation, filtration, oxidation, ozonation, advanced oxidation processes, UV photolysis, and sonolysis (Wu et al. 2008, Zhou et al. 2008) are available for the treatment of textile dye effluents. Additionally, unconventional methods, like nano-, micro-, and ultrafiltration and reverse osmosis are employed for dye removal.

Although these methods of dye treatment may remove color from dye wastewaters, serious concerns about further toxicity add complications to their applications. Secondary pollution problems of sludge and leachates cannot be avoided via physico-chemical methods. The generated sludge is generally disposed of in landfills, where they may also cause serious damage to the environment. Further, the advanced methods are not affordable for many processors. As mentioned earlier, current legislation is ignored because of these reasons and limitations, and the problem persists.

13.4.2 Biological Methods for Dye Removal from Effluents and Their Limitations

To resolve the problems and limitations of physico-chemical methods, modern bioremediation techniques are frequently advocated. The enormous potential of microbes and their metabolic capacities provide tools which can utilize and transform dyes into inoffensive endproducts. Enzymatic machineries of the employed microbes can bring about complete mineralization or degradation and removal of dyes from textile wastewaters. In addition, a number of cost-effective advantages, like forming less sludge, make them more environmentally friendly (Khandare and Govindwar 2015a).

A very broad array of microbial species have been demonstrated in the last few decades to have capacities for degradation of textile dyes (Reema et al. 2011). Various classes of fungi (Saratale et al. 2006, 2009a,b), actinomycetes (Machado et al. 2006), yeasts (Jadhav and Govindwar 2006), algae (Khataee et al. 2010a,b, 2013a), and bacteria (Kurade et al. 2011, Saratale et al. 2011) have been studied for the potential for textile dye degradation.

13.5 Emergence of Plant-Based Technologies for Textile Effluent Treatment

Of late, phytoremediation of pesticides, chlorinated solvents, polychlorinated biphenyls, radionuclides, poisonous gases, crude oil, landfill leachates, polyaromatic hydrocarbons, petroleum, explosives, munitions, and heavy metals have methodically been tested as a prospective tool for pollutant removal from the environment (Pilon-Smits 2005). Of all these pollutants, removal of heavy metals is the most meticulously demonstrated and proven method via phytoremediation. Phytoremediation of textile dyes has remained a less explored area of investigation. It is evident from the available literature that plants can be an invaluable alternative for removal of textile dyes. The extraordinary metabolism possessed by plants provides them incomparable capabilities for pollutant removal (Govindwar and Kagalkar 2010). Bioaugmentaion and phytostimulation by virtue of the presence of rhizospheric microbes give more natural assistance to phytoremediation candidate plants. Plants with fibrous and deep root systems and also rapid growth may prove appropriate for phytoremediation of dyestuffs (Khandare et al. 2011a, 2013a). In the future, basic information about the metabolic processes involved in phytoremediation will provide more avenues and additional insights for development and engineering of plants.

13.6 Understanding Phytoremediation of Dyes through Various Modes

Although plants appear to be complex living beings, they adopt very specialized mechanisms to deal with various pollutants like textile dyes. Indigenous defense mechanisms are triggered, and the toxicants are neutralized or eliminated by virtue of increased expression of various oxidoreductase and antioxidant enzyme genes. Tissues from different plant parts perform various roles for uptake of contaminants and to achieve further metabolism. To understand these mechanisms, it is imperative to study changes at the histological level and the biochemical status of enzymes involved in pollutant removal. As mentioned earlier, the advanced phytoremediation processes rely on manipulations at the genetic and precisely biochemical levels. Basic information regarding various modes of phytoremediation is available. As plants are metabolically complex systems, none of the pollutant-handling modes is exclusive, and a number of different mechanisms can simultaneously work. Major plant mechanisms of pollutant treatment are described below.

13.6.1 Plant Mechanisms of Phytoremediation

Nature has relentlessly used this green and eco-friendly technology since long ago for treatment of contaminants and to help to maintain water quality, to prevent soil erosion, and most importantly, to remove contaminants from soil, air, and water. There are various subprocesses that constitute phytoremediation, and these may work together or individually (Figure 13.3).

13.6.1.1 Phytoextraction

Phytoextraction is the extraction of pollutants from soil, water, or air by plants through accumulation and compartmentalization. When plants are used to take up contaminants

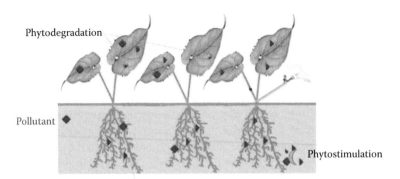

FIGURE 13.3
Possible fates of pollutants during phytoremediation.

and accumulate them in harvestable biomass, this is known as phytoextraction or phyto-accumulation. The phytoextraction process has become popular worldwide during the last 20 years. The process is more useful for harvesting heavy metals, like nickel, zinc, copper, lead, cadmium, chromium, and arsenic, than for organic contaminants (Roy et al. 2005).

Nowadays, phytoremediation is rising as a phytomining technology for commercial phytoextraction of valuable heavy metals, like gold and silver, from the soil matrix (Wilson-Corral et al. 2012). Some strategies, such as overexpression or insertion of genes related to uptake, transport, partitioning storage, and binding proteins (including regulatory transcription factors and organ-specific promoters), can be applied for enhancement of the process. Textile effluents to some extent also contain heavy metals, which can be removed by candidate plants. Use of aquatic vegetation such as water hyacinths has been shown to remove heavy metals from textile wastewater (Sanmuga and Senthamil 2014).

13.6.1.2 Phytostabilization

In phytostabilization, reduced mobility of pollutants toward groundwater or their dispersion in soil or water via enhanced precipitation or sequesteration to roots occurs. In this process, plants carry out immobilization of contaminants by adsorption or accumulation and provide a zone around the roots where contaminants and their derivatives are precipitated and stabilized. The phytostabilization process, unlike phytoextraction, focuses mainly on transformation of contaminants into more stable derivatives and sequestration of contaminants in soil near the root but not actually in plant tissues. Some strategies, such as amendments of binders or sequesters and microbial populations suitable for the purpose, can be applied for the enhancement of the process. The root-associated microflora of phytoremediation candidate plants have been shown to enhance pollutant stability, increasing the bioavailability of dyes to the plants. Sulfonated anthraquinones, which are raw materials for synthesis of a number of synthetic textile dyes, are removed by *R. rhabarbarum*, with accumulation as the prime mechanism.

13.6.1.3 Phytostimulation

In phytostimulation, release of plant exudates and enzymes into the rhizosphere stimulates the microbial and fungal degradation of organic pollutants; therefore, this process is also known as rhizospheric degradation, i.e., a process that exhibits plant-bacteria interactions for removal of contaminants. A plant rhizosphere is supposed to be the oasis for bacterial growth where stimulated growth of root-associated microflora leads to enhanced degradation of pollutants in a synergistic way. Some strategies, such as overexpression or insertion of genes producing such microbial stimulants, can be applied for the enhancement of the process. Bacteria-associated phytoremediation of textile dyes and real effluents was achieved using *Z. angustifolia*, *P. grandiflora*, *G. pulchella*, and *P. crinitum* in independent studies (Khandare et al. 2012, 2013, Kabra et al. 2013, Watharkar et al. 2015).

13.6.1.4 Phytodegradation

In phytodegradation, partial or total degradation of complex organic molecules within the plants takes place. In this process, *ex planta* and *in planta* enzymes of plants carry out partial or complete breakdown of contaminants and convert them to relatively less toxic compounds. Phytodegradation is a well-known process for the recovery of soil polluted with hydrocarbons, herbicides, pesticides, explosives, etc. Some strategies, such as overexpression or insertion of genes related to uptake, transport, degradation, and metabolism or of transcription factors can be applied for the enhancement of the process. *T. angustifolia*, commonly called narrow-leaved cattails, a well-known hyperextractor of dissolved nutrients, has shown efficient uptake of Reactive red 141. Transmission electron microscopic imaging of plant roots and leaves revealed the accumulation of dye in tissues after 28 days of exposure (Nilratnisakorn et al. 2007).

13.6.1.5 Phytovolatization

In phytovolatization, uptake, transport, and volatilization of volatile organics take place through stomata. In this process, plants take up toxic pollutants and release them in volatile form via their leaves. Sometimes, phytotransformation of highly volatile compounds to less volatile compounds is performed by plants and then they are released into the air (Pilon-Smits 2005). Some strategies, such as insertion and overexpression of genes related to uptake, transport, degradation, metabolism, and volatilization as well as transcription can be applied for enhancement of this process.

13.6.1.6 Rhizofiltration

In rhizofiltration, plant roots are used to absorb or adsorb pollutants from water and aqueous waste streams. This type of phytoremediation involves the filtration of polluted water by the massive root biomass of plants and simultaneous removal of toxic substances and excess nutrients by adsorption as well as absorption phenomena. Some strategies, such as manipulation for desired and extensive root systems and higher uptake of the pollutants, can be applied for the enhancement of the process. Partial rhizofiltration of real textile effluent samples was carried out in developed phytoreactors in independent studies using the plants *P. grandiflora*, *G. pulchella*, and *T. angustifolia* (Kabra et al. 2012, Khandare et al. 2013b). Rhizofiltration of dye effluents was also studied and demonstrated in a reactor lagoon in a study exploring the rhizofiltration method using *S. molesta* (Chandanshive et al. 2016).

13.6.2 Plant Stress Response Mechanisms through Enzymatic Systems for Textile Dyes

Plants have the potential to transform organic compounds and convert them into simpler forms which are less toxic to life forms. Thus, after treatment with plants, these treated compounds can be safely released into the environment. Different kinds of mechanisms (Figure 13.4) have been developed by plants for protection from various abiotic stresses,

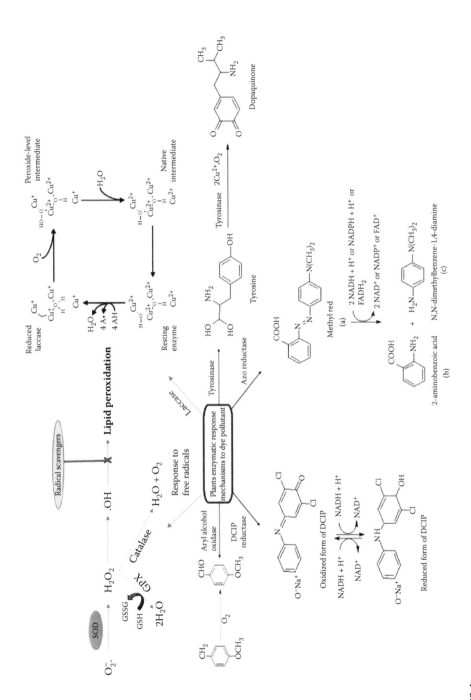

FIGURE 13.4
Plant enzymatic response mechanisms to dye pollutants.

including exposure to xenobiotic compounds in their vicinity (Page and Schwitzguébel 2009). Plants have very diverse and complex metabolic systems by which they are able to treat toxic pollutants that are not degradable by a variety of microorganisms. Plants carry well-organized enzymatic mechanisms that can also take up aromatic compounds as substrates and treat them (Aubert and Schwitzguébel 2004). Enzymes like catalase (CAT), superoxide dismutase (SOD), and glutathione *S*-transferase (GST) are induced in response to oxidative stress generated after exposure to any abiotic stress, including that from toxic textile dyes. Various oxidoreductive enzymes, including cytochrome P450, jointly take part in treatment and detoxification of dyes through complete mineralization. Accurate detection and quantification of every dye metabolite shows the role of each of these oxidoreductive enzymes in degradation pathways and gives important information to improve our understanding. Plant systems, through their intrinsic enzymatic systems, overcome the stress created by exposure to xenobiotic compounds. Reactive oxygen species (ROS) are increased by enzymes like NADPH oxidase as a part of a general plant defense mechanism that results in activation of antioxidant enzymes. SOD is known to catalyze the conversion of ROS into H_2O_2, which later is transformed into water and oxygen by the action of CAT and peroxidases.

Plants possess three different phases for the detoxification of a toxicant or contaminant. In the first phase of action, xenobiotics are oxidized by cytochrome P450 and peroxidases. In the second phase of action, conjugation of oxidized xenobiotics and glutathione takes place by the action of GST. In the third phase of action, these conjugated products are translocated into vacuoles differentially (Carias et al. 2008). Cytochrome P450 monooxygenases, a multigene family of enzymes, is involved in detoxification of these xenobiotic compounds. Cytochrome P450 plays an important role in metabolism of various plant compounds involved in cell signalling and defense mechanisms. Dioxygenase was reported earlier to add oxygen to sulfonated anthraquinones, which results in a double bond with the sulfonate group and removes it subsequently (Schwitzguébel et al. 2002). Translocation sites were investigated by using the capillary electrophoresis method with rhubarb, in which sulfonated anthraquinones were found to translocate and accumulate primarily in the leaves. The presence of newly formed metabolites from the parent molecule was also investigated, and these were detected in further studies, confirming the biotransformation of these compounds (Aubert and Schwitzguébel 2004).

GST is another enzyme that takes part in detoxification of contaminants and has been studied in plants. GST functions to covalently link glutathione to a wide-ranging variety of reactive hydrophobic and electrophilic substrates and results in polar and less reactive moieties. A pilot-scale constructed wetland with *P. australis*, used to degrade Acid Orange 7, increased the GST enzyme activity (Carias et al. 2008). Genetic-level studies were carried out in *P. australis* for the analysis of antioxidant enzyme expression after exposure to Acid Orange 7. That study revealed increased gene expression levels for CAT isoforms, Cu/Zn and Mn SOD, and glutathione peroxidase. A sudden increase in the production of ROS after dye exposure causes overexpression of these genes that are responsible for the antioxidant action on ROS (Davies et al. 2009). Increases in the levels of dehydroascorbate reductase enzyme, which is responsible for regulation of the ascorbate glutathione pathway, were also found in *P. australis* when it was exposed to the dye Acid Orange 7. Significant increases in GST activity revealed the compartmentalization and distribution of dye in plant cells (Carias et al. 2008).

Peroxidases have been studied with a focus on their significant role in degradation of textile dyes by plants. Various oxidoreductases were assayed in three different plants, namely, cress, mustard, and alfalfa, and these plants showed that peroxidases were the dominating enzymes in roots, shoots, and exudates (Gramss and Rudeschko 1998). Lignin peroxidases (LiP) are the key enzymes in oxidation of lignin structures of wood, and the reaction yields an intermediate cation radical that undergoes spontaneous fission. Studies with lignin model-like compounds have revealed that LiP attacks the C_α and C_β of its propyl side chain that is present in the aryl glycerol β-aryl ether substructure of lignin (Sarkanen et al. 1991). LiP, a heme group-containing enzyme, is useful in exploiting a wide range of organic chemical species as electron donor substrates and hydrogen peroxide for oxidation of the compound. First, cleavage of the hydrogen peroxide molecule takes place, resulting in the formation of a water molecule followed by incorporation of the oxygen atom into an initial intermediate compound. In further steps, the enzyme is reduced and subsequently regenerated. The azo dyes may act as electron donors for these enzymes (Davies et al. 2005). Numerous plant species have been reported to contain peroxidases in their tissues. Addition of H_2O_2 in growth medium was found to enhance the degradation of Remazol brilliant blue R by *Rumex crispus* (Takahashi et al. 2005). Roots of *B. malcolmii*, *A. amellus*, *Portulaca grandiflora*, *Z. angustifolia*, *G. pulchella*, *T. flagelliforme*, *Petunia grandiflora*, *I. hederifolia*, *S. portulacastrum*, and *N. cochenillifera* showed significant induction in activities of intracellular LiP during degradation of different textile dyes (Kagalkar et al. 2009, 2010, Kabra et al. 2011a, Khandare et al. 2011a,b, 2012, Adki et al. 2012, Patil et al. 2012, Watharkar et al. 2013a, Rane et al. 2014).

Laccases are from a class of polyphenol oxidases that are able to catalyze the oxidation of different substituted phenolic compounds, consuming O_2 as an electron acceptor in the reaction. Laccases attack hydroxyl groups of ortho- and para-substituted mono- and polyphenolic substrates. Formation of aromatic amines by transfer of one electron and removal of an H atom leads to release of free radicals, which can further undergo demethylation, depolymerization, repolymerization, or quinone formation (Abadulla et al. 2000). Laccases have been shown to possess enormous potential in decolorization of textile industry dyes and degradation of wastes from the textile industry. Bacterial and fungal laccases are well known and were reported early on for their potential in dye decolorization. Tyrosinase has been comprehensively assessed for its role and involvement in microbial dye degradation. Comprehensive studies with plants for dye removal, conversely, are yet to be performed. Root cells of *B. malcolmii* and *Portulaca grandiflora* tissue cultures showed strong induction in activities of tyrosinase while degrading Direct Red 5B (Kagalkar et al. 2009, Khandare et al. 2011a). *T. patula* hairy roots when exposed to Reactive Red 198 were also found to show significant induction in extracellular and intracellular enzyme activities of tyrosinase (Patil et al. 2009).

A number of enzymes has also been reported to show induction of activities while degrading textile dyes. Activities of azo reductase, dichlorophenol indophenol reductase, and riboflavin reductase were found to increase significantly, to varying extents, during Direct Red 5B degradation by *B. malcolmii* (Kagalkar et al. 2009). The status of enzymatic inductions and dye removal mechanisms are found to be highly unpredictable in different plant species. *G. pulchella* showed variable extents of veratryl alcohol oxidase, LiP, dichlorophenol indophenol reductase, and tyrosinase activities when exposed independently to Red HE3B, Brilliant Blue R, Rubine GFL Scarlet RR, and Navy Blue 2R. Induction of enzyme activities with respect to time and differential patterns was seen during treatment of dye mixtures. Enzymatic activities of LiP and veratryl alcohol oxidase were found to increase constantly during 34 to 96 h of exposure, whereas tyrosinase activity was determined to be enhanced only after 24 h and its activity started to decrease later, showing its role in the degradation process (Kabra et al. 2012).

13.7 Purified Plant Enzymes for Degradation of Textile Dyes

Enzymes are effectively used to achieve environmental cleanup. Direct involvement and catalytic action of plant enzymes in treatment of textile dye decolorization have been reported in a number of studies. However, the purification processes generally have high production costs and concerns about stability and low yields for their use in remediation processes. Agricultural substrates with low costs can be used for reducing the cost of purification of enzymes. Furthermore, technical efforts to elevate the percetages of recovery and purification should be made (Shaffiqu et al. 2002). Immobilization could be another alternate method, as separation of reactants and substrate becomes possible, and this would allow use of enzymes repeatedly. Repeated usage would reduce overall costs of a method (Mohan et al. 2005). Immobilized horseradish peroxidase with calcium alginate and a polyacrylamide gel entrapment method at an optimum pH of 4 was found to decolorize Azo Direct Yellow 12 completely, and it performed better than free enzyme (Maddhinni et al. 2006).

13.8 Results of Screening Candidate Plants for Phytoremediation of Dyes

Phytoremediation of textile dyes is a new and innovative method for textile effluent treatment. Various approaches for laboratory-scale studies have been used. Studies on textile dye removal have mostly focused on the use of wild macrophytes. The macrophyte *P. australis* was reported earlier for the decolorization of Acid Orange 7 as the model dye (Davies et al. 2005). *T. angustifolia* (narrow-leaved cattail) showed potential in effective treatment of Reactive Red 41 (100–300 mg/L) and could achieve around 60% decolorization (Nilratnisakorn et al. 2007). *Lemna minor* was also reported to decolorize methylene blue dye, with a prominent mechanism of accumulation (Reema et al. 2011). Aquatic plants such as *Eichhornia* spp. have also been found to decolorize and metabolize Black HY, Direct Dark Blue 6B, and Congo Red (Anjana and Thanga 2011).

Harmful textile dyes were also removed by a narrow range of herbaceous plant species. *Z. angustifolia*, *Brassica juncea*, *B. malcolmii*, *T. patula*, *T. flagelliforme*, *A. amellus*, *G. pulchella*, *Sesuvium portulacastrum*, *Gaillardia grandiflora*, *Rosmarinus officinalis*, *Petunia grandiflora*, *L. minor*, *Azolla filiculoides*, *Thymus vulgaris*, *Portulaca grandiflora*, *Hydrocotyle vulgaris*, and *I. hederifolia* are some examples which have been proposed for the removal of textile effluents and dyes (Zheng and Shetty 2000, Davies et al. 2005, 2009, Nilratnisakorn et al. 2007, Patil et al. 2009, 2012, Govindwar and Kagalkar 2010, Kabra et al. 2011a, 2011b, Khandare et al. 2011a, 2011b, 2012, Khataee et al. 2012, 2013b, Vafaei et al. 2012, 2013, Watharkar et al. 2013a, 2013b, Watharkar and Jadhav 2014, Rane et al. 2014). Table 13.1 provides a list of the potential different wild and indigenous plants and their dye remediation performances. Hydroponic systems provide another substitute for cultivation of plants for experimental purposes. These plants that are not grown on soil offer benefits, as most of the root system can be made available for remediation. Hydroponic systems are cost effective compared to tissue cultures, and experimentation is close to field applications. Studies on hydroponically grown duckweed, *L. minor*, with artificial neural network modeling has provided promising results, and the plant was found to attain substantial decolorization of the dye Acid Blue 92 (Khataee et al. 2012). In independent studies with hydroponics, Acid Blue 92,

TABLE 13.1

Phytoremediation Performances of Different Wild and Indigenous Plants for Textile Dyes and Effluents

Sr. No.	Name of Plant	Dye or Effluent	Concn (mg/L)	Decolorization Time (h)	% Decolorization	Reference(s)
1	*Salvinia molesta*	Rubin GFL	100	72	97	Chandanshive et al. 2016
2	*Ipomoea aquatica*	Brown 5R	200	72	94	Rane et al. 2016
3	*Alternanthera philoxeroides*	Remazol Red	70	72	100	Rane et al. 2015
4	*Pogonatherum crinitum*	Effluent	NA[a]	288	74	Watharkar et al. 2015
5	*Nasturtium officinale*	Acid Blue 92	20	96	78	Torbati et al. 2015
6	*Ipomoea hederifolia*	Scarlet RR	50	60	96	Rane et al. 2014
7	*Typha angustifolia*	Reactive Blue 19	75	144	70	Mahmood et al. 2014
8	*Bouteloua dactyloides*	Effluent	NA	24	92	Vijayalakshmidevi & Muthukumar 2014
9	*Petunia grandiflora*	Brillaint Blue G	20	36	86	Watharkar et al. 2013a, 2013b
10	*Azolla filiculoide*	Basic Red 46, Acid Blue 92	20	144, 168	90 and 80	Vafaei et al. 2012, Khataee et al. 2013a
11	*Lemna minor*	Methylene blue, Acid Blue 92	10, 2500	144	98 and 77	Reema et al. 2011, Khataee et al. 2012
12	*Portulaca grandiflora*	Reactive Blue 172	20	40	98	Khandare et al. 2011
13	*Glandularia pulchella*	Green HE4B, Remazol Orange 3R	20	48, 96	92 and 100	Kabra et al. 2011a, 2011b
14	*Aster amellus*	Remazol Red, Remazol Orange 3R	20	60, 72	96 and 100	Kabra et al. 2011b, Khandare et al. 2011c
15	*Typhonium flagelliforme*	Brilliant Blue R	20	96	65	Kagalkar et al. 2010
16	*Blumea malcolmii*	Malachite green, Red HE4B, Methyl orange, Reactive Red 2, Direct Red 5B	20	72	96, 76, 88, 80, 42	Kagalkar et al. 2009
17	*Phragmites australis*	Acid Orange 7	750, 100	8 times/day, 144 h	68, 98	Davies et al. 2009, Ong et al. 2010

Source: Adopted and modified from Khandare, R., Govindwar, S., *Biotechnology Advances* 33:1697–1714, 2015.

[a] NA, not available.

Basic Red 46, and Reactive Blue 19 dyes were found to be removed and metabolized effectively by *Nasturtium officinale, A. filiculoides,* and *T. angustifolia,* respectively (Khataee et al. 2013b, Mahmood et al. 2014). Phytoremediation performances of various plants in hydroponic culture are included in Table 13.1 along with initial concentrations of dye, the percent dye removal, and the time required for the individual treatments.

Rumex hydrolapatum, R. rhabarbarum, Rumex acetosa, and *Apium graveolens* were screened for the removal of sulfonated anthraquinones in hydroponic solutions (Aubert and Schwitzguébel 2004). There are some disadvantages in the use of hydroponic systems for phytoremediation. The water absorption capacity becomes a key factor during studies utilizing hydroponics systems. Among the tested plants, *R. acetosa* possessed a lower water absorption capacity and normal metabolism. The transpiration was also observed to decline compared to that of plants in soil (Aubert and Schwitzguébel 2004). In another experiment with hydroponically grown *Rumex* spp., collection of leaves of the same age, growth stage, and development was found to be an unmanageable task (Page and Schwitzguébel 2009). Hydroponically grown grasses, namely, *Eleocharis calva, Cyperus javanicus, Fimbristylis cymosa,* and *Pennisetum purpureum,* were observed to decolorize Poly R-478 (20 mg/L) up to 59, 61, 66, and 88%, respectively, within 1–4 weeks (Paquin et al. 2006). In the same report, herbaceous plants like *Rumex crispus* and *Hibiscus furcellatus* were also shown to decolorize Poly R-478 up to 70 and 55%, respectively.

Microbial contaminants in association with nutritional status, photoperiod, and other environmental factors can change the enzyme activities of plants possessing a phytoremediation potential (Govindwar and Kagalkar 2010). To overcome these issues with hydroponically or wetland-grown plants, tissue culture-based technologies have been the focus of some researchers. Although tissue culture entail high costs, it can be an effective tool to realize the actual role of plants in pollutant treatment. Taking advantage of plants in tissue cultures or hydroponics is indeed different than the real soil and rhizospheric environment. These *in vitro* built systems therefore must go along with or be followed by on-site field trials for testing their sustainability and viability.

13.9 *In Vitro* Techniques to Study Phytoremediation of Textile Dyes

Plant tissue cultures have been found to be useful tools to understand the role of individual plants in phytoremediation of pollutants. Various types of cultures, like whole plants, cell suspensions, and hairy roots, have frequently been exploited in phytoremediation experimentation. Tissue culture techniques offer a big advantage in understanding phytoremediation potentials, metabolic capabilities, and toxicity tolerance by the model or selected plant. Although there is a large amount of articles and texts available on phytoremediation and decolorization of dyes, much less information has been published concerning basic information on mechanistic and metabolism or biotransformation of dyes. Tissue culture technologies provide an additional advantage of growing plants under controlled conditions and establishing them with large numbers of propagating plants. These *in vitro*-grown plants can be made available throughout the year in any season, whereas wild plants have a limited natural life span. *In vitro* tissue culture techniques involve a microbe-free environment and thus can be utilized further to discriminate the actual responses and potentials of plants. Tissue culture techniques offer controlled conditions in terms of light, nutrients, humidity requirements, differentiation patterns, and hormone levels

(Doran 2009). Exploring plant tissue culture techniques for phytoremediation of various toxicants, on the other hand, has limitations in terms of the actual load and rate of flow, i.e., hydraulics of the pollutant during on-site phytoremediation are entirely different and fluctuate regularly. Use of tissue culture-grown plants can be used for providing basic information on mechanisms of metabolism during phytoremediation processes. However, it is clearly inadvisable to use plant tissue culture for remediation purposes.

13.9.1 Whole-Plant *In Vitro* Cultures for Dye Degradation

Application of *in vitro*-grown whole plantlets seems to be best suited for understanding the role of plants and their actual mechanism of phytoremediation of dyes. In a whole-plantlet form, translocation of a pollutant from roots to various shoot parts of a plant can show the fate of metabolic products of a pollutant. Factual insights into phytoremediation mechanisms can thus be studied by using whole-plant *in vitro* cultures. Although roots are the first tissues to come in contact with dyes after exposure, further metabolism of a dye can occur partially in the shoot tissues (Nilratnisakorn et al. 2007). Standardization of different parameters can also be done with the use of whole plants. Whole-plant *in vitro* systems are used in initial optimization studies, and this primary information is useful in obtaining important information for further pilot-scale applications. A few *in vitro*-grown plant species, like *B. malcolmii, T. flagelliforme, Portulaca grandiflora, T. patula, S. portulacastrum, G. pulchella, G. grandiflora, Petunia grandiflora, Z. angustifolia, Nopalea cochenillifera,* and *I. hederifolia,* have been utilized for degradation of dyes like Brilliant Blue R, Remazol Red, Remazol Black B, Reactive Red 198, Direct Red 5B, Navy Blue HE2R, Green HE4B, and Reactive Blue 160 (Kagalkar et al. 2009, 2011, Patil et al. 2009, 2012, Khandare et al. 2011a, 2012, 2013a, Adki et al. 2012, Kabra et al. 2013, Watharkar et al. 2013a,b, Rane et al. 2014). As whole-plant systems possess greater tolerance to a pollutant, they are more suitable and favorable for treatment of dyes.

13.9.2 Hairy Root Cultures for Dye Degradation

As a plant takes up nutrients from it roots and then transports it to other organs, pollutants also first come in contact with roots and tissues that are present in the vicinity of roots. Understanding the phenomenon of the primary interaction of the roots with pollutants is crucial, as subsequent metabolism clearly relies on initial uptake. Hairy root culture systems have previously been employed for phytoremediation of pharmaceuticals, radionuclides, heavy metals, trinitrotoluene, polychlorobiphenyl compounds, and phenolics (Guillon et al. 2006). Phytoremediation studies (Patil et al. 2009) showed the potential of *T. patula* hairy roots in degradation of Reactive Red 198. In the same study, hairy roots of *Nicotiana tabacum, Solanum xanthocarpum,* and *Solanum indicum* decolorized Reactive Red 198 (20 mg/L) up to 95, 96, and 86%, respectively, within 12–30 days. Methyl Orange were also found to be decolorized in an independent study with *B. juncea* hairy root cultures. Practically, use of hairy roots for remediation of textile dyes at higher concentrations and on larger scales nevertheless remains uncertain (Govindwar and Kagalkar 2010).

13.9.3 Callus and Suspension Cultures for Dye Degradation

Leaf wax, epidermis, endodermis, cuticles, and bark are the regulatory parts of plants that control and regulate penetration and the entry of foreign substances from the environment. Use of callus and suspension cell cultures favors the overall rise in phytoremediation of contaminant, as these cultures are devoid of barriers possessed by whole plants and thus the cells come into direct contact with pollutants and act on them. As callus and suspension cultures are not differentiated, their translocation processes are indeterminate and hence the uptake of external compounds is more uniform than in whole plants (Doran 2009). Uptake of pollutants is generally higher in free cells than in complete plants, as they provide uniform exposure to the contaminant. Suspension cells are relatively homogenous and can be used easily to standardize various procedures. This improves the chances of reproducibility of results, compared with those with a complete plantlet (Doran 2009). Cell suspension culture of *Rheum palmatum* has been used for the study of accumulation of sulfonated anthraquinones (Duc et al. 1999). Kagalkar et al. (2011) showed the use of suspension cells of *B. malcolmii* for degradation of a triphenylmethane dye Malachite Green. Cell cultures of *N. cochenillifera* have also been shown to independently transform various toxic textile dyes, like Red HE7B, Orange M2R, Malachite Green, Green HE4BD, Navy Blue HE2R, Toluene Blue, Golden Yellow HER, and Methyl Orange (Adki et al. 2011). Phytoremediation potentials of various plant tissue cultures in treatment of textile dyes with initial dye concentrations and the percent decolorization are listed in Table 13.2.

13.10 Need for Large-Scale Applications of Phytoremediation Technology through Constructed Wetlands

A great deal of research has been performed on phytoremediation of a variety of pollutants at the laboratory-scale level, but only a few studies have been applied on actual contaminated sites. The development of constructed wetlands, especially for phytoremediation of textile wastewater or to remediate a polluted soil matrix, can take us a step closer to make this application as efficient phytotechnology with potent plant species on the actual sites of contamination, as shown in Table 13.3. Experiments performed in the laboratory are carried out under controlled conditions. The behavior and efficiency of the system when applied at the actual site of contamination may be nearer to experimental or pilot-scale results or may be completely variable due to other environmental factors.

13.11 Constructed Wetlands

A constructed wetland or wet park is an artificially constructed wetland used for the purpose of restoring habitat for treatment of wastewater and land reclamation after ecological disturbances. CWs for wastewater treatment are engineered wetlands which receive and purify wastewater of various types, based on the naturally occurring treatment processes. Moreover, CWs are engineered systems specially designed and created to utilize the natural processes concerning wetland vegetation, soils, and the associated microbial assemblages to help in treating wastewaters within a more controlled environment (Vymazal 2010).

Constructed wetlands have been employed widely for the treatment of municipal, industrial, and agricultural wastewater, as well as for urban storm water. This technology is now accepted universally, based on its high nutrient absorption capacity, simplicity, low

TABLE 13.2

Phytoremediation Performance of Various Tissue Cultures for Removal of Textile Dyes

Sr. No.	Plant Used	Tissue Culture Type	Dye or Effluent (Conc, mg/L)	Decolorization Time (hours)	% Decolorization	Reference(s)
1	*Ipomoea hederifolia*	Whole plant	Scarlet RR (50)	96	90	Rane et al. 2014
2	*Gaillardia grandiflora*	Whole plant	Simulated dye mixture	36	62	Watharkar and Jadhav 2014
3	*Physalis minima*	Hairy roots	Reactive Black (830)	120	76	Jha et al. 2014
4	*Glandularia pulchella*	Whole plant	Scarlet RR (50)	48	97	Kabra et al. 2013
5	*Petunia grandiflora*	Whole plant	Brilliant Blue G and Navy Blue RX (20)	36	86, 80	Watharkar et al. 2013a, 2013b
6	*Tagetes patula*	Whole plant	Reactive Blue (160)	96	90	Patil et al. 2013
7	*Zinnia angustifolia*	Whole plant	Remazol Black B (20)	72	94	Khandare et al. 2012
8	*Portulaca grandiflora*	Whole plant	Reactive Blue (172), Direct Red 5B (20)	40, 96	98, 92	Khandare et al. 2011a, 2013a
9	*Nopalea cochenillifera*	Cell culture	Malachite Green, Red HE7B (40)	91, 65	168	Adki et al. 2012
10	*Sesuvium portulacastrum*	Whole plant	Green HE4B (30)	120	70	Patil et al. 2012
11	*Brassica juncea*	Hairy roots	Methyl Orange (20)	96	92	Telke et al. 2011
12	*Blumea malcolmii*	Cell suspension	Malachite Green (20)	24	93	Kagalkar et al. 2011
13	*Typhonium flagelliforme*	Whole plant	Brilliant Blue R	96	80	Kagalkar et al. 2010
14	*Blumea malcolmii*	Whole plant	Direct Red 5B (20)	72	60	Kagalkar et al. 2009
15	*Tagetes patula*	Hairy roots	Reactive Red 198 (110)	288	95	Patil et al. 2009

Source: Adopted and modified from Khandare, R., Govindwar, S. *Biotechnology Advances* 33:1697–1714, 2015.

TABLE 13.3

Constructed Wetlands from Around the World for Textile Wastewater Treatment

Type of CW	Size	Plant(s) and Bacteria Utilized	Reference
CW for textile wastewater	Length 10 m, height 1 m, width 6 m (60,000-L volume)	*Ipomoea aquatica*	Rane et al. 2016
CW for textile wastewater	Length 7 m, height 2 m, width 5 m (52,500-L volume)	*Salvinia molesta*	Chandanshive et al. 2016
Vertical subsurface flow augmented bioreactor (EXP)[a]	Length 46 cm, height 18 cm, width 29 cm	*Portulaca grandiflora* augmented with *Pseudomonas putida*	Khandare et al. 2013a
Vertical subsurface flow augmented bioreactor (EXP)	Length 40 cm, height 20 cm, width 30 cm	*Glandularia pulchella* augmented with *Pseudomonas monteilii* ANK	Kabra et al. 2013
Hydroponic continuous flow phytotunnel system (EXP)	Length 1.6 m, diameter 1 cm	*Portulaca grandiflora*	Khandare et al. 2013b
CW for textile wastewater (EXP)	1000 mL	*Eichhornia crassipes*	Shivkumar et al. 2013
CW with bagasse and sand media (EXP)	20 L	*Phragmites karka* augmented with *Bacillus* sp.	Goyal et al. 2009
Horizontal subsurface flow CW	10 × 5 m	*Phragmites australis*	Davies and Cottingham 1992
Engineered wetland systems	200 L	Cattail, coco yam	Mbuligwe 2004
Hybrid CW (EXP)	VF beds: 5 × 4 × 0.6 m HF bed: 8 × 5 × 0.5 m	*Phragmites australis*	Bulc and Ojstršek 2007
Up Flow CW (EXP)	Diameter 18 cm, height 70 cm	*Phragmites australis*	Ong et al. 2008
Vertical Flow CW (pilot scale)	0.96 m² × 0.87 m	*Phragmites australis*	Carias et al. 2008
Vertical Flow CW (pilot scale)	Length 0.8 m, height 0.6 m, width 0.6 m	Narrow-leaved cattails	Nilratnisakorn et al. 2009

[a] EXP, experimental CW.

construction, operational, and maintenance costs, low energy demand, process stability, and low sludge production, and they provide potential places for generating biodiversity (Korkusuz et al. 2005). Properly designed and constructed artificial wetland ecosystems are tremendously efficient at utilizing and cleaning nutrient-rich wastewaters (Mitsch and Gosselink 1993). In addition, this process has earned greater than ever acceptance for many types of bioremediation, including mining and agribusiness wastewater (Hammer 1989, U.S. EPA 1993, Reed et al. 1995).

13.11.1 Types of CWs

CW for wastewater treatment can be classified according to the wetland hydrology employed, i.e., free water surface flow and subsurface flow CW systems (Figure 13.5).

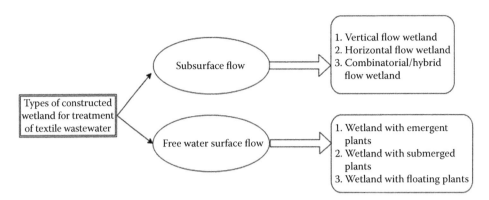

FIGURE 13.5
Classification of constructed wetlands for wastewater treatment.

13.11.1.1 Free Water Surface Flow CW

Free water surface flow CWs are typically classified as (i) a free-floating system and (ii) a horizontal surface flow system. Surface flow systems require less land area for water treatment and are generally suitable for wildlife habitats. Free water surface flow CWs comprise shallow channels or basins, with a sealed bottom to prevent wastewater seepage to the underlying aquifer. They contain a soil layer of up to 40 cm, where the macrophytes (usually emergent, but also submerged or floating) are planted. The water flows nearly horizontally at low velocity above the soil layer along the system, creating a water column depth of 20–40 cm (Vymazal et al. 2006) or even up to 80 cm (Akratos et al. 2006, Crites et al. 2006); therefore, the water is exposed to the atmosphere and to partial sunlight. The water level is maintained with an appropriate outlet level control arrangement. The water flows through the wetland bed and comes into contact with the soil grains and plant parts, thus enabling a series of physical, biological, and chemical processes to take place, which contribute to the degradation and removal of various pollutants.

- Free floating system: This system is generally planted with various vegetation forms like macrophyte and floating leaved and submerged macrophytes (Figure 13.6).

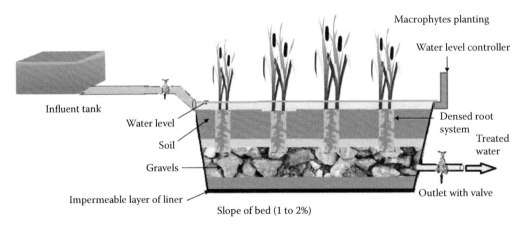

FIGURE 13.6
Schematic diagram of a constructed wetland with free surface water.

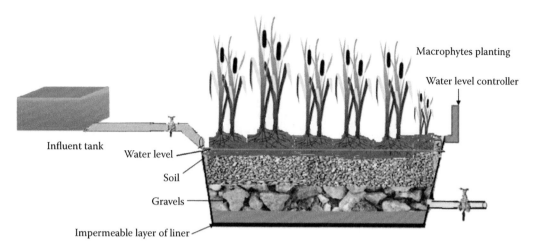

FIGURE 13.7
Diagram of a constructed wetland with horizontal subsurface flow.

- Horizontal surface flow system: This system consists of a basin or channels with some type of barrier to prevent seepage, soil to support the roots of the emergent vegetation, and water at a relatively shallow depth flowing through the system (Figure 13.7). The water surface is open to the atmosphere, and the projected flow path of water through the system is horizontal (Rani et al. 2011).

13.11.1.2 Subsurface Flow CW

Subsurface flow CWs can be further classified on the basis of the flow direction, i.e., horizontal or vertical (Vymazal and Kröpfelová 2008). A subsurface flow system is generally planted with rooted emergent plants (Brix and Schierup 1989). Subsurface flow wetlands move effluent through a gravel or sand medium on which plants are rooted. The medium support the root structure of the emergent vegetation (Rani et al. 2011). The design of these systems assumes that the water level in the bed will remain below the top of the rock or gravel media. Subsurface flow wetlands are advantageous over free surface water system as they do not cause odor problems and do not support growth of mosquitoes, whose populations can be a problem in surface flow constructed wetlands, because the water surface is directly exposed to environment.

- Horizontal subsurface flow CW: In a subsurface flow system, the effluent can move horizontally, parallel to the surface. Davies and Cottingham (1992) described the use of a horizontal subsurface flow system for a textile wastewater system.
- Vertical subsurface flow: In this type of CW, the effluent moves vertically from the planted layer down through the substrate and out. Phytoremediation of textile effluents containing azo dye by using *Phragmites australis* in a vertical flow intermittent feeding constructed wetland has been performed (Davies et al. 2005) (Figure 13.8).

13.11.1.3 Hydroponic System

Hydroponics is also called as aquaculture, a method of growing plants using mineral nutrient solutions prepared in water specifically in the absence of soil. Terrestrial plants

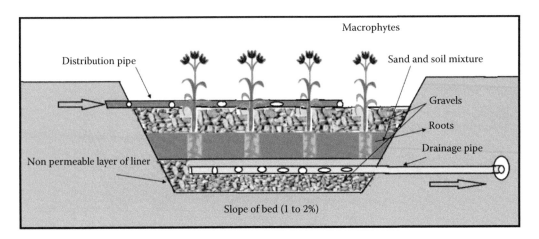

FIGURE 13.8
Constructed wetlands with vertical flow.

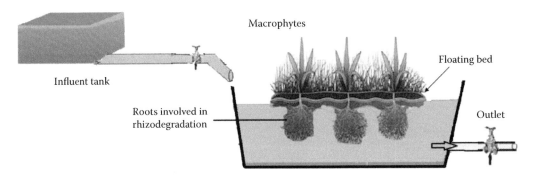

FIGURE 13.9
Schematic representation of a floating treatment wetland.

may be developed with their roots in the mineral nutrient solution only or in an inert medium, such as perlite, gravel, biochar, mineral wool, expanded clay pebbles or coconut husk. Use of hydroponics for wastewater treatment is a novel strategy in the bioremediation field. Hydroponics has the advantage that it prevents the pollution of soil, as it does not require soil as a substratum and as a nutrient for the growth of plants (Figure 13.9).

- Static solution system: This type of hydroponic system utilizes plant roots are growing in contact with wastewater at static condition and bioremediation of the desired pollutants is performed.
- Continuous flow solution system: This is hydroponic system where plant roots deepen in continuous flowing wastewater for the treatment. Khandare et al. (2013) have developed a cost-effective phyto-tunnel system for treatment real textile effluent using plant *Portulaca grandiflora*.

13.11.1.4 Hybrid Systems

Hybrid systems are "optimized" constructed wetlands to achieve a better treatment outcome by using advantages of individual systems (Figure 13.10). Most hybrid constructed

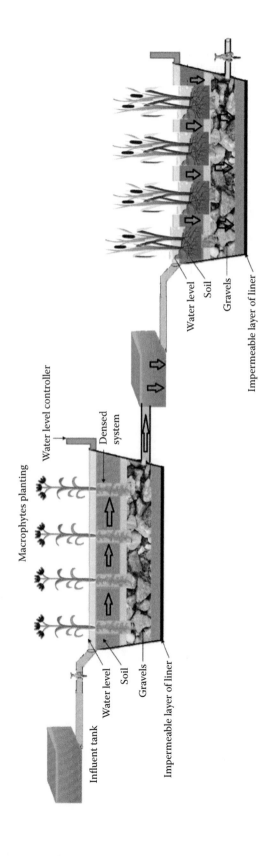

FIGURE 13.10
Combined horizontal and vertical subsurface flow system.

wetlands combine vertical subsurface flow and horizontal subsurface flow systems (Vymazal 2005). The concept is to exploit the advantages of the one type to counterbalance the disadvantages of the other. Thus, the fact that horizontal subsurface flow systems have lower nitrification capacity due to limited oxygen transfer capacity can be offset with vertical flow CWs which are more effective in nitrification (higher oxygen transfer capacity) (Vymazal et al. 1998). On the other hand, horizontal subsurface flow CWs provide good conditions for denitrification, contrariwise to vertical flow CWs. Dye rich textile wastewater has been treated with hybrid system combined with horizontal and vertical subsurface flow planted with *Phragmites australis* (Bulc and Ojstrsek 2008).

13.11.1.5 On-Site Application for Treatment of Primary Treated Textile Effluent by Lagoon System

A total of 60,000 L of effluent treatment for 8 days by *I. aquatica* was carried out (Figure 13.11). The effluent after treatment in a lagoon for 192 h revealed significant reductions in important environmental safety parameters. The effluent before treatment had a specific pungent odor which completely disappeared after the treatment in the lagoon. Other important parameters, like organic carbon, total nitrogen, total phosphorus, ferrous, chlorides, and sulfates were reduced after treatment. Metals from textile effluent were also removed by *I. aquatica* effectively. The results obtained with the lagoon system were encouraging, as healthy growth of *I. aquatica* was observed, as well as increased root lengths, formation of denser root mats, and increased vegetative growth.

FIGURE 13.11
A macrophyte, *Ipomoea aquatica*, in a lagoon of approximately 60,000 L.

13.12 Commercial Feasibility of Phytoremediation Projects

In the Untied States, phytoremediation of various pollutants involving all the contaminant types has been demonstrated and carried out commercially on a pilot scale at nearly 200 sites (Glass et al. 1999, Shekhar et al. 2004). A rising concern over for a safe and sustainable environment has created a huge need globally for such eco-friendly techniques within a practical commercial setup. Several universities, government bodies, research institutes, and private companies are cooperating with each other to develop large-scale economically viable projects for cleanup of toxicants contaminating various sites. Such efforts and practices are, however, confined to developed countries which are getting better public perception and pressure for the sustainable eco-friendly developmental projects. Other parts of the world, especially most developing countries, are yet to be adequately alerted to the cause of the environmental pollution and the sustainable development which is a task ahead. It is evident, that phytotechnology will receive momentum throughout the world, as there is no better options ahead of us to treat contaminated water, air, and land sites.

Contaminated sites are creating a high risk of health hazards to humans and other living beings and also damaging the green cover and plant productivity enormously by polluting the land. Large-scale phytoremediation of contaminated sites has been achieved for organic xenobiotics, radionuclides, and heavy metals (Schwitzguébel et al. 2002, Schwitzguébel 2004). Developing a commercial phytoremediation strategy needs attention to both preharvest steps (e.g., contaminant-level plant selection, agro-climatic suitability of phytoremediators, decontamination rates, monitoring, groundwater capture zone, transpiration rate, and required cleanup time) and postharvest processing steps (e.g., harvestable biomass collection, underground residues disposal, leftovers, and removal of the contaminated plant materials). Phytoremediation techniques can be applied to a broad range of toxicants, with minimal environmental disturbances and less secondary air or water waste compared to other traditional physico-chemical methods. The organic contaminants may ideally be degraded and mineralized to CO_2 and H_2O, reducing environmental toxicity. It is always beneficial when treating large volumes of water, air, or land to have low to moderate concentrations of the contaminants. During land reclamation using phytoremediation, the top soil is left in a usable condition and may be developed for agricultural use, as the soil remains intact at the site after contaminants have been removed, in contrast to conventional methods. Rhizosphere amendments with chelators and enhancers, bacteria, and mycorrhizae have been used to enhance bioavailability of the contaminants to the remediating plants for large-scale remediation strategies (Chaudhary et al. 1998, Khan et al. 2000, Thangavel and Subburaam 2004, Schwitzguébel 2004). Rhizosphere manipulations to deal with various layers and depths of the contaminants and to provide subsurface aeration, for example, have been provided in some systems developed by companies dealing with this technology. Though hybrid poplar willows (*Salix* sp.), clover, Indian mustard, alpine pennycress (*Thlaspi* sp.), grasses, ferns (*Pteris vittala*), perennial ryegrass, sunflowers, geraniums, red root pig weed, etc., have been plants of choice for many commercial phytoremediation systems, several new plants with higher efficiency and better suitability for phytoremediation can be found with the extensive phytoprospecting of new contaminated sites. In addition, genetically modified superior quality phytoremediators can be developed to handle specific situations. A large number of large-scale demonstration and treatment projects have established the commercial viability of phytoremediation through constructed wetlands as a sustainable, ecological, and viable treatment technology at the present and for the future.

13.13 Concluding Remarks

Phytoremediation, the use of plants for cleaning up contaminated sites is an *in situ*, solar-powered remediation technology that requires minimal site disturbance and maintenance, resulting in a low cost and high public acceptance. Plants possess inherent enzymatic and uptake processes that can break down or degrade organic pollutants, and they contain and stabilize contaminants by acting as filters or traps. Plant-associated microflora as well uncultivable microorganisms can be cultivated in soil source for dye degradation. Plants from actual sites of contamination have an aquatic habitat possessing a creeping and layering root system that forms interwoven dense mats over the surface of the water. Use of macrophytes appears to be more logical for treatments of large volumes of effluents. Large biomasses of macrophytes are additionally advantageous as they facilitate greater accumulation and subsequent degradation of the dyes. Macrophytes can grow normally in aquatic environments and are able to handle fluctuating dye loads. Also, because it is costly to produce root biomass through transformation for environmental cleanup purposes, the use of adventitious roots appears to be a novel and promising approach to improve phytoremediation.

Studies at actual sites of dye disposal and full pilot-scale studies find more avenues of research in the field of textile dye removal from wastewaters. In order to look for better replacements, combinatorial plant–plant or plant–microbe systems should be undertaken on larger scales and on fields. Phytoremediation using garden ornamental plants needs to be motivated, as this aspect adds aesthetics to treatment systems; bacterial augmentation can be used to enhance the treatment efficacies of the plants. Moderate prior physico-chemical treatments of textile effluents before actual exposure to plants are attractive alternative, which would contribute to the longevity of this phytotechnology and enhance remediation. Tree vegetation in high-rate transpiration systems also holds possibilities in treatment of huge volumes textile effluents. Such more approaches of field application and trials for on-site treatments are needed for finding solutions for dye effluent management.

References

Abadulla, E., Tzanov, T., Costa, S., Robra, K., Cavaco-Paulo, A., Gubitz, G. 2000. Decolorization and detoxification of textile dyes with a laccase from *Trametes hirsuta*. *Applied and Environmental Microbiology* 66:3357–3362.

Abhilash, P.C., Powell, J.R., Singh, H.B., Singh, B.K. 2012. Plant-microbe interactions: Novel applications for exploitation in multipurpose remediation technologies. *Trends in Biotechnology* 30:416–420.

Adki, V., Shedbalkar, U., Jagtap, U., Jadhav, J., Bapat, V. 2011. Detoxification of a carcinogenic paint preservative by *Blumea malcolmii* Hook cell cultures. *Journal of Hazardous Materials* 191:150–157.

Adki, V., Jadhav, J., Bapat, V. 2012. Exploring the phytoremediation potential of cactus (*Nopalea cochenillifera* Salm. Dyck.) cell cultures for textile dye degradation. *International Journal of Phytoremediation* 14:554–569.

Aken, B. 2008. Transgenic plants for phytoremediation: Helping nature to clean up environmental pollution. *Trends in Biotechology* 5:225–227.

Akratos, C.S., Tsihrintzis, V.A., Pechlivanidis, I., Sylaios, G.K., Jerrentrup, H. 2006. A free water surface constructed wetland for the treatment of agricultural drainage entering Vassova Lagoon, Kavala, Greece. *Fresenius Environmental Bulletin* 15:1553–1562.

Alexandros, S., Christos, A., Vassilios, T. 2014. *Vertical Flow Constructed Wetlands.* Amsterdam: Elsevier Science.

Anjana, S., Thanga, V. 2011. Phytoremediation of synthetic textile dyes. *Asian Journal of Microbiology Biotechnology and Environmental Science* 13:30–39.

Aubert, S., Schwitzguébel, J. 2004. Screening of plant species for the phytotreatment of wastewater containing sulphonated anthraquinones. *Water Research* 38:3569–3575.

Bae, J., Freeman, H. 2005. Aquatic toxicity evaluation of copper complexed direct dyes to the *Daphnia magna*. *Dye Pigment* 73:126–132.

Birhanli, A., Ozmen, M. 2005. Evaluation of the toxicity and teratogenity of six commercial textile dyes using the frog embryo teratogenesis assay: *Xenopus*. *Drug and Chemical Toxicology* 1:51–65.

Bloemberg, G.V., Lugtenberg, J. 2001. Molecular basis of plant growth promotion and biocontrol by rhizobacteria. *Current Opinion in Plant Biology* 4:343–350.

Bulc, T., Ojstrsek, A. 2008. The use of constructed wetland for dye-rich textile wastewater treatment. *Journal of Hazardous Materials* 155:76–82.

Carias, C., Novais, M., Martins-Dias, S., Novais, J. 2008. Are *Phragmites australis* enzymes involved in the degradation of the textile azo dye acid orange 7? *Bioresource Technology* 99:243–251.

Chandanshive, V.V., Rane, N.R., Gholave, A.R., Patil, S.M., Byong-Hun, J., Govindwar, S.P. 2016. Efficient decolorization and detoxification of textile industry effluent by *Salvinia molesta* in lagoon treatment. *Environmental Research* 150:88–96.

Chaudhary, T.M., Hayes, W.J., Khan, A.G., Khoo, C.S. 1998. Role of plants, mycorrhizae and phytochelators in heavy metal contaminated land remediation. *Chemosphere* 21:197–207.

Chen, H., Burns, L. 2006. Environmental analysis of textile products. *Clothing Textile Research Journal* 24:248–261.

Chequer, F., Angeli, J., Ferraz, E., Tsuboy, M., Marcarini, J., Mantovani, M., Oliveira, D. 2009. The azo dyes disperse red 1 and disperse orange 1 increase the micronuclei frequencies in human lymphocytes and in HepG2 cells. *Mutation Research* 676:83–86.

Couto, S.R. 2009. Dye removal by immobilised fungi. *Biotechnology Advances* 27:227–235.

Crites, R.W., Middlebrooks, J., Reed, S.C. 2006. *Natural Wastewater Treatment Systems.* Boca Raton, FL: CRC Press.

Davies, L., Carias, C., Novais, J., Martins-Dias, S. 2005. Phytoremediation of textile effluents containing azo dye by using *Phragmites australis* in a vertical flow constructed intermittent feeding constructed wetland. *Ecological Engineering* 25:594–605.

Davies, L., Ferreira, R., Carias, C., Novais, J. 2009. Integrated study of the role of *Phragmites australis* in azo-dye treatment in a constructed wetland: From pilot to molecular scale. *Ecological Engineering* 5:961–970.

Davies, T., Cottingham, P. 1992. The use of constructed wetlands for treating industrial effluent, pp. XXX. *Proceedings of the 3rd International Conference on Wetland Systems in Water Pollution Control.* Sydney, Australia: IAWQ and Australian Water and Wastewater Association.

Doran, P. 2009. Application of plant tissue cultures in phytoremediation research: Incentives and limitations. *Biotechnology Bioengineering* 103:60–76.

Duc, R., Vanek, T., Soudek, P., Schwitzguébel, J. 1999. Accumulation and transformation of sulfonated aromatic compounds by rhubarb cells (*Rheum palmatum*). *International Journal of Phytoremediation* 1:255–271.

Ferraz, E., Grando, M., Oliveira, D. 2011. The azo dye Disperse Orange 1 induces DNA damage and cytotoxic effects but does not cause ecotoxic effects in *Daphnia similis* and *Vibrio fischeri*. *Journal of Hazardous Materials* 192:628–633.

Gerhardt, K.E., Huang, X.D, Glick, B.R., Greenberg, B.M. 2009. Phytoremediation and rhizoremediation of organic soil contaminants: Potential and challenges. *Plant Science* 176:20–30.

Glass, J.D. 1999. *US and International Markets for Phytoremediation 1999–2000*.

Glick, B.R. 2010. Using soil bacteria to facilitate phytoremediation. *Biotechnology Advances* 28:367–374.

Glick, B.R. 2003. Phytoremediation: Synergistic use of plants and bacteria to clean up the environment. *Biotechnology Advances* 21:383–393.

Govindwar, S., Kagalkar, A. 2010. *Phytoremediation technologies for the removal of textile dyes: An overview and future prospectus*, pp. 1–53. New York: Nova Science Publishers Inc.

Gramss, G., Rudeschko, O. 1998. Activities of oxidoreductase enzymes in tissue extracts and sterile root exudates of three crop plants, and some properties of the peroxidase component. *New Phytologist* 138:401–409.

Guillon, S., Trémouillaux-Guiller, J., Pati, P.K., Rideau, M., Gantet, P. 2006. Hairy root research: Recent scenario and exciting prospects. *Current Opinion in Plant Biology* 9:341–346.

Hammer, D. 1989. *Constructed Wetlands for Wastewater Treatment: Municipal, Industrial and Agricultural*. Chelsea, MA: Lewis Publishers.

IARC. 1982. *Monographs on the Evaluation of the Carcinogenic Risk of Chemicals to Humans, Suppl. 4. Chemicals, Industrial Processes and Industries Associated with Cancer in Humans*. Lyon: IARC.

Kabra, A., Khandare, R., Govindwar, S. 2013. Development of a bioreactor for remediation of textile effluent and dye mixture: A plant-bacterial synergistic strategy. *Water Research* 47:1035–1048.

Kabra, A., Khandare, R., Kurade, M., Govindwar, S. 2011a. Phytoremediation of a sulphonated azo dye Green HE4B by *Glandularia pulchella* (Sweet) Tronc. (moss verbena). *Environmental Science and Pollution Research* 18:1360–1373.

Kabra, A., Khandare, R., Waghmode, T., Govindwar, S. 2011b. Differential fate of metabolism of a sulfonated azo dye remazol orange 3r by plants *Aster amellus* Linn., *Glandularia pulchella* (Sweet) Tronc. and their consortium. *Journal of Hazardous Materials* 190:424–431.

Kabra, A., Khandare, R., Waghmode, T., Govindwar, S. 2012. Phytoremediation of textile effluent and mixture of structurally different dyes by *Glandularia pulchella* (Sweet) Tronc. *Chemosphere* 87:265–272.

Kagalkar, A., Jadhav, M., Bapat, V., Govindwar, S. 2011. Phytodegradation of the triphenylmethane dye Malachite Green mediated by cell suspension cultures of *Blumea malcolmii* Hook. *Bioresource Technology* 102:10312–10318.

Kagalkar, A., Jagtap, U., Jadhav, J., Govindwar, S., Bapat, V. 2010. Studies on phytoremediation potentiality of *Typhonium flagelliforme* for the degradation of Brilliant Blue R. *Planta* 232:271–285.

Kagalkar, A., Jagtap, U., Jadhav, J., Bapat, V., Govindwar, S. 2009. Biotechnological strategies for phytoremediation of the sulphonated azo dye direct red 5b using *Blumea malcolmii* Hook. *Bioresource Technology* 100:4104–4110.

Khan, A., Kuek, C., Chaudhary, T., Khaoo, C., Hayes, W. 2000. Role of plants, mycorrhizae and phytochelators in heavy metal contaminated land remediation. *Chemosphere* 21:197–207.

Khandare, R., Govindwar, S. 2015a. Microbial degradation mechanism of textile dye and its metabolic pathway for environmental safety, pp. 399–439. *Environmental Waste Management*, Chandra, R. (ed.). Boca Raton: Taylor & Francis Group.

Khandare, R., Govindwar, S. 2015b. Phytoremediation of textile dyes and effluents: Current scenario and future prospects. *Biotechnology Advances* 33:1697–1714.

Khandare, R., Kabra, A., Kadam, A., Govindwar, S. 2013a. Treatment of dye containing wastewaters by a developed lab scale phytoreactor and enhancement of its efficacy by bacterial augmentation. *International Biodeterioration and Biodegradation* 78:89–97.

Khandare, R., Watharkar, A., Kabra, A., Kachole, M., Govindwar, S. 2013b. Development of a low-cost, phyto-tunnel system using *Portulaca grandiflora* and its application for the treatment of dye-containing wastewaters. *Biotechnology Letters* 36:47–55.

Khandare, R., Rane, N., Waghmode, T., Govindwar, S. 2012. Bacterial assisted phytoremediation for enhanced degradation of highly sulfonated diazo reactive dye. *Environmental Science and Pollution Research* 19:1709–1718.

Khandare, R., Kabra, A., Kurade, M., Govindwar, S. 2011b. Phytoremediation potential of *Portulaca grandiflora* Hook. (Moss-Rose) in degrading a sulfonated diazo reactive dye Navy Blue HE2R (Reactive Blue 172). *Bioresource Technology* 102:6774–6777.

Khandare, R., Kabra, A., Tamboli, D., Govindwar, S. 2011a. The role of *Aster amellus* Linn. in the degradation of a sulfonated azo dye Remazol Red: A phytoremediation strategy. *Chemosphere* 82:1147–1154.

Khataee, A., Dehghan, G., Zarei, M., Fallah, S., Niaei, G., Atazadeh, I. 2013a. Degradation of an azo dye using the green macroalga *Enteromorpha* sp. *Journal of Chemical Ecology* 29:37–41.

Khataee, A., Movafeghi, A., Vafaei, F., Lisar, S.Y.S., Zarei, M. 2013b. Potential of the aquatic fern *Azolla filiculoides* in biodegradation of an azo dye: Modeling of experimental results by artificial neural networks. *International Journal of Phytoremediation* 15:729–742.

Khataee, A., Movafeghi, Torbati, S., Salehi Lisar, S.Y., Zarei, M. 2012. Phytoremediation potential of duckweed (*Lemna minor* L.) in degradation of C.I. Acid Blue 92: Artificial neural network modeling. *Ecotoxicology and Environmental Safety* 80:291–298.

Khataee, A., Dehghan, G., Ebadi, A., Zarei, M., Pourhassan, M. 2010a. Biological treatment of a dye solution by macroalgae *Chara* sp.: Effect of operational parameters, intermediates identification and artificial neural network modeling. *Bioresource Technology* 101:2252–2258.

Khataee, A., Zarei, M., Pourhassan, M. 2010b. Bioremediation of Malachite Green from contaminated water by three microalgae: Neural network modeling. *CLEAN* 38:96–103.

Korkusuz, E., Beklioglu, M., Demirer, G. 2005. Comparison of the treatment performances of blast furnace slag-based and gravel-based vertical flow wetlands operated identically for domestic wastewater treatment in Turkey. *Ecological Engineering* 24:187–200.

Kurade, M., Waghmode, T., Govindwar, S. 2011. Preferential biodegradation of structurally dissimilar dyes from a mixture by *Brevibacillus laterosporus*. *Journal of Hazardous Materials* 192:1746–1755.

LeFevre, G., Hozalski, R., Novak, P. 2013. Root exudates enhanced contaminant desorption: An abiotic contribution to the rhizosphere effect. *Environmental Science and Technology* 47:11545–11553.

Maddhinni, V., Vurimindi, H., Yerramilli, A. 2006. Degradation of azo dye with horseradish peroxidase (HRP). *Journal of the Indian Institute of Science* 86:507–514.

Mahmood, Q., Masood, F., Bhatti, Z.A., Siddique, M., Bilal, M., Yaqoob, H. 2014. Biological treatment of the dye Reactive Blue 19 by cattails and anaerobic bacterial consortia. *Toxicological and Environmental Chemistry* 96:530–541.

Mbuligwe, S. 2005. Comparative treatment of dye-rich wastewater in engineered wetland systems (EWSs) vegetated with different plants. *Water Research* 39:271–280.

Mittal, A., Mittal, J., Kurup, L., Singh, A. 2006. Process development for the removal and recovery of hazardous dye erythrosine from wastewater by waste materials: Bottom ash and de-oiled soya as adsorbents. *Journal of Hazardous Materials* 138:95–105.

Mitsch, W., Gosselink, J. 1993. *Wetlands*, 2nd ed. New York: Van Nostr and Rheinhold.

Mohan, S., Prasad, K., Rao, N., Sarma, P. 2005. Acid azo dye degradation by free and immobilized horseradish peroxidase (HRP) catalyzed process. *Chemosphere* 58:1097–1105.

Nilratnisakorn, S., Thiravetyan, P., Nakbanpote, W. 2007. Synthetic reactive dye wastewater treatment by narrow-leaved cattails (*Typha angustifolia* Linn.): Effects of dye, salinity and metals. *Science of the Total Environment* 384:67–76.

Page, V., Schwitzguébel, J. 2009. The role of cytochromes P450 and peroxidases in the detoxification of sulphonated anthraquinones by rhubarb and common sorrel plants cultivated under hydroponic conditions. *Environmental Science and Polluttion Research* 16:805–816.

Paquin, D., Sun, W., Tang, C., Li, Q. 2006. A phytoremediation study: Selection of tropical and other vascular plants for decolorization of Poly R-478 dye. *Remediation Journal* 16:97–107.

Patil, P., Desai, N., Govindwar, S., Jadhav, J., Bapat, V. 2009. Degradation analysis of Reactive Red 198 by hairy roots of *Tagetes patula* L. (Marigold). *Planta* 230:725–735.

Pilon-Smits, E. 2005. Phytoremediation. *Annual Review in Plant Biology* 56:15–39.

Rane, N., Chandanshive, V., Khandare, R., Gholave, A., Yadav, S., Govindwar, S. 2014. Green remediation of textile dyes containing wastewater by *Ipomoea hederifolia* L. *RSC Advances* 4:6623–36632.

Rane, N., Chandanshive, V., Watharkar, A., Khandare, R., Patil, T., Pawar, P., Govindwar, S. 2015. Phytoremediation of sulfonated Remazol Red dye and textile effluents by *Alternanthera philoxeroides*: An anatomical, enzymatic and pilot scale study. *Water Research* 83:271–281.

Rani, R., Din, M., Yusof, M., Chelliapan, S. 2011. Overview of subsurface constructed wetlands application in tropical climates. *Universal Journal of Environmental Research and Technology* 2:103–114.

Reed, S., Crites, R., Middlebrooks, E. 1995. *Natural Systems for Waste Management and Treatment*, 2nd ed. McGraw-Hill, New York.

Reema, R., Saravanan, P., Kumar, M., Renganathan, S. 2011. Accumulation of methylene blue dye by growing *Lemna minor*. *Separation Science and Technology* 46:1052–1058.

Roy, S., Labelle, S., Meht, P., Mihoc, A., Fortin, N., Masson, C., Leblanc, R., Chateauneuf, G., Sura, C., Gallipeau, C., Olsen, C., Delisle, S., Labrecque, M., Greer, C. 2005. Phytoremediation of heavy metal and PAH-contaminated brownfield sites. *Plant and Soil* 272:277–290.

Sanmuga, P., Senthamil, P. 2014. Water hyacinth (*Eichhornia crassipes*): An efficient and economic adsorbent for textile effluent treatment. A review. *Arabian Journal of Chemistry* 10(Suppl. 2):S3548–S3558.

Saratale, G., Kalme, S., Govindwar, S. 2006. Decolorisation of textile dyes by *Aspergillus ochraceus* (NCIM-1146). *Indian Journal of Biotechnology* 5:407–410.

Saratale, R., Saratale, G., Chang, J., Govindwar, S. 2009a. Ecofriendly degradation of sulfonated diazo dye C.I. Reactive Green 19A using *Micrococcus glutamicus* NCIM-2168. *Bioresource Technolology* 100:3897–3905.

Saratale, R., Saratale, G., Chang, J., Govindwar, S. 2009b. Decolorization and biodegradation of textile dye Navy blue HER by *Trichosporon beigelii* NCIM-3326. *Journal of Hazardous Materials* 166:1421–1428.

Saratale, R., Saratale, G., Chang, J., Govindwar, S. 2011. Bacterial decolorization and degradation of azo dyes: A review. *Journal of the Taiwan Institute of Chemical Engineers* 42:138–157.

Sarkanen, S., Razals, R., Piccariellos, T., Yamamotoa, E., Lewiss, N. 1991. Lignin peroxidase: Toward a clarification of its role *in vivo*. *Journal of Biological Chemistry* 266:3636–3643.

Schwitzguebel, J., Van der Lelie, D., Glass, D., Vangronsveld, J., Baker, A. 2002. Phytoremediation: European and American trends, successes, obstacles, and needs. *Journal of Soils and Sediments* 2:91–99.

Shaffiqu, T., Roy, J., Nair, R., Abraham, T. 2002. Degradation of textile dyes mediated by plant peroxidases. *Applied Biochemistry and Biotechnology* 102–103:315–326.

Sharma, S., Kalpana, Arti, Shweta, Suryavath, V., Singh, P. et al. 2007. Toxicity assessment of textile dye wastewater using swiss albino rats. *Australasian Journal of Ecotoxicology* 13:81–85.

Srivastava, S., Singh, N., Srivastava, A., Sinha, R. 1995. Acute toxicity of Malachite Green and its effects on certain blood parameters of a catfish, *Heteropneustes fossilis*. *Aquatic Toxicology* 31:241–247.

Stammati, A., Nebbia, C., Angelis, I., Albo, A., Carletti, M., Rebecchi, C., Zampaglioni, F., Dacasto, M. 2005. Effects of Malachite Green (MG) and its major metabolite, Leucomalachite Green (LMG), in two human cell lines. *Toxicology in Vitro* 19:853–858.

Stępniewska, Z., Kuźniar, A. 2013. Endophytic microorganisms-promising applications in bioremediation of greenhouse gases. *Applied Microbiology and Biotechnology* 97:9589–9596.

Takahashi, M., Tsukamoto, S., Kawaguchi, A., Sakamoto, A., Morikawa, H. 2005. Phytoremediators from abandoned rice field. *Plant Biotechnology* 22:167-170.

Thangavel, P., Subbhuraam, C. 2004. Role of hyperaccumulators in metal contaminated soils. *Proceedings of the Indian National Science Academy B* 70:109–130.

Tsuboy, M., Angeli, J., Mantovani, M., Knasmuller, S., Umbuzeiro, G., Ribeiro, L. 2007. Genotoxic, mutagenic and cytotoxic effects of the commercial dye CI Disperse Blue 291 in the human hepatic cell line HepG2. *Toxicology in Vitro* 21:1650–1655.

Uera, R., Paz-Alberto, A., Sigua, G. 2007. Phytoremediation potentials of selected tropical plants for ethidium bromide. *Environmental Science and Pollution Research* 14:505–509.

U.S. EPA. 1993. *Subsurface Flow Constructed Wetlands for Wastewater Treatment: A Technology Assessment.* EPA 832-R 93-008. Washington, D.C.: U.S. EPA Office of Water.

Vafaei, F., Khataee, A., Movafeghi, A., Lisar, S., Zarei, M. 2012. Bioremoval of an azo dye by *Azolla filiculoides*: Study of growth, photosynthetic pigments and antioxidant enzymes status. *International Biodeterioration and Biodegradation* 75:194–200.

Vafaei, F., Movafeghi, A., Khataee, A., Zarei, M., Salehi, L. 2013. Potential of *Hydrocotyle vulgaris* for phytoremediation of a textile dye: Inducing antioxidant response in roots and leaves. *Ecotoxicology and Environmental Safety* 93:128–134.

Vymazal, J. 2005. Horizontal sub-surface flow and hybrid constructed wetlands for wastewater treatment. *Ecological Engineering* 25:478–490.

Vymazal, J. 2010. Constructed wetlands for wastewater treatment water. *Water* 2:530–549.

Vymazal, J., Kröpfelová, L. 2008. *Wastewater Treatment in Constructed Wetlands with Horizontal Sub-Surface Flow.* Springer: Dordrecht, The Netherlands.

Vymazal, J., Greenway, M., Tonderski, K., Brix, H., Mander, U. 2006. Constructed wetlands for wastewater treatment, pp. 69–94. *Wetlands and Natural Resource Management*, Verhoeven, J.T.A., Beltman, B., Bobbink, R., Whigham, D.F. (eds.). Berlin: Spinger-Verlag.

Watharkar, A., Khandare, R., Waghmare, P., Jagadale, A., Govindwar, S., Jadhav, J. 2015. Treatment of textile effluent in a developed phytoreactor with immobilized bacterial augmentation and subsequent toxicity studies on *Etheostoma olmstedi* fish. *Journal of Hazardous Materials* 283:698–704.

Watharkar, A., Jadhav, J. 2014. Detoxification and decolorization of a simulated textile dye mixture by phytoremediation using *Petunia grandiflora* and *Gaillardia grandiflora*: A plant–plant consortial strategy. *Ecotoxicology and Environmental Safety* 103:1–8.

Watharkar, A., Rane, N., Patil, S., Khandare, R., Jadhav, J. 2013. Enhanced phytotransformation of Navy Blue RX dye by *Petunia grandiflora* Juss. with augmentation of rhizospheric *Bacillus pumilus* strain PgJ and subsequent toxicity analysis. *Bioresource Technolology* 142:246–254.

Wright, L. 1976. Effect of malachite green and formaline on the survival of largemouth bass eggs and fry. *Progressive Fish Culturist* 38:155–157.

Yahagi, T., Degawa, M., Seino, Y., Matsushima, T., Nagao, M., Sugimura, T., Hashimoto, Y. 1975. Mutagenicity of carcinogenic azo dyes and their derivatives. *Cancer Letter* 1:91–96.

14

Phytotoxicity: An Essential Tool in Ecological Risk Assessment

Rajesh Kumar Sharma, Bhanu Pandey, and Shivani Uniyal

CONTENTS

14.1 Introduction

Man's influence on the ecosphere has been extremely wide and complex and most often has led to irreversible alteration. A growing capability of man to alter his surroundings and to control several natural processes is a source of drastic changes altering the balance of fragile natural systems based upon the symmetric flow of elements and energy. All anthropogenic changes perturb the natural equilibrium of each ecosystem that has been formed evolutionarily over eras. These changes often result in a deterioration of the natural human environment (Salati and Moore 2010). Though the impacts of anthropogenic activities on the environment date back to the Neolithic Period, degradation of ecosystems due to pollutants became increasingly acute during the last decades of the twentieth century.

Population growth, increased urbanization, and advanced standards of human lifestyle have contributed to an increase in both variety and quantity of solid wastes generated by agricultural, industrial, and domestic activities. Industrial wastes have contributed more than 85% of solid-waste generation globally (Bhada-Tata and Hoornweg 2016). Ever-expanding consumption of materials, energy, and space by growing populations is associated with a rising flux of anthropogenic chemicals toxic to the environment. Many toxic chemicals, although applied or introduced to confined locations, become widely dispersed—even to the "ends of the earth." The advancement of recent technology has brought a dramatic increase in the production and consumption of chemicals. In a few cases, the benefits of chemical use have been followed by unexpected adverse effects. Although some of these chemicals are not of direct environmental concern, numerous compounds have been continuously introduced in large quantities. These activities are releasing a variety of extraneous materials and energy into the environment in a continuous and uninterrupted manner.

Anthropogenic pollution is not new, as humans started contaminating the environment with an awareness of controlling fire and smelting metals. The characteristics and distribution of contaminants in the environment has altered in late history as new compounds have been formed. Most early anthropogenic pollution was localized, while early metal smelting, even 2000 years ago, resulted in hemispheric-scale pollution. Due to industrial development, populations concentrated in cities, which led to increased air pollution due to the burning of fossil fuels, and water pollution due to organic pollutants in the form of sewage. These alterations caused illness and disease in humans and killed fish as well as other living organisms in rivers (Davis et al. 2002).

14.2 Phytotoxicity

Phytotoxicity is normally associated with the phenomenon where a potentially harmful substance is accumulated in the plant tissue to a level affecting optimal growth and development of the plant. It can also be described as detrimental deviations from normal growth and appearance pattern of plants in response to a given substance (OECD 2006). According to Chang et al. (1992), the positive confirmation of an incidence of phytotoxicity requires that (i) plants have sustained injuries, (ii) a potentially phytotoxic metal has accumulated in the plant tissue, (iii) the observed abnormalities are not due to other disorders of plant growth, and (iv) the biochemical mechanisms that cause metal toxicity to be harmful to plants are observed during the course of growth.

The current guidelines on metal phytotoxicity are based on total metal concentrations, which are inadequate. Depending on soil type, plants can tolerate significantly higher concentrations without detrimental risk to their lives. The phytoavailability of metals is determined by the nature of the metal species, their interaction with soil colloids, the soil characteristics (e.g., soil pH, clay, organic matter content, and type), moisture content and duration of contact with the surface binding these metals. Soil characteristics also determine availability to plants by controlling the speciation of the elements, temporary binding by particle surfaces (adsorption–desorption processes), precipitation reactions, and availability in soil solution. Variability in metal bioavailability in soil suggests that total metal concentration may not be an appropriate and sensitive indicator for phytotoxicity (Naidu et al. 2003). An appropriate strategy for the guideline may be a two-tiered system—that is, (i) an assessment of total metal concentration followed by (ii) phyto- or bioavailability assessment using a chemical extraction technique. The bioaccumulations of heavy metals in plants have shown the phytotoxic effects on many plant species as reviewed by Sharma and Agrawal (2005) and Nagajyoti et al. (2010).

14.3 Plant Species Used to Determine Phytotoxicity

The most commonly used plant species for the determination of phytotoxicity is garden cress (*Lepidium sativum* L.) as it is easy to manage and shows fast germination and growth. Many other plant species have also been used, among which horticultural crops such as tomato (*Lycopersicum esculentum* Mill), carrot (*Daucus carota* L.), cucumber (*Cucumis sativus* L.), cabbage (*Brassica oleracea*), radish (*Raphanus sativus* L.), lettuce (*lactuca sativa* L.), and bean (*Phaseolus vulgaris* L); cereals such as barley (*Hordeum vulagre* L.), Italian rye grass (*Lolium multiflorum* Lam.), rice (*Oryza sativa* L.), wheat (*Treticum aestivum* L.), rye (*Secale cereal* L.), soya (*Glycine max* L.), or corn (*Zea mays* L.); and sunflowers (*Helianthum annuus* L.), petunia (*Petunia hybrida* L.), and amaranthus (*Amaranthus tricolor* L.) are common. Various studies have been carried out to determine phytotoxicity using these species (Sharma and Agrawal 2006). OECD guideline 208 assesses the potential effects of a test item (i.e., toxic chemical) on seedling emergence and early growth of seedlings of some terrestrial plants such as *Allium cepa* (onion), *Zea mays* (corn), *Brassica napus* (oilseed rape), *Cucumis sativus* (cucumber), *Glycinev max* (soybean), *Helianthus annuus* (sunflower), *Solanum lycopersicon* (tomato), or *Stellaria media* (common chickweed) following exposure to the toxic chemical in or on the soil (OECD 2006).

14.3.1 Approaches for Phytotoxicity

Hazard assessment of contaminated environmental samples, toxicants, and pesticides has relied on a few standardized tests (Krewski et al. 2010). The connection between inherent hazard, based on limited laboratory determinations of toxicity, and ecological risk often poses unacceptable levels of uncertainty. The tools used to obtain the preliminary data for ecological risk assessments (EcoRAs) are classical sampling methods and standardized laboratory toxicity tests (Landis et al. 2003). The combination of toxicity and exposure is constrained by the power of the primary data (Kapustka 1996). The principal deficiencies of standardized tests vis-à-vis EcoRAs relate to exposure conditions, relevant endpoints, interspecies extrapolation, and lab-to-field extrapolation. Fundamental changes that widen the types and ranges of "standardized" tests and expand the suite of measurement endpoints are needed. The next generation of standardized tests must be developed in light of risk assessment requirements and expectations if they are to be effective.

Short-term plant toxicity tests initially were developed from simple measures used in plant physiology and weed science (Chaignon and Hinsinger 2003). The tests have been ratifying estimates of single chemical and mixed chemical effects. Recently, they have been used to assess soil contamination. They are used to test soils brought to the laboratory for ecological assessment of terrestrial waste (Suter 2016) and for comparative hazard ranking between chemicals. The seed germination assay, often approached as testing a sensitive and critical stage in the plant life cycle, is comparatively not sensitive to many toxic substances.

Two factors are responsible for insensitivity: (i) various chemicals may not be taken into the seed, and (ii) the embryonic plant used seed storage materials for its nutritional requirements and is effectively isolated from the environment. Finally, from an ecological point of view, seed germination is comparatively insignificant for perennial plants. Even for nondomesticated annuals, extremely low percentages of seed germination are typical (Suter 2016). Nevertheless, EcoRAs depend widely on standard toxicity data. In simple form, the EcoRA considers the inherent toxicity of individual chemicals on ecological resources, evaluates the likelihood of exposure appropriate to specified pathways in site-specific situations or prescribed scenarios, and combines the toxicity and exposure information to quantify the risk to ecological resources (Suter 2016). Under the latest guidelines, the EcoRA focus has been extended to include nonchemical stressors (e.g., physical disturbance, temperature, drought, or herbivory). Incorporating standard toxicity information into EcoRAs has been difficult. In an ideal situation, toxicity and exposure data would be available for the taxa and endpoints of concern. Usually, neither toxicity nor exposure data are available for the taxa or the endpoints of concern.

14.3.2 Phytotoxicants

Several types of compounds can induce toxicity in plants by inhibiting their growth, seed germination, root elongation, photosynthetic activity, and biomass production (Table 14.1). Thus, specific plant species can be characterized for a particular contaminant based on the phytotoxicity induced and thus can aid in ecological risk assessment.

14.3.2.1 Cyanotoxins

Ever-increasing pollution has quantified the toxin-producing cyanobacteria all through the world. Cyanobacteria are known to synthesize secondary metabolites such as peptides, glycosides, and macrolides (Sun 2016), which can cause phytotoxicity. It has been documented that cyanotoxins inhibit growth of the green alga *Chlorella pyrenoidosa* (Ikawa et al. 2001), reduce photosynthetic oxygen production change pigment pattern in aquatic macrophytes such as *C. demersum* and *Myriophyllum spicatum* (Pflugmacher 2002), inhibit root hair growth and modify the cortical cell shape of *Arabidopsis thaliana* L. at 3 nM (Smith et al. 1994).

14.3.2.2 Nanoparticles

Engineered nanoparticles (NPs) like zinc oxide (ZnO) NPs are widely used as electronic sensors and solar voltaics (Ma et al. 2013). Their accidental or deliberate release into the

TABLE 14.1

Overview of Phytotoxic Compounds and Their Reported Toxicity to Plants

S. No.	Type	Toxicants	Plant	Impact On Plant	Ref.
1.	Cyanobacterial toxin	Microcystin-LR	*P. australis*	Potent and specific inhibitors of protein phosphatases 1 and 2A	Hastie et al. 2005
		Cylindrospermopsin (CYN)	*Nicotiana tabacum*	Inhibition of pollen germination; partial inhibition of protein production in the germinating pollen tubes following exposure to 138 µg mL^{-1} of toxin	Metcalf et al. 2004
2.	Nanoparticles	Copper nanoparticles	*Phaseolus radiates, Triticum aestivum*	Reduced seedling growth rate	Lee et al. 2010
		Carbon nanotubes	*Lycopersicon esculentum*	Affected root elongation at high concentration	Canas et al. 2008
3.	Inorganic	Cr	*Salvinia minima*	Reduction in the concentration of all photosynthetic pigments	Brent Nichols et al. 2000
		Cu, Zn, Cd	*P. oceanica*	Generation of lipid peroxidation	Hamoutene et al. 1996
		Al^{3+}	*Hordeum vulgare*	Reduced Ca^{2+} influx	Nichol and Oliveiram 1995
4.	Organic	Doxycycline	*Triticum aestivum*	Inhibition of photosynthesis and reduction in chlorophyll content	Opris et al. 2013
		Tetracycline	*Lolium perenne*	Reduction in root biomass and phosphorus assimilation reduction	Wei et al. 2009
		Sulfadiazine	*Zea mays*	Reduction in plant growth, root alterations	Michelini et al. 2012
		Sulfamethazine	*Phragmites australis*	Decreased root activity and chlorophyll content	Liu et al. 2013
		Crude oil, VDC dispersant, dispersed oil (1:5)	*Zostera capricorni*	Toxicity at 0.25% crude oil concentration in laboratory; less effect in field enclosures	Macinnis-Ng and Ralph 2003
		Tapis crude oil	*Zostera capricorni*	Temporary toxicity at 0.4% concentration	Wilson and Ralph 2008

environment can cause a potential hazard. Several plant species have been reported to accumulate and posses a phytotoxic effect in the presence of nanoparticles (López-Moreno et al. 2010). However, the phytotoxic effect seems to be dose dependent. ZnO nanoparticles have been reported to cause a significant reduction in the biomass and root elongation in *Lolium perenne* (ryegrass) at a level of 1000 mg L^{-1} and inhibition of seed germination in *Arabidopsis thaliana* (Lee et al. 2010). Similarly, Ma et al. (2010) reported CeO_2 nanoparticles induced phytotoxicity in *P. subcapitata* cells and suggested that local aggregation of CeO_2 nanoparticles around *P. subcapitata* cells resulted in nutrient depletion and shading, which contributed to the potential phytotoxicity.

14.3.2.3 Organic Compounds

The widespread use of pesticides in agriculture for increased crop production can cause direct and indirect hazards to the environment (Uniyal et al. 2016). Some pesticides are persistent in nature and thus provide multiple routes for chronic (and acute, in some cases) exposure of nontargeted organisms (Uniyal et al. 2016). Some pesticides can exert a phytotoxic effect on the basis of enantioselectivity. In a study, enantioselective phytotoxicity of the enantiomers of the herbicide napropamide was investigated (Qi et al. 2015). Results indicated that (−)-napropamide caused greater toxicity to soybean and cucumber in terms of root, shoot, and fresh weight than the racemate and (+)-napropamide.

Some organic compounds can exert toxic effects on plants and thereby aid in their associated ecological risk assessment. For instance, *Arabidopsis thaliana* was reported to show a significant decrease in seed germination, as well as reduction in fresh weight and chlorophyll content in response to 2′,3′,7′,8′-tetrachlorinated dibenzo-*p*-dioxin (TCDD) exposure (Hanano et al. 2014). Similarly, Yan et al. (2015) observed the toxic effects of di-*n*-butyl phthalate (DBP) on algae species—namely, *Chlamydomonas reinhardtii*, *Pseudokirchneriella subcapitata*, and *Gymnodinium breve*. Ecological risk assessment study showed an apparent ecological risk of DBP in the middle Yellow River, the Xuanwu Lake, and the Yuehu Lake of China. Oils and dispersants are also reported to induce phytotoxicity in some plant species. For instance, the phytotoxic effect of diesel fuel oil at a concentration of 0.2%, 2%, and 20% in *Salicornia fragilis* was evaluated (Meudec et al. 2007). Results indicated that oiled sediments were lethal at 2% and 20% and the effects are more pronounced when in contaminated soil than in shoot coating.

Pharmaceutical compounds can also pose a severe threat to the environment by affecting nontarget species. Veterinary drugs can cause potential ecological hazards as they represent a group of chemical contaminants that have biological effects at low concentrations (Arnold et al. 2013). Florfenicol, a veterinary drug belonging to the phenicols, has been reported to inhibit growth of *Lemna minor* (Kolodziejska et al. 2013). Another compound—namely, trimethoprim—was observed to affect *Cichaorium endivia* by inhibiting seed germination (Liu et al. 2009).

14.3.2.4 Inorganic Pollutants

Heavy-metal pollution can pose a serious hazard to health and environment by causing prenatal and developmental defects and can also induce biological effects on aquatic organisms (Uniyal et al. 2016). After discharge from various sources, heavy metals may go into the soil, vegetation, and water, depending on their density. Subsequent to their deposition in different systems, metals cannot be degraded and, as a result, persist in the environment, causing human health complications through inhalation, ingestion, and skin absorption. Heavy metals have freely been used by humans for quite a long time in metal alloys and pigments for paints, cement, paper, rubber, and other materials and their use is rising even now in most parts of the globe despite their well-known unfavorable effects. Acute exposure to metals leads to nausea, anorexia, vomiting, gastrointestinal abnormalities, and dermatitis. Heavy-metal toxicity can also cause decline in central nervous system function and disturb blood composition, kidneys (Reglero et al. 2009), lungs (Kampa and Castanas 2008), livers (Sadik 2008), and other main organs. Additionally, the continuing exposure to heavy metals may gradually impair physical, muscular, and neurological processes similarly to Alzheimer's disease (Kampa and Castanas 2008), Parkinson's disease (Kwakye et al. 2016), muscular dystrophy, and multiple sclerosis (Turabelidze et al. 2008). High exposure also leads to obstructive lung disease and has been related to lung cancer and harm to human respiratory systems. In addition, some metals, like Cu, Se, and Zn (trace elements), are important in maintaining the metabolism of the human body. Cu, for example, is an essential element to human life, but in higher doses, it can cause anemia, liver and kidney damage, and stomach and intestinal irritation.

Contamination of soils by heavy metals followed by human consumption through different agencies like foods, feeds, water, etc. (Khan et al. 2008) has become one of the most serious problems as threatened the precious human health. Therefore, an urgent and collective attempt to clean up the contaminants from surroundings is needed so that the risk of metal toxicity can be minimized. The concern resulting from the potential exposure of populations vulnerable to toxicants has, however, forced workers in various disciplines to act jointly in order to develop methodologies so that the actual effect of heavy metals on both the varying environment and human health could be assessed (Eriyamremu et al. 2005).

Health risk assessment is, in fact, the method of assessing the prospect of damage caused to human beings resulting from exposure to contaminants at a site. Therefore, both the deleterious effects of pollutants and the ways that people may be exposed to these substances are evaluated. In this context, for evaluating the risk due to heavy metals, various workers apply different approaches (Li et al. 2014). However, the role of both risk assessors and decision makers in the evaluation process is most important to understanding the risk assessment. Usually, two approaches can be applied for evaluating the risk of a specific toxicant to any individual population: direct (biological) and indirect (environmental monitoring). Different human bioindicators, like urine and plasma, human hair, and adipose tissue, can be used for observation. Even though these sources may provide real and direct information about how the population is exposed to toxicants, they are variable and depend largely on personal characteristics, such as dietary habits, smoking, weight, etc., rather than on low-level environmental exposures (Zaidi et al. 2012). The chemical analysis of the toxic concentrations originating from different sources like air, soil, vegetation, sediment, etc. may be an interesting indirect methodology for human health risk assessment. However, in order to make chemical methods more viable and effective, this analysis should be complemented with biological and toxicological methods (Gruiz 2005). Considering these, it is generally believed that health risk assessment may play an important role in protecting humans from the effects of heavy metals.

Plants show a phytotoxic response to inorganic contaminants depending on their chemical stage, dose, and physical factors. Heavy metals such as Cd, Pb, Ni, and Cu can induce phytotoxicity in plants at high concentrations. Heavy metals at higher concentrations cause severe damage to various metabolic activities, leading consequently to the death of plants. However, some plant species have the ability to survive in heavily contaminated soils (Deng et al. 2004). Metal at remarkably high concentrations has been documented to damage plant species by (i) reducing physiologically active enzymes (Lemire et al. 2013), (ii) inactivating photosystems, and (iii) damaging mineral metabolism (Gadd 2010). At high concentration, Cd can drastically slow down the radical growth in lettuce (Salvatore et al. 2008) and the growth of *S. nigrum* L., and can decrease concentration of photosynthetic pigments. Cd has also been reported to cause structural disturbances in a planktonic community (Moe et al. 2013) and integrity and functionality of a cell membrane by inducing extensive lipid peroxidation; thus, it can change cell function in an irreversible manner (Azzi et al. 2017). Cobalt (Co^{2+}) has been observed to inhibit vegetative growth, chlorophyll content, and photosynthetic efficiency of *Lemna minor* (duckweed) (Sree et al. 2015).

Sandmann and Bflger (1980) have pointed out the importance of lipid peroxidation by metal (e.g., Cu) stress. Under nutrient-deficient soil, the solubility of organic carbon and concurrently the mobility of heavy metals are amplified. Dissolved soil organic matter has major effects on the transformation of heavy metals through the rise of heavy-metal solubility, root growth, and plant uptake (Kim and Kim 2016). Copper and Pb accumulation in maize (*Zea mays* L.) and soybean (*Glycine max* L.), as affected by application of plant nutrients in soil such as N, P, and K (Xie et al. 2011), resulted in decrease in stomatal conductance, photosynthesis, and biomass. Application of cadmium caused reduction in the net rate of photosynthesis, stomatal conductance, and biomass in pak choi (Chinese cabbage) and mustard (Chen et al. 2013) but higher total chlorophyll content in tomato and decreased total biomass (Rehman et al. 2011). Accumulation of Zn and Cd in roots, petioles, and leaves of *Potentilla griffithii* increased considerably with the addition of these metals separately, while the addition of Zn showed reduced root Cd accumulation and increased Cd concentration in petioles and leaves (Qiu et al. 2010). The defensive effect of Mg against Cd toxicity could in part be due to the maintenance of Fe or to the increase in antioxidative capacity, detoxification, and protection of the photosynthetic apparatus (Hermans et al. 2011).

14.4 Mechanism of Phytotoxicity in Plants

Contaminants proceed in a precise manner to induce phytotoxicity in plants. For instance, arsenic can induce oxidative stress in plants at high concentrations (Figure 14.1). Aluminum can interfere with cell division and cause cross-linking of pectins, thereby increasing the rigidity of the cell wall (Durner 2013). It can reduce the rate of DNA replication and root respiration, and affect sugar phosphorylation. It can induce phosphorous deficiency by fixing phosphorous in less available forms in soils and interferes with uptake and transport of essential nutrients such as Ca, Mg, K, P, and Fe. Mercury (Hg) can induce phytotoxicity through the change of the permeability of cell membranes. It can react with sulphohydryl and phosphate groups, trigger replacement of essential ions, and disrupt the process involving critical or nonprotected proteins (Patra et al. 2004).

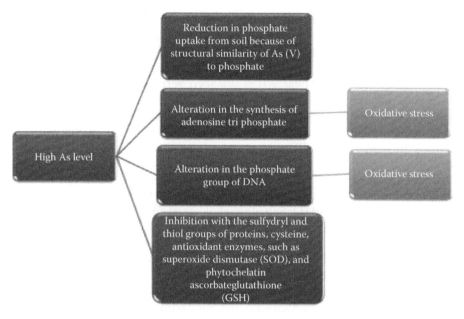

FIGURE 14.1
Mechanism of As toxicity in plants.

14.5 Standardized Methods for Phytotoxicity Tests

The literature contains a mixture of standard phytotoxicity laboratory tests, laboratory experiments, greenhouse experiments, field experiments, and field surveys. Since the objectives of these procedures differ, the information hardly can be applied directly in an EcoRA. Critical information might not be reported; alternatively, the information may be too comprehensive to readily permit comparison between different sources (Hayes and Kruger 2014). Specialist decisions and computer-aided study are often required before data can be included into EcoRAs.

14.5.1 Terrestrial Plant Tests

The most extensively used acute phytotoxicity tests concerning vascular plants are the seed germination and the root elongation tests. The seed germination test has been used widely since standardized protocols were introduced (Di Salvatore et al. 2008). Previously sorted seed lots are exposed to test toxicants in a soil matrix. Site soil or test chemicals are mixed with control soils in a logarithmic series. Germination is measured 5 days after initiating the test. The effective concentration of the test soil to give a 50% decrease of seed germination is used for calculation of the EC_{50} (direct soil toxicity test). Species usually used are selected to cover four to five types of plants. Alfalfa, clover, beet, lettuce, corn, cucumber, mustard, oats, perennial ryegrass, pinto bean, soybean, sorghum, radish, and wheat have been documented widely.

An alteration of the seed germination tests was developed for field use (Nwosu et al. 1991). On-site containers were kept under a canopy and shaded from the sun and rain. Test performance was evaluated against companion laboratory tests. Biologically significant differences were obtained between field and laboratory methods with red clover, cucumber, and lettuce, but not with wheat. An *in situ* seed germination test requires particular consideration to assure best quality control criteria. The major benefit of the test is the reduction of shipment, handling effort and the accompanying costs.

The root elongation test was developed as an indirect toxicity test. Roots are exposed to water extracts and soluble test soil constituents potentially toxic to the growing roots. After incubation in a chamber controlled by temperature and moisture, root length is measured. The EC_{50} of the test group is calculated as the concentration of the extract that inhibits root length of test samples by half that of the control samples. Preference seems to have been given to lettuce as a test species (Di Salvatore et al. 2008). Other species that have different root morphology, development patterns, and carbon allocation patterns may provide more ecologically relevant data for risk assessments.

The early seedling growth assay overcomes deficiencies of the seed germination and root elongation tests. The duration of the test provides for exposure well into the autotrophic stage of plant development. Exposure occurs in soil (either synthetic soil mix or an environmental sample), which provides a better approximation of field conditions than the root elongation test.

Life cycle bioassays are used to assess sublethal responses of plants to toxic chemicals. Exposure may be either acute or chronic. The endpoints used to quantify the effects of toxic chemicals include morphological and phenological measurements, which can be accomplished easily in a greenhouse, growth chamber, or field conditions. This system also allows examination of the roots for morphological impact. Two plant-related genera have been used in developing rapid life cycle tests: *Arabidopsis* (Ratsch et al. 1986) and *Brassica* (Shimabuku et al. 1991). *Arabidopsis* is well characterized physiologically and genetically and is ideally suited for laboratory assays. Technical impediments arise from the prostrate growth habit and tiny seed size. The small seeds virtually preclude measures of any parameter involving seed counts (e.g., percentage germination, reproductive success).

The rapid cycling *Brassica* has been developed by the Crucifer Genetics Cooperative of the University of Wisconsin, Madison. This group of plants is gaining popularity as a model system, especially with molecular biologists and geneticists. The advantages of *Brassica* compared to *Arabidopsis* include the upright growing habit and large seed size. Relatively large variations in many growth parameters may limit the utility of some potential endpoints. However, the short life cycle permits up to 10 generations in a year. This offers a good opportunity to investigate nonlethal effects of considerable ecological import (e.g., reproductive potential). Legitimate questions regarding the representativeness of these mustards as surrogates for other plants continue to slow acceptance of the life cycle bioassays.

Recently, the Office of Prevention, Pesticides and Toxic Substances (OPPTS) of the US Environmental Protection Agency (EPA) has developed a series of ecological effects test guidelines (USEPA 1991). These guidelines combine current standardized bioassays used under the Toxic Substances Control Act (TSCA) and the Federal, Insecticide, Fungicide, and Rodenticide Act (FIFRA). Group D of these guidelines is designed to estimate the effects of toxicants on nontarget plants and includes tests for terrestrial plant species (Table 14.2).

Several different types of bioassays exist, including soil or hydroponics, foliar, petri dish, and tissue culture assays. Soil bioassays incorporate the toxicant into the soil, exposing the plant through the roots or as a seed. In foliar bioassays the plant is exposed by application

TABLE 14.2

Ecological Effects Test Guidelines, Group D—Nontarget Plants Test Guidelines, Developed by the Office of Prevention, Pesticides, and Toxic Substances (OPPTS)

Office of Prevention, Pesticides, and Toxic Substances	Environmental Protection Agency	Name
850.4000	712-C-96–151	Background—nontarget plant testing
850.4025	712-C-96–152	Target area phytotoxicity
850.4100	712-C-96–153	Terrestrial plant toxicity, tier I (seedling emergence)
850.4150	712-C-96–163	Terrestrial plant toxicity, tier I (vegetative vigor)
850.4200	712-C-96–154	Seed germination/root elongation toxicity test
850.4225	712-C-96–363	Seedling emergence, tier II
850.4230	712-C-96–347	Early seedling growth toxicity test
850.4250	712-C-96–364	Vegetative vigor, tier II
850.4300	712-C-96–155	Terrestrial plants field study, tier III
850.4400	712-C-96–156	Aquatic plant toxicity test using *Lemna* spp., tiers I and II
850.4450	712-C-96–157	Aquatic plants field study, tier III
850.4600	712-C-96–158	Rhizobium–legume toxicity
850.4800	712-C-96–159	Plant uptake and translocation test

of the toxicant directly onto the leaves. Plants can be grown in either soil or hydroponic systems for foliar assays. Petri dish bioassays are generally used for seed-germination studies, where seeds of the test species are directly exposed to a substrate (moist filter paper, sand, or glass beads) containing the toxicant. However, plant tissues may also be exposed in petri dishes. Tissue culture assays use *in vitro* cultivation of plant cells or tissues for phytotoxicity, biochemical, and mode-of-action studies.

Various plant responses (Table 14.3) can be measured in terrestrial phytotoxicity bioassays depending on the design of the experiment. Overall, effects of toxicants on the plants

TABLE 14.3

Measured Responses to Toxicants in Terrestrial Phytotoxicity Tests

Response	Ref.
Seedling germination	Rivera et al. 2013
Seedling emergence	Rivera et al. 2013
Plant growth	USEPA 1991
Chlorosis	Kaur et al. 2015
Cupping of the leaves	Boutin et al. 2000
Photosynthesis	Park et al. 2016
Respiration	Truelove and Davis 1986
CO_2 assimilation	Truelove and Davis 1986
Ethylene and ACC	Mueller et al. 2000
Protein content	Cummins et al. 1997
Fatty acid composition	Verdoni et al. 2001
Glutathione (GSH)	Mehra and Tripathi 2000
Enzyme activities	Forlani et al. 2000
Glutathione S-transferase activity	Forlani et al. 2000
Phtyochelatins	Mehra and Tripathi 2000
Transpiration	Wilson et al. 1999

may be analyzed using seedling germination, seedling emergence, plant growth measured by fresh weight, shoot or root length change over the duration of the test, or physical observations such as chlorosis, yellowing of the leaves, cupping of the leaves, etc. Biochemical endpoints, such as photosynthesis, respiration, CO_2 assimilation, protein synthesis, enzyme activities, and lipid composition, may be used (Verdoni et al. 2001). Several review papers can be found on photosynthesis, ethylene synthesis, phytochelatins (Mehra and Tripathi 2000), glutathione, glutathione-S-transferases, and other enzymes (Table 14.3).

14.5.2 Aquatic Macrophyte Tests

Evaluation of wetland plants can be achieved best with aquatic macrophyte tests. Duckweed has been used to characterize single toxicant dose–response relationships (Naumann et al. 2007). For some effluents, duckweed was more sensitive than daphnia or fish for determining effluent toxicity (Liu et al. 2014). Bioassay endpoints include reduction of frond production, reduction of root length, biomass, ^{14}C uptake, total Kjeldahl nitrogen, and chlorophyll. Reduction of chlorophyll pigments can be more sensitive than frond production (Wang 1986). The ease of culture and bioassay methods have favored the use of duckweed to evaluate contaminants in water and saturated soils.

Nevertheless, it is better to use rooted aquatic plants to evaluate sediment toxicity. *Hydrilla verticillata* (hydrilla), a common aquatic angiosperm in the southeastern United States, is easy to culture and handle, tolerant of a broad range of environmental conditions, and has a fast growth rate. The most reproducible and toxicant-related endpoints are new root growth and peroxide activity. Uptake and translocation of chemicals to shoots may be an important route of chemical mobility in the environment. Since it is an exotic species, however, it must not be used in the field for *in situ* bioassays. Other plants that have similar growth and culture characteristics to hydrilla include *Elodea canadensis, Myriophyllum spicatum*, and *Potomogeton pectinatus*.

14.5.3 Endpoints

A measurement endpoint in phytotoxicity is characteristic of terrestrial plants that can be scored either qualitatively or quantitatively. Measurement endpoints should have definable levels of precision and accuracy; relate to important physiological, morphological, or ecological features; and exhibit a graded response to one or more agents. These become the unprocessed data used to estimate the plant performance in a given test. Quantitative data normally are the most precious, though numerous attributes, particularly signs of morbidity, can be scored subjectively and yield helpful information. Standardized test methods identify primary endpoints to be scored. In general, these methods support the collection of additional data such as observations of wilting, chlorosis, or other descriptions of plant health. Much of the primary phytotoxicity information appears in the agronomy and physiology literature, which preceded standardization of measurement endpoints.

14.6 Need for Ecological Risk Assessment

Toxic substances have an effect on the structure and functioning of the ecosystem (Pandey et al. 2014). Plants are the basis of food for the whole system (including human) and have further roles in terms of nutrient cycles, land protection and equilibrium of gases in the atmosphere. Different types of toxic compounds, such as heavy metals and pesticides, affect the growth of flora and fauna. The toxic compounds also affect aboveground and foliar invertebrates, which are the food for other organisms and perform crucial roles as a pest controller, saprophages, pollinators, and detrivores (Hladun et al. 2013).

The effects of such chemicals on food, soil, air, and water surfaces have noticeable economic and/or social consequences (Handy et al. 2008). Accumulation of these toxic chemicals in food and in the food chain is a matter of concern related to the consumption of this food by humans and domestic animals (Dorne and Fink-Gremmels 2013). The aquatic ecosystems, including seas, estuaries, rivers, and lakes, get toxicants due to anthropogenic activities from terrestrial sources. In aquatic organisms, there is a high affinity for muscles, liver, and gills to accumulate mercury, cadmium, and lead, respectively (Dural et al. 2007). It is well established that vertebrates and invertebrates are capable of accumulating heavy metals (Cd, Pb, and Zn) from the aquatic environment (Kim and Kim 2016). Pesticides affect crustaceans, fishes, and mollusks. Echinoderms are very sensitive to oil pollution (Mearns et al. 2015). Pesticides and heavy metals accumulate in aquatic organisms through food chain concentration.

Toxicants can cause impairment in humans when they are absorbed and reach the organs that are targets of their toxicity. This can occur through ingestion, inhalation, and absorption through the skin. In animals, toxicants often affect one or more target organs (i.e., the lungs, skin, or gastrointestinal tract) (Croom 2016). The toxicants can cross the external environment of the lung, skin, or gastrointestinal tract into the bloodstream. Various parts of the human body are designed to absorb chemicals quickly. Once absorbed in the body, toxins are metabolized, transported, or excreted and during the metabolic process; they have unfavorable biochemical effects and result in the manifestation of poisoning. These processes are described by kinetics and dynamic phases of toxicants. A toxic substance goes through absorption, temporary storage, metabolism, distribution, and excretion during the kinetic phase (Manahan 2002). If a toxicant's leftovers are unchanged throughout the metabolic process, it is available for further biochemical interactions as an active parent compound. A number of toxicants can be detoxified and excreted or transformed to a nontoxic metabolite (Manahan 2002). In the dynamic phase, a toxicant or toxic metabolite interacts in cells, tissues, or organs in the body, leading to various types of toxic effects.

The issue of the hazardous effect of a substance on an organism was relegated to the curiosity of physiologists until study related to toxicology arose in the early 1800s (Otter 2016). Hazard is commonly defined as "a property or situation that in particular circumstances could lead to harm" (Patnaik 2007). All the chemical toxicants are hazardous; however, the effect of hazardous substances depends on several environmental conditions, and therefore assessment of the probability or chance that a hazardous substance can affect the living organism is more important from an ecological point of view.

The risk is the chance of undesirable effects caused by specific conditions in an organism, a population, or an ecological system. Risk assessment for the probable effects of any substance entering the environment that may impair people must sum up exposures through all routes in order to determine the total exposure and then the possible effect. Risk assessment often culminates in the development of a model to predict toxic effects using environmental and long-term data. In addition, models may not be transferable from one site to another, because no two sites have identical characteristics. The challenge of environmental toxicology now is to identify the common principles that might allow extrapolation and prediction of toxic effects on the environment. Risk assessment has grown to be a commonly used approach in exploring environmental problems. It is used to investigate risks of very different natures.

Risk assessment and management approaches to environmental issues are increasingly being used at all levels of policy and regulation development. The techniques have a wide range of application, including the design of regulation (in determining societally "acceptable" risk levels, which may form the basis of environmental standards); providing a basis for site-specific decisions (land-use planning or sites of hazardous installations); prioritization of environmental risks (determination of which chemicals to regulate first); and comparison of risks (to enable comparisons to be made between the resources being allocated to the control of different types of risk, or to allow risk substitution decisions to be made).

In the early twelfth century, the use of living organisms in determining the level of response of organisms to contamination was proposed. To date, several living organisms like bacteria, algae, and higher plants have been used in ecological risk assessment studies (Zhang et al. 2015). Plants can be used effectively in ecological risk assessment as they have several advantages over other living organisms. For instance, they are sedentary, highly sensitive to environmental changes, and produce a rapid response to environmental contaminants in comparison to other organisms (Jäger 2016). Phytotoxicity data for different contaminants can be generated and used for developing water quality criteria to protect aquatic life, toxicity evaluation of effluents from municipal and industrial sources, and registration and re-registration of commercial chemicals (Smilanick et al. 2014). Several types of toxic compounds, like cyanotoxins, nanoparticles, heavy metals, etc., are being continuously released into the environment either naturally or by anthropogenic means; plants can act as an effective tool to measure consequent ecological risk.

14.7 Development of EcoRA

EcoRA is essential for appraising a broad spectrum of harmful impacts of toxic compounds on different ecosystems and therefore highly recommended for environmental decision making (Chen et al. 2013). In the late 1980s, the US EPA proposed specific guidelines for environmental risk assessment for different toxicants, such as carcinogenic compounds, and developed a series of technical documents (Qiuying and Jingling 2014). The development process of environmental risk assessment started with the embryonic phase, which lasted for 30–60 years of the twentieth century. This period was focused on a qualitative approach for toxicant identification (Figure 14.2). It was followed by a peak phase, which aimed at the quantitative methodology for risk assessment and lasted for 70–80 years of the twentieth century. After this, the improvement phase was initiated, in which the concept of ecological risk assessment was proposed in the period of 90 years of the twentieth century (US EPA 1991). From the late 1990s to early 2000s, the development stage of regional ecological risk assessment, in which risk assessment was proposed on a large scale (Xu and Liu 2009), started.

14.7.1 Process

Risk assessment is the process of determining the probability and extent of adverse effects on organisms' subsequent exposure to toxic material. EcoRA from contaminants (e.g., heavy metals) can be useful in identifying exposure pathways for plant and human

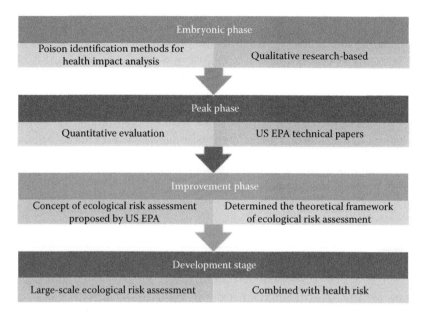

FIGURE 14.2
Developmental process of EcoRA.

populations; such assessments are significant for setting permissible limits of chemical toxicants in the environment. The risk assessment follows four basic steps that are briefly explained here.

14.7.1.1 Hazard Identification

Hazard identification deals with defining the nature of the possible cause of environmental health harms. It needs to assure that we have a clear knowledge of the ecological hazard. This is anything from connections with other living organisms, the immediate environment, or interactions among living organisms and the environment that poses the threat.

14.7.1.2 Exposure Assessment

The condition of a chemical contacting the outer boundary of an organism is exposure. The majority of the time, a chemical is contained in soil, water, air, a product, or a carrier medium; the chemical concentration at the point of contact is the exposure concentration. Exposure over a period of time can be showed by a time-dependent profile of the exposure concentration. Ecological exposure occurs when a chemical reduces growth of an organism and in a form that is bioavailable. The lower the bioavailability is, the lower is the risk that will be posed by the contaminant. Bioavailability differs among the species and can change under different environmental conditions. This step mainly deals with determining the concentration of a contaminating agent in the environment as well as estimating the intensity, duration, and frequency of organism exposure to a particular toxin. Exposure to a particular toxin can occur via inhalation, ingestion of food, or through the skin. Contaminant sources, release mechanisms, and transformation steps are all important aspects of exposure assessment.

14.7.1.3 Dose–Response Assessment

Dose–response assessment involves quantifying the adverse effects arising from exposure to a hazardous toxin based on the level of exposure. Sometimes, it is hard to show a relationship between the dose of a toxicant and the response being produced by it.

14.7.1.4 Risk Characterization

In this step, potential impacts of a toxicant are estimated based on the severity of its effects and level of exposure to a particular toxicant. Once risks are characterized, regularity options are evaluated in a process called risk management, which attempts to put together technical, legal, political, social, and economic issues.

14.8 Methods for Risk Assessment

Different methods like hazard quotient (HQ), health risk index (HRI), and daily dietary index (DI) for estimation of the heavy-metal concentrations in the human body following consumption of contaminated vegetables are summarized briefly in this section.

14.8.1 Hazard Quotient

This is a ratio of average daily dose (ADD) to the reference dose (RfD). According to this, if the HQ ratio is less than 1, there is no risk to population but if the ratio is equal to or greater than 1, then the ADD of a particular metal is greater than RfD, indicating that the population is likely to have health risks due to that metal and therefore requires a toxicity management option. This risk assessment method has been used by many workers (Chary et al. 2008) and has been found to be convincing and precise. For calculating the HQ, the following equation is used:

$$HQ = \frac{[W_{Plant}] \times [M_{plant}]}{R_f D \times B}. \tag{14.1}$$

where
   W_{plant} is the dry weight of contaminated plant material consumed (mg kg^{-1})
   M_{plant} is the concentration of metal in contaminated plant material (mg kg^{-1})
   RfD is the food reference dose for the metal (mg d^{-1})
   B is the average body mass (kg)

The values of RfD for metals are taken from the Integrated Risk Information System and the Department of Environment, Food and Rural Affairs. By the use of the HQ, the risk to human health of consuming metal-contaminated vegetables and grains has been assessed by different researchers (Yang et al. 2011). For example, the risk to human health, expressed as HQ, was generally highest for cadmium-, followed by lead-, zinc-, nickel-, and copper-accumulating leafy vegetables like *Gynandropsis gynandra* (spiderwisp), *Amaranthus dubius* (spleen amaranth), *Cucurbita maxima* (pumpkin), and *Vigna unguiculata* (cowpea) grown in metal-contaminated gardens of farmers in Uganda (Nabulo et al. 2010). Nevertheless, it was evident that urban farming of leafy vegetables could be safely continued on a good number of sites, subject to site-specific assessment of soil metal load, cautious choice of vegetable types, and adoption of washing in clean water prior to cooking.

In the same way, the health risk of metals such as lead, cadmium, nickel, and chromium via consumption of bell peppers and greenhouse cucumbers produced in Iran by total noncancer hazard quotient (THQ) and cancer risk assessment measures was studied (Aghili et al. 2009). The individual metal THQ values showed no cancer health effects linked with consumption of a single metal via intake of greenhouse cucumbers or bell peppers. The THQ for all population groups that consumed greenhouse cucumbers and bell peppers was below a unit. This value showed a low possibility of any noticeable risk. Among metals, Cd was recognized as the prominent risk factor for the consumers. The cancer risk assessment for Pb for Qom (Qom Province, Iran) mature population groups via consumption of cucumber and bell peppers was greater than a unit. Higher lead and cadmium levels in greenhouse vegetables were found to be a major concern that requires immediate consideration. In another study, the THQ was estimated to assess the noncarcinogenic health risk from individual heavy metals (e.g., Hg, Pb, Cd, Zn, and Cu) and combined heavy metals due to food intake by adults and children in the industrial area of Huludao City, China (Zheng et al. 2007). TQH for a single heavy metal following intake of individual foodstuff (e.g., cereal, sea product, vegetable, fruit, eggplant, and bean) in the industrial area of Huludao was 1. The relative HIs for Hg, Pb, Cd, Zn, and Cu were 1.7%, 11.7%, 24%, 23.4%, and 39.6% for adults and 1.5%, 11.7%, 21.8%, 26%, and 38.8% for children, respectively.

14.8.2 Daily Intake of Metal (DIM)

Since many foods or food products do contain heavy metals if collected from metal-contaminated environments, their daily intake by humans needs to be evaluated consistently for comparison as suggested by the US EPA. The DIM is calculated using the following formula:

$$DIM = \frac{[C_{metal} \times C_{factor} \times C_{food\ intake}]}{[B_{average\ weight}]}. \tag{14.2}$$

where
C_{metal} is heavy-metal concentrations in plants (mg kg^{-1})
C_{factor} is the conversion factor
$C_{food\ intake}$ is the daily intake of vegetables

The conversion factor of 0.085 is to convert fresh vegetable weight to dry weight (Rattan et al. 2005). Following this method, Yang et al. (2011) measured the concentration and daily intake of heavy metals (DIM) like Pb, Zn, Mn, Cu, Cd, and Cr in market vegetables in Chongqing, a major city of southwest China. Also, the potential health risk to local consumers was assessed by calculating THQ. The observed values for Pb and Cd were higher than the safety limit fixed by FAO/ WHO and Chinese regulations, signifying that market vegetables were critically contaminated by tested metals. The DIM values for Pb, Mn, and Cd were also more than the international guideline basis, and therefore consumers were at higher health risk. The individual THQ for Pb and Cd in pak choi and Cd in mustard and the combined THQ for all metals in each vegetable species (not including lettuce) were above the threshold of 1, implying a clearly adverse effect on health.

14.8.3 Health Risk Index

By using DIM values and reference oral dose, the HRI is calculated as HRI ¼ DIM/RDf; if the calculated HRI value is less than 1, the exposed population is considered safe (IRIS 2003). By the use of this method, the health risks of heavy metals (chromium, copper, and zinc) in edible seeds of crops grown in sewage-irrigated soils situated in the Langfang region of Hebei Province, China, were assessed. HRI values for every heavy metal apart from copper following intake of the edible seeds were less than a unit, indicating that the crops grown in sewage-irrigated soil did not cause any health risk to humans and were thus considered harmless for human intake (Chen et al. 2013). In a similar study, Khan et al. (2008) studied the health risks of heavy metals in crops grown in soil irrigated with wastewater. Significant amounts of heavy metals were built up in wastewater-irrigated soils collected from Beijing, China. Heavy-metal concentrations were significantly higher in crop plants grown in wastewater-irrigated soils compared to those observed for crops grown in untreated soil. Crop plants grown in wastewater-irrigated soil had heavy metals higher than the acceptable limits set by the State Environmental Protection Administration (SEPA) in China and the WHO. In addition, both adult and child populations tested in this study had significant amounts of the metals when they were allowed to consume crops grown in wastewater-irrigated soil. The HRI values were, however, less than 1, which suggests that there were no health risks for these groups, even when they consumed contaminated vegetables. In a follow-up study, the HRI values were also found to be less than 1 for crops such as *Brassica rapa* (common mustard), *Spinacia oleracae* L. (spinach), *Lycopersicum esculantum* (tomato), *Mentha viridis* (spearmint), *Coriandum sativum* (coriander), and *Lactuca sativa* (lettuce) cultivated in wastewater-irrigated soil enriched with Zn. However, such crops had a chance of causing health risks when grown in wastewater-irrigated soil containing higher concentrations of Mn (Jan et al. 2010). Risk to human health by heavy metals like Cd, Cu, Pb, Ni, and Cr after consuming vegetables and cereal crops collected from wastewater-irrigated sites was assessed by Singh et al. (2010). When analyzed, it was observed that all the collected samples from wastewater-irrigated sites had significantly higher concentrations of metals compared to those grown with clean water only. Of the various metals determined, the levels of Cd, Pb, and Ni were above the "safe" limits of Indian and WHO/FAO standards in all the vegetables and cereals. Furthermore, the higher metal pollution index and HRI values suggested that human populations who may consume these food materials collected from wastewater-irrigated sites are likely to experience health-related problems.

Considering the threat of heavy metals to soil fertility, food safety, and human health, a numbers of efforts are being directed to reducing/completely alleviating metal toxicity to all forms of life. However, further efforts are required to carefully monitor and regulate the discharge of properly treated by-products of different industries or to avoid the use of substances that otherwise contain heavy metals in agricultural practices. By doing this, the uptake of metals by various plants and, hence, by the food materials could be avoided. Moreover, methods should be developed to rapidly identify the presence of toxic substances in consumable food items so that a suitable strategy is adopted in time to eliminate human health problems. Therefore, concerted efforts from the public, scientists, and policy makers are required to achieve a common goal of a safe and secure environment for better living around the world.

14.8.4 Ecological Risk Potential Index

Hakanson (1980) developed a method for the ecological risk potential assessment of heavy metals, which is based on following equation:

$$C_f^i = \frac{C_i^s}{C_r^i}; E_r^i = T_f^i \times C_f^i; ERi = \sum_{i=1}^{n} E_r^i \tag{14.3}$$

where
  C_f^i is the contamination level of heavy metal
  C_i^s is the concentration of heavy metal on sediment
  C_r^i is the reference value of heavy metal in study location
  E_r^i is the ecological risk potential of heavy metal
  T_f^i is the toxicity response factor of heavy metal
  ERi (risk index) is the ecological risk potential of environment

To determine the ecological potential risk of heavy metals and ecological risk criteria of the environment, an ecological risk potential index was proposed (Gan et al. 2000). This index provides comparative data for potential toxicity of different heavy metals and is thus quite useful for ecological risk assessment. According to this index, ecological risk criteria for heavy metal are designated as low, moderate, considerable, very high, and disastrous with values of $E_r^i < 30$, $30 < E_r^i < 50$, $50 < E_r^i < 100$, $100 < E_r^i < 150$, $E_r^i > 150$, respectively (Gan et al. 2000). Similarly, ecological risk criteria for the environment are considered as low, moderate, considerable, very high, and disastrous with values of Eri < 100, 100 < ERi < 150, 150 < ERi < 200, 200 < ERi < 300, ERi > 300, respectively (Gan et al. 2000).

14.9 Conclusion

A variety of physiological and biochemical responses are shown by plant species in response to toxicants present in the environment. Plants frequently show reduction in root growth, seed germination, biomass production, and photosynthetic pigment content. A number of cellular changes, such as reduced rate of DNA replication, an increase in rigidity of the cell wall, and changes in the permeability of the cell membrane, have also been exhibited by plants as a phytotoxicity response. The plants showing these types of phytotoxic responses offer encouraging possibilities to be used in early detection of deteriorated quality of the environment and thus could be efficiently used in ecological risk assessment.

References

Aghili, F., Khoshgoftarmanesh, A.H., Afyuni, M., Schulin, R. 2009. Health risks of heavy metals through consumption of greenhouse vegetables grown in central Iran. *Human and Ecological Risk Assessment* 15:999–1015.

Arnold, K.E., Boxall, A.B., Brown, A.R., Cuthbert, R.J., Gaw, S., Hutchinson, T.H., Jobling, S., Madden, J.C., Metcalfe, C.D., Naidoo, V., Shore, R.F., Smits, J.E., Taggart, M.A., Thompson, H.M. 2013. Assessing the exposure risk and impacts of pharmaceuticals in the environment on individuals and ecosystems. *Biology Letters* 9(4):20130492.

Azzi, V., Kanso, A., Kazpard, V., Kobeissi, A., Lartiges, B., El Samrani, A. 2017. Lactuca sativa growth in compacted and non-compacted semi-arid alkaline soil under phosphate fertilizer treatment and cadmium contamination. *Soil and Tillage Research* 165:1–10.

Bhada-Tata, P., Hoornweg, D. 2016. Solid waste and climate change. In *State of the World* (pp. 239–255). Island Press/Center for Resource Economics.

Boutin, C., Hing-Biu, L., Peart, E.T., Batchelor, P.S., Maguire, R.J. 2000. Effects of the sulfonylurea herbicide metsulfuron methyl on growth and reproduction of five wetland and terrestrial plant species. *Environmental Toxicology and Chemistry* 19(10):2532–2541.

Brent Nichols, P., Couch, J.D., Al-Hamdani, S.H. 2000. Selected physiological responses of *Salvinia minima* to different chromium concentrations. *Aquatic Botany* 68:313–319.

Canas, J.E., Long, M., Nations, S., Vadan, R., Dai, L., Luo, M., Ambikapathi, R., Lee, E.H., Olszyk, D. 2008. Effects of functionalized and nonfunctionalized single-walled carbon nanotubes on root elongation of select crop species. *Environmental Toxicology Chemistry* 27:1922–1931.

Chaignon, V., Hinsinger, P. 2003. A biotest for evaluating copper bioavailability to plants in a contaminated soil. *Journal of Environmental Quality* 32(3):824–833.

Chang, A.C., Granato, T.C., Page, A.L. 1992. A methodology for establishing phytotoxicity criteria for chromium, copper, nickel and zinc in agricultural land application of municipal sewage sludges. *Journal of Environmental Quality* 21:521–536.

Chary, N.S., Kamala, C.T., Raj, D.S.S. 2008. Assessing risk of heavy metals from consuming food grown on sewage irrigated soils and food chain transfer. *Ecotoxicology and Environmental Safety* 69:513–524.

Chen, S., Chen, B., Fath, B.D. 2013. Ecological risk assessment on the system scale: A review of state-of-the-art models and future perspectives. *Ecological Modelling* 250:25–33.

Croom, E.L. 2016. The role of toxicokinetics and toxicodynamics in developmental and translational toxicology. In *Translational Toxicology* (pp. 45–81). Springer International Publishing.

Cummins, I., Cole, D.J., Edwards, R. 1997. Purification of multiple glutathione transferases involved in herbicide detoxification from wheat (*Triticum aestivum* L.) treated with the safener fenchlorazole-ethyl. *Pesticide Biochemistry and Physiology* 59(1):35–49.

Davis, D.L., Bell, M.L., Fletcher, T. 2002. A look back at the London smog of 1952 and the half century since. *Environmental Health Perspectives* 110:A734–A735.

Deng, H., Ye, Z.H., Wong, M.H. 2004. Accumulation of lead, zinc, copper and cadmium by 12 wetland plant species thriving in metal-contaminated sites in China. *Environmental Pollution* 132(1):29–40.

Di Salvatore, M., Carafa, A.M., Carratù, G. 2008. Assessment of heavy metals phytotoxicity using seed germination and root elongation tests: A comparison of two growth substrates. *Chemosphere* 73(9):1461–1464.

Dorne, J.L.C.M., Fink-Gremmels, J. 2013. Human and animal health risk assessments of chemicals in the food chain: Comparative aspects and future perspectives. *Toxicology and Applied Pharmacology* 270(3):187–195.

Dural, M., Göksu, M.Z.L., Özak, A.A. 2007. Investigation of heavy metal levels in economically important fish species captured from the Tuzla lagoon. *Food Chemistry* 102(1):415–421.

Durner, E.F. 2013. *Principles of horticultural physiology*. CABI.

Eriyamremu, G.E., Asagba, S.O., Ojeaburu, A. 2005. Evaluation of lead and cadmium levels in some commonly consumed vegetables in the Niger-Delta oil area of Nigeria. *Bulletin of Environmental Contamination and Toxicology* 75:278–283.

Forlani, G., Lejczak, B., Kafarski, P. 2000. The herbicidally active compound N-2-(5-chloro-pyridyl)-aminomethylene-bisphosphonic acid acts by inhibiting both glutamine and aromatic amino acid biosynthesis. *Australian Journal of Plant Physiology* 27:677–683.

Gadd, G.M. 2010. Metals, minerals and microbes: Geomicrobiology and bioremediation. *Microbiology* 156:609–643.

Gan, J.L., Jia, X.P., Lin, Q., Li, C.H., Wang, Z.H., Zhou, G.J., Wang, X.P., Cai, W.G., Lu, X.Y. 2000. A primary study on ecological risk caused by the heavy metals in coastal sediments. *Journal of Fisheries of China* 24:533–538.

Gruiz, K. 2005. Soil testing triad and interactive ecotoxicity tests for contaminated soil. In *Soil Remediation*, vol 6 (pp. 45–70), Fava, F., Canepa, P. (eds). INCA, Venice.

Gybina, A.A., Prohaska, J.R. 2008. Copper deficiency results in AMP-activated protein kinase activation and acetylCoA carboxylase phosphorylation in rat cerebellum. *Brain Research* 1204:69–76.

Hakanson, L. 1980. An ecological risk index for aquatic pollution control, a sedimentological approach. *Water Research* 14:975–1001.

Hamoutene, D., Romeo, M., Gnassia, M., Lafaurie, M. 1996. Cadmium effects on oxidative metabolism in a marine seagrass: *Posidonia oceanica*. *Bulletin of Environmental Contamination and Toxicology* 56(2):327–334.

Hanano, A., Almousally, I., Shaban, M. 2014. Phytotoxicity effects and biological responses of *Arabidopsis thaliana* to 2,3,7,8-tetrachlorinated dibenzo-p-dioxin exposure. *Chemosphere* 104:76–84.

Handy, R.D., Kammer, F., Lead, J.R., Hassellov, M., Owen, R., Crane, M. 2008. The ecotoxicology and chemistry of manufactured nanoparticles. *Ecotoxicology* 17:287–314.

Hastie, C.J., Borthwick, E.B., Morrison, L.F., Codd, G.A., Cohen, P.T.W. 2005. Inhibition of several protein phosphatases by a non-covalently interacting microcystin and a novel cyanobacterial peptide, nostocyclin. *Biochemica Biophysica Acta* 1726:187–193.

Hayes, A.W., Kruger, C.L. (eds). 2014. *Hayes' Principles and Methods of Toxicology*. CRC Press, Boca Raton, FL.

Hermans, C., Chen, J., Coppens, F., Inze, D., Verbruggen, N. 2011 Low magnesium status in plants enhances tolerance to cadmium exposure. *New Phytologist* 192(2):428–436.

Hladun, K.R., Parker, D.R., Tran, K.D., Trumble, J.T. 2013. Effects of selenium accumulation on phytotoxicity, herbivory, and pollination ecology in radish (*Raphanus sativus* L.). *Environmental Pollution* 172:70–75.

Ikawa, M., Sasner, J.J., Haney, J.F. 2001. Activity of cyanobacterial and algal odor compounds found in lake waters on green alga *Chlorella pyrenoidosa* growth. *Hydrobiologia* 443:19–22.

Integrated Risk Information System report (IRIS). 2003. http://www.epa.gov/iris/

Jäger, H.J. 2016. Biochemical diagnostic tests for the effect of air pollution on plants. *Gaseous Air Pollutants and Plant Metabolism* 333.

Jan, F.A., Ishaq, M., Khan, S., Ihsanullah, I., Ahmad, I., Shakirullah, M. 2010. A comparative study of human health risks via consumption of food crops grown on wastewater irrigated soil (Peshawar) and relatively clean water irrigated soil (lower Dir). *Journal of Hazardous Materials* 179:612–621.

Kabata-Pendias, A., Pendias, H. 2001. *Trace Elements in Soil and Plants*. CRC, Boca Raton, FL.

Kampa, M., Castanas, E. 2008. Human health effects of air pollution. *Environmental Pollution* 151(2):362–367.

Kapustka, L.A. 1996. Plant ecotoxicology: The design and evaluation of plant performance in risk assessments and forensic ecology. In *Environmental Toxicology and Risk Assessment: Fourth Volume* (pp. 110–121), La Point, T.W., Price, F.T., Little, E.E. (eds). ASTM STP 1262, American Society for Testing and Materials, Philadelphia, PA.

Kaur, M., Aggarwal, N.K., Dhiman, R. 2015. Screening of phytotoxicity of *Alternaria macrospora* MKP1 against *Parthenium hysterophorus* L. *Archives of Phytopathology and Plant Protection* 48(17–20):890–897.

Khan, S., Cao, Q., Zheng, Y.M., Huang, Y.Z., Zhu, Y.G. 2008. Health risks of heavy metals in contaminated soils and food crops irrigated with wastewater in Beijing, China. *Environmental Pollution* 152:686–692.

Kim, H.T., Kim, J.G. 2016. Uptake of cadmium, copper, lead, and zinc from sediments by an aquatic macrophyte and by terrestrial arthropods in a freshwater wetland ecosystem. *Archives of Environmental Contamination and Toxicology* 71(2):198–209.

Kolodziejska, M., Maszkowska, J., Bialk-Bielinska, A., Steudte, S., Kumirska, J., Stepnowski, P., Stolte, S. 2013. Aquatic toxicity of four veterinary drugs commonly applied in fish farming and animal husbandry. *Chemosphere* 92(9):1253–1259.

Krewski, D., Acosta, J.D., Andersen, M., Anderson, H., Bailar, III J.C., Boekelheide, K., Brent, R., Charnley, G., Cheung, V.G., Green, J.S., Kelsey, K.T. 2010. Toxicity testing in the 21st century: A vision and a strategy. *Journal of Toxicology and Environmental Health, Part B* 13(2–4):51–138.

Kwakye, G.F., McMinimy, R.A., Aschner, M. 2016. Disease–toxicant interactions in Parkinson's disease neuropathology. *Neurochemical Research* 1–15.

Landis, W., Sofield, R., Yu, M.H., Landis, W.G. 2003. *Introduction to Environmental Toxicology: Impacts of Chemicals upon Ecological Systems*. CRC Press, Boca Raton, FL.

Lee, C.W., Mahendra, S., Zodrow, K., Li, D., Tsai, Y.C., Braam, J. 2010. Developmental phytotoxicity of metal oxide nanoparticles to Arabidopsis thaliana. *Environmental Toxicology and Chemistry* 29(3):669–675.

Lemire, J.A., Harrison, J.J., Turner, R.J. 2013. Antimicrobial activity of metals: Mechanisms, molecular targets and applications. *Nature Reviews Microbiology* 11(6):371–384.

Li, Z., Ma, Z., van der Kuijp, T.J., Yuan, Z., Huang, L. 2014. A review of soil heavy metal pollution from mines in China: Pollution and health risk assessment. *Science of the Total Environment* 468:843–853.

Liu, F., Ying, G.G., Tao, R., Zhao, J.L., Yang, J.F., Zhao, L.F. 2009. Effects of six selected antibiotics on plant growth and soil microbial and enzymatic activities. *Environmental Pollution* 157(5):1636–1642.

Liu, J., Liu, L., Li, Z., Yu, L., Yang, G., Lu, Z., Wang, X. 2014. An alternative fish assay for sustainable industrial wastewater toxicity assessment and management: The Case of Northeast China. *International Journal Water Research* 2(1): 33-42.

Liu, L., Liu, Y.H., Liu, C.X., Wang, Z., Dong, J., Zhu, G.F., Huang, X. 2013. Potential effect and accumulation of veterinary antibiotics in *Phragmites australis* under hydroponic conditions. *Ecological Engineering* 53:138–143.

López-Moreno, M.L., Rosa, G.D.L., Hernández-Viezcas, J.A., Peralta-Videa, J.R., Gardea-Torresdey, J.L. 2010. X-ray absorption spectroscopy (XAS) corroboration of the uptake and storage of CeO$_2$ nanoparticles and assessment of their differential toxicity in four edible plant species. *Journal of Agriculture, Food, and Chemistry* 58:3689–3693.

Ma, H., Williams, P.L., Diamond, S.A. 2013. Ecotoxicity of manufactured ZnO nanoparticles— A review. *Environmental Pollution* 172:76–85.

Ma, X., Lee, J.G., Deng, Y., Kolmakov, A. 2010. Interactions between engineered nanoparticles (ENPs) and plants: Phytotoxicity, uptake and accumulation. *Science of the Total Environment* 408:3053–3061.

Macinnis-Ng, C.M.O., Ralph, P.J. 2003. In-situ impact of petrochemicals on the photosynthesis of the seagrass *Zostera capricorni*. *Marine Pollution Bulletin* 46:1395–1407.

Manahan, S.E. 2002. *Toxicological Chemistry and Biochemistry*. CRC Press, Boca Raton, FL.

Mearns, A.J., Reish, D.J., Oshida, P.S., Ginn, T., Rempel-Hester, M.A., Arthur, C., Rutherford, N., Pryor, R. 2015. Effects of pollution on marine organisms. *Water Environment Research* 87(10):1718–1816.

Mehra, R.K., Tripathi, R.D. 2000. Phytochelatins and metal tolerance. In *Environmental Pollution and Plant Responses*, Agrawal, S.B., Agrawal, M. (eds). Lewis Publishers, Boca Raton, FL.

Metcalf, J.S., Barakate, A., Codd, G.A. 2004. Inhibition of plant protein synthesis by the cyanobacterial hepatotoxin, cylindrospermopsin. *FEMS Microbiology Letters* 235:125–129.

Meudec, A., Poupart, N., Dussauze, J., Deslandes, E. 2007. Relationship between heavy fuel oil phytotoxicity and polycyclic hydrocarbon contamination in *Salicornia fragilis*. *Science of the Total Environment* 38:146–156.

Michelini, L., Reichel, R., Werner, W., Ghisi, R., Thiele Bruhn, S. 2012. Sulfadiazine uptake and effects on *Salix fragilis* L. and *Zea mays* L. plants. *Water, Air, & Soil Pollution* 223(8):5243–5257.

Moe, S.J., De Schamphelaere, K., Clements, W.H., Sorensen, M.T., Van den Brink, P.J., Liess, M. 2013. Combined and interactive effects of global climate change and toxicants on populations and communities. *Environmental Toxicology and Chemistry* 32(1):49–61.

Müller, R., Sisler, E.C., Serek, M. 2000. Stress induced ethylene production, ethylene binding, and the response to the ethylene action inhibitor 1-MCP in miniature roses. *Scientia Horticulturae* 83(1):51–59.

Nabulo, G., Young, S.D., Black, C.R. 2010. Assessing risk to human health from tropical leafy vegetables grown on contaminated urban soils. *Science of the Total Environment* 408:5338–5351.

Nagajyoti, P.C., Lee, K.D., Sreekanth, T.V.M. 2010. Heavy metals, occurrence and toxicity for plants: A review. *Environmental Chemistry Letters* 8:199–216.

Naidu, R., Oliver, D., McConnell, S. 2003. Heavy metal phytotoxicity in soil. In *Proceedings of the Fifth National Workshop on the Assessment of Site Contamination* (pp. 235–241). Langley, A., Gilbey, M., Kennedy, B. (eds). National Environment Protection Council Service Corporation.

Naumann, B., Eberius, M., Appenroth, K.J. 2007. Growth rate based dose–response relationships and EC-values of ten heavy metals using the duckweed growth inhibition test (ISO 20079) with *Lemna minor* L. clone St. *Journal of Plant Physiology* 164(12):1656–1664.

Nichol, B.E., Oliveira, L.A. 1995. Effects of aluminium on the growth and distribution of calcium in roots of an aluminium-sensitive cultivar of barley (*Hordeum vulgare*). *Canadian Journal of Botany* 73:1849–1858.

Nwosu, J.U., Ratsch, H.C., Kapustka, L.A. 1991. A protocol for on-site seed germination test. In *Plants for Toxicity Assessment: Second Volume* (pp. 333–340), Gorsuch, J.W., Lower, W.R., Lewis, M.A., Wang, W. (eds). ASTM STP 1115, American Society for Testing and Materials, Philadelphia, PA.

OECD-OCDE (Organisation for Economic Cooperation and Development). 2006. OECD Guideline for testing of chemicals test no 208. *Terrestrial Plant Test—Seedling Emergence and Seedling Growth Test*. OECD Publishing, Paris.

Opris, O., Copaciu, F., Loredana Soran, M., Ristoiu, D., Niinemets, Ü., Copolovici, L. 2013. Influence of nine antibiotics on key secondary metabolites and physiological characteristics in *Triticum aestivum*: Leaf volatiles as a promising new tool to assess toxicity. *Ecotoxicology and Environmental Safety* 87:70–79.

Otter, C. 2016. Artificial Britain: Risk, systems and synthetics since 1800. In *Governing Risks in Modern Britain* (pp. 79–103). Palgrave Macmillan, UK.

Pandey, B., Agrawal, M., Singh, S. 2014. Coal mining activities change plant community structure due to air pollution and soil degradation. *Ecotoxicology* 23(8):1474–1483.

Park, J., Brown, M.T., Depuydt, S., Kim, J.K., Won, D., Han, T. 2016. Comparing the acute sensitivity of growth and photosynthetic endpoints in three *Lemna* species exposed to four herbicides. *Environmental Pollution* 220(PartB):818–827.

Patnaik, P. 2007. *A Comprehensive Guide to the Hazardous Properties of Chemical Substances*. John Wiley & Sons.

Patra, M., Bhowmik, N., Bandopadhyay, B., Sharma, A. 2004. Comparison of mercury, lead and arsenic with respect to genotoxic effects on plant systems and the development of genetic tolerance. *Environmental and Experimental Botany* 52(3):199–223.

Patra, M., Sharma, A. 2000. Mercury toxicity in plants. *Botanical Review* 66(3):379–422.

Pflugmacher, S. 2002. Possible allelopathic effects of cyanotoxins, with reference to microcystin-LR, in aquatic ecosystems. *Environmental Toxicology* 17:407–413.

Qi, Y., Liu, D., Zhao, W., Liu, C., Zhou, Z., Wang, P. 2015. Enantioselective phytotoxicity and bioactivity of the enantiomers of the herbicide napropamide. *Pesticide Biochemistry and Physiology* 125:38–44.

Qiu, R.L., Thangavel, P., Hu, P.J., Senthilkumar, P., Ying, R.R., Tang, Y.T. 2010. Interaction of cadmium and zinc on accumulation and sub-cellular distribution in leaves of hyperaccumulator *Potentilla griffithii*. *Journal of Hazardous Materials* 186:1425–1430.

Qiuying, C., Jingling, L. 2014. Development process and perspective on ecological risk assessment. *Acta Ecologica Sinica* 34:239–245.

Ratsch, H.C., Johndro, D.J., McFarlane, J.C. 1986. Growth inhibition and morphological effects of several chemicals in *Arabidopsis thaliana* (L.) Heynh. *Environmental Toxicology and Chemistry* 5:55–60.

Rattan, R.K., Datta, S.P., Chhonkar, P.K., Suribabu, K., Singh, A.K. 2005. Long term impact of irrigation with sewage effluents on heavy metal contents in soils, crops and ground water—A case study. *Agriculture, Ecosystems & Environment* 109:310–322.

Reglero, M.M., Taggart, M.A., Monsalve-González, L., Mateo, R. 2009. Heavy metal exposure in large game from a lead mining area: Effects on oxidative stress and fatty acid composition in liver. In: *Environmental Pollution* 157(4):1388–1395.

Rehman, F., Khan, F.A., Varshney, D., Naushin, F., Rastogi, J. 2011. Effect of cadmium on the growth of tomato. *Biology and Medicine* 3:187–190.

Rivera, E.B., Milla, O.V., Huang, W.J., Ho, Y.S., Chiu, J.Y., Chang, H.Y. 2013. Rice germination as a bioassay to test the phytotoxicity of MSWI bottom ash recycling wastewater. *Journal of Hazardous, Toxic, and Radioactive Waste* 17(2):140–145.

Sadik, N.A. 2008. Effects of diallyl sulfide and zinc on testicular steroidogenesis in cadmium-treated male rats. *Journal of Biochemical and Molecular Toxicology* 22(5):345–353.

Salati, S., Moore, F. 2010. Assessment of heavy metal concentration in the Khoshk River water and sediment, Shiraz, Southwest Iran. *Environmental Monitoring and Assessment* 164(1–4):677–689.

Salvatore, M.D., Carratu, G., Carafa, A.M. 2009. Assessment of heavy metals transfer from a moderately polluted soil into the edible parts of vegetables. *Journal of Food, Agriculture, and Environment* 7:683–688.

Sandmann, G., Bflger, P. 1980. Copper-mediated lipid peroxidation processes in photosynthetic membranes. *Plant Physiology* 66:797–800.

Sharma, R.K., Agrawal, M. 2005. Biological effects of heavy metals: An overview. *Journal of Environmental Biology* 26(3/4):301–313.

Sharma, R.K., Agrawal, M. 2006. Single and combined effects of cadmium and zinc on carrots: Uptake and bioaccumulation. *Journal of Plant Nutrition* 29(10):1791–1804.

Sharma, R.K., Agrawal, M., Marshall, F.M. 2007. Heavy metals contamination of soil and vegetables in suburban areas of Varanasi, India. *Ecotoxicology and Environmental Safety* 66:258–266.

Shimabuku, R.A., Ratsch, H.C., Wise, C.M., Nwosu, J.U., Kapustka, L.A. 1991. *Methodology for a New Plant Life-Cycle Bioassay Featuring Rapid Cycling Brassica*, (pp. 365–375), vol. 2, Gorsuch, J.W., Lower, W.R., St. John, K.R. (eds). ASTM STP 1115, American Society for Testing and Materials, Philadelphia, PA.

Singh, A., Sharma, R.K., Agrawal, M., Marshall, F.M. 2010. Health risk assessment of heavy metals via dietary intake of foodstuffs from the wastewater irrigated site of a dry tropical area of India. *Food Chemistry and Toxicology* 48:611–619.

Smilanick, J.L., Mansour, M., Sorenson, D. 2014. Performance of fogged disinfectants to inactivate conidia of *Penicillium digitatum* within citrus degreening rooms. *Postharvest Biology and Technology* 91:134–140.

Smith, R.D., Wilson, J.E., Walker, J.C., Baskin, T.I. 1994. Protein-phosphatase inhibitors block root hair growth and alter cortical cell shape of *Arabidopsis* roots. *Planta* 194:516–524.

Sree, K.S., Keresztes, A., Mueller-Roeber, B., Brandt, R., Eberius, M., Fischer, W., Appenroth, K.J. 2015. Phytotoxicity of cobalt ions on the duckweed *Lemna minor*—Morphology, ion uptake, and starch accumulation. *Chemosphere* 131:149–156.

Sun, M. 2016. Investigation on the production of secondary metabolites from anoxygenic phototrophic bacteria (Doctoral dissertation, Christian-Albrechts-Universität Kiel).

Suter, II G.W. 2016. *Ecological Risk Assessment*. CRC Press, Boca Raton, FL.

Truelove, B., Davis, D.E. 1986. Chapter XVII The measurement of photosynthesis and respiration using whole plants or plant ofigans. *Research Methods in Weed Science* 325.

Turabelidze, G., Schootman, M., Zhu, B.P., Malone, J.L., Horowitz, S., Weidinger, J., Williamson, D., Simoes, E. 2008. Multiple sclerosis prevalence and possible lead exposure. *Journal of the Neurological Sciences* 269(1):158–162.

Uniyal, S., Paliwal, R., Sharma, R.K., Rai, J.P.N. 2016. Degradation of fipronil by *Stenotrophomonas acidaminiphila* isolated from rhizospheric soil of *Zea mays*. *3 Biotech* 6:48.

US EPA. 1991. Ecological assessment of superfund sites: An overview, Publication, pp. 9345–9351.

Verdoni, N., Mench, M., Cassagne, C., Bessoule, J.J. 2001. Fatty acid composition of tomato leaves as biomarkers of metal contaminated soils. *Environmental Toxicology and Chemistry* 20(2):382–388.

Wang, W. 1986. The effect of river water on phytotoxicity of Ba, Cd and Cr. *Environmental Pollution* 33(b):193–204.

Wei, X., Wu, S.C., Nie, X.P., Yediler, A., Wong, M.H. 2009. The effects of residual tetracycline on soil enzymatic activities and plant growth. *Journal of Environmental Science and Health B* 44(5):461–471.

Wilson, K.G., Ralph, P.J. 2008. A comparison of the effects of Tapis crude oil and dispersed crude oil on subtidal Zostera capricorni. In *Proceedings International Oil Spill Conference* (pp. 859–864).

Wilson, P.C., Whitwell, T., Klaine, S.J. 1999. Phytotoxicity, uptake, and distribution of [C-14] simazine in Canna hybrida 'Yellow King Humbert'. *Environmental Toxicology and Chemistry* 18:1462–8.

Xie, W., Zhou, J., Wang, H., Liu, Q., Xia, J., Lv, X. 2011. Cu and Pb accumulation in maize (*Zea mays* L.) and soybean (*Glycine max* L.) as affected by N, P and K application. *African Journal of Agricultural Research* 6:1469–1476.

Xu, L.Y., Liu, G.Y. 2009. The study of a method of regional environmental risk assessment, *Journal of Environmental Management* 90:3290–3296.

Yan, Z., Wang, W., Zhou, J., Yi, X., Zhang, J., Wang, X., Liu, Z. 2015. Screening of high phytotoxicity priority pollutants and their ecological risk assessment in China's surface waters. *Chemosphere* 128:28–35.

Yang, Q.W., Xu, X., Liu, S.J., He, J.F., Long, F.Y. 2011. Concentration and potential health risk of heavy metals in market vegetables in Chongqing, China. *Ecotoxicology, Environment and Safety* 74:1664–1669.

Zaidi, A., Wani, P.A., Khan, M.S. (eds). 2012. *Toxicity of Heavy Metals to Legumes and Bioremediation*. Springer Science & Business Media.

Zhang, D., Hua, T., Xiao, F., Chen, C., Gersberg, R.M., Liu, Y., Stuckey, D., Ng, W.J., Tan, S.K. 2015. Phytotoxicity and bioaccumulation of ZnO nanoparticles in *Schoenoplectus tabernaemontani*. *Chemosphere* 120:211–219.

Zheng, N., Wang, Q., Zhang, X., Zheng, D., Zhang, Z., Zhang, S. 2007. Population health risk due to dietary intake of heavy metals in the industrial area of Huludao City, China. *Science of the Total Environment* 387:96–104.

15

Vermicomposting of Lignocellulosic Waste: A Biotechnological Tool for Waste Management

Kavita Sharma and V. K. Garg

CONTENTS

15.1 Introduction

Remarkable acceleration in solid waste generation has been observed all over the world, which is more alarming in developing nations compared to developed ones. Currently, global waste generation is about 1.3 billion tons/year and that is likely to increase by 2.2 billion tons/year by 2025 (Halbach 2013). Main sources of solid wastes are various domestic, industrial, commercial, and agricultural activities that generate organic as well as inorganic residues. Lignocellulosic wastes are a major fraction of the biodegradable portion of solid wastes. Sources of lignocellulosic wastes are various agricultural practices and agro-based industries (paper pulp industries, timber industries, palm oil mills, sugar mills, etc). Basic components of the lignocellulose are lignin, cellulose, and hemicelluloses (Saha 2003). The complex nature of these structural components of lignocellulosic waste makes them useful raw materials for biotechnological application. Lignocellulosic waste has served as substrate for the production of several products, such as chemicals, acids, enzymes, medicines, and, more often, biofuels like ethanol (Mtui 2009). But the complexity of chemical linkage between cellulose, hemicelluloses, and lignin resists degradation of lignocellulosic waste (Xiao et al. 2012). So conversion of lignocellulosic waste into some value-added product requires mechanical, chemical, physical, and biological pretreatment. These pretreatment methods are complex and enhance the cost of product recovery from lignocellulosic waste. Hence, an alternative technology is required for the utilization of lignocellulosic waste.

The most commonly used techniques for waste management are open dumping, landfilling, incineration, pyrolysis, and gasification. But these techniques have faced several hurdles in terms of legislation, acceptance by the public, environmental concern, and financial issues (Singh et al. 2011). In such a scenario, policy makers, scientists, and urban local bodies (ULBs) are in search of sustainable and economically viable solutions to solid waste management. In recent years, vermicomposting has emerged as a viable solution for the management of nontoxic and biodegradable solid wastes.

Various species of earthworms have efficiently been utilized for the vermicomposting of cattle manure, domestic waste, complex agroindustrial residues, and sewage sludge. Huge quantities of lignocellulosic waste as phytomass generated annually may be utilized as substrate to produce vermicompost. This chapter discusses various waste management techniques—specifically, vermitechnology for lignocellulosic waste utilization. A brief account of various studies on vermicomposting of lignocellulosic waste is also presented.

15.2 Lignocellulosic Waste

Lignocellulosic waste mainly refers to phytomass composed of cellulose, hemicellulose, and lignin. These are polymers of carbohydrates forming major structural units of all plant materials. Generally, lignocellulosic waste comprises ~40%–50% cellulose, 20%–30% hemicellulose, and 10%–25% lignin (Anwar et al. 2014). However, composition of these structural units varies depending on the origin of the biomass. Lignocellulosic waste can be categorized as forestry residues (sawdust, paper mill waste, wood chips, grasses, waste paper, etc.), agricultural residues (crop straw, stover, peelings, cobs, stalks, nutshells, nonfood seeds, bagasse, etc.), domestic wastes of lignocellulosic nature (vegetable waste, fruit

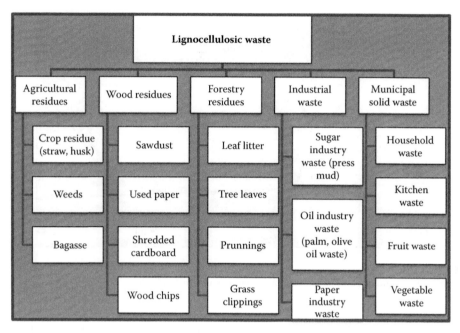

FIGURE 15.1
Classification of lignocellulosic waste.

waste, food industry residues), and municipal solid wastes (Sanchez 2009). Broad classification of lignocellulosic waste based on its origin is given in Figure 15.1.

Lignocellulosic resources can be converted into value-added bioproducts such as biofuels, enzymes, acids, compost, etc., by various processing methods. But a major problem in the conversion of lignocellulosic biomass into value-added products is multistep processing that includes pretreatment (mechanical, chemical, or biological) followed by enzymatic hydrolysis and fermentation processes (Wyman 1999). Degradation of the lignocellulosic waste also takes more time because of its structural complexity (Buswell and Odier 1995). So, there is a need to explore alternative eco-friendly and economically cheaper methods for the proper utilization of lignocellulosic waste. Vermicomposting offers potential opportunities for economic utilization of lignocellulosic wastes.

15.3 Waste Management Methods

Various factors influencing the waste management systems of a nation, include its socioeconomic conditions, climate, etc. In the present era, waste management systems are changing all over the globe. Figure 15.2 presents an overview of various waste management technologies prevalent worldwide. Land scarcity has caused several countries to move toward recycling processes (Lim et al. 2016).

The concept of sustainable development and environmental protection gives rise to a new concept—that is, the "zero waste" concept. Zero waste is utilization of organic residues produced from diverse fields like agriculture, forestry, municipalities, and industries.

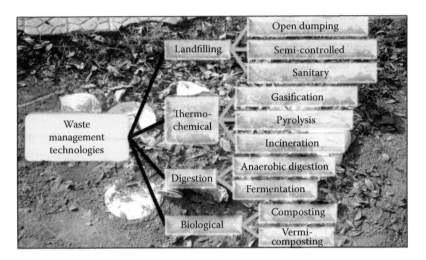

FIGURE 15.2
Various physical-, chemical-, and biological-waste management technologies.

Biological processes such as composting and vermicomposting have been widely recognized in converting organic materials into nutrient-rich fertilizers and soil conditioners.

15.3.1 Landfilling

Landfilling is a common and dominant method for the disposal of solid wastes (Laner et al. 2012). In developing nations, more than 90% of solid waste in cities and towns is directly disposed off in an unscientific manner. In this method, waste is deposited in specially designed areas with an impervious layer (man-made or natural). Landfills are classified into three main types: open landfills, semicontrolled landfills, and sanitary landfills (Singh et al. 2011). In India, there is no proper management for covering, discharge of leachate, and emission of landfill gases. Sanitary landfills avoid the problem of open spreading as they are engineered disposal alternatives, but they release obnoxious gases. Many problems are associated with landfilling that adversely affect the environment and create aesthetic hazards. Leachate carries organic and inorganic contaminants, contributing to greenhouse gas emission (Romero et al. 2013). Greenhouse gas emissions by developing countries are estimated to account for about 29% of global emissions and are projected to increase about 64% by 2030 (Lim et al. 2016). These sites also contribute to the pollution load of surface and groundwaters, adversely affecting the aquatic flora and fauna. In addition, most landfills require proper maintenance and continuous care after their closure, requiring extra costs.

15.3.2 Thermochemical Conversion

15.3.2.1 Gasification and Pyrolysis

Pyrolysis is the thermal degradation of waste in the absence of air to produce gas (often termed syngas), liquid (pyrolysis oil), or solid (char, mainly ash and carbon). Pyrolysis generally takes place between 400°C and 1,000°C. However, gasification takes place at higher temperatures of 1,000°C–1,400°C in a controlled supply of oxygen. These techniques use heat and an oxygen-starved environment to convert biomass into other forms. Syngas is mainly composed of carbon monoxide and hydrogen (85%), with smaller quantities of carbon dioxide, nitrogen, methane, and various other hydrocarbon gases. These processes produce a mixture of combustible and noncombustible gases as well as pyroligenous liquid. All of these products have a high heat value and can be utilized. These processes also reduce the emission of sulfur dioxide and particulate matter.

15.3.2.2 Incineration

Incineration is a thermal waste management process in which raw or unprocessed waste is used as fuel; the process is carried out in the presence of a sufficient quantity of air to oxidize the waste. Incineration is the combustion of wastes under controlled conditions at 850°C–1,100°C in an enclosed structure; waste is converted to carbon dioxide, water, and noncombustible materials with solid residues. Incineration can reduce the volume of waste by 90%; however the drawbacks associated with incineration include high capital, technical and operation costs. Incineration is also questioned for destroying not only raw materials, but also all the energy, water, and other natural resources used to produce it. Furthermore, misuse of incineration poses risks to the environment and human health because of a less positive energy balance compared to recycling. The most important pollutants with incineration are particles, acidic gases and aerosols, metals, and organic compounds.

15.3.3 Anaerobic Digestion

Anaerobic digestion is biological decomposition of organic materials in the absence of oxygen to produce a mixture of methane and carbon dioxide (biogas) and a digestate (slurry) with high nutrient value. The anaerobic digestion of the organic waste occurs in three different stages: hydrolysis, acidogenesis, and methanogenesis. It has widely been investigated as a waste management technique for the production of energy in various countries. Lignocellulosic wastes are the most suitable organic wastes for methanogenesis. The acid-forming bacteria during the conversion process utilize the amount of oxygen remaining in the medium and make the environment anaerobic.

15.3.4 Recycling

Recycling is another approach to reducing waste by the process of separating, collecting, and preparing waste substrates in a way in which they could be remanufactured into useful material. Recycling is the most desirable option for waste management. A hindrance to successful recycling practice, especially in developing countries, is the lack of better collection services and facilities for sorting and processing (Hoornweg and Bhada-Tata 2012).

15.3.5 Composting and Vermicomposting

Composting is a natural phenomenon of biological decomposition of organic wastes driven by the soil microfauna under aerobic or anaerobic conditions. Similarly, vermicomposting is also a biological decomposition process that is driven by earthworms and assisted by microorganisms. The end products obtained after these processes—composts and vermicomposts, respectively—are used as nutrient-rich organic fertilizers. Management of waste employing composting and vermicomposting is highly favored due to technical simplicity and economics. Economically cheaper nature of these techniques also makes them feasible for developing countries. These biological decomposition processes can be considered as a sustainable waste management strategy, which is in line with the zero waste concept.

15.4 Vermicomposting Technology

15.4.1 Vermicomposting

Vermicomposting is a decomposition process involving interactions between earthworms and microorganisms. It is an innovative and cost-effective green biological tool used for the recycling of various wastes into stabilized, finely divided, peat-like material called vermicompost (Ndegwa and Thompson 2001). Earthworms play the role of mechanical blender by comminuting and fragmenting the waste substrates. Earthworms are accompanied by microorganisms for the biochemical degradation of the organic matter that alter the biological, physical, and chemical properties of waste material. Earthworm gut acts as a reactor and during passage of organic fragments increases its surface area for further microbial action and bacteria-rich excrements as vermicast. Vermicompost is used as organic manure as it has an ideal C:N ratio, high porosity and water-holding capacity, and, most importantly, nutrients in a form readily taken up by plants. Vermicompost application in agricultural fields also eliminates pathogens as it contains antibacterial properties.

15.4.2 Substrates Used as Feedstock for Vermitechnology

Earthworms are considered voracious feeders and they can consume waste up to half of their body weight per day (Munroe 2007). Naturally, earthworms feed mainly on dead and decaying organic matter. In captivity, any type of organic matter generated from plants and animals can be used as feed for earthworms. Considerable work has been carried out on vermicomposting of various organic materials starting from simple waste to complex waste. Different waste residues such as animal excreta, agricultural residues, domestic waste, sewage sludge, paper waste, industrial sludge, coconut husk, palm oil residues, vinasses, etc., have been used as earthworm feed in various research trials. However, feedstock needs precomposting before vermicomposting. Bedding material to be used for vermicomposting must have high absorbency, good bulking potential, and a high protein/nitrogen ratio. Examples of common bedding/bulking substrates are various animal manures (cow, buffalo, rabbit, goat, sheep, horse, etc.), wood residues, sawdust, municipal solid waste, industrial waste, and agricultural waste.

15.4.3 Vermicomposting Procedure

The process of vermicomposting occurs in two different phases: active and maturation. Earthworms process waste by modifying its physical state and microbial composition (Lores et al. 2006) during the active phase. In the maturation phase, earthworms move toward the fresh layers of waste. During this phase, the workload is mainly taken on by the microbes that take over the decomposition of the earthworm processed waste (Dominguez and Edwards 2011). Vermicomposting is a bio-oxidative process that occurs under mesophilic

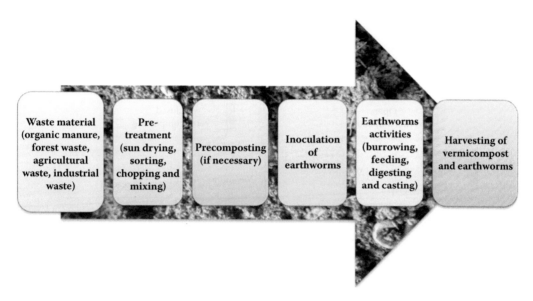

FIGURE 15.3
Stepwise procedure for the process of vermicomposting.

conditions aided by the biochemical action of microorganisms (Dominguez and Edwards 2010). Vermicomposting depends on several factors, including waste material quality, pH, temperature, moisture, aeration, etc., and the earthworm species used. The stepwise procedure (Figure 15.3) for vermicomposting is as follows:

1. **Waste collection:** The first requirement for vermicomposting is collection of nontoxic biodegradable waste to be used as feedstock for earthworms.
2. **Pretreatment:** Waste may require pretreatment such as sorting, drying, grinding, and chopping for crop straw, sewage sludge, wigs, cobs, woody residues, etc.
3. **Precomposting:** It is essential to make the feedstock palatable to earthworms and release toxic gases, if any. This may take 3–4 weeks.
4. **Earthworm inoculation:** Worms are inoculated to the feedstock after precomposting.
5. **Earthworm action on waste:** Earthworms degrade the feedstock by various activities such as digestion, fragmentation, and excretion and convert it into nutrient-rich vermicompost.
6. **Harvesting:** At last peat-like, homogenized, odor-free vermicompost is obtained and then collected and stored for agronomic purposes. Earthworms can be utilized for animal feed or subjected to further vermicomposting of other feed stocks.

15.4.4 Types of Vermicomposting

Several types of vermicomposting systems have been studied by different researchers to provide better habitats for earthworms. Different types, from low-cost simple technology to high-cost complex systems, are available. Munroe (2007) suggested three basic types of vermicomposting systems at a commercial level: windrows, beds or bins, and flow-through reactors. However, vermitechnology divides into two broad types: open and

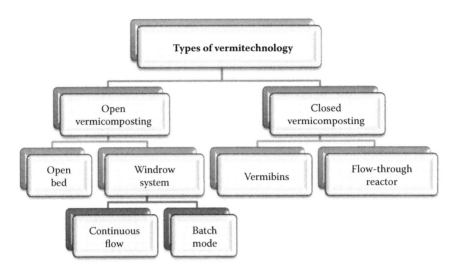

FIGURE 15.4
Types of vermitechnology systems.

closed vermicomposting systems. Figure 15.4 presents a broad classification of vermitechnology systems.

15.4.4.1 Open Vermicomposting System

1. **Open bed:** In this traditional method, feedstocks are placed in open places in the form of pits, heaps, or windrows. This type of system is applicable mainly on a large scale and is weather dependent. Dominguez and Edwards (2011) have classified open vermicomposting systems as low cost systems.

2. **Windrow system:** In this system mixtures of different feedstocks are placed in long and narrow piles. The most common types of windrow systems are batch flow and continuous flow systems (Edwards and Bohlen 1996). Batch flows, also known as static pile windrows, are piles of mixed bedding and feedstock inoculated with earthworms and left for the process of vermicomposting to be completed. Continuous flow windrows are similar to batch flows except that in a continuous flow windrow, waste is not mixed and placed as a batch but deposited as a continuous flow system.

15.4.4.2 Closed Vermicomposting System

Several closed vermicomposting systems also have been proposed:

1. **Vermibins:** Bins are closed containers of small to big size that are mainly used in urban areas.

2. **Flow-through reactors:** In this type of system, earthworms are allowed to live in a closed box; feedstock is added from the top and vermicompost is collected from the bottom.

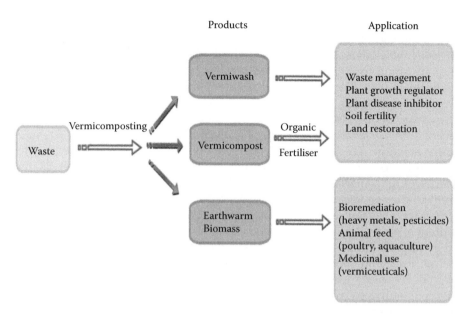

FIGURE 15.5
Products of vermitechnology and their applications in various fields.

15.4.5 Application of Vermicomposting Technology

Vermicomposting is a multifaceted technology and its application has expanded into different fields, including waste management, soil fertility improvement, land reclamation, bioremediation, feed supplement in aquaculture and poultry, vermiceuticals, etc. (Figure 15.5).

Earthworms are widely used as feed material to promote the fishery and poultry industries and even for manufacture of protein food for human consumption. Earthworms are rich in proteins (65%) with amino acid profiles similar to fish with 70%–80% high-quality essential amino acids lysine and methionine. Mohanta et al. (2016) reported 52% whole body protein and 18% lipid contents of earthworm, *Eisenia fetida*, used as feed for *Labeo rohita*. In addition, earthworm tissues contain a preponderance of long chain fatty acids and an adequate mineral content (Hansen and Czochanska 1975). They also have an excellent range of vitamins, being rich in niacin and vitamin B12. Studies are available on the use of earthworm powder as a feed ingredient for making fish feed (Ngoc et al. 2016) or fish fed with dried, frozen, or live earthworms (Kostecka and Paczka 2006).

15.5 Different Species of Earthworms

About 4,000 species of earthworms are distributed all over the world, and most of them are terrestrial. Based on ecological and trophic (feed) functions, earthworms are classified as epigeic, anecic, and endogeic (Bhatnagar and Palta 1996). Epigeics are smaller in size, with a short life cycle but high reproduction rate and regeneration. They reside in the superficial soil surface within litters feeding on the surface litter and mineralize them. Feeding habits of these earthworms are phytophagous; they are highly tolerant and efficient biodegraders. Examples are *Eisenia fetida, Lumbricus rubellus, Lumbricus astaneus, Bimastus minusculus, Bimastus eiseni, Dendrodrilus rubidus,* and *Dendrobaena veneta.* Endogeics are small to large with a medium life cycle and reproduction rate. They are moderately tolerant to disturbance and are geophagous feeders on particulate organic matter and soil. Endogeics are mainly used for soil improvement as they efficiently utilize energy from poor soils. Examples are *Aporrectodea caliginosa, Aporrectodea trapezoides, Aporrectodea rosea, Pontoscolex corethrurus, Allolobophora chlorotica,* and *Aminthas* sp. Aneceic earthworms are large-sized, dorsally pigmented worms with low reproductive rates. They are very sensitive to disturbance and have a phytogeophagous feeding habit. Examples are *Lumbricus terrestris, Lumbricus polyphemus,* and *Aporrectodea longa.*

Different studies on vermicomposting have reported a range of earthworm species utilized for vermiconversion of wastes (Elvira et al. 1997). Earthworms of different species and different ecological niches differ in their ability to fragment various organic wastes (Lattaud et al. 1998). From temperate regions, the two most suitable species reported are *Dendrobaena veneta* and *Lumbricus rubellus.* However, in the Tropics, *Eisenia fetida, Eudrilus eugeniae,* and *Perionyx* species are potentially most widely used for stabilizing organic waste. Various authors have studied the survival, growth, mortality, and reproduction of different earthworm species under laboratory conditions as well as on a commercial scale. Suthar (2007b) studied the decomposition efficiency of *Perionyx sansibaricus* (Perrier) for agriculture waste, farm yard manure, and urban solid waste. Gajalakshmi et al. (2001) used the manure worm *Eudrilus eugeniae* for vermicomposting because of its high growth rate and reproduction in different types of organic wastes. Liu et al. (2005) studied the ability of *Eisenia fetida* to reduce copper and cadmium availability in sewage sludge. Results of these studies offered different conclusions: Some species were more prolific, some grew more rapidly, and some gained large biomass quickly, contributing in different ways to vermicompost and earthworm production. This may mainly be due to species-specific differences in the biology and ecology of these earthworms.

Researchers have also used combinations of different species of worms (polyculture) for waste degradation (Kaviraj and Sharma 2003). A comparative study was conducted by Kaviraj and Sharma (2003) using exotic (*Eisenia fetida*) and local (*Lempito mauritii*) species of earthworm. Suthar (2007a) also compared decomposition efficiencies of monoculture and polyculture vermibins using epigeic (*Eisenia fetida*) and anecic (*Lempito mauritii*) earthworms and suggested that polyculture is more appropriate for rapid recycling of organic wastes.

15.6 Vermicomposting of Lignocellulosic Waste

Various authors have attempted the vermicomposting of different lignocellulosic waste. A brief review of such studies is given next.

15.6.1 Forestry Residues

Leaf litter, dried leaves, and other plant residues create major disposal problems in forestry systems. Recycling of this waste could be a good source of organic matter and nutrient flux back to the environment, contributing to soil fertility (Chaudhuri et al. 2016). Forest leaf litter can be converted into organic manure for sustainable soil fertility management programs. Leaf litter has become the most common and widely used feedstock for vermicomposting. Table 15.1 encapsulates various studies conducted on the vermicomposting of leaf litter.

Mango leaf litter was used as feed for composting and vermicomposting by Gajalakshmi et al. (2005). Results showed that earthworms grow and reproduce well in mango leaf litter and different earthworm density produced different amounts of vermicast. Vermicomposting of neem (*Azadirachta indica*) leaves was conducted in the reactors using *Eudrilus eugeniae* (Gajalakshmi and Abbasi 2004). Earthworms converted up to 7% of the neem leaves into vermicompost per day and also grew and reproduced more rapidly in the neem-fed vermireactors. Use of precomposted leaf litter of acacia (*Acacia auriculiformis*) was studied by Ganesh et al. (2009) employing *Eudrilus eugeniae* and *Lampito mauritii* worm species. The results showed that vermicompost output was dependent on surface area-to-volume ratio.

Suthar and Gairola (2014) used leaf litter of silver oak and bamboo spiked with cow dung for vermicomposting. The results showed an increase in electrical conductivity,

TABLE 15.1

Vermicomposting of Leaf Litter Employing Various Species of Earthworms

Sr. No.	Waste (Leaf Litter/ Plant Leaves)	Earthworm Species	Bulking Agent	Duration (in days)	Ref.
1.	Acacia	*L. mauritii, E. eugeniae*	CD (cow dung)	49	Ganesh et al. 2009
2.	Leaves of Indian mast tree, pearl millet, corn grass	*L. mauritii, P. ceylanesis*	CD	120	Karmegam and Daniel 2009
3.	Parthenium, neem leaves	*E. fetida*	CD	45	Sivakumar et al. 2009
4.	Silkworm litter	*E. fetida*	–	–	Rajasekar and Karmegam 2009
5.	Avaram, Sababul	*E. fetida, E. eugeniae*	CD	60	Sivasankari et al. 2010
6.	Leaf litter	*E. eugeniae*	CD	25	Ravindran et al. 2010
7.	Mango leaf litter	*P. ceylanensis*	CD	60	Prakash and Karmegam 2010
8.	Sago industry waste	*E. fetida*	CD, poultry manure	45	Subramanian et al. 2010
9.	Babul leaves, dhaincha leaves, paddy straw, wheat straw, sawdust	*E. fetida, E. eugeniae, P. excavates*	CD	120	Indrajeet et al. 2010
10.	Acacia, eucalyptus leaf litter	*E. fetida*	CD	45	Srivastava et al. 2011
11.	Mango, guava leaves	*E. eugeniae*	CD	60	Vasanthi et al. 2013
12.	Silver oak and bamboo leaf litter	*E. fetida*	CD	70	Suthar and Gairola 2014
13.	Cashew leaf litter	*P. excavatus*	CD, sheep and horse dung	60	Parthasarathi et al. 2016

ash content, nitrogen phosphorus, and calcium, as well as a decrease in the C:N ratio. Furthermore, vermicomposting increased the microbial populations of bacteria (423.33–684.0 colony-forming units [CFUs] $\times$ 10^5 g^{-1}), fungi (22.0–36.67 CFUs $\times$ 10^5 g^{-1}), and actinomycetes (107.67–141.67 CFUs $\times$ 10^5 g^{-1}). Parthasarathi et al. (2016) used cashew leaf litter as lignocellulosic waste resources for vermicomposting after mixing with animal dung (horse, cow, sheep) using the epigeic earthworm *Perionyx excavatus* (Perrier). The study revealed that lignin, cellulose, and hemicellulose content in the final product decreased compared to initial substrates. This may be due to joint action of various lignocellulolytic microflora present in the earthworm intestine.

15.6.2 Food Waste

Management of food waste is a challenge for scientists and policy makers. Hanc and Pliva (2013) reported that about one-third of food is lost or wasted; this represents 1.3 billion tons. Various studies conducted in past years indicate that vermicomposting can be a viable option for the management of food waste. Vermicomposting of food industry waste was carried out after spiking with cow dung, biogas plant slurry, and poultry droppings as bulking material in different ratios (Garg et al. 2012). Hanc and Pliva (2013) made an attempt to vermicompost kitchen biowaste (precomposted and raw) after mixing with paper and wood chips. The results showed that mixture of paper and kitchen waste is a suitable feedstock for earthworms and that vermicomposting finally enhances the nutrient regime and its availability to plants. Paper waste seems to be a very suitable feed for earthworms because they are able to produce cellulose enzymes for its decomposition.

Fornes et al. (2012) used tomato waste for composting, vermicomposting, and a combination of both processes. Compost and vermicompost obtained were found suitable for horticultural purposes due to their manurial value. Garg et al. (2006) reported conversion of different types of organic substrates (kitchen waste, agroresidues, institutional and industrial wastes including textile industry sludge and fibers) into vermicompost. The percentage of nitrogen–phosphorus–potassium (NPK) in vermicompost was higher than in feedstock. Based on these studies it can be concluded that NPK content in vermicompost depends on the feedstock quality.

15.6.3 Fruit and Vegetable Waste

Fruit and vegetable wastes constitute a large portion of the total biodegradable fraction of wastes (Huang et al. 2016). High biodegradability and moisture content of fruit and vegetable wastes make them ideal feedstock for earthworms. Various studies conducted on the vermicomposting of fruit and vegetable wastes are given in Table 15.2.

Jadia and Fulekar (2008) developed a hydro-based operating bioreactor for aeration and turning of waste during the vermicomposting of vegetable waste. Vermicompost produced this way had high nutrient content. Huang et al. (2014) studied the effect of earthworms on fragmenting fresh fruit and vegetable wastes for 35 weeks. Quantitative polymerase chain reaction (PCR) revealed enhanced bacterial and fungal densities in vermicompost. Huang et al. (2016) investigated vermistabilization of fresh fruit and vegetable wastes (banana peels, watermelon peels, cabbage, lettuce, and potato) and analyzed biological properties of the end product as a microbial population using quantitative PCR. Earthworms' activity on waste enhances bacterial diversity with significant greater populations of actinobacteria and ammonia oxidizing bacteria compared to control. However, vermicompost quality was variable depending on raw waste.

TABLE 15.2

Management of Fruit and Vegetable Waste Using Vermitechnology in the Past

Sr. No.	Waste (Fruit/ Vegetable Waste)	Earthworm Species	Bulking Agent	Duration (in days)	Ref.
1.	Fruit waste	*E. fetida,* *E. eugeniae*	–	45	Laxshmi Prabha et al. 2007
2.	Vegetable waste	*E. fetida*	–	140	Jadia and Fulekar 2008
3.	Pineapple waste	*E. eugeniae*	Soil leaf litter	140	Mainoo et al. 2009
4.	Vegetable waste, biogas slurry	*Allolobophora Parva*	CD, biogas slurry	75	Suthar 2009b
5.	Vegetable waste	*E. fetida*	CD, wheat straw, biogas slurry	105	Suthar, 2009a
6.	Vegetable waste	*E. fetida,* *E. eugeniae,* *P. excavates*	CD	150	Chauhan et al. 2010
7.	Tomato-fruit waste	*E. fetida*	–	150	Fernandez Gomez et al. 2010a
8.	Greenhouse vegetable waste	*E. andrei*	CD, Straw	84	Fernandez Gomez et al. 2010b
9.	Vegetable waste	*E. fetida,* *E. eugeniae*	CD	45	Shanmuga and Lakshmi Prabha 2011
10.	Vegetable waste	*E. fetida,* *E. eugeniae,* *P. excavatus*	CD, saw dust	45	Khwairakpam and Kalamdhad 2011
11.	Grape marc	*E. andrei*	–	15	Brandona et al. 2011
12.	Vegetable waste	*E. fetida*	CD	105	Garg and Gupta 2011
13.	Vegetable waste	*L. rubellus*	CD	21	Palaniappan and Thiruganasambadam 2012
14.	Market refused fruits	*E. eugeniae*	Rice husk	63	Lim et al. 2012
15.	Fruit, vegetable waste	*E. fetida*	Horse manure, wheat straw	80	Asadollahfardi and Mohebi 2012
16.	Fruit, vegetable waste, pruning waste	*E. andrei,* *E. fetida*	–	42	Lleo et al. 2013
17.	Fruit waste, vegetable waste, leaves	*L. rubellus*	CD, mango foliage, sawdust	60	Sethuraman and Kavitha 2013
18.	Fruit and vegetable waste	*E. fetida*	–	35	Huang et al. 2014
19.	Fruit and vegetable waste	*E. fetida*	–	28	Huang et al. 2016
20.	Vegetable waste and rice straw	*E. fetida*		60	Hussain et al. 2016d

Hussain et al. (2016d) conducted a study of various combinations of vegetable market waste, rice straw, and cow dung employing *Eisenia fetida* and *Perionyx excavatus*. The main aim of the work was to identify and isolate the N-fixing and P-solubilizing bacteria from earthworm gut. Results revealed that the microbial population increased during vermicomposting and that isolated strains of bacteria significantly improved soil health and crop growth.

15.6.4 Municipal Solid Waste

Global generation of municipal solid waste is projected to increase to 70% by 2025, with an average annual increase of +5.3%. Of municipal solid waste, 40% is organic in nature, so this fraction of waste can be vermicomposted. Various attempts that have been made to vermicompost the municipally solid waste are reviewed in this section.

Villar et al. (2016) studied the degradation of fresh and composted municipal sewage sludge employing *Eisenia andrei*. Composting followed by vermicomposting enhanced the waste communition correlated with greater decrease in the phosphorus and carbon cycling enzymes. Sahariah et al. (2015) explored the possibility of vermicomposting of municipal solid waste after mixing it with cow dung. Results revealed that vermicomposting enhanced NPK content of the final product and reduced the heavy-metal content. Yang et al. (2014) studied the vermicomposting of sewage sludge and reported an increase in nitrate, electrical conductivity, and degree of humification, as well as reduction in pH and water-extractable carbon. Guar gum industry waste also proved to be a suitable feedstock for vermicomposting. Elvira et al. (1997) reported the vermicomposting of solid paper mill sludge with sewage sludge mixed in different ratios and reported that a 3:2 ratio resulted to highest growth rate of worms. However, the efficiency of worms at higher sludge proportion in feedstock decreased due to unfavorable conditions (Yadav and Garg 2010). Madan and Yadav (2012) used distillery sludge mixed with agricultural waste, municipal waste, and poultry waste. Three different combinations of distillery sludge with different wastes in ratios of 1:1, 1:3, and 3:1 were prepared. The results showed that macronutrients increased after vermicomposting but that organic carbon decreased.

15.6.5 Agroindustrial Waste

Intensive agriculture and food processing technologies have led to generation of huge quantities of agroindustrial residues and wastes—proper management of which using vermicomposting could be a boon for waste recycling and soil fertility (Table 15.3). Agroindustrial biomass comprising lignocellulosic waste is inexpensive, renewable, and abundant and provides a natural resource for waste recycling.

15.6.5.1 Agricultural Waste

Agricultural waste is a by-product generated from various agricultural activities such as collection and processing of phytomass. Numbers of studies have shown that vermicomposting of agricultural waste gives a product rich in nutrients with higher humic and fulvic acids. Pramanik et al. (2016) evaluated the possibility of recycling prunings of the tea plant (*Camellia sinensis* L.) to nutrient-rich vermifertilizer via vermicomposting. The study further explored the vermicomposting technique for replacing cattle manure addition with partially decomposed plant residues (weeds). Vermicomposting decreased total organic carbon content and increased total concentrations of major nutrients (NPK) in the final vermicompost. The vermiprocess also enhanced cellulolytic microbial populations and cellulase enzyme activity, which in turn improved microbial properties of the vermicompost.

15.6.5.2 Crop Residue

Vermitechnology is an eco-friendly technique in the recycling of crop residues (straw, cobs, husk, stalks) left after harvesting. A large proportion of the crop nutrient input during

TABLE 15.3

Studies on Vermicomposting of Lignocellulosic Waste from Various Agroindustrial Sources

Sr. No.	Mixed Agroindustrial Residues	Earthworm Species	Bulking Substrate	Time Period (days)	Ref.
1.	Kitchen waste, agroresidue, institutional waste	*E. fetida*	CD and soil	100	Garg et al. 2006
2.	Grasses, aquatic weeds, municipal solid waste (MSW)	*E. fetida*	CD	85	Pramanik et al. 2007
3.	Crop residue, vegetable waste, leaf litter	*P. sansibaricus*	CD	150	Suthar 2007b
4.	Green forages (grasses, green leaves of vegetables, herbs and plant materials)	*E. fetida*		85	Tejada et al. 2008
5.	Water hyacinth, sugar cane bagasse, rice husk	*E. fetida*	CD	90	Saini et al. 2008
6.	Olive pomace	*E. andrei*	CD	240	Plaza et al.2008
7.	Indian mast tree	*P. ceylanensis*	CD	45	Prakash et al. 2008
8.	Residues of wheat straw, millets, and pulse	*E. fetida*	CD + sheep manure	150	Suthar 2009a
9.	Coffee grounds, kitchen waste	*L. rubellus*	CD	49	Adi and Noor 2009
10.	Bagasse, coir	*E. fetida*	CD	60	Pramanik 2010
11.	Vinasse	*E. eugeniae, E. fetida*	Fly ash	70	Pramanik and Chung 2011
12.	Flower waste	*E. eugeniae*	CD	41	Shouche et al. 2011
13.	Coffee husk, Enset waste (*Enset ventricosum*), khat waste (*Catha edulis*), vegetable waste	*E. andrei*	CD	90	Degefe et al. 2012
14.	Tomato crop waste	*E. fetida*	Almond shell	198	Fornes et al. 2012
15.	Kitchen waste	*E. fetida*	Wood chips and paper	150	Hanc and Pliva 2013
16.	Water hyacinth, parthenium, food industry sludge	*E. fetida*	CD	91	Yadav and Garg 2013
17.	Garden, kitchen waste	*E. fetida*			Wani et al. 2013
18.	Flower, farmyard, kitchen wastes	*E. fetida*	CD	120	Singh et al. 2013.
19.	Grasses	*E. eugeniae*	Pig manure	100	Nweke 2013
20.	Coir pith	*E. eugeniae*	CD	60	Nattudurai et al. 2013
21.	Leaf litter, vegetable waste, coffee waste, flower waste, May flower waste	*E. eugeniae*	CD	90	Mujeebunisa et al. 2013
22.	Temple waste, kitchen waste, farmyard waste	*E. fetida*	CD	120	Singh et al. 2013
23.	Vine shoots, vinasse, olive cakes	*E. fetida*	Goat manure	105	Castillo et al. 2013
24.	Grass clippings	*E. fetida*	Horse manure, sludge	90	Manakova et al. 2014
25.	Mushroom residues	*E. fetida*	CD	120	Song et al. 2014

(Continued)

TABLE 15.3 (CONTINUED)

Studies on Vermicomposting of Lignocellulosic Waste from Various Agroindustrial Sources

Sr. No.	Mixed Agroindustrial Residues	Earthworm Species	Bulking Substrate	Time Period (days)	Ref.
26.	Apple pomace, wheat straw	*Eisenia spp.*	Beef manure	30	Hanc and Chadimova 2014
27.	Rice straw, water hyacinth, Ipomoea	*E. eugeniae*	CD		Mahantaa et al. 2014
28.	Flower waste	*E. fetida*	CD	80	Kumar et al. 2015
29.	Municipal solid waste	*Metaphire posthuma, E. fetida*	CD	60	Sahariah et al. 2015
30.	Fly ash	*E. fetida*	CD, paper waste mixture	70	Mupambwa et al. 2016
31.	Filter cake	*E. fetida*	Cattle manure	120	Busato et al. 2016
32.	Palm oil mill by-product	*E. eugeniae*	Rice straw	28	Lim and Wu 2016
33.	Coconut husk	*E. eugeniae*	Pig slurry, poultry manure	120	Swarnam et al. 2016
34.	Chicken manure and shredded paper waste	*E. fetida*	–	49	Ravindran and Mnkeni 2016
35.	Municipal sewage sludge	*E. andrei*	Wood chips	112	Villar et al. 2016

cultivation remains in the form of plant residues. About 30%–35% of applied nitrogen and phosphate and 70%–80% of potassium remain in the residues of food crops that could be an attractive option as feedstock for earthworms. Vermicomposting of apple pomace waste mixed with straw produced nutrient-rich vermicompost and worm biomass (Hanc and Chadimova 2014). Lim et al. (2011) subjected soybean husk to vermicomposting for 63 days and obtained nutrient-rich vermicompost. Rice husk, an abundant agricultural waste resulting from the rice-milling process, was vermicomposted by Lim et al. (2012). To achieve the objective, rice husk was amended with market-refused fruits (i.e., banana, papaya, and honeydew) and subjected to vermicomposting using *Eudrilus eugeniae*. In this way two types of waste disposal problems (rice husk and fruit waste) were solved. Calcium, potassium, phosphorus, and carbon increased after vermicomposting. When all the treatments were compared, rice husk mixed with papaya waste produced superior quality vermicompost with higher nutritional status.

Earthworms have also been used as bioremediation agents in some studies (Singh and Kalamdhad 2013, Song et al. 2014). Swarnam et al. (2016) vermicomposted coconut husk after mixing it with either pig slurry or poultry manure. They reported that it is possible to convert coconut husk into vermicompost by substitution with animal manure. Best results were obtained with pig slurry mixed in 80:20 ratio reporting higher nutrient recovery (35%–43%) of both total and available forms of nutrients compared to the control. The fertilizing index for different treatments varied from 4.0 to 4.7, indicating high manurial value of the end product. Song et al. (2014) reported mushroom residues as a potential vermicomposting substrate. Earthworm activities accelerated organic matter mineralization (reduction in C:N ratio and increase in total concentrations of N, P, and K) and humification (e.g., increase in humic acid concentration, humification ratio, and humification index).

Availability of heavy metals decreased due to bioaccumulation by earthworms. This may be due to formation of stable metal–humus complexes and the humification process.

15.6.5.3 Weeds

Weeds are unwanted lignocellulosic phytomass that grow very fast in cropland, forests, grasslands, and aquatic bodies. Many weed species are reported that grow at alarming rates and infest millions of hectares of natural and cultivated lands in different regions of the world. Weeds compete with other useful plants for space, water, and nutrients and hence destroy biodiversity, causing ecological as well as economic loss. Furthermore, negative allelopathy and toxicity due to toxic compounds harm the surrounding soil and may affect other plants. Vermicomposting is one of the promising alternatives for the management of weeds as these weeds are also good sources of organic matter and nutrients. Various authors have reported the vermicomposting of various weeds—for example, water hyacinth (Singh and Kalamdhad 2013, Kumar et al. 2015), lantana (Hussain et al. 2016a), *Salvinia* (Hussain et al. 2016b), *Parthenium* (Hussain et al. 2016c). Table 15.4 gives an overview of various studies related to vermicomposting of weeds.

Abbasi et al. (2015) developed a new concept of high-rate vermicomposting of phytomass without any precomposting or manure supplementation. Moreover, the vermicomposting was achieved at a rate three to four times faster than the rate possible until now. For high-rate vermicomposting, a reactor was designed with a high surface area-to-volume and low-aspect ratio and high earthworm-to-feed ratio. This ensures maximum earthworm and substrate contact followed by mixing and aeration of feedstock and easy cast deposition. Tauseef et al. (2013) designed and patented a novel continuously operating

TABLE 15.4

Various Species of Weeds Used as Feedstock for Earthworms during Vermicomposting

Sr. No.	Waste (Weed Species)	Earthworm Species	Bulking Agent	Duration (in days)	Ref.
1.	Water hyacinth	*E. fetida*	CD	147	Gupta et al. 2007
2.	Parthenium, water hyacinth, cannabis	*E. fetida*	CD		Chauhan and Joshi 2010
3.	Water hyacinth	*P. sansibaricus*	Pig manure	56	Zirbes et al. 2011
4.	Parthenium	*E. fetida*	CD	126	Yadav and Garg 2011
5.	Citronella plant waste	*E. eugeniae*	CD	105	Deka et al. 2011
6.	Water hyacinth	*E. fetida*	Poultry manure, shredded cardboard	60	Patil et al. 2012
7.	Parthenium	*E. eugeniae*	CD	45	Rajiv et al. 2013
8.	Lantana	E. fetida	CD	60	Suthar and Sharma 2013
9.	Water hyacinth	*E. fetida*	CD, sawdust	45	Singh and Kalamdhad 2013
10.	Water weed (macrophytes)	*E. fetida*	CD	60	Najar and Khan 2013
11.	Lantana	*E. fetida*	–	–	Hussain et al. 2016a
12.	Salvinia	*E. fetida*	–	–	Hussain et al. 2016b
13.	Parthenium	*E. fetida*	–	–	Hussain et al. 2016c
14.	Salvinia natans	*E. fetida*	CD, sawdust	45	Singh and Kalamdhad 2016

vermireactor system for large-scale vermicomposting of phytomass. Singh and Kalamdhad (2013) reported vermicomposting of water hyacinth after mixing with cattle manure and sawdust using *Eisenia fetida*. The main finding of the study was that vermicomposting reduced the bioavailability of heavy metals (Zn, Ni, Pb, Cd, and Cr) in the vermicompost.

The possibility of recycling Guatemala grass (*Tripsacum andersonii*), a weed in humid subtropical regions, was studied by Pramanik et al. (2016). Results revealed that a high rate of vermicomposting was achieved with partially decomposed weeds and that this could replace cattle manure as a mixing substrate. Najar and Khan (2013) vermicomposted freshwater weeds (macrophytes) using *Eisenia fetida* in various reactors. Five mixtures prepared using *Azolla pinnata*, *Trapa natans*, and *Ceratophyllum demersum* free-floating macrophyte and submerged macrophyte were given as feed to earthworms. Results of the study reported increments in pH, EC, N, and K and decreases in total organic carbon (TOC) and C:N ratio after vermicomposting. Singh and Kalamdhad (2016) studied bioavailability and leachability of various nutrients and heavy metals during vermicomposting of *Salvinia natans*. Results revealed that vermicomposting enhanced the nutrient content. Bioremediation of metals made vermicompost a safe organic manure.

Fourier transform infrared (FTIR) spectra of vermicomposts revealed nearly complete degradation of phenols and sesquiterpene lactones, which are responsible for the allelopathic impacts of lantana (Hussain et al. 2016a). More complex and larger biomolecules are converted into simpler compounds. Rajiv et al. (2013) reported production of parthenin-free vermicompost from *Parthenium hysterophorus*. Hussain et al. (2016c) also explored the potential effects of *Parthenium*-derived vermicompost on germination and growth of common vegetables (green gram, okra, and cucumber). Based on results, it was concluded that negative allelopathic effects of *Parthenium* were completely destroyed during vermicomposting. In this study also FTIR spectra indicated complete disintegration of phenol and the sesquiterpene lactones and reduction in the lignin content.

15.6.6 Industrial Waste

Enormous quantities of liquid, gaseous, or solid wastes are generated by various industries. Vermicomposting has been employed to stabilize the distillery plant sludge spiked with cow dung using *Perionyx excavates* by Suthar and Singh (2008). Lower proportions of distillery sludge in the feedstock resulted in higher earthworm biomass production and better reproduction performance.

Busato et al. (2016) subjected filter cake to vermiprocessing and obtained vermicompost with higher nitrogen content, humic acid content, and enzymes. The results showed a positive correlation of phosphatase with TOC, pH, and water-soluble phosphorus. However, phosphatase activities negatively correlated with humic acid content.

Olive oil cake is lignocellulosic residue left after oil extraction from crude olive cake. Benitez et al. (2005) subjected dry olive cake mixed with municipal biosolids to vermicomposting and extracted humic substances and three hydrolytic enzymes (β-glucosidase, phosphatase, and urease). Addition of municipal biosolids to olive cake increased microbial activity three times higher than adding olive cake alone did. Lim and Wu (2016) used vermicomposting as a suitable technique for recycling of decanter cake (produced as a waste in palm oil industries). Decanter cake and rice straw were subjected to vermiprocessing to convert into organic fertilizer. Various chemical (pH, EC, organic matter content, and C:N ratio) and instrumental (FTIR spectroscopy, thermogravimetric analysis) characterization proved that vermicompost prepared from waste was safe and beneficial for agricultural purposes. Management of food industry sludge mixed with biogas plant slurry via vermicomposting was reported by Yadav and Garg (2010). Results indicated that mixture of 30%–40% food industry sludge with biogas plant slurry can be successfully used as a feedstock for vermicomposting.

15.6.7 Paper Waste

Waste paper mixed with dairy manure was subjected to vermicomposting by Mupondi et al. (2010). In this study, vermicomposting was compared with combined thermophilic composting and vermicomposting. The results showed that both methods were good for biodegradation of dairy manure and paper waste mixtures having a C:N ratio of 30. Differences were observed in the pathogen reduction, where the combining method eliminated *Escherichia coli* completely but vermicomposting only reduced the pathogen population.

Arumugam et al. (2015) reported the vermicomposting of disposable paper cups mixed with cow dung using different ratios employing *Eisenia fetida*. Physicochemical analysis of vermicompost revealed that the best quality vermicomposting was obtained when paper cup/cow dung were mixed in a 1:1 ratio. Mupambwa et al. (2016) evaluated the potential of vermicomposting in the management of fly ash, paper waste inoculated with specialized microbial culture along with *Eisenia fetida*. Results of the study revealed that vermicomposting enhanced microbial and fungal populations and significantly enhanced extractable phosphorus content.

Recently, Ravindran and Mnkeni (2016) experimented with vermicomposting by using shredded waste paper and chicken manure. Results revealed that if feedstock has a C:N ratio of 40 a higher quality vermicompost is obtained. Earthworm gut and cast consisting of phosphate-solubilizing, cellulolytic, amylolytic, proteolytic, and lignocellulolytic microflora helped in decomposition of ligncellulosic compounds (Parthasarathi et al. 2016). Combined activity of microflora in the gut of worm and inoculated lignocellulolytic fungi intensified cellulolysis and lignolysis. Hussain et al. (2016d) isolated N-fixing and P-solubilizing strains of *Serratia* and *Bacillus* from intestines of two earthworm species (*Eisenia fetida* and *Perionyx excavatus*). The isolated bacteria were compared to commercial biofertilizers and significant improvement in soil health and crop growth was reported.

Alidadi et al. (2016) investigated the dynamics of vermicompost maturity prepared from municipal solid waste containing sawdust, boxwood leaves, and cardboard. The study reported that degradation percentage of lignin increased after the 50th day of vermicomposting. The increment may be attributed to the production of cellulase and lignocellulasse enzymes by earthworm gut, which converted cellulose substances to lignin (Benitez et al. 1999).

15.7 Conclusion

Vermicomposting is an eco-friendly technology that involves the use of earthworms as versatile natural bioreactors to harness beneficial microflora, thus converting organic waste into valuable products. Various studies have proved that vermicomposting of lignocellulosic waste can be a good recycling technique to return nutrients to soil. Various lignocellulosic wastes, such as sugarcane bagasse, rice and wheat husks, forest residues, wood, leaves, etc., can serve as an excellent source of plant nutrients. Hence, vermicomposting of these renewable resources offers a sustainable and waste recycling option for waste remediation and organic agricultural practices. Vermicomposting enables quick degradation of ligninocellulolytic materials (lignin, cellulose, hemicellulose) and helps to preserve available resources and soil fertility.

References

Abbasi, S.A., Nayeem Shah, M., Abbasi, T. 2015. Vermicomposting of phytomass: Limitations of the past approaches and the emerging directions. *Journal of Cleaner Production* 93:103–114.

Adi, A.J. and Noor, Z.M. 2009. Waste recycling: Utilization of coffee grounds and kitchen waste in vermicomposting. *Bioresource Technology* 100:1027–1030.

Alidadi, H., Hosseinzadeh, A., Najafpoor, A.A., Esmaili, H., Zanganeh, J., Takab, M.D., Piranloo, F. G. 2016. Waste recycling by vermicomposting: Maturity and quality assessment via dehydrogenase enzyme activity, lignin, water soluble carbon, nitrogen, phosphorous and other indicators. *Journal of Environmental Management* 182:134–140.

Anwar, Z., Gulfraz, M., Irshad, M. 2014. Agro-industrial lignocellulosic biomass a key to unlock the future bio-energy: A brief review. *Journal of Radiation Research and Applied Sciences* 7:163–173.

Arumugam, K., Ganesan, S., Muthunarayanan, V., Vivek, S., Sugumar, S., Munusamy, V. 2015. Potentiality of *Eisenia fetida* to degrade disposable paper cups—An ecofriendly solution to solid waste pollution. *Environment Science and Pollution Research* 22:2868–2876.

Asadollahfardi, G. and Mohebi, A. 2012. Fruits and vegetables residue vermicomposting using earthworm *Eisenia fetida*. *Journal of International Environment and Applied Science* 7(2):261–265.

Benitez, E., Nogales, R., Elvira, C., Masciandaro, G., Ceccanti, B., 1999. Enzyme activities as indicators of the stabilization of sewage sludges composting with *Eisenia foetida*. *Bioresource Technology* 3:297–303.

Benitez, E., Sainz, H. and Nogales, R. 2005. Hydrolytic enzyme activities of extracted humic substances during the vermicomposting of a lignocellulosic olive waste. *Bioresource Technology* 96:785–790.

Bhatnagar, R.K., and Palta, R.K. 1996. *Earthworm—Vermiculture and Vermicomposting*. Kalyani Publishers, New Delhi.

Brandona, M.G., Lazcanob, C., Loresc, M., Dominguez, J. 2011. Short-term stabilization of grape marc through earthworms. *Journal of Hazardous Materials* 187:291–295.

Busato, J.G., Papa, G., Canellas, L.P., Adani, F., Oliveira, A.L., Leao, T.P. 2016. Phosphatase activity and its relationship with physical and chemical parameters during vermicomposting of filter cake and cattle manure. *Journal of Science in Food and Agriculture* 9 6:1223–1230.

Buswell, J.A., and Odier, E. 1995. Lignin biodegradation. *Critical Reviews in Biotechnology* 6:1–60.

Castillo, J.M., Romero, E., Nogales, R. 2013. Dynamics of microbial communities related to biochemical parameters during vermicomposting and maturation of agro industrial lignocelluloses wastes. *Bioresource Technology* 146:345–354.

Chaudhuri, P.S., Paul, T.K. Dey, A., Datta, M., Dey, S.K. 2016. Effects of rubber leaf litter vermicompost on earthworm population and yield of pineapple (*Ananas comosus*) in West Tripura, India. *International Journal of Recycling Organic Waste in Agriculture* 5:93–103.

Chauhan, A. and Joshi, P.C. 2010. Composting of some dangerous and toxic weeds using *Eisenia fetida*. *Journal of American Science* 6(3):1–6.

Chauhan, A., Kumar, S., Singh, A.P., Gupta, M. 2010. Vermicomposting of vegetable wastes with cowdung using three earthworm species *Eisenia foetida*, *Eudrilus eugeniae* and *Perionyx excavatus*. *Natural Science* 8(1).

Degefe, G., Mengistu, S., Dominguez, J. 2012. Vermicomposting as a sustainable practice, to manage Coffee Husk, Enset waste (*Enset ventricosum*), Khat waste (*Catha edulis*), and Vegetable waste amended with Cowdung using an epigeic earthworm, *Eisenia andrei*, (Bouch' 1972). *International Journal of PharmTech Research* 4(1):15.

Deka, H., Deka, S., Baruah, C., Das, J., Hoque, S., Sarma, H., Sarma, N. 2011. Vermicomposting potentiality of *Perionyx excavatus* for recycling of waste biomass of java citronella—An aromatic oil yielding plant. *Bioresource Technolology* 102:11212–11217.

Dominguez, J. and Edwards, C.A. 2010. Relationships between composting and vermicomposting. In *Vermiculture Technology: Earthworms, Organic Wastes, and Environmental Management*, Edwards, C.A., Arancon, N.Q., Sherman, R.L. (eds). CRC Press: Boca Raton, FL, pp. 11–26.

Dominguez, J. and Edwards, C.A. 2011. Biology and ecology of earthworm species used for vermicomposting. In *Vermiculture Technology—Earthworms, Organic Wastes, and Environmental Management*, CRC Press, Boca Raton, FL, pp. 27–40.

Edwards, C.A. and Bohlen, P.J. 1996. *Biology and Ecology of Earthworms*. Chapman & Hall, London.

Elvira, C., Sampedro, L., Dominguez, J. and Mato, S. 1997. Vermicomposting of wastewater sludge from paper pulp industry with nitrogen rich materials. *Soil Biology and Biochemistry* 29:759–762.

Fernandez Gomez, M.J., Nogales, R., Insam, H., Romero, E., Goberna, M. 2010a. Continuous-feeding vermicomposting as a recycling management method to revalue tomato-fruit wastes from greenhouse crops. *Waste Management* 30(12):2461–2468.

Fernandez Gomez, M.J., Romero, E., Nogales, R. 2010b. Feasibility of vermicomposting for vegetable greenhouse waste recycling. *Bioresource Technology* 101(24):9654–9660.

Fornes, F., Mendoza-Hernandez, D., Garcia-de-la-Fuente, R., Abad, M., Belda, R.M. 2012. Composting versus vermicomposting: A comparative study of organic matter evolution through straight and combined processes. *Bioresource Technology* 118:296–305.

Gajalakshmi, S. and Abbasi, S.A. 2004. Neem leaves as a source of fertilizer cum pesticide vermicompost. *Bioresource Technology* 92:291–296.

Gajalakshmi, S., Ramasamy, E., Abbasi, S. 2001. Potential of two epigeic and two anecic earthworm species in vermicomposting of water hyacinth. *Bioresource Technology* 76:177–181.

Gajalakshmi, S., Ramasamy, E.V. and Abbasi, S.A. 2005. Composting–vermicomposting of leaf litter ensuing from the trees of mango (*Mangifera indica*). *Bioresource Technology* 96:1057–1061.

Ganesh, P.S., Gajalakshmi, S. and Abbasi, S.A. 2009. Vermicomposting of the leaf litter of acacia (*Acacia auriculiformis*): Possible roles of reactor geometry, polyphenols, and lignin. *Bioresource Technology* 100:1819–1827.

Garg, P., Gupta, A., Satya, S. 2006. Vermicomposting of different types of waste using *Eisenia foetida*: A comparative study. *Bioresource Technology* 97:391–395.

Garg, V.K. and Gupta, R. 2011. Optimization of cow dung spiked pre-consumer processing vegetable waste for vermicomposting using *Eisenia fetida*. *Ecotoxicology and Environment Safety* 74:19–24.

Garg, V.K., Suthar, S., Yadav, A. 2012. Management of food industry waste employing vermicomposting technology. *Bioresource Technology* 126:437–444.

Gupta, R., Mutiyar, P.K., Rawat, N.K., Saini, M.S., Garg, V.K. 2007. Development of a water hyacinth based vermireactor using an epigeic earthworm *Eisenia foetida*. *Bioresource Technology* 98:2605–2610.

Halbach, T.R. 2013. International trends in solid waste handling: 2013. Solid waste management & recycling export roundtable. Department of Soil, Water and Climate, University of Minnesota, Carlson School of Management Minneapolis, Minneapolis, MN, March 14, 2013.

Hanc, A., and Chadimova, Z. 2014. Nutrient recovery from apple pomace waste by vermicomposting technology. *Bioresource Technology* 168:240–244.

Hanc, A. and Pliva, P. 2013. Vermicomposting technology as a tool for nutrient recovery from kitchen bio-waste. *Journal of Material Cycles and Waste Management* 15:431–439.

Hansen, R.P. and Czochanska, Z. 1975. The fatty acid composition of the lipids of earthworms. *Journal of the Science of Food and Agriculture* 26:961–971.

Hoornweg, D. and Bhada-Tata, P. 2012. *What a Waste: A Global Review of Solid Waste Management.* World Bank, Washington, DC.

Huang, K., Li, F., Wei, Y., Fu, X. and Chen, X. 2014. Effects of earthworms on physiochemical properties and microbial profiles during vermicomposting of fresh fruit and vegetable wastes. *Bioresource Technology* 170:45–52.

Huang, K., Xia, H., Li, F., Wei, Y., Cui, G., Fu, X and Chen, X. 2016. Optimal growth condition of earthworms and their vermicompost features during recycling of five different fresh fruit and vegetable wastes. *Environmental Science and Pollution Research* 23:13569–13575.

Hussain, N., Abbasi, T., and Abbasi, S.A. 2016a. Transformation of toxic and allelopathic lantana into a benign organic fertilizer through vermicomposting. *Spectrochimica Acta Part A: Molecular and Biomolecular Spectroscopy* 163:162–169.

Hussain, N., Abbasi, T., Abbasi, S.A. 2016b. Vermiremediation of an invasive and pernicious weed salvinia (*Salvinia molesta*). *Ecological Engineering* 91:432–440.

Hussain, N., Abbasi, T. and Abbasi, S.A. 2016c. Vermicomposting transforms allelopathic parthenium into a benign organic fertilizer. *Journal of Environmental Management* 180:180–189.

Hussain, N., Singh, A., Saha, S., Kumar, M.V.S., Bhattacharyya, P., Bhattacharya, S.S. 2016d. Excellent N-fixing and P-solubilizing traits in earthworm gut-isolated bacteria: A vermicompost based assessment with vegetable market waste and rice straw feed mixtures. *Bioresource Technology* 222:165–174.

Indrajeet, Rai, S.N., Singh, J. 2010. Vermicomposting of farm garbage in different combination. *Journal of Recent Advance in Applied Science* 25:15–18.

Jadia, C.D. and Fulekar, M.H. 2008. Vermicomposting of vegetable waste: A biophysicochemical process based on hydro-operating bioreactor. *African Journal of Biotechnology* 7(20):3723–3730.

Karmegam, N. and Daniel, T. 2009. Growth, reproductive biology and life cycle of the vermicomposting earthworm *Perionyx ceylanensis* Mich. (Oligochaeta: Megascolecidae). *Bioresource Technology* 100:4790–4796.

Kaviraj and Sharma, S. 2003. Municipal solid waste management through vermicomposting employing exotic and local species of earthworms. *Bioresource Technology* 90:169–173.

Khwairakpam, M. and Kalamdhad, A.S. 2011. Vermicomposting of vegetable wastes amended with cattle manure. *Research Journal of Chemical Sciences* 1(8):49–56.

Kostecka, J. and Pączka, G. 2006. Possible use of earthworm *Eisenia fetida* (Sav.) biomass for breeding aquarium fish. *European Journal of Soil Biology* 42:S231–S233.

Kumar, M.S., Rajiv, P., Rajeshwari, S., Venckatesh, R. 2015. Spectroscopic analysis of vermicompost for determination of nutritional quality. *Spectrochim Acta A:Molecular and Biomolecular Spectroscopy* 135:252–255.

Lakshmi Prabha, M., Indira, A., Jeyaraaj, R. and Srinivasa Rao, D. 2007. Comparative studies on the levels of vitamins during vermicomposting of fruit wastes by *Eudrilus eugeniae* and *Eisenia fetida*. *Applied Ecology and Environmental Research* 5(1):57–61.

Laner, D., Crest, M., Scharff, H., Morris, J.W.F., Barlaz, M.A. 2012. A review of approaches for the long-term management of municipal solid waste landfills. *Waste Management* 32:498–512.

Lattaud, C., Locati, S., Mora, P., Rouland, C., Lavelle, P. 1998. The diversity of the digestive enzymes in the geophagous earthworm. *Applied Soil Ecology* 9:189–195.

Lim, S.L. and Wu, T.Y. 2016. Characterization of matured vermicompost derived from valorization of palm oil mill by product. *Journal of Agriculture and Food Chemistry* 64:1761–1769.

Lim, S.L., Lee, L.H., Wu, T.Y. 2016. Sustainability of using composting and vermicomposting technologies for organic solid waste biotransformation: Recent overview, greenhouse gases emissions and economic analysis. *Journal of Cleaner Production* 111:262–278.

Lim, S.L., Wu, T.Y., Sim, E.Y.S., Lim, P.N., Clarke, C. 2012. Biotranformation of rice husk into organic fertilizer through vermicomposting. *Ecological Engineering* 41:60–64.

Liu, X., Hu, C., Zhang, S. 2005. Effects of earthworm activity on fertility and heavy metal bioavailability in sewage sludge. *Environment International* 31:874–879.

Lleo, T., Albacete, E., Barrena, R., Font, X., Artola, A., Sanchez, A. 2013. Home and vermicomposting as sustainable options for biowaste management. *Journal of Cleaner Production* 47:70–76.

Lores, M., Gomez Brandon, M., Perez-Diaz, D., Dominguez, J. 2006. Using FAME profiles for the characterization of animal wastes and vermicomposts. *Soil Biology and Biochemistry* 38:2993–2996.

Madan, S. and Yadav, A. 2012. Vermicomposting of Distillery sludge with different wastes by using *Eisenia fetida. Advances in Applied Science Research* 3(6):3844–3847.

Mahantaa, K., Jhaa, D.K., Rajkhowab, D.J., Kumar, M. 2014. Microbial enrichment of vermicompost prepared from different plant biomasses and their effect on rice (*Oryza sativa* L.) growth and soil fertility. *Biological Agriculture and Horticulture* 28:241–250.

Mainoo, N.O., Barrington, S., Whalen, J.K., Sampedro, L. 2009. Pilot-scale vermicomposting of pineapple wastes with earthworms native to Accra, Ghana. *Bioresource Technology* 100(23):5872–5875.

Manakova, B., Kuta, J., Svobodova, M., Hofman, J. 2014. Effects of combined composting and vermicomposting of waste sludge on arsenic fate and bioavailability. *Journal of Hazardous Materials* 280:544–551.

Mohanta, K.N., Subramanian, S., Korikanthimath, V.S. 2016. Potential of earthworm (*Eisenia foetida*) as dietary protein source for rohu (*Labeo rohita*) advanced fry. *Cogent Food & Agriculture* 2:1138594.

Mtui, Y.S. 2009. Recent advances in pretreatment of lignocellulosic wastes and production of value added products. *African Journal of Biotechnology* 8(8):1398–1415.

Mujeebunisa, M., Divya, V., Aruna, D. 2013. Effect of leaf litter, vegetable waste, coffee waste, flower waste and may flower waste added vermicompost on weight gain in *Eudrilus eugeniae. Asian Journal of Plant Science and Research* 3(6):1–4.

Munroe, G. 2007. *Manual of On-farm Vermicomposting and Vermiculture.* Pub. of Organic Agriculture Centre of Canada, p. 39.

Mupambwa, H.A., Ravindran, B., Mnkeni, P.N.S. 2016. Potential of Effective micro-organisms and *Eisenia fetida* in enhancing vermi-degradation and nutrient release of fly ash incorporated into cow dung–paper waste mixture. *Waste Management* 48:165–173.

Mupondi, L.T., Mnkeni, P.N.S., Muchaonyerwa, P. 2010. Effectiveness of combined thermophilic composting and vermicomposting on biodegradation and sanitization of mixtures of dairy manure and waste paper. *African Journal of Biotechnology* 9(30):4754–4763.

Najar, I.A. and Khan, A.B. 2013. Management of fresh water weeds (macrophytes) by vermicomposting using *Eisenia fetida. Environmental Science and Pollution Research* 20(9):6406–6417.

Nattudurai, G., EzhilVendan, S., Ramachandran, P.V., Lingathurai, S. 2013. Vermicomposting of coirpith with cowdung by *Eudrilus eugeniae* Kinberg and its efficacy on the growth of *Cyamopsis tetragonaloba* (L) Taub. *Journal of Saudi Society of Agriculture Science* 13(1):23–27.

Ndegwa, P.M. and Thompson, S.A. 2001. Integrating composting and vermicomposting in the treatment of bioconversion of biosolids. *Bioresource Technology* 76:107–112.

Ngoc, T.N., Pucher, J., Becker, K., Focken, U. 2016. Earthworm powder as an alternative protein source in diets for common carp (*Cyprinus carpio* L.). *Aquaculture Research* 47:2917–2927.

Nweke, I.A. 2013. Plant nutrient release composition in vermicompost as influenced by *Eudrilus eugeniae* using different organic diets. *Journal of Ecological Natural Enviroment* 5:346–351.

Palaniappan, S. and Thiruganasambadam, K. 2012. Investigation on application of synthetic nutrients for worm growth rate vermicomposting. *Journal of Urban and Environmental Engineering* 6(1):30–35.

Parthasarathi, K., Balamurugan, M., Prashija, K.V., Jayanthi, L., Basha, S.A. 2016. Potential of *Perionyx excavatus* (Perrier) in lignocellulosic solid waste management and quality vermifertilizer production for soil health. *International Journal of Recycling Organic Waste in Agriculture* 5:65–86.

Patil, J.H., Sanil, P.H., Malini, B.M., Manoj, V., Deepika, D., Chaitra, D. 2012. Vermicomposting of water hyacinth with poultry litter using rotary drum reactor. *Journal of Chemical and Pharmaceutical Research* 4(5):2585–2589.

Plaza, C., Nogales, R., Senesi, N., Benitez, E., Polo, A. 2008. Organic matter humification by vermicomposting of cattle manures alone and mixed with two-phase olive pomace. *Bioresource Technology* 99:5085–5089.

Prakash, M. and Karmegam, N. 2010. Dyanmics of nutrients and microflora during vermicomposting of mango leaf litter using *P. ceylanensis*. *International Journal of Global Environmental Issues* 10(3/4):339–353.

Prakash, M., Jayakumar, M., Karmegam, N. 2008. Physico-chemical characteristics and fungal Flora in the casts of the earthworm, *Perionyx ceylanensis* Mich. Reared in *Polyalthia longifolia* leaf litter. *Journal of Applied Science and Reserach* 4(1):53–57.

Pramanik, P. 2010. Changes in microbial properties and nutrient dynamics in bagasse and coir during vermicomposting: Quantification of fungal biomass through ergosterol estimation in vermicompost. *Waste Management* 30:787–791.

Pramanik, P. and Chung, Y.R. 2011. Changes in fungal population of fly ash and vinasse mixture during vermicomposting by *Eudrilus eugeniae* and *Eisenia fetida*: Documentation of cellulase isozymes in vermicompost. *Waste Management* 31:1169–1175.

Pramanik, P., Ghosh, G.K., Ghosal, P.K., Banik, P. 2007. Changes in organic C, N, P and K and enzyme activities in vermicompost of biodegradable organic wastes under liming and microbial inoculants. *Bioresource Technology* 98:2485–2494.

Pramanik, P., Safique, S., Jahan, A., Bahagat, R.M. 2016. Effect of vermicomposting on treated hard stem leftover wastes from pruning of tea plantation: A novel approach. *Ecological Engineering* 97:410–415.

Rajasekar, K. and Karmegam, N. 2009. Efficiency of the earthworm, *Eisenia fetida* (Sav.) in vermistabilization of silkworm litter mixed with leaf litter. *International Journal of Applied Environmental Science* 4(4):481–486.

Rajiv, P., Rajeshwaria, S., Yadav, R.H., Rajendran, V. 2013. Vermiremediation: Detoxification of Parthenin toxin from Parthenium. *Journal of Hazardous Materials* 262:489–495.

Ravindran, B. and Mnkeni, P.N. 2016. Bio-optimization of the carbon-to-nitrogen ratio for efficient vermicomposting of chicken manure and waste paper using *Eisenia fetida*. *Environmental Science and Pollution Research* 23:16965–16976.

Ravindran, B., Sravani, R., Mandal, A.B., Contreras-Ramos, S.M., Sekaran, G. 2010. Intrumental evidence for biodegradation of tannery waste during vermicomposting process using *Eudrilus eugeniae*. *Journal of Thermal Analysis and Calorimetry*. http://dx.doi.org/ 10.1007/s10973-011-2081-9.

Romero, C., Ramos, P., Costa, C., Marquez, M.C. 2013. Raw and digested municipal waste compost leachate as potential fertilizer: Comparison with a commercial fertilizer. *Journal of Cleaner Production* 59:73–78.

Saha, B.C. 2003. Hemicellulose bioconversion. *Journal of Industrial Microbiology Biotechnology* 30(5):279–291.

Sahariah, B., Goswami, L., Kim, K.H., Bhattachatyya, P. and Bhattacharya S.S. 2015. Metal remediation and biodegradation potential of earthworm species on municipal solid waste a parallel analysis between *Metaphire posthuma* and *Eisenia fetida*. *Bioresource Technology* 180:230–236.

Saini, V.K., Sihag, R.C., Sharma, R.C., Gahlawat, S.K., Gupta, R.K. 2008. Biodegradation of water hyacinth, sugarcane bagasse and rice husk through vermicomposting. *Intersciences Enterprises Ltd*. 1478–9876.

Sethuraman, T.R. and Kavitha, K.V. 2013. Vermicomposting of Green waste using earthworm *Lumbricus rubellus*. *National Environment Pollution Technology* 12(2):371–374.

Shanmuga, M.P. and Lakshmi Prabha, M. 2011. Compartive studies on enzymatic levels of vegetable wastes decomposed by *Eudriles eugeniae* and *Eisenia fetida*. *Advance Techniques in Biology* 11(4):13–15.

Shouche, S., Pandey, A., Bhati, P. 2011. Study about the changes in physical parameters during vermicompsting of floral wastes. *Journal of Environmental Research and Development* 6(1) (July–September).

Singh, A., Jain, A., Sharma, B.K., Abhilash, P.C., Singh, H.B. 2013. Solid waste management of temple floral offerings by vermicomposting using *Eisenia fetida*. *Waste Management* 33:1113–1118.

Singh, J. and Kalamdhad, A.S. 2013. Assessment of bioavailability and leachability of heavy metals during rotary drum composting of green waste (Water hyacinth). *Ecological Engineering* 52:59–69.

Singh, R.P., Singh, P., Araujoc, S.F.M., Ibrahima, H.M., Sulaiman, O. 2011. Management of urban solid waste: Vermicomposting a sustainable option. *Resources, Conservation and Recycling* 55:719–729.

Singh, S. and Kalamdhad, A.S. 2016. Transformation of nutrients and heavy metals during vermicomposting of the invasive green weed *Salvinia natans* using *Eisenia fetida*. *International Journal of Recycling Organic Waste in Agriculture* 5:205–220.

Sivakumar, S., Kasthuri, H., Prabha, D., Senthil, K.S., Subbhuram, C.V., Song, Y.C. 2009. Efficiency of composting parthenium plant and neem leaves in the presence and absence of an oligocahete, *Eisenia fetida*. *Iranian Journal of Environmental Health Science and Engineering* 6(3):201–208.

Sivasankari, B., Anitha, W., Daniel, T. 2010. Effect of application of vermicompost prepared from leaf materials on growth of *Vigna unguiculata* L. Walp. *Journal of Pure and Applied Microbiology* 4(2):895–898.

Song, X., Liu, M., Wu, D. Qi, L., Ye, Jiao, J., Hu, F. 2014. Heavy metal and nutrient changes during vermicomposting animal manure spiked with mushroom residues. *Waste Management* 34:1977–1983.

Srivastava, P.K., Singh, P.C., Gupta, M., Sinha, A., Vaish, A., Shukla, A., Singh, N., Tewari, S.K. 2011. Influence of earthworm culture on fertilization potential and biological activities of vermicomposts prepared from different plant wastes. *Journal of Plant Nutrition and Soil Science* 174(3):420–429.

Subramanian, S., Sivarajan, M., Saravanapriya, S. 2010. Chemical changes during vermicomposting of sago industry solid waste. *Journal of Hazardous Material* 179:318–322.

Suthar, S. 2007a. Vermicomposting potential of *Perionyx sansibaricus* (Perrier) in different waste materials. *Bioresource Technology* 98(6):1231–1237.

Suthar, S. 2007b. Influence of different food sources on growth and reproduction performance of composting epigeics: *Eudrilus eugeniae*, *Perionyx excavatus* and *Perionyx sansibaricus*. *Applied Ecology and Environmental Research* 5:79–92.

Suthar, S. 2009a. Vermicomposting of vegetable-market solid waste using *Eisenia fetida*: Impact of bulking material on earthworm growth and decomposition rate. *Ecological Engineering* 35:914–920.

Suthar, S. 2009b. Potential of *Allolobophora parva* (Oligochaeta) in vermicomposting. *Bioresourse Technology* 100:6422–6427.

Suthar, S. and Gairola, S. 2014. Nutrient recovery from urban forest leaf litter waste solids using *Eisenia fetida*. *Ecological Engineering* 71:660–666.

Suthar, S. and Sharma, P. 2013. Vermicomposting of toxic weed *Lantana camara* biomass: Chemical and microbial properties changes and assessment of toxicity of end product using seed bioassay. *Ecotoxicology and Environmental Safety* 95:179–187.

Suthar, S. and Singh, S. 2008. Feasibility of vermicomposting in biostabilization of sludge from a distillery industry. *Science of the Total Environment* 394:237–243.

Swarnam, T.P., Velmurugan, A., Pandey, S.K., Roy, S.D. 2016. Enhancing nutrient recovery and compost maturity of coconut husk by vermicomposting technology. *Bioresource Technology* 207:76–84.

Tauseef, S.M., Abbasi, T., Banupriya, D., Vaishnavi, V., Abbasi, S.A. 2013. HEVSPAR: A novel vermireactor system for treating paper waste. *Official Journal of Patent Office* 24:12726.

Tejada, M., Gonzalez, J.L., Hernandez, M.T., Garcia, C. 2008. Agricultural use of leachates obtained from two different vermicomposting processes. *Bioresource Technology* 99:6228–6232.

Vasanthi, K., Chairman, K., Ranjit Singh, A.J.A. 2013. Vermicomposting of leaf litter ensuing from the trees of Mango (*Mangifera indica*) and Guava (*Psidium guujuvu*) leaves. *International Journal of Advanced Research* 1(3):33–38.

Villar, I., Alves, D., Perez-Diaz, D., Mato, S. 2016. Changes in microbial dynamics during vermicomposting of fresh and composted sewage sludge. *Waste Management* 48:409–417.

Wani, K.A., Mamta, Rao, R.J. 2013. Bioconversion of garden waste, kitchen waste and cow dung into value-added products using earthworm *Eisenia fetida*. *Saudi Journal of Biological Sciences* 20:149–154.

Wyman, C.E. 1999. Biomass ethanol: Technical progress, opportunities and commercial challenges. *Annual Review of Energy and the Environment* 24:189–226.

Xiao, W., Wang, Y., Xia, S., Ma, P. 2012. The study of factors affecting the enzymatic hydrolysis of cellulose after ionic liquid pretreatment. *Carbohydrate Polymers* 87:2019–2023.

Yadav, A. and Garg, V.K. 2010. Bioconversion of Food Industry Sludge into value-added product (vermicompost) using epigeic earthworm *Eisenia fetida. World Review of Science, Technology and Sustainable Development* 7(3):225–238.

Yadav, A. and Garg, V.K. 2011. Vermicomposting e an effective tool for the management of invasive weed *Parthenium hysterophorus. Bioresource Technology* 102:5891–5895.

Yadav, A. and Garg, V.K. 2013. Nutrient recycling from industrial solid wastes and weeds by vermiprocessing using earthworms. *Pedosphere* 23:668–677.

Yang, J., Lv, B., Zhang, J., Xing, M. 2014. Insight into the roles of earthworm in vermicomposting of sewage sludge by determining the water-extracts through chemical and spectroscopic methods. *Bioresource Technology* 154:94–100.

Zirbes, V.L., Renard, Q., Dufey, J., Khanh Tu, P., Duyet, H.N. 2011. Volarisation of water hyacinth in vermicomposting using an epigeics earthworm *P. exacavatus. Biotechnology, Agronomy, Society and Environment* 15(1):85–93.

16

Phytocapping Technology for Sustainable Management of Landfill Sites

Sunil Kumar and Abhishek Khapre

CONTENTS

16.1 Introduction

Landfills are the most economical and easiest means of disposing of waste. In India, most of the municipal solid waste (MSW) ends up in landfills. In developed countries, landfills are highly engineered structures that receive a wide range of materials (e.g., paper, unrecyclable plastics). In developing countries, the proportion of adequately designed landfill is lower, and serious human health and environmental problems are ominous. India is among the highest 10 solid-waste generators in the world. Although there has been a significant increase in approaches to sustainable waste management focusing on reduction, reuse, and recycling of solid waste in recent years, disposal to landfill will inevitably remain the most widely used waste management method globally. In India, the majority of local communities and associated municipalities have generated landfill sites, with an excess of 8,000 landfill sites expected nationwide (Asnani 2015). Landfill design in India is guided by notifications from the government from time to time. Guidelines for the adoption of liners and covers for MSW landfills have been issued by the Ministry of Environment and Forests (MoEF 2005) and Ministry of Urban Affairs (MUA 2000). Similarly, guidelines for liners and covers for hazardous waste (HW) landfills were issued by the CPCB in 2004 (CPCB 2004). But in most Indian cities, land filling is a common practice that adversely affects the environment and public health.

16.1.1 Adverse Effects of Landfills

With the various advantages of landfill practice, some adverse effects are also of concern: (i) groundwater contamination by the leachate generated in the waste dump, (ii) surface water contamination by the run-off from the waste dump, (iii) bad odor, pests, rodents, and wind-blown litter in and around the waste dump, (iv) fires within the waste dump, (v) erosion and stability problems, (vi) acidity to surrounding soil, and (vii) release of greenhouse gases (GHGs), such as methane (CH_4), nitrous oxide (N_2O), carbon dioxide (CO_2), etc.

In landfills, MSW is degraded by the microbes in anaerobic conditions. This produces CH_4 and CO_2. CH_4 is one of the most important GHGs. As a result of human activities, CH_4 emission concentrations in the atmosphere have increased from 715 ppb during the preindustrial age to 1,732 ppb in the early 1990s and 1,774 ppb in 2005 (IPCC 2007). Although the CH_4 concentration in the atmosphere is much lower than that of CO_2, its global warming potential is 21 times greater than that of CO_2 (IPCC 2007). A study by Lamb et al. (2013) showed that the global CH_4 concentration is approximately 1.8 ppmv, which represents a doubling during the last 200 years (IPCC 2007). However, a robust engineering design and operation is required due to the generation of leachate, which would potentially contaminate nearby ground- and surface water.

16.2 Environmental Issues of Landfills

Waste disposal is one of the major environmental issues facing society today. Recycling of materials is of increasing interest to minimize the use of nonrenewable environmental resources, such as metals, plastics, etc., but also to minimize the amount of waste added to landfills. Landfills have very long life spans, usually requiring maintenance and management for several decades. Environmental issues are associated with all phases of a landfill's lifetime, ranging from construction, operation, and post closure. Landfills have adverse environmental impacts via production of landfill leachate and generation of landfill gases (predominantly CH_4).

16.2.1 Leachate Generation and Characteristics

Landfill caps improve aesthetics; prevent waste from blowing across the landscape; reduce odor; control insects, rodents, and flies; reduce fire hazards; and, most importantly, minimize infiltration of water into the waste. Landfill leachate is generated when excessive water percolates from the surface through to the bottom of the landfill. Conventional caps rely on the reduction of hydraulic conductivity to minimize entry of water into the buried waste. However, often this is not achieved due to the formation of micro- and macrocracks as the landfill ages under drying and wetting cycles associated with seasonal changes in rainfall and relative humidity. Landfill leachate is contaminated liquid that is generated as a result of interaction with solid waste that may contain a wide variety of dissolved and colloidal material.

Leachate characteristics in a landfill vary widely based on the type of dumped waste and age of the landfill. The leachate generation is caused principally due to percolation of precipitated water through waste deposited in a landfill. Due to the contact with solid-waste mass, the percolating water becomes contaminated and, if it flows out of the waste material, it is termed leachate. Additional leachate volume is produced during this decomposition of carbonaceous material, producing a wide range of other materials including CH_4 and CO_2, a complex mixture of organic acids, aldehydes, alcohols, and simple sugars. (Raj 2011).

16.2.2 Landfill Gas Emission

Recently, GHG emissions from landfills have been widely discussed due to their linkage to climate change. Landfill gas (LFG) generated due to anaerobic decomposition of organic fractions in buried waste consists of 55%–60% CH_4 and 40%–45% CO_2 (Kumar 2010). The total amount of LFG that can be produced from a landfill is determined by the fraction of degradable organic carbon (DOC) in the waste (Sun et al. 2013). However, due to the inefficiency of landfills as anaerobic digesters, the production of LFG is slow and often continues for decades. As a major component of LFG, CH_4 is a significant GHG produced by landfills, due to its potential global-warming contribution.

CH_4 in LFG is a significant source of anthropogenic CH_4 emissions and accounts for more than half of GHG emissions from the waste sector (Bogner 2006). Landfills rank as the third major anthropogenic source of CH_4 emissions after rice paddies and ruminant manure. A total of 40–60 metric tons of CH_4 is emitted from landfills worldwide, accounting for approximately 11%–12% of global anthropogenic CH_4 emissions (Ritzkowski and Stegmann 2010). CH_4 gas migration from landfills to the surrounding environment negatively affects both humankind and the environment. Gas explosion disasters due to LFG migration resulting from variations in atmospheric pressure were reported in the village of Loscoe in England in 1986 and at Skellingsted Landfill in Denmark (Christophersen and Kjeldsen 2001).

Mitigating CH_4 emissions from landfills can significantly reduce GHG emissions from waste sectors and attenuate the global CH_4 budget. Furthermore, under global carbon trading initiatives, this can be a significant cost saving in landfill operations (Venkatraman and Ashwath 2005). Considerable efforts have been invested to reduce CH_4 emissions from landfills, especially in developed countries.

16.2.3 Mitigation of Landfill Gas

The prospect of effectively mitigating CH_4 emissions from landfills is promising. In comparison with other anthropogenic CH_4 sources (e.g., natural gas systems, rice cultivation, etc.), landfill CH_4 emissions are unique in many ways, such as small scale, high intensity, and closeness to residential areas. These characteristics can facilitate efficient technologies to reduce emissions. However, there are many situations in which LFG extraction and energy recovery are financially and technically impractical, especially for small, rural, or older landfills. In modern landfill practice, a gas extraction/collection system is the single most important measure to reduce emissions (Bogner 2006). The efficiency of a gas extraction system can reach over 90% by means of modern engineering construction and management. The overall efficiency of CH_4 recovery is considered poor for the entire life span of a landfill in spite of having a gas extraction system, as there is still a significant amount of LFG that might escape during active landfilling or after gas extraction turnoff (Huber-Humer et al. 2008).

16.3 Options for Capping Landfills

CH_4 oxidation by landfill cover soils is considered an important secondary measure to mitigate landfill CH_4 emissions for both long-term performance and sustainable development of landfills. Biotic systems that utilize enhanced CH_4 oxidation capacities to mitigate landfill CH_4 emissions have been a contemporary research topic. Biotic systems are a viable option in attenuating CH_4 emissions from landfills. One of the main criteria of interest to environmental regulators in measuring the performance of a landfill cover is the quantity of water draining through the cover into the buried waste (Sun et al. 2013).

Conventionally, the materials considered to be most suitable for the construction of landfill covers have been impermeable barriers commonly constructed of compacted clay layers. However, there is enough evidence to suggest that the barrier function of a compacted clay cover can deteriorate with time, as the clay is subjected to cracking under cycles of repeated drying and wetting (Venkatraman and Ashwath 2006).

16.3.1 Disadvantages

The main disadvantages of these systems include

1. The system may not ensure the prevention of surrounding gas migration.
2. Systems have limited operational control.
3. Systems are limited by material demand.
4. There is a risk of CH_4 overload and formation of extracellular polysaccharides.
5. The system is more expensive if it is installed with gas venting systems.

To overcome these drawbacks, an alternative design of landfills with vegetated soil covers is proposed. These are called as phytocaps ("phyto" = "plant") and are a novel approach that is a relatively low-cost technique for reducing LFG emissions by oxidizing CH_4 in landfill cover.

16.4 Phytocapping Technology

The failure of conventional caps over the medium to long term has resulted in increasing interest in and adoption of an alternative to impermeable liners. The clay content, bulk density, nutrient content, and other poor soil factors (pH, salinity) make the productive growth of plants difficult due to low water availability and poor soil structure and aeration. A definition of "phytocapping" is lacking, although it may be considered a subsection of the intensively researched area of phytoremediation. Cunningham and Lee (1995) defined *phytoremediation* as "the use of green plants to remove, contain, or render harmless environmental contaminants." In landfill soils where contamination has occurred or that are irrigated with landfill leachate, phytoremediation of soil contaminants may be necessary or a part of the phytocapping technique. Phytocapping as a subdiscipline to phytoremediation may be defined as the use of higher plants to minimize percolation into the waste of landfill sites to stabilize and mitigate offsite migration of soil or gaseous environmental contaminants (Reichenauer et al. 2011). The role that plants play in phyto-capping thus relates specifically to the environmental management of landfill sites, and includes associated microflora, particularly in the rhizosphere of soils (Venkatraman and Ashwath 2009).

Phytocapping, particularly in the United States, is often known by different terms. These include broad terms, such as *alternative earthen covers* or alternative covers, *agronomic covers*, or *natural covers*. The term *alternative landfill covers* has received some acceptance, particularly in the United States as a result of the Alternative Covers Assessment Project. Other terms that are commonly used include *evapotranspiration covers* or ET covers (Hauser et al. 2001).

Phytocapping minimizes water entry through the cover by a variety of processes. These include canopy evaporation, in which part of the rain directly evaporates into the atmo-sphere without reaching the ground surface, water storage in the soil matrix after rainfall and/or canopy throughfall reaches the ground, and plant evapotranspiration. The per-formance of phytocapping technology impinges critically on properly accounting for the local climate, water storage capacity, and plant influences on the hydrological cycle, par-ticularly transpiration rate during critical wet periods.

16.4.1 Approaches to Phytocapping

The approach to phytocapping landfills depends primarily on the final end use that is desired. Factors that may influence the decision on end use may include legislative directives on the appropriateness of plant species on the landfill, the climate, and the location relative to urban land use. The majority of landfills in India have not been designed specifically for phytocapping technology. Phytocapping ideally requires a soil cover that is not compacted and possesses loamy-like textures. Landfill designs not explicitly including vegetative cover tend to possess high clay content with high bulk density. The typical design of the landfill "cover" is the addition of a low-permeability clay lining that is compacted to reduce the hydraulic conductivity. The specifications may differ. An additional soil material is placed above the clay liner. Despite the difficulty in obtaining soils of adequate nutrient content for the growth of plants, and the likelihood of high clay content of cover materials, the texture of landfill covers is often a silty loam to clay loam (Ashwath and Venkatraman 2015).

Phytocapping of landfills generally takes two approaches. Phytocappingcan be used to revegetate the area with endemic plant species for the creation of a wildlife corridor or habitat for local fauna. The alternative option is to establish plantations of high biomass plant species for production of plant-based products. Methane emission from closed and future landfill cover soils is among the most critical environmental issues associated with landfill sites. It hasbeen shown that methanotrophic bacteria native to most soils are able to oxidize CH_4 to CO_2 and water. As a result, the design of "biocovers" has been proposed to maximize the effectiveness of LFG oxidizing microbial communities and other trace gas contaminants (e.g., TCE, BTEX). Although only an emerging area, there is increasing interest in the use of plants to assist microbial communitiesto oxidize CH_4 and other LFGs. The design of biocovers has involved the use of waste materials to maximize pore space and nutrients, but to date has not incorporated the potential for plant growth for assisted LFG oxidation.

16.4.2 Mechanisms and Processes of Phytocapping

There are many advantages to using the phytocapping system for landfill remediation. The primary benefits include a sustainable and green technology; reduction in the cost of establishment as compared to conventional capping; providing community benefits, such as creation of parklands and biodiversity conservation centers; and taking advantage of the opportunities to use these sites for carbon sequestration and, in some instances, production of plants that can be used in the cut-flower industry or in bioenergy production. The major disadvantage in phytocapping is that this system may not be effective if the evapotranspiration rates of the sites where it will be established are not greater than the precipitation (e.g., cool and humid locations). Selection of plant species requires an in-depth understanding of site conditions, local weather patterns, and local soil properties. It is also possible that the plant species could access buried wastes, thus leading to plant death and/or dispersal of contaminants into the environment, if those chemicals are present in the waste at large concentrations.

16.5 Concluding Remarks

Landfill gases are produced by the decomposition of biodegradable fractions of MSW. Biodegradation occurs rapidly if the water comes into contact with buried wastes in prevailing anaerobic conditions. To minimize the percolation of water in landfills, clay capping, a conventional approach, is still practiced in India. Clay capping has been proven unsuccessful in most of the landfills in India, resulting in the formation of leachate and contamination of the groundwater in areas near landfills. A gas extraction system is an expensive technique, especially in India, where the amount of waste generation results in the huge and mostly unscientific disposal of wastes. Thus, a new cost-effective technique, "phytocapping," to mitigate landfill gases and to minimize percolation of water into the landfill is catching the interest of researchers. But, varying climatic conditions in India comprise the biggest challenge for its field trial. In addition, the investigation of how soil chemical properties change due to vegetation can help in the selection of plants for the design of phytocaps under different climatic conditions.

References

Asnani, P.U. 2015. *Solid waste management*. In India Infrastructure Report: Urban Infrastructure, 3(2015), 160–189. New Delhi: Oxford University Press.

Ashwath, N., and Venkatraman, K. 2015. Phytocapping of municipal landfills. *Environmental Research Journal* 9(1), 43–77.

Bogner, J. 2006. Garbage and global change. *Waste Management* 26(5), 451–452.

CPCB. 2004. *Management of Municipal Solid Waste*. Central Pollution Control Board, Ministry of Environment and Forests, New Delhi, India.

Christophersen, M., and Kjeldsen, P. 2001. Lateral gas transport in soil adjacent to an old landfill: Factors governing gas migration. *Waste Management & Research* 19(6), 579–594.

Cunningham, S.D., and Lee, C.R. 1995. Phytoremediation: Plant-based remediation of contaminated soils and sediments. In H.D. Slipper and R.F. Turco (Eds.), *Bioremediation* (pp. 145–156). Madison, WI: Soil Science Society of America Inc.

Hauser, V.L., Weand, B.L., and Gill, M.D. 2001. Natural covers for landfills and buried waste. *Journal of Environmental Engineering* 127, 768–775.

Huber-Humer, M., Gebert, J., and Hilger, H. 2008. Biotic systems to mitigate landfill methane emissions, *Waste Management & Research* 26(1), 33–46.

IPCC. 2007. Summary for Policymakers. In Climate Change 2007: Mitigation. Contribution of Working Group III to the Fourth Assessment Report of the Intergovernmental Panel on Climate Change.

Kumar, S. 2010. Effective municipal solid waste management in India. In S. Kumar (Ed.) *Waste Management* (pp. 1–8). ISBN 978-953-7619-84-8.

Lamb, T.D., Venkatraman, K., Bolan, N., Ashwath, N., Choppala, G., and Naidu, R. 2013. Phytocapping: An alternative technology for the sustainable management of landfill sites. *Critical Review in Environmental Science and Technology* (6)44, 2014.

Ministry of Environment and Forests (MoEF). 2005. www.envfor.nic.in as on 16th June 2005 116.

Ministry of Urban Affairs (MUA) Annual Report. 2000. Ministry of Urban Affairs and Employment, GOI.

Raj, K. 2011. Dissertation Report on Anaerobic Treatment of Municipal Solid Waste, pp. 1–135.

Reichenauer, T.G., Watzinger, A., Riesing, J., and Gerzabek, M.H. 2011. Impact of different plants on the gas profile of a landfill cover. *Waste Management* 31(5), 843–853.

Ritzkowski, M., and Stegmann, R. 2010. Generating CO2-credits through landfill in situ aeration. *Waste Management* 30(4) 702–706.

Sun, J., Yuen, S., Bogner, J., Chen, D., and Asadi, M. 2013. Evaluation of phytocaps as biotic systems to mitigate landfill methane emissions. *14th International Waste Management and Landfill Symposium*. CISA Publisher.

Venkatraman, K., and Ashwath, N. 2005. Phytocapping: An alternative technique to reduce leachate and methane generation from municipal landfills. In *Environmental Research Event Hobart, Australia 2005*.

Venkatraman, K., and Ashwath, N. 2006. Phytocapping: Can it mitigate methane emissions from municipal landfills, In *2nd International Conference on Environmental Science and Technology, Houston, Texas, USA*.

Venkatraman, K., and Ashwath, N. 2009. Can phytocapping technique reduce methane emission from municipal Landfills. *Proceedings of International Environmental Research Event* 1–11.

17

Plant–Endophytic Bacterial Diversity for Production of Useful Metabolites and Their Effect on Environmental Parameters

Ajit Kumar Passari, Vineet Kumar Mishra, Zothanpuia, and Bhim Pratap Singh

CONTENTS

17.1 Introduction

Plants are commonly associated with a diverse group of microorganisms that help them in several aspects. One such group, endophytic bacteria, resides inside the plant tissues either in symbiotic or mutualistic association without causing any adverse effect on plant health (Schulz and Boyle 2006). Endophytes gain access to their host plants mainly through the roots; they reside within plant cells or in the intercellular spaces or they colonize the plant systematically by vascular system and spread in the whole host plant body. After gaining access in the plant tissues, the endophytes are known to produce various bioactive compounds (Strobel et al. 2004). Of the nearly 300,000 plant species that subsist on the earth, each individual plant is host to one or more endophytes (Strobel and Daisy 2003). Endophytes have been well accepted to produce growth-promoting metabolites, insect and pest repellents, and antimicrobials against plant pathogens and to protect the host during stress conditions (Staniek et al. 2008, Rai et al. 2006).

In contrast with rhizosphere and phyllosphere bacteria, endophytic bacteria more closely interact with their host. In plant–endophyte interactions, endophytes can live inside plant tissues and supply nutrients, where endophytes can directly or indirectly improve plant growth and health (Mastretta et al. 2006). Direct plant growth–promoting mechanisms may depend on various plant growth regulators such as auxins, cytokinins and gibberellins, 1-aminocyclopropane-1-carboxylate (ACC) deaminase activity, nitrogen fixation, and unavailable nutrient supply via phosphate solubilization. Endophytic bacteria can help plant growth indirectly against several plant pathogens through competition for space and nutrients by production of hydrolytic enzymes, antibiosis, and induction of plant defense mechanisms.

Endophytes have been considered for the improvement of plant growth potential and also effectiveness in phytoremediation. Phytoremediation (the use of plants and their associated microorganisms to remove pollutants from the environment) is an *in situ* highly promising technology that involves minimal site disturbance and maintenance resulting in low cost and high public acceptance. The currently available conventional remediation methods are expensive and often environmentally invasive. Phytoremediation has emerged as a promising alternative, especially in the treatment of large, pollution-contaminated areas. However, large-scale phytoremediation is still facing difficulty with the levels of contaminants (toxic for the organisms involved in remediation) along with evapotranspiration of volatile organic pollutants from soil or groundwater to the atmosphere (Ryan et al. 2008, Stępniewska and Kuźniar 2013).

Many endophytes have played a role as vectors for xenobiotic degradation as well as having the capacity to exhibit resistance to heavy metals/antimicrobials and degrade organic compounds in the plant/soil niche. These endophytes must be investigated to improve phytoremediation (Siciliano et al. 2001, Barac et al. 2004, Germaine et al. 2004, 2006, Porteous-Moore et al. 2006, Ryan et al. 2007). Selected or engineered microorganisms have also been used to enhance the rate of phytoremediation (Barac et al. 2004, Germaine et al. 2004, 2006, Ryan et al. 2007). Several researchers have suggested that endophytic microorganisms live inside both specific plant tissues and the root cortex or the xylem, enhancing the phytoremediation process (Khan and Dotty 2011, Li et al. 2012). Moreover, they colonize the plant by the vascular or apoplast system.

Endophytes can utilize various metabolic pathways for the degradation of bioremediation, which can be used for assimilation of methane, nitrogen fixation, bioremediation of pollutants (e.g., pesticides, herbicides, insecticides, petrochemicals, polychlorobiphenyls, phenols/chlorophenols), and biotransformation of organic substances (e.g., propylene to epoxypropane) and production of chiral alcohols (Gai et al. 2009, Kim et al. 2012).

Phytoremediation of organic contaminants mainly depends on plants and their associated microorganisms, such as mycorrhizal fungi and bacteria. Soil contaminants, especially organic xenobiotics with a log K_{ow} in the range of 0.5 and 3.5 and weak electrolytes (weak acids and bases such as herbicides), are easily taken up by plants (Trapp et al. 1994, Trapp 2000). Plant–bacteria interaction can enhance contaminant degradation in the rhizosphere, while root-associated bacteria can play a major role in plant growth in contaminated soil. Siciliano et al. (2001) examined the ability of host plants to selectively enhance the occurrence of endophytes having pollutant catabolic genes contaminated with different pollutants and found that the enhancement of catabolic genotypes is both plant and contaminant dependent (Siciliano et al. 2001).

This chapter demonstrates the use of molecular markers for identification of endophytic bacteria associated with plants and the potential for a plant–endophyte system to improve phytoremediation of organic contaminants, toxic metals, and gas pollutants, especially methane and carbon dioxide from the atmosphere.

17.2 Isolation of Endophytic Bacteria

Several methods and procedures, including plant sampling, surface sterilization, and use of various media, have been employed by researchers for the isolation of endophytic bacteria. The isolation of bacteria from plants is a very crucial and important trend that depends on various factors such as species of the host plant, sampling season, host–endophyte interactions, age and type of tissue, geographical and habitat distribution, surface sterility, selective media, and culture conditions (Zhang et al. 2006). Hallman et al. (2006) and Coombs Franco (2003) have reported detailed information about methods and procedures as well as sampling of plants and media used for the isolation of bacteria.

Firstly, surface sterilization is the most essential and crucial step in the endophytic isolation. The commonly used sterilizing agents include ethanol (70%–95%), sodium hypochlorite (3%–10%), and hydrogen peroxide. Hallmann et al. (2006) reported that Tween-20, Tween-80, and Triton X-100 can be used as surfactants to increase the effect of surface sterilization procedures. Hallmann et al. (2006) and Lodewyckx et al. (2002) have described a five-step procedure that indicates that addition of sodium thiosulfate solution after treatment of sodium hypochlorite (NaOCl) can improve cultivation of microorganisms on media plates as well as it suppresses the negative effects of remaining NaOCl on plant surfaces, which may kill the endophytes or suppress their growth. After the treatment, the plant tissues are rinsed with distilled water and can be soaked in 10% NaHCO$_3$ to inhibit endophytic fungal growth that could otherwise overgrow in the plates (Nimnoi et al. 2010), followed with the validation of surface sterilization to prove nonexistence of any epiphytic microbial growth and whether the surface sterilization is complete or not. The sterilization procedure (especially, the time) should be standardized for each plant, as the sensitivity of the procedure differs with plant species, age, and organs (Qin et al. 2011).

Pretreatment of plant tissue at 80°C to 100°C for 15–30 minutes is also an important step for the isolation of endophytic bacteria, especially the phylum Actinobacteria (Coombs and Franco 2003, Qin et al. 2009). The tissue samples were divided into small fragments by using a blender and inoculated onto respective culture media. Alternatively, the tissue sample can also be ground or homogenized using a mortar pestle with phosphate buffer or extraction solution and used as a gradient dilution method (Hallmann et al. 1997a) for the recovery of an endophytic bacterial population. At the same time, Hallmann et al. (1997b) reported another method of maceration, vacuum, and pressure bomb techniques, which is more useful in the isolating of Gram-positive bacteria. Furthermore, Qin et al. (2009) and Ikeda et al. (2009) recommended combined enzymatic hydrolysis and differential centrifugation bacterial cell enrichment methods, which can be used to isolate rare organisms. Antifungal antibiotics such as nystatin and cycloheximide are supplemented with media to inhibit the fungal growth. After that, the plates are incubated at 37°C ± 2°C for 2–7 days, and then individual colonies are selected based on microscopic and morphologic study. Pure cultures are obtained by repeatedly streaking the isolates on fresh nutritional media plates (Golinska et al. 2015) (Figure 17.1).

Media composition and culture conditions play an important role for growth of the bacteria. A variety of growth media have been described by certain authors for the isolation of endophytic bacteria that include chitin vitamin B (Hayakawa 1990), tap water yeast extract (TWYE) (Crawford et al. 1993), yeast extract casamino acid (YECA) (Mincer et al. 2002), tryptic soya agar media, nutrient agar media, Luria-Bertani agar media, and yeast malt extract agar (ISP2) media (Passari et al. 2016a).

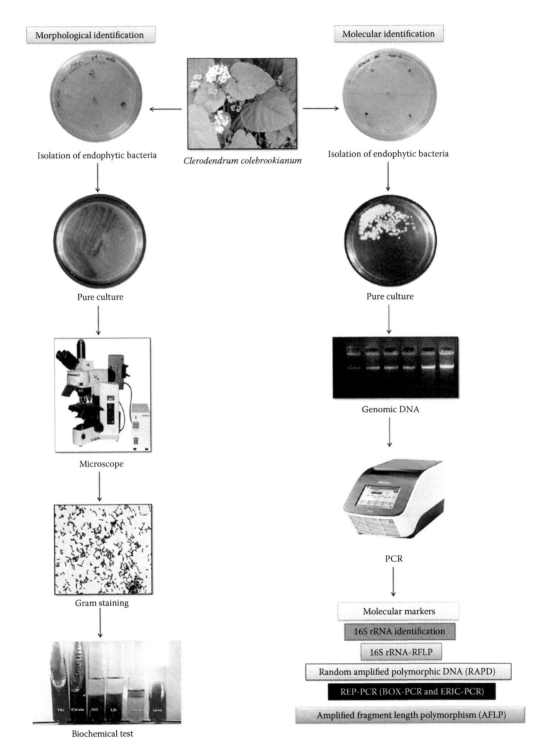

FIGURE 17.1
Different techniques for morphological identification and genetic profiling of endophytic bacteria.

17.3 Occurrence of Endophytic Bacteria

Approximately 300,000 plant species are present on the earth and each plant can harbor a range of endophytes from monocotyledonous to dicotyledonous plants (Strobel and Daisy 2003, Kelemu et al. 2011, Liu et al. 2012, Ma et al. 2013). The presence of endophytes has been proven in several plants, such as beets, cabbage, bananas, tomatoes, and rice roots (Cao et al. 2005, Altalhi 2009). Recently, Passari et al. (2016a) reported that 73 endophytic bacterial isolates were recovered from different tissues of *C. colebrookianum* and assessed for their diversity, *in vitro* screening for their plant growth promoting (PGP) activities, and phytohormone production. Vega et al. (2005) reported that several genera—*Bacillus, Burkholderia, Clavibacter, Curtobacterium, Escherichia, Micrococcus, Pantoea, Pseudomonas, Serratia,* and *Stenotrophomonas*—were recovered from berry seeds. Medicinal plants are reported to harbor larger numbers of endophytes (three endophytic organisms per plant) as compared to weeds (2.4 endophytes per plant) (Ting et al. 2009).

17.4 Plant Colonization with Bacterial Endophytes

Plant colonization is completely dependent on absorption of soil aggregates, biodiversity of plants, and their microbial population as well as their physiology (Hardoim et al. 2008). Conrath et al. (2006) suggested that colonization of microbes involves various factors such as plant genotype, growth stage, physiological status, type of plant tissues, and some soil environmental conditions, as well as some agricultural practices. However, the microbial metabolic pathways of colonization may play a significant role as determinants of endophytic diversity. For example, bacterial endophytes may be evolved from the rhizosphere, which can be attracting microbes in the presence of root exudates (Compant et al. 2010, Philippot et al. 2013). Bacterial endophytes can enter into roots via colonization of root hairs (Mercado-Blanco and Prieto 2012), whereas, in a few cases, stem and leaf also produce exudates that attract microorganisms (Hallmann 2001). Hence, only adapted bacteria can exist and enter the plant via stomata, wounds, and hydathodes. There are several routes for endophytes to colonize within the plants as described by Hallmann (2001) and James et al. (2002). Bacteria can migrate from the rhizoplane to the cortical cell layer, where the plant endodermis forms a wall for further colonization. Then, bacteria invade the endodermis and the xylem vascular system becomes the main transport for colonization in plant tissues (Mercado-Blanco and Lugtenberg 2014).

In order to colonize the plant, the bacteria can find their way through cracks formed at the emergence of lateral roots or at the zone of elongation and during differentiation of the root. Lytic enzymes produced by these bacteria might also contribute to more efficient penetration and colonization. Dong et al. (2001) showed that cells of *Klebsiella* sp. strain Kp342 aggregate at lateral-root junctions of wheat and alfalfa. Similarly, *Gluconacetobacter diazotrophicus* and *Herbaspirillum seropedicae* also colonize lateral-root junctions in high numbers (James and Olivares 1997). Possible infection and colonization sites have been illustrated by Rosenblueth and Martinez-Romero (2006). Some rhizospheric bacteria can colonize the internal roots and stems, showing that these bacteria are a source for endophytes (Germaine et al. 2004); however, phyllosphere bacteria may also be a source of endophytes (Hallmann et al. 1997). It has been proposed that cellulolytic and pectinolytic enzymes produced by endophytes are involved in the infection process (Hallmann et al. 1997); as in *Klebsiella* strains, pectate lyase has been implicated in participating during plant colonization (Kovtunovych et al. 1999). The cell wall–degrading enzymes endogluconase and polygalacturonase seem to be required for the infection of *Vitis vinifera* by *Burkholderia* sp. (Compant et al. 2005).

17.5 Diversity of Bacterial Endophytes and Phylogenetic Analysis

Molecular marker–based identification and classification is not only authenticated but also easy and less time consuming. DNA is used for the identification of microbial populations as well as to emphasize their evolution by the help of phylogenetic studies. Molecular markers can be used to check bacterial diversity based on polymerase chain reaction (PCR) and non-PCR molecular methods (Vieira et al. 2007). A molecular marker should follow some criteria—for example, it should be polymorphic, have codominant inheritance, be evenly and frequently distributed in a genome, be reproducible and economical, and be fast and accurate. The development of molecular markers has a widely significant role in clinical medicine, the food industry, and microbiology to detect early diseases and also help in forensic science (Vieira et al. 2007, Liu et al. 2012).

The 16S rRNA and the intergenic spacer between these 16S–23S rRNA gene sequences have been widely used for the identification and characterization of bacteria as well as to search highly conserved regions; this has been mostly used to study their diversity as well as phylogenetic relationships between closely related species (Eichler et al. 2006). Consequently, the 16S rRNA gene is highly useful to differentiate the bacterial population up to the species level (Liu et al. 2012). The 16S rRNA gene sequence shows evolutionary distance and relationships between bacterial populations (Lloyd-Jones et al. 2005). Additionally, the 16S rRNA gene is so universal in bacteria that it has the capacity to evaluate all bacteria from phyla level to species level (Woese et al. 1985, Woese 1987). Because of the unique characteristics of 16S rRNA, it has applications in several fields. This gene sequence of each bacterium is species specific and thus can be used for identification and characterization of all prokaryotic bacteria (Clarridge 2004). Phylogenetic study of all endophytic bacterial strains based on 16S rRNA sequencing is performed to study their bacterial phylogenies as well as their taxonomy (Yin et al. 2008).

Endophytic bacteria have been obtained from different tissues (root, stem, leaves, bark, petiole, flower, and fruit) of vegetable plant species like wheat, rice, potato, black pepper, maize, zinger, carrots, and tomato (Nejad and Johnson 2000, Surette et al. 2003, Aravind et al. 2009, Dawwam et al. 2013, Ikeda et al. 2009, Ngoma et al. 2001, Jasim et al. 2014, Ji et al. 2014, Passari et al. 2016b). Aravind et al. (2009) reported 74 endophytic bacterial strains from root and stem of the *Piper nigrum* L. plant of India that were identified using a 16S rRNA gene sequence. These strains belong to six genera—that is, *Pseudomonas* spp. (20 strains), *Serratia* (1 strain), *Bacillus* sp. (22 strains), *Arthrobacter* sp. (15 strains), *Micrococcus* sp. (7 strains), *Curtobacterium* sp. (1 strain), and eight unidentified strains. Seven endophytic bacterial isolates were isolated from healthy roots of sweet potato (*Ipomoea batatas* L.) and, based on their PGP activity, two strains (P31 and P35) were identified by 16S rDNA sequencing analysis as *Bacillus cereus* and *Achromobacter xylosoxidans*. These two strains could be recommended as biofertilizers to sustain agriculture (Dawwam et al. 2013).

Qin et al. (2009) reported that 2,174 endophytic actinobacteria were isolated from different medicinal plants in the Xishuangbanna tropical rainforest of China and they were divided into 10 different suborders and 32 genera, including at least 19 new taxa. Among them, *Streptomyces* was the dominant genus, which was reported as endophyte. Some rare actinobacteria genera like *Dietzia*, *Blastococcus*, *Dactylosporangium*, *Actinocorallia*, *Jiangella*, *Promicromonospora*, *Oerskovia*, *Microtetraspora*, and *Intrasporangium* were also reported (Qin et al. 2009). Most of these rare species had less than 97% similarity with the nearest types of strains based on the analysis of the 16S rRNA gene sequences (Qin et al. 2009).

Ji et al. (2014) reported isolation of 576 endophytic bacteria from the leaves, stems, and roots of 10 varieties of rice cultivars. Among them, 12 strains were selected based on the presence of the *nif* gene and were identified by 16S rDNA gene sequence analysis. They belonged to four genera—that is, four species of *Klebsiella* followed by three species of *Bacillus* and *Microbacterium* and two species of *Paenibacillus*. Passari et al. (2016a) isolated and identified 73 endophytic bacteria from different tissues (root, stems, and leaves) of *C. colebrookianum* and, based on 16S rRNA gene sequencing, they phylogenetically analyzed them by using BOX-PCR fingerprinting and 16S rDNA sequencing results. All the isolates were classified into 11 families, belonging to *Bacillaceae* (32.8%) followed by *Alcaligenaceae* (1.36%), *Micrococcaceae* (31.5%), *Xanthomonadaceae* (8.2%), *Pseudomonadaceae* (4.1%), *Comamonadaceae* (1.36), *Phyllobacteriaceae* (2.73%), *Flavobacteriaceae* (2.73%), *Sphingomonadaceae* (4.1%), *Amaranthaceae* (1.36%), and *Bradyrhizobiaceae* (1.36%). A few isolates were identified as uncultured bacterium clones and bacterium based on the 16S rRNA sequence similarity with the NCBI-BLASTn tool. All the isolates showed 99%–100% similarity and a neighbor joining phylogenetic tree was constructed using the Mega 5.05 version. Two isolates, *Pseudomonas psychrotolerans* and *Labrys wisconsinensis*, were reported for the first time as endophytes (Passari et al. 2016a) (Figures 17.1 and 17.2).

A total of 50 endophytic bacteria were obtained from the roots of three different plants (*Amaranthus hybridus*, *Solanum lycopersicum*, and *Cucurbita maxima*) and eight isolates were selected for 16S rDNA gene sequencing based on their different phenotypic and morphological characteristics (Ngoma et al. 2011). Three strains each belonged to the genera *Pseudomonas* and *Achromobacter*; two strains belonged to the genus *Stenotrophomonas* (Ngoma et al. 2011).

An endophytic bacterial population was identified based on 16S rRNA gene analysis and characterized using different molecular markers to discriminate between the species, using various DNA segment presences in the bacterial genome and identification of bacterial species. Various molecular techniques, such as random amplification of polymorphic DNA (RAPD)-PCR and amplification of repetitive extragenic palindromic-PCR (REP-PCR), such as enterobacterial repetitive intergenic consensus-PCR (ERIC-PCR) and BOX-PCR, have been progressively used for identification and comparative analysis (Ishii and Sadowsky 2009, Kumar et al. 2014, Passari et al. 2015b). RAPD methodologies have been successfully used to study genetic diversity of several genera in endophytic bacteria.

The genetic diversity of 32 *Erwinia/Pantoea* and 16 *Agrobacterium* sp. strains was assessed with the RAPD technique. There was a high level of genetic polymorphism among all the

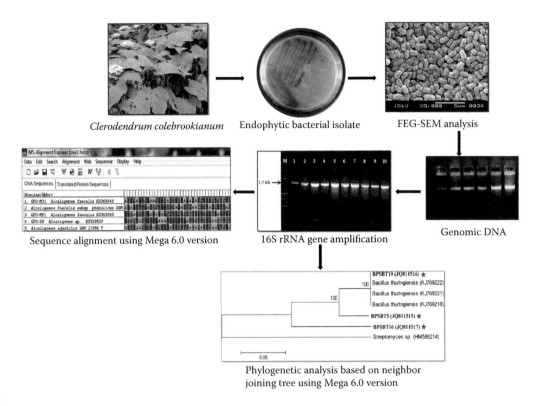

FIGURE 17.2
Identification and phylogenetic analysis of endophytic bacteria based on 16S rRNA gene analysis.

isolates and there was positive correlation between the clusters of the strains (Procópio et al. 2009). Enterobacterial repetitive intergenic consensus polymerase chain reaction (ERIC PCR) is also used to study the genetic relationship between endophytic bacteria. Various families of interspersed repetitive sequences are present in the genomes of bacterial species (e.g., the enterobacterial repetitive intergenic consensus (ERIC) and the BOX element) (Lupski and Weinstock 1992, Rademaker and de Bruijn 1997). Kumar et al. (2014) reported that REP-PCR fingerprinting is a tremendous tool to characterize and discriminate the strains at the genus level that showed close genetic relation among the genera and to differentiate genera *B. subtilis*, *B. pumulis*, *B. megaterium*, and *B. amyloliquefaciens*, respectively (Figure 17.2).

17.6 Role of Endophytes in Phytoremediation

The endophytic bacteria reside inside the plant tissues in either symbiotic or mutualistic relationship with the host plant—contrary to rhizospheric bacteria that live around the plant roots. Endophytes promote plant growth as well as provide resistance against plant fungal pathogens, drought, and even herbivores (Selosse et al. 2004). Reinhold-Hurek and Hurek (1998) reported that some endophytes are diazotrophic and can supply fixed nitrogen to the host plant, which might be beneficial for the plant. Many researchers have suggested the benefits of rhizospheric bacteria in phytoremediation (Anderson et al. 1993, Abou-Shanab et al. 2003). However, endophytic bacteria have several advantages over rhizospheric bacteria (Doty 2008, Ryan et al. 2008, Rajkumar et al. 2009). A rhizospheric population is more diverse and there is a high rate of competition between microbes due to selective metabolism of the pollutant. But, endophytic bacteria live inside the plant tissues and their population is controlled by the plant itself (Doty 2008). Transgenic plants are reported to have increased metabolism of the pollutant from solution compared to nontransgenic plants (Doty et al. 2008).

The endophytic microbial community might also change depending on the climatic conditions of a place and other environmental variables. For example, bacteria isolated from plants growing in a petroleum-contaminated soil have shown degradation of petroleum (Siciliano et al. 2001). Interestingly, some plants selectively supplemented bacteria to remove pollutants, while other plants growing in the same location were incapable of doing so (Taghavi et al. 2005, McGuinness and Dowling 2009). The genes encoding enzymes involved in petroleum degradation—alkane monooxygenase and naphthalene dioxygenase—were more prevalent in bacteria from the root interior than in those from the surrounding soil (Doty 2008, Ryan et al. 2008).

17.6.1 2,4-Dinitrophenol (DNP) Degradation by Endophytic Bacteria

Endophytic bacteria *Bacillus* sp. has decreased cadmium levels up to 94% in the presence of metabolic inhibitors *N,N'*-dicyclohexylcarbodiimide or 2,4-dinitrophenol (DNP) (Guo et al. 2010). Similarly, Wan et al. (2012) reported that DNP mixed with endophytic bacteria, *Serratia nematodiphila* LRE07, has improved growth activity in *Solanum nigrum* L. in the presence of cadmium (Figure 17.3 and Table 17.1).

17.6.2 2,4,6-Trinitrotoluene (TNT), Hexahydro-1,3,5-trinitro-1,3,5-triazine (HMX), and Hexahydro-1,3,5 trinitro-1,3,5-triazine (RDX) Degradation

Van Aken et al. (2004) reported that a methylotrophic bacterium—*Methylobacterium populum* sp. nov. strain BJ001 obtained from poplar trees (*P. deltoides* × *P. nigra* DN34)—has the capacity to degrade energetic compounds such as 2,4,6-trinitrotoluene (TNT), hexahydro-1,3,5-trinitro-1,3,5-triazine (HMX), and hexahydro-1,3,5-trinitro-1,3,5-triazine (RDX). They observed that 60% of RDX and HMX were mineralized to carbon dioxide (CO_2) within 2 months. Germaine et al. (2004) reported that three bioremediation potential strains of *Pseudomonas* sp. isolated from xylem sap of poplar trees can degrade xenobiotic compounds. Similarly, the poplar endophyte *R. tropiciby populus* strain PTD1 could remove RDX readily from solution (Doty et al. 2009). Moreover, three strains isolated from Cerrado plants showed potential for the degradation of different fractions of petroleum, diesel oil, and gasoline (De-Oliveira et al. 2012) (Figure 17.3 and Table 17.1).

17.6.3 BTEX Degradation

Toluene is one of the four components of benzene, toluene, ethylbenzene, and xylene (BTEX) contamination. Barac et al. (2004) reported that the biotechnologically engineered bacterial endophyte *Burkholderia cepacia* G4 has increased plant tolerance to toluene and reduced the transpiration of toluene to the environment. These researchers confirmed that the endophyte

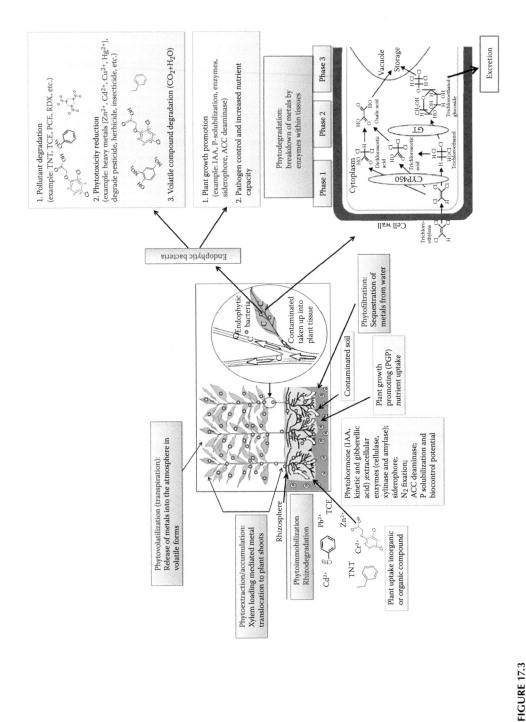

FIGURE 17.3

Plants and their associated endophytic bacteria interaction to degrade a pollutant using the phytoremediation process.

TABLE 17.1

Xenobiotic Compounds Degraded by Endophytic Bacteria Associated with Plants

Sl. No	Compound Name	Structure	Endophytic Bacteria	Ref.
1.	2,4-Dinitrophenol (DNP)		*Bacillus* sp.	Guo et al. (2010)
2.	2,4,6-Trinitrotoluene (TNT)		*Methylobacterium populum* sp. nov. strain BJ001	Van Aken et al. (2004)
3.	Hexahydro-1,3,5-trinitro-1,3,5-triazine (RDX)		*Methylobacterium populum* sp. nov. strain BJ001	Van Aken et al. (2004)

(Continued)

TABLE 17.1 (CONTINUED)

Xenobiotic Compounds Degraded by Endophytic Bacteria Associated with Plants

Sl. No	Compound Name	Structure	Endophytic Bacteria	Ref.
4.	BTEX	Benzene, o-Xylene, Toluene, m-Xylene, Ethylbenzene, p-Xylene	*Burkholderia cepacia* G4	Barac et al. (2004)
5.	2,4-Dichlorophenoxyacetic acid		*Pseudomonas putida* VM1441	Germaine et al. (2009)
6.	Trichloroethylene (TCE)		*Pseudomonas putida* W619	Weyens et al. (2009)
7.	Volatile organic compounds	Oxidizing methane (CH$_4$) in the atmosphere	*Sphagnum* sp.	Raghoebarsing et al. (2005)

Burkholderia cepacia G4 strain has the capacity to enhance phytoremediation by decreasing toxicity and increasing degradation of the xenobiotic compounds. The recombinant strain induced strong (up to 50%–70%) degradation of toluene (Figure 17.3 and Table 17.1).

17.6.4 2,4-Dichlorophenoxyacetic Acid Degradation

Germaine et al. (2009) described that a *Pisum sativum* plant inoculated with genetically modified *Pseudomonas putida* VM1441 (pNAH7) had the ability to degrade 40% of 2,4-dichlorophenoxyacetic acid from the soil (Table 17.1).

17.6.5 Trichloroethylene (TCE) Degradation

For the first time, the TCE-degrading root endophyte strain *Pseudomonas putida* W619-TCE was isolated from poplar trees growing on a trichloroethylene (TCE)-contaminated site showed 90% reduction of TCE evapotranspiration under field conditions as described by Weyens et al. (2009). The members of the poplar's endogenous endophytic population developed TCE metabolic activity due to horizontal gene transfer (Weyens et al. 2009). Kang et al. (2012) reported a novel endophyte, *Enterobacter* sp. PDN3, obtained from the hybrid poplar (*Populus deltoids* × *P. nigra*) that showed high tolerance to TCE up to 55.3 μM (Kang et al. 2012). This approach is given for development of the effectiveness of phytoremediation of volatile organic contaminants (Figure 17.3 and Table 17.1).

17.6.6 2,4-Dichlorophenoxyacetic Acid (2,4-D) Degradation

Pea plants (*Pisum sativum*) inoculated with a *Pseudomonas* endophyte had the ability to degrade the organochlorine herbicide 2,4-dichlorophenoxyacetic acid (2,4-D) according to Germaine et al. (2006). They showed no accumulation of the herbicide into their tissues or phytotoxicity in the presence of inoculated plants exposed to 2,4-D, whereas uninoculated plants exhibited significant accumulation of 2,4-D and indicated signs of toxicity, including decrease in biomass, leaf abscission, and callus growth on their roots. The relationship between plant and endophyte could be beneficial to increase the phytoremediation of herbicide-contaminated substrates and decrease the concentrations of toxic herbicide residues in crop plants. This study clearly suggested that the inoculations of endophytic bacteria might be useful to increase phytoremediation potential (Figure 17.3 and Table 17.1).

17.6.7 Volatile Organic Compound (CH_4 and CO_2) Degradation

Many researchers have targeted the bioremediation of volatile organic compounds like methane and carbon dioxide depending on the composition of vegetation (Chen and Murrell 2010, Parmentier et al. 2011, Goraj et al. 2013, López et al. 2013). Recently, endophytic methanotrophic bacteria *Sphagnum* sp. was recovered from moss tissues that are capable of oxidizing methane at atmospheric levels (Raghoebarsing et al. 2005). For example, endophytic bacteria *Methylocella palustris* and *Methylocapsa acidiphila* can oxidize methane to carbon dioxide, which is used by *Sphagnum* plants during photosynthesis (Raghoebarsing et al. 2005, Stępniewska et al. 2013). This finding has severely changed the carbon cycle in ecosystems as well as the global carbon cycle. Thus, methanotrophic endophytes can play a role as a natural methane filter that can decrease 50% of CH_4 and CO_2 emission from peat lands (Kip et al. 2012).

Goraj et al. (2013) reported that some potential plant methanotrophic bacteria have the ability to reduce 77% of methane emission, depending on the season and the host plant.

Molecular diversity has indicated that the dominant endophytic bacteria belong to the genera *Burkholderia, Pseudomonas, Flavobacterium, Serratia,* and *Collimonas*. These bacteria having a main objective of development of plant growth promotion as well as being used for the productivity of crops in agriculture fields (Shcherbakov et al. 2013). Additionally, the endophytic bacteria inhabiting *Sphagnum* sp. can be inoculated in plants inhabiting an artificial wetland system, which can be utilized to treat mixed contaminations—for example, heavy metals, different organic contaminations, and greenhouse gases. They could be effective for developing an efficient metal-removal system (Li et al. 2012) (Figure 17.3 and Table 17.1).

Endophytic microorganisms can play a major role in phytoremediation. Hence, it is essential to search for endophytes associated with plants from unique and harsh habitats for further understanding of their contributions to the degradation of xenobiotic compounds.

17.7 Concluding Remarks

Molecular techniques have been well known for the identification of microorganisms based on the phylogenetic evolutionary relationship among them. It has been reported that particular bacteria should be identified by two or more suitable genetic markers for more reliable and authenticated results. The 16S rRNA gene is the most widely used molecular marker, especially for the identification and molecular characterization of an endophytic bacterial population. However, RAPD, REP-PCR (BOX-PCR and ERIC-PCR) genotypic fingerprinting techniques are more reliable tools compared to 16S rRNA gene sequences for distinguishing the endophytic bacterial isolates.

The development of plant–endophyte interaction for the removal of contaminated soils and groundwater is a very significant area. Endophytic bacteria can play a vital role in phytoremediation of organic contaminant cases. Endophytic bacteria can help their host plant with metal-sequestration systems or the ability to produce natural metal chelators by degrading the contaminants that are absorbed by plants. This results in (a) toxicity due to the accumulation of the original compound, (b) evapotranspiration of volatile contaminants, and (c) reduction of metal toxicity for the host plant or increasing metal translocation to the aerial parts.

In the phytoremediation of toxic metals, endophytic bacteria protect the plant from toxicity of the pollutant, enhance the rate of phytoremediation, and promote the growth of the plant. In the future, we have to search potential metal-resistant endophytic bacteria that could be inoculated at the field level to promote the production of biomass and bioenergy crops in conjunction with phytoremediation of soil contamination. A stress-tolerant plant can be targeted to isolate endophytic microbes with the ability to degrade multiple types of metal or various organic contaminants by using bioremediation processes.

Acknowledgments

BPS is thankful to the Department of Biotechnology (DBT), government of India, New Delhi, for funding the NER Twinning project. The authors are thankful to the Department of Biotechnology for establishment of the DBT-BIF center and DBT-state biotech hub in the department, which was used for the present study.

References

Abou-Shanab, R.A., Angle, J.S., Delorme, T.A., Chaney, R.L., van Berkum, P., Moawad, H., Ghanem, K. and Ghozlan, H.A. 2003. Rhizobacterial effects on nickel extraction from soil and uptake by *Alyssum murale*. *New Phytologist* 158(1):219–224.

Altalhi, A.D. 2009. Plasmids profiles, antibiotic and heavy metal resistance incidence of endophytic bacteria isolated from grapevine (*Vitis vinifera* L.). *African Journal of Biotechnology* 8:5873–5882.

Anderson, T.A., Guthrie, E.A. and Walton, B.T. 1993. Bioremediation in the rhizosphere: Plant roots and associated microbes clean contaminated soil. *Environmental Science and Technology* 27(13): 2630–2636.

Aravind, A., Kumar, A., Eapen, S.J. and Ramana, K.V. 2009. Endophytic bacterial flora in root and stem tissues of black pepper (*Piper nigrum* L.) genotype: Isolation, identification and evaluation against *Phytophthora capsici*. *Letters in Applied Microbiology* 48:58–64.

Barac, T., Taghavi, S., Borremans, B., Provoost, A., Oeyen, L., Colpaert, J.V., Vangronsveld, J. and van der Lelie, D. 2004. Engineered endophytic bacteria improve phytoremediation of watersoluble, volatile, organic pollutants. *Nature Biotechnology* 22:583–588.

Cao, L., Qui, Z., You, J., Tan, H. and Zhou, S. 2005. Isolation and characterization of endophytic *Streptomycete* antagonists of *Fusarium wilt* pathogen from surface-sterilized banana roots. *FEMS Microbiology* 247:147–152.

Chen, Y. and Murrell, J.C. 2010. Methanotrophs in moss. *Nature Geoscience* 3:595–596.

Clarridge, J.E. 2004. Impact of 16S rRNA gene sequence analysis for identification of bacteria on clinical microbiology and infectious diseases. *Clinical Microbiology Review* 17:840–862.

Compant, S., Clement, C. and Sessitsch, A. 2010. Plant growth-promoting bacteria in the rhizo- and endosphere of plants: Their role, colonization, mechanisms involved and prospects for utilization. *Soil Biology Biochemistry* 42:669–678.

Compant, S., Reiter, B., Sessitsch, A., Nowak, J., Clement, C. and Barka, E.A. 2005. Endophytic colonization of *Vitis vinifera* L. by a plant growth promoting bacterium, *Burkholderia* sp. strain PsJN. *Applied and Environmental Microbiology* 71:4951–4959.

Conrath, U., Beckers, G.J.M., Flors, V., Garcia-Agustin, P., Jakab, G., Mauch, F., Newman, M.A., Pieterse, C.M.J., Poinssot, B., Pozo, M.J., Pugin, A., Schaffrath, U., Ton, J., Wendehenne, D., Zimmerli, L. and Mauch-Mani, B. 2006. Priming: Getting ready for battle. *Molecular Plant–Microbe Interaction* 19:1062–1071.

Coombs, J.T. and Franco, C.M.M. 2003. Isolation and identification of actinobacteria isolated from surface-sterilized wheat roots. *Applied and Environmental Microbiology* 69:5603–5608.

Crawford, D.L., Lynch, J.M., Whipps, J.M. and Ousley, M.A. 1993. Isolation and characterization of actinomycete antagonists of a fungal root pathogen. *Applied and Environmental Microbiology* 59(11):3889–3905.

Dawwam, G.E., Elbeltagy, A., Emara, H.M., Abbas, I.H. and Hassan, M.M. 2013. Beneficial effect of plant growth promoting bacteria isolated from the roots of potato plant. *Annals of Agricultural Sciences* 58:195–201.

De-Oliveira, N.C., Rodrigues, A.A., Alves, M.I.R., Filho, N.R.A., Sadoyama, G. and Vieira, J.D.G. 2012. Endophytic bacteria with potential for bioremediation of petroleum hydrocarbons and derivatives. *African Journal of Biotechnology* 11:2977–2984.

Dong, Y., Glasner, J.D., Blattner, F.R. and Triplett, E.W. 2001. Genomic interspecies microarray hybridization: Rapid discovery of three thousand genes in the maize endophyte *Klebsiella pneumonia* 342, by microarray hybridization with *Escherichia coli* K12 open reading frams. *Applied and Environmental Microbiology* 67:1911–1921.

Doty, S.L. 2008. Enhancing phytoremediation through the use of transgenics and endophytes. *New Phytologist* 179(2):318–333.

Doty, S.L., Oakley, B., Xin, G., Kang, J.W., Singleton, G., Khan, Z., Vajzovic, A. and Staley, J.T. 2009. Diazotrophic endophytes of native black cottonwood and willow. *Symbiosis* 47:23–33.

Eichler, S., Christen, R., Holtje, C., Westphal, P., Botel, J., Brettar, I., Mehling, A. and Hofle, M.G. 2006. Composition and dynamics of bacterial communities of a drinking water supply system as assessed by RNA and DNA based 16S rRNA gene fingerprinting. *Applied and Environmental Microbiology* 72:1858–1872.

Gai, C.S., Lacava, P.T., Quecine, M.C., Auriac, M.C., Lopes, J.R.S., Araújo, W.L., Miller, T.A. and Azevedo, J.L. 2009. Transmission of *Methylobacterium mesophilicum* by *Bucephalogonia xanthophis* for paratransgenic control strategy of citrus variegated chlorosis. *Journal of Microbiology* 47:448–454.

Germaine, K., Keogh, E., Borremans, B., Lelie, D., Barac, T., Oeyen, L., Vangronsveld, J., Moore, F.P., Campbell, C.D., Ryan, D. and Dowling, D.N. 2004. Colonisation of poplar trees by gfp expressing bacterial endophytes. *FEMS Microbiology Ecology* 48:109–118.

Germaine, K., Liu, X., Cabellos, G., Hogan, J., Ryan, D. and Dowling, D.N. 2006. Bacterial endophyte enhanced phytoremediation of the organochlorine herbicide 2,4-dichlorophenoxyacetic acid. *FEMS Microbiology Ecology* 57:302–310.

Germaine, K.J., Keogh, E., Ryan, D. and Dowling, D.N. 2009. Bacterial endophyte-mediated naphthalene phytoprotection and phytoremediation. *FEMS Microbiol Lett.* 296:226–234.

Golinska, P., Wypij, M., Agarkar, G., Rathod, D., Dahm, H. and Rai, M. 2015. Endophytes actinobacteria of medicinal plants: Diversity and bioactivity. *Antonie van Leeuwenhoek* 108(2):267–289.

Goraj, W., Kuźniar, A., Urban, D., Pietrzykowska, K. and Stępniewska, Z. 2013. Influence of plant composition on methane emission from Moszne peatland. *Journal of Ecological Engineering* 14:53–57.

Guo, H., Luo, S., Chen, L., Xiao, X., Xi, Q., Wei, W., Zeng, G., Liu, C., Wan, Y., Chen, J. and He, Y. 2010. Bioremediation of heavy metals by growing hyperaccumulaor endophytic bacterium *Bacillus* sp. L14. *Bioresource Technology* 101:8599–8605.

Hallmann, A., Benhamou, N., and Kloepper, J.W. 1997. Bacterial endophytes in cotton: Mechanisms of entering the plant. *Canadian Journal of Microbiology* 43:577–582.

Hallmann, J. 2001. Plant interactions with endophytic bacteria. In *Biotic interactions in plant-pathogen associations,* eds. M.J. Jeger and N.J. Spence, 87–119. Wallingford, United Kingdom: CABI Publishing.

Hallmann, J., Berg, G. and Schulz, B. 2006. Isolation procedures for endophytic microorganisms. In *Microbial root endophytes,* eds. B.J.E. Schulz, C.J.C. Boyle and T.N. Sieber, 299–314. New York: Springer.

Hallmann, J., Kloepper, J.W. and Rodríguez-Kábana, R. 1997a. Application of the scholander pressure bomb to studies on endophytic bacteria of plants. *Canadian Journal of Microbiology* 43:411–416.

Hallmann, J., Quadt-Hallmann, A., Mahaffee, W.F. and Kloepper, J.W. 1997b. Bacterial endophytes in agricultural crops. *Canadian Journal of Microbiology* 43:895–914.

Hardoim, P.R., van Overbeek, L.S. and van Elsas, J.D. 2008. Properties of bacterial endophytes and their proposed role in plant growth. *Trends in Microbiology* 16:463–471.

Hayakawa, M. 1990. Selective isolation methods and distribution of soil actinomycetes. *Actinomycetologica* 4:103–112.

Ikeda, S., Kaneko, T., Okubo, T., Rallos, L.E.E., Eda, S., Mitsui, H., Sato, S., Nakamura, Y., Tabata, S. and Minamisawa, K. 2009. Development of a bacterial cell enrichment method and its application to the community analysis in soybean stems. *Microbial Ecology* 58:703–714.

Ishii, S. and Sadowsky, M.J. 2009. Applications of the rep-PCR DNA fingerprinting technique to study microbial diversity, ecology and evolution. *Environmental Microbiology* 11:733–740.

James, E.K., Gyaneshwar, P., Mathan, N., Barraquio, Q.L., Reddy, P.M., Iannetta, P.P.M., Olivares, F.L. and Ladha, J. K. 2002. Infection and colonization of rice seedlings by the plant growth-promoting bacterium *Herbaspirillum seropedicae* Z67. *Molecular Plant–Microbe Interaction* 15:894–906.

James, E.K. and Olivares, F.B. 1997. Inoculation and colonization of sugar cane and other graminaceous plants by endophytic diazotrophs. *Critical Review in Plant Sciences* 17:77–119.

Jasim, B., Joseph, A.A., John, J., Mathew, J. and Radhakrishnan, E.K. 2014. Isolation and characterization of plant growth promoting endophytic bacteria from the rhizome of *Zingiber officinale*. *3 Biotechnology* 4:197–204.

Ji, S.H., Gururani, M.A. and Chun, S.C. 2014. Isolation and characterization of plant growth promoting endophytic diazotrophic bacteria Korean rice cultivars. *Microbiology Research* 169:83–98.

Kang, J.W., Khan, Z. and Doty, S.L. 2012. Biodegradation of trichloroethylene by an endophyte of hybrid poplar. *Applied and Environmental Microbiology* 12:3504–3507.

Kelemu, S., Fory, P., Zuleta, C., Ricaurte, J., Rao, I. and Lascano, C. 2011. Detecting bacterial endophytes in tropical grasses of the *Brachiaria* genus and determining their role in improving plant growth. *African Journal of Biotechnology* 10:965–976.

Khan, Z. and Dotty, S. 2011. Endophyte-assisted phytoremediation. *Current Opinion in Plant Biology* 12:97–105.

Kim, T.U., Cho, S.H., Han, J.H., Shin, Y.M., Lee, H.B. and Kim, S.B. 2012. Diversity and physiological properties of root endophytic actinobacteria in native herbaceous plants of Korea. *Journal of Microbiology* 50:50–57.

Kip, N., Fritz, C., Langelaan, E.S., Pan, Y., Bodrossy, L., Pancotto, V., Jetten, M.S.M., Smolders, A.J.P. and Op den Camp, H.J.M. 2012. Methanotrophic activity and diversity in different *Sphagnum magellanicum* dominated habitats in the southernmost peat bogs of Patagonia. *Biogeosciences* 9:47–55.

Kovtunovych, G., Lar, O., Kamalova, S., Kordyum, V., Kleiner, D. and Kozyrovska, N. 1999. Correlation between pectate lyase activity and ability of diazotrophic *Klebsiella oxytoca* VN 13 to penetrate into plant tissues. *Plant Soil* 215:1–6.

Kumar, A., Kumar, A. and Pratush, A. 2014. Molecular diversity and functional variability of environmental isolates of *Bacillus* species. *SpringerPlus* 3:312.

Li, H.Y., Wei, D.Q., Shen, M. and Zhou, Z.P. 2012. Endophytes and their role in phytoremediation. *Fungal Diversity* 54:11–18.

Liu, W., Li, L., Khan, M.A. and Zhu, F. 2012. Popular molecular markers in bacteria. *Molecular Genetics, Microbiology and Virology* 27(3):103–107.

Lloyd-Jones, G., Laurie, A.D. and Tizzard, A.C. 2005. Quantification of the *Pseudomonas* population in New Zealand soils by fluorogenic PCR assay and culturing techniques. *Journal of Microbiological Methods* 60:217–224.

Lodewyckx, C., Vangronsveld, J., Porteous, F., Moore, E.R.B., Taghavi, S., Mezgeay, M. and Lelie, D.V.D. 2002. Endophytic bacteria and their potential applications. *Critical Review in Plant Sciences* 21(6):583–606.

López, J.C., Quijano, G., Souza, T.S., Estrada, J.M., Lebrero, R. and Muñoz, R. 2013. Biotechnologies for greenhouse gases (CH4, N2O and CO2) abatement: State of the art and challenges. *Applied Microbiology and Biotechnology* 97:2277–2303.

Lupski, J.R. and Weinstocks, G.M. 1992. Short, interspersed repetitive DNA sequences in prokaryotic genomes. *Journal of Bacteriology* 174:4525–4529.

Ma, L., Cao, Y.H., Cheng, M.H., Huang, Y., Mo, M.H., Wang, Y., Yang, J.Z. and Yang, F.X. 2013. Phylogenetic diversity of bacterial endophytes of *Panax notoginseng* with antagonistic characteristics towards pathogens of root-rot disease complex. *Antonie van Leeuwenhoek*, doi: 10.1007/s10482-012-9810-3.

Mastretta, C., Barac, T., Vangronsveld, J., Newman, L., Taghavi, S. and van der Lelie, D. 2006. Endophytic bacteria and their potential application to improve the phytoremediation of contaminated environments. *Biotechnology & Genetic Engineering Review* 23:175–207.

McGuinness, M. and Dowling, D. 2009. Plant associated bacterial degradation of toxic organic compounds in soil. *International Journal of Environmental Research and Public Health* 6(8):2226–2247.

Mercado-Blanco, J. and Lugtenberg, B.J.J. 2014. Biotechnological applications of bacterial endophytes. *Current Biotechnology* 3:60–75.

Mercado-Blanco, J. and Prieto, P. 2012. Bacterial endophytes and root hairs. *Plant Soil* 361:301–306.

Mincer, T.J., Jensen, P.R., Kauffman, C.A. and Fenical, W. 2002. Widespread and persistent populations of a major new marine actinomycete taxon in ocean sediments. *Applied and Environmental Microbiology* 68:5005–5011.

Nejad, P. and Johnson, P.A. 2000. Endophytic bacteria induce growth promotion and wilt disease suppression in oilseed rape and tomato. *Biological Control* 18:208–215.

Ngoma, L., Esau, B. and Babalola, O.O. 2001. Isolation and characterization of beneficial indigenous endophytic bacteria for plant growth promoting activity in Molelwane farm, Mafikeng, South Africa. *African Journal of Biotechnology* 12:4105–4114.

Nimnoi, P., Pongsilp, N. and Lumyong, S. 2010. Endophytic actinomycetes isolated from *Aquilaria crassna* Pierre ex Lec and screening of plant growth promoters production. *World Journal of Microbiology & Biotechnology* 26:193–203.

Parmentier, F.J.W., van Huissteden, J., Kip, N., Op den Camp, H.J.M., Jetten, M.S.M., Maximov, T.C. and Dolman, A.J. 2011. The role of endophytic methane-oxidizing bacteria in submerged Sphagnum in determining methane emissions of Northeastern Siberian tundra. *Biogeosciences* 8:1267–1278.

Passari, A.K., Chandra, P., Zothanpuia, Mishra, V.K., Leo, V.V., Gupta, V.K., Kumar, B. and Singh, B.P. 2016b. Detection of biosynthetic gene and phytohormone production by endophytic actinobacteria associated with *Solanum lycopersicum* and their plant-growth-promoting effect. *Research in Microbiology* 167:692–705.

Passari, A.K., Mishra, V.K., Gupta, V.K., Yadav, M.K., Saikia, R. and Singh, B.P. 2015b. *In vitro* and *in vivo* plant growth promoting activities and DNA fingerprinting of antagonistic endophytic actinomycetes associates with medicinal plants. *PLoS ONE* 10(9):e0139468.

Passari, A.K., Mishra, V.K., Leo, V.V., Gupta, V.K. and Singh, B.P. 2016a. Phytohormone production endowed with antagonistic potential and plant growth promoting abilities of culturable endophytic bacteria isolated from *Clerodendrum colebrookianum* Walp. *Microbiology Research* 193:57–73.

Philippot, L., Raaijmakers, J.M., Lemanceau, P. and van der Putten, W.H. 2013. Going back to the roots: The microbial ecology of the rhizosphere. *Nature Reviews Microbiology* 11:789–799.

Porteous-Moore, F., Barac, T., Borremans, B., Oeyen, L., Vangronsveld, J., van der Lelie, D., Campbell, D. and Moore, E.R.B. 2006. Endophytic bacterial diversity in poplar trees growing on a BTEX-contaminated site: The characterisation of isolates with potential to enhance phytoremediation. *Systematic and Applied Microbiology* 29:539–556.

Procópio, R.E.L., Araújo, W.L., Maccheroni Jr, W. and Azevedo, J.L. 2009. Characterization of an endophytic bacterial community associated with *Eucalyptus* spp. *Genetics and Molecular Research* 8(4):1408–1422.

Qin, S., Li, J., Chen, H.H., Zhao, G.Z. and Zhu, W.Y. 2009. Isolation, diversity and antimicrobial activity of rare actinobacteria from medicinal plants of tropical rain forests in Xishuangbanna, China. *Applied and Environmental Microbiology* 75:6176–6186.

Qin, S., Xing, K., Jiang, J., Xu, L. and Li, W. 2011. Biodiversity, bioactive natural products and biotechnological potential of plant-associated endophytic actinobacteria. *Applied Microbiology and Biotechnology* 89:457–473.

Rademaker, J.L.W. and de Bruijn, F.J. 1997. Characterization and classification of microbes by rep-PCR genomic fingerprinting and computer assisted pattern analysis. In *DNA Markers: Protocols, Applications and Overviews*, eds. G. Caetano-Anollés and P.M. Gresshoff, 151–171. New York, NY: John Wiley & Sons.

Raghoebarsing, A.A., Alfons, J.P., Smolders, A.J.P., Schmid, M.C., Rijpstra, W.I.C., Wolters-Arts, M., Derksen, J., Jetten, M.S.M., Schouten, S., Damste, J.S.S., Lamers, L.P.M., Roelofs, J.G.M., Op den Camp, H.J.M. and Strous, M. 2005. Methanotrophic symbionts provide carbon for photosynthesis in peat bogs. *Nature* 436:1153–1156.

Rai, M., Agarkar, G. and Rathod, D. 2006. Multiple applications of endophytic *Colletotrichum* species occurring in medicinal plants. In *Novel Plant Bioresources: Applications in Food, Medicine and Cosmetics, Novel Plant Bioresources*, ed. A. Gurib-Fakim, 227–236. Chichester: Wiley.

Rajkumar, M., Ae, N. and Freitas, H. 2009. Endophytic bacteria and their potential to enhance heavy metal phytoextraction. *Chemosphere* 77:153–160.

Reinhold-Hurek, B. and Hurek, T. 1998. Life in grasses: Diazotrophic endophytes. *Trends in Microbiology* 6:139–144.

Rosenblueth, M. and Martinez-Romero, E. 2006. Bacterial endophytes and their interactions with hosts. *Molecular Plant–Microbe Interaction* 19:827–837.

Ryan, R.P., Germaine, K., Franks, A., Ryan, D.J. and Dowling, D.N. 2008. Bacterial endophytes: Recent developments and applications. *FEMS Microbiology Letters* 278:1–9.

Ryan, R.P., Ryan, D.J., Sun, Y.C., Li, F.M., Wang, Y. and Dowling, D.N. 2007. An acquired efflux system is responsible for copper resistance in *Xanthomonas* strain IG-8 isolated from China. *FEMS Microbiol Lett.* 268:40–46.

Schulz, B. and Boyle, C. 2006. Microbial root endophytes. In *What Are Endophytes*, ed. T.N. Sieber, 9, 1–13. Berlin: Springer.

Selosse, M.A., Baudoin, E. and Vandenkoornhuyse, P. 2004. Symbiotic microorganisms, a key for ecological success and protection of plants. *Comptes Rendus Biologies* 327:639–648.

Shcherbakov, A.V., Bragina, A.V., Kuzmina, E.Y., Berg, C., Muntyan, A.N., Makarova, N.M., Malfanova, N.V., Cardinale, M., Berg, G., Chebotar, V.K. and Tikhonovich, I.A. 2013. Endophytic bacteria of *Sphagnum mosses* as promising objects of agricultural microbiology. *Microbiology* 82:306–315.

Siciliano, S., Fortin, N., Himoc, N., Wisse, G., Labelle, S., Beaumier, D., Ouellette, D., Roy, R., Whyte, L.G., Banks, M.K., Schwab, P., Lee, K. and Greer, C.W. 2001. Selection of specific endophytic bacterial genotypes by plants in response to soil contamination. *Applied and Environmental Microbiology* 67:2469–2475.

Staniek, A., Woerdenbag, H.J. and Kayser, O. 2008. Endophytes: Exploiting biodiversity for the improvement of natural product-based drug discovery. *Journal of Plant Interactions* 3:75–98.

Stępniewska, Z. and Kuźniar, A. 2013. Endophytic microorganisms-promising applications in bioremediation of greenhouse gases. *Applied Microbiology and Biotechnology* 97:9589–9596.

Stępniewska, Z., Kuźniar, A., Pytlak, A. and Szymczycha, J. 2013. Detection of methanotrophic endosymbionts in *Sphagnum* sp. originating from Moszne peat bog (East Poland). *African Journal of Microbiology Research* 7:1319–1325.

Strobel, G. and Daisy, B. 2003. Bioprospecting for microbial endophytes and their natural products. *Microbiology and Molecular Biology Reviews* 67:491–502.

Strobel, G., Daisy, B., Castillo, U. and Harper, J. 2004. Natural products from endophytic microorganisms. *Journal of Natural Products* 67:257–268.

Surette, M.A., Sturz, A.V., Lada, R.R. and Nowak, J. 2003. Bacterial endophytes in processing carrots (*Daucus carota* L. var. *sativus*): Their localization, population density, biodiversity and their effects on plant growth. *Plant Soil* 253:381–390.

Taghavi, S., Barac, T., Greenberg, B., Borremans, B., Vangronsveld, J. and van der Lelie, D. 2005. Horizontal gene transfer to endogenous endophytic bacteria from Poplar improves phytoremediation of toluene. *Applied and Environmental Microbiology* 71:8500–8505.

Ting, A.S.Y., Mah, S.W. and Tee, C.S. 2009. Prevalence of endophytes antagonistic towards *Fusarium oxysporum* f. sp. *Cubense* Race 4 in various plants. *American–Eurasian Journal of Sustainable Agriculture* 3:399–406.

Trapp, S. 2000. Modeling uptake into roots and subsequent translocation of neutral and ionisable organic compounds. *Pest Manage Science* 56:767–778.

Trapp, S., McFarlane, J.C. and Matthies, M. 1994. Model for uptake of xenobiotics into plants-validation with bromacil experiments. *Environmental Toxicology Chemistry* 13:413–422.

Van Aken, B., Tehrani, R. and Schnoor, J.L. 2004. Biodegradation of nitrosubstituted explosives 2,4,6-trinitrotoluene, hexahydro-1,3,5- trinitro-1,3,5-triazine, and octahydro-1,3,5,7-tetranitro-1,3,5-tetrazocine by a photosymbiotic *Methylobacterium* sp. associated with poplar tissues (*Populus deltoids ×nigra* DN34). *Applied and Environmental Microbiology* 70:508.

Vieira, J., Mendes, M.V., Albuquerque, P., Moradas-Ferreira, P. and Tavares, F. 2007. A novel approach for the identification of bacterial taxa-specific molecular markers. *Letters in Applied Microbiology* 44:506–512.

Wan, Y., Luo, S., Chen, J., Xiao, X., Chen, L., Zeng, G., Liu, C. and He, Y. 2012. Effect of endophyte-infection on growth parameters and Cd-induced phytotoxicity of Cd-hyperaccumulator *Solanum nigrum* L. *Chemosphere* 89:743–750.

Weyens, N., van der Lelie, D., Artois, T., Smeets, K., Taghavi, K., Newman, L., Carleer, R. and Vangronsveld, J. 2009. Bioaugmentation with engineered endophytic bacteria improves contaminant fate in phytoremediation. *Environmental Science and Technology* 43:9413–9418.

Woese, C.R. 1987. Bacterial evolution. *Microbiology Review* 51(2):221–271.

Woese, C.R., Stackebrandt, E., Macke, T.J. and Fox, G.E. 1985. A phylogenetic definition of the major eubacterial taxa. *Systematic and Applied Microbiology* 6:143–151.

Yin, H., Cao, L., Xie, M., Chen, Q., Qiu, G., Zhou, J., Wu, L., Wang, D. and Liu, X. 2008. Bacterial diversity based on 16S rRNA and gyrB genes at Yinshan mine, China. *Systematic and Applied Microbiology* 31(4):302–311.

Zhang, H.W., Song, Y.C. and Tan, R.X. 2006. Biology and chemistry of endophytes. *Nature Product Reports* 23:753–771.

18

Phytoremediation of Industrial Pollutants and Life Cycle Assessment

Ram Chandra, Vineet Kumar, Sonam Tripathi, and Pooja Sharma

CONTENTS

18.1 Introduction

As a result of various industrial activities over the past two centuries, soil and water contamination is a major environmental issue worldwide. The cleanup of contaminated soil and water is a cost-intensive and technically complex procedure; various physical and chemical methods (i.e., coagulation, flocculation, ozonolysis, electrokinesis, chemical oxidation/reduction, etc.) are already being used for soil and water treatment. In contrast to these traditional remediation approaches, a number of researchers and organizations have proposed the adoption of less invasive alternative remediation options ("gentle" remediation technologies, the so called "green remediation") based on life cycle assessment (LCA) in order to conserve resources and minimize environmental impacts (USEPA 2008, Hanl et al. 2009). LCA is a methodology for assessing all the environmental impacts associated with a product, process, or activity, by identifying and evaluating all the resources consumed and all emissions and wastes released into the environment (Rebitzer et al. 2004, Parkes et al. 2015). It is the systematic approach of looking at a product's complete life cycle, from raw materials to final disposal of the product. During the last century, it was mainly used in industrial fields but, nowadays, most researchers use it widely to assess the impact of products, processes, and activities on the environment (Nie et al. 2011).

In an LCA-contaminated site, impacts have normally been referred to as primary, secondary, and tertiary effects. Phytoremediation is widely viewed as an ecologically responsible alternative to the environmentally destructive physical remediation methods currently practiced, given that it is based on the use of green plants to extract, sequester, and/or detoxify pollutants (Chandra et al. 2015). It has been increasingly used as a more sustainable approach for the remediation of contaminated sites. In the LCA method, the inputs, outputs, and potential environmental impacts of a product system are compiled and evaluated through the product's life span (Khasreen et al. 2009). Although LCA has of late been more widely applied in agricultural than industrial fields, only a few studies have documented and used LCA to determine environmental impacts associated with phytoremediation patterns. A study in China, based on LCA, showed that intercropping maize with suitable plants (peanut and soybean) could be a convenient, effective approach to phytoremediation of a field affected by nitrogen pollution (Nie et al. 2011).

In view of life cycle analysis of industrial waste-contaminated sites, phytoremediation technologies can be described as assessment of industrial pollutants to the level of discharge to a contaminated site where potential plants are growing, with a capability to degrade these pollutants to some limit by the technique called "phytoremediation." Phytoremediation technology is applicable to a broad range of contaminants, including metals and radionuclides, as well as organic compounds like chlorinated solvents, polychlorobiphenyls, polycyclic aromatic hydrocarbons, pesticides/insecticides, explosives, and surfactants. Phytoremediation includes a variety of remediation techniques that include many treatment strategies leading to contaminant degradation, removal (through accumulation or dissipation), or immobilization (Chandra et al. 2015, Chandra and Kumar 2017a). Research is underway to understand the role of phytoremediation to remediate perchlorate, a contaminant that has been shown to be persistent in surface and groundwater systems.

Phytoremediation may be used to clean up contaminants found in soil and groundwater. For radioactive substances, chelating agents are sometimes used to make the contaminants amenable to plant uptake. The phytoremediation process is accomplished with integrated functions of both plants and microbes. Root zone/rhizofiltration technology is very effective in remediation of pollutants present in wastewater. The assessment of environmental impacts is a very important aspect in soil management and LCA can be a very useful tool. In fact, LCA has been used for studying a wide range of remediation technologies such as land filling, stabilization/solidification, etc. Other important aspects are that existing sustainability assessment in this area uses LCA methodology to compare remediation technologies and that LCA can help in the decision-making process. According to the ISO standard, a typical LCA study consists of four stages (i) goal, scope, and definition of LCA; (ii) life cycle inventory analysis; (iii) life cycle impact assessment; and (iv) results interpretation. The four phases of LCA study are illustrated in Figure 18.1.

The assessment process also includes identifying and quantifying energy and materials used and wastes released to the environment, assessing their environmental impact, and evaluating opportunities for improvement as illustrated in Figure 18.2.

The purpose of this chapter is to present an analytical framework for using the LCA approach to assess technologies that can be applied to the remediation process of sites contaminated with organic and inorganic pollutants. In this context, LCA of industrial waste through a systematic process, along with the phytoremediation, can be studied by step-by-step strategies for different types of pollutants (organic, inorganic metallic, and inorganic nonmetallic).

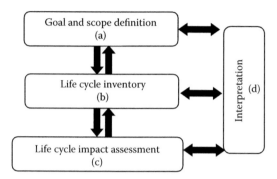

FIGURE 18.1
Four stages of life cycle assessment method.

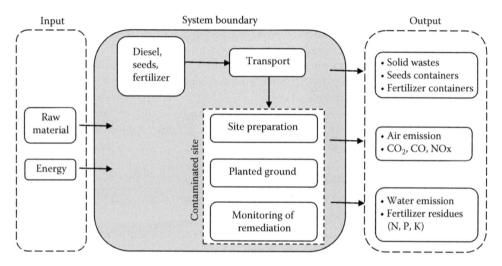

FIGURE 18.2
Life cycle assessment processes.

18.2 Contaminated Sites and Industrial Pollutants

Contamination of soils, groundwater, sediments, surface water, and air with hazardous and toxic chemicals is one of the major problems facing the industrialized world today (Chandra and Kumar 2017a). Wastewater may become contaminated by the accumulation of heavy metals and metalloids through emissions from rapidly expanding industrial areas; mine tailings; disposal of high metal wastes, leaded gasoline, and paints; land application of fertilizers; animal manures; sewage sludge; pesticides; wastewater irrigation; coal combustion residues; spillage of petrochemicals; and atmospheric deposition. Heavy metals constitute an ill-defined group of inorganic chemical hazards, and those most commonly found at contaminated sites are lead (Pb), chromium (Cr), arsenic (As), zinc (Zn), cadmium (Cd), copper (Cu), mercury (Hg), and nickel (Ni) (Barakat 2011). Soil is the major sink for heavy metals released into the environment by anthropogenic activities; unlike organic contaminant oxide by microbial action, most metals do not undergo microbial or chemical degradation, and their total concentration in soils persists for a long time after their introduction. The presence of toxic metals in soil can severely inhibit the biodegradation of organic contaminants and heavy-metal contamination of soil may pose risks and hazards to humans and the ecosystem through direct ingestion or contact with contaminated soil, the food chain (soil–plant–human or soil–plant–animal–human), drinking of contaminated ground water, reduction in food quality (safety and marketability) via phytotoxicity, reduction in land usability for agricultural production causing food insecurity, and land tenure problems.

The adequate protection and restoration of soil ecosystems contaminated by heavy metals require their characterization and remediation. Contemporary legislation respecting environmental protection and public health, at both national and international levels, is based on data that characterize chemical properties of environmental phenomena, especially those that reside in our food chains. While soil characterization would provide an insight into heavy-metal speciation and bioavailability, attempts at remediation of heavy metal–contaminated soils would entail knowledge of the source of contamination, basic chemistry, and environmental and associated health effects (risks) of these heavy metals' ecology. Risk assessment is an effective scientific tool that enables decision makers to manage sites thus contaminated in a cost-effective manner while preserving public and ecosystem health. Heavy metals occur naturally in the soil environment from the pedogenetic processes of weathering of parent materials at levels that are regarded as trace (1,000 mg kg^{-1}) and rarely toxic. Due to man's disturbance and acceleration of nature's slowly occurring geochemical cycle of metals, most soils of rural and urban environments may accumulate one or more of the heavy metals' background values high enough to cause risks to human health, plants, animals, ecosystems, or other media. The heavy metals essentially become contaminants in the soil environment because

1. Their rates of generation via manmade cycles are more rapid relative to natural ones.
2. They become transferred from mines to random environmental locations where higher potentials of direct exposure occur.
3. The concentrations of the metals in discarded products are relatively high compared to those in the receiving environment.
4. The chemical form (species) in which a metal is found in the receiving environmental system may render it more bioavailable.

A number of metal ions are micronutrients essential for the healthy growth of plants. However, very high concentrations of some of these may harm the plants. Soil may contain metals that are not nutrients, and some may be toxic to the plant or animals that feed on the plant. These harmful metals are often referred to as "heavy-metal pollutants." A heavy metal is one with a relatively high density, typically greater than 5 g cm^{-3}, or with a high relative atomic mass. Arsenic and selenium are often called "semimetals" because their properties are somewhere between those of metals and those of nonmetals. But there is neither proper listing nor inventory of various pollutants and the environmental impact has not been assessed so far. Moreover, several complex industrial pollutants are persistent in nature and have carcinogenic or mutagenic effects (Tables 18.1 and 18.2). Furthermore, several industrial discharges act as endocrine disrupting chemicals and severely affect aquatic flora and fauna (Table 18.3). But, unfortunately no national or statewide accurate data are available regarding life and effect of these pollutants. Broadly, the environmental pollutants may be classified into five different categories: (i) inorganic metallic, (ii) inorganic nonmetallic, (iii) natural organic, (iv) industrial by-products from natural compounds, and (v) gaseous. See Tables 18.1 through 18.4.

TABLE 18.1

Inorganic Metallic Pollutants Discharged into the Environment from Various Industries and Their Toxic Effects

Heavy Metals	Source	Toxicity
Cd	Pulp paper mill waste, distillery waste, metal pipes, fertilizers, dental alloys, batteries, copper refineries, fungicides, pesticides	Degenerative bone disease, kidney damage, renal disorder, human carcinogen
Cr	Pulp paper mill waste, distillery waste, electroplating, leather tanning, textile industries	DNA damage, carcinogenic, allergy reaction, irritant dermatitis, gastrointestinal hemorrhage (bleeding), hemolysis, (red blood cell destruction), acute renal failure
Cu	Pulp paper mill waste, distillery waste, fungicides, insecticides	Liver damage, Wilson disease, insomnia
Pb	Pulp paper mill waste, distillery waste, insecticides, tobacco smoke, textiles	Damage to fetal brain, diseases of the kidneys, circulatory system, and nervous system
Ni	Pulp paper mill waste, distillery waste	Dermatitis, nausea, chronic asthma, coughing, human carcinogen, cardiovascular damage (heart and blood vessels)
Zn	Pulp paper mill waste, distillery waste	Depression, lethargy, neurological signs, increased thirst, disruption, central nervous system damage, gastrointestinal, respiratory (from the nose to the lungs)
Mn	Pulp paper mill effluent, distillery waste	Central nervous system, cardiovascular (heart and blood vessels), liver, nervous system, and respiratory damage
Fe	Pulp paper mill waste, distillery waste	Bloody vomiting, liver failure, dilation blood vessels
As	Pulp paper mill waste, tannery waste	Skin manifestations, visceral cancers, vascular disease, carcinogens, dermal, gastrointestinal, hepatic, and respiratory damage

TABLE 18.2

Inorganic Nonmetallic Pollutants Discharged from Various Industries into the Environment and Their Toxic Effects

Pollutants	Source	Toxicity
Na^+	Pulp paper mill waste, distillery waste, tannery waste	Hypertension, stress, premenstrual syndrome, osteoporosis, urinary stones
P	Pulp paper mill waste, distillery waste, fertilizer industry	Hyperphosphatemia or high blood phosphate levels, chlorosis
Cl_2	Pulp paper mill waste, distillery waste	Eye tearing, nose and throat irritation, chest pain or retrosternal burning, muscle weakness, abdominal discomfort, nausea and vomiting (with the smell of chlorine in emesis)
S	Pulp paper mill waste, distillery waste	Irritation to mucous membranes, carcinogenicity, reproductive and developmental toxicity, neurotoxicity, and acute toxicity
N	Pulp paper mill waste, distillery waste	Chest tightness; chest pain; diaphoresis (sweating); eye, nose, or throat irritation; dyspnea (shortness of breath); light-headedness or headache
C	Pulp paper mill waste, distillery waste	Coal workers' pneumoconiosis (CWP), inflammation, fibrosis, necrosis
Cl^-	Pulp paper mill waste, distillery waste	Carcinogenic, binding tendency with other pollutants

TABLE 18.3

Organic Compounds Identified by GC-MS Analysis Discharged from Various Industries

Industries	Organic Pollutants	Ref.
Pulp and paper	Pentadecanone; 2-pentadecanone; benzene dicarboxilic acid; benzene acitic acid; gloxylic acid; propanoic acid; acetic acid; 1,2-benzene dicarboxylic acid; hexadecanoic acid; 2-methoxy-4-(1-propenyl) phenol; octadecanoic acid; 1,2 benzene carboxylic acid; 1-phenyl-1-nonyne; 4,5-octanediol,3,6-dimethyl; diphenylthiocarbazide	Chandra et al. 2017; Chandra and Singh 2012
Distillery	Benzoic acid, 3,4-bis[(TMS)oxy], TMS ester; vanillypropionic acid, bis(TMS); benzeneacetic acid, α,4-bis[(TMS) oxy]-methyl ester; valeric acid, 5-methoxy, TMS ester; 2-hydroxysocaproic acid, TMS ether; benzenepropanoic acid, α-[(TMS)oxy], TMS ester; benzoic acid, 4-[(TMS)oxy], TMS ester; 2-hydroxyheptanoic acid 2TMS; trimethylsilyl 3,5 dimethoxy-4-(TMS oxy)benzoate; silane, (preg-5-ene-3β,11β,17,20β-tetraylteraoxy)tetrakis (trimethyl)	Chandra and Kumar 2017b, c
Pesticides	Hexachlorocyclohexane, chlorpyrifos, parathion, glyphosate, coumaphos, ethoprophos, diazinon, fenitrothion	Singh and Walker 2006
Tannery	Hexanoic acid; benzene propanoic acid; acetylthiocarbamic acid; 3-methoxy-4-benzaldehyde; benzoic acid; butanedioic acid; benzene; 2-hydroxy-3-methyl-butanoic acid; ethanol; 2-hydroxy-3-methyl-butanoic acid	Chandra et al. 2011

TABLE 18.4

Gaseous Pollutants Emitted from Various Industries into the Environment and Their Toxic Effects

Pollutants	Source	Toxicity
H_2s	Pulp paper mill waste, distillery waste	Nausea, headaches, delirium, disturbed equilibrium, tremors, convulsions, and skin and eye irritation
SO_2	Pulp paper mill waste, distillery waste	Decreased respiration, inflammation or infection of the airways, and destruction of areas of the lung; irritates skin and mucous membranes of the eyes, nose, throat, and lungs; inflammation and irritation of the respiratory system, worsening of asthma attacks, and aggravation of existing heart disease in sensitive groups
Cl_2	Pulp paper mill waste	Carcinogenic, binding tendency with other pollutants
H_2O_2	Pulp paper mill waste	Inhalation, ingestion
CH_3OH	Pulp paper mill waste	Vision loss, neurologic manifestations headache, nausea, vomiting, or epigastria pain
Na_2S	Pulp paper mill waste	Toxicity to humans, including carcinogenicity, reproductive and developmental toxicity, neurotoxicity, and acute toxicity
CH_4S	Pulp paper mill waste	Inhalation, ingestion
$NaClO_3$	Pulp paper mill waste	Long-term toxicity of sodium chlorate to birds resulted in reduced egg production and fertility
Na_2CO_3	Pulp paper mill waste	Skin or eye irritation; oral, chest, or abdominal pain; shock difficulty; breathing; stomach upset

Notes: H_2S: hydrogen sulfide; SO_2: sulfur dioxide; Cl_2: chlorine dioxide; H_2O_2: hydrogen peroxide; CH_3OH: methanol; Na_2S: sodium sulfide; CH_4S: methanethiol; $NaClO_3$: sodium chlorate; Na_2CO_3: sodium carbonate.

18.3 Goal, Scope, and Definition of LCA

The first step of an LCA study defines the reason for the study and the intended use of the results (Rebitzer et al. 2004). The comparison between two systems is justified only if they have the same function (i.e., the same functional unit), as well as the same limits and hypotheses. For LCA studies in the agricultural sector, this could be, for instance, to investigate the environmental impacts of different intensities in crop production or to analyze the advantages and disadvantages of intensive or extensive arable farming systems. Furthermore, this step describes the system under investigation, its function, and its boundaries. The goal of LCA is to compare the full range of environmental effects assignable to products and services by quantifying all inputs and outputs of material flows and assessing how these material flows affect the environment. This information is used to improve processes, support policy, and provide a sound basis for informed decisions. Finally, it sets the boundaries for the assessment, which include a boundary between the technosphere (the technological system) and the ecosphere (the environment) (i.e., where an environmental flow crosses this boundary and becomes an emission), a system boundary defining the studied system and the included and excluded processes, and a temporal boundary defining the time frame for the inventory of environmental flows.

The scope should describe the depth of the study and show that the purpose can be fulfilled with the actual extent of the limitations. In the scope of an LCA, the functional unit and the system boundaries should be considered and described. The functional unit is a key element of the LCA that must be clearly defined. The definition of functional unit is a critical step in LCA because it determines the reference flows and dictates the upstream and downstream process alternatives to be included in the study. This enables comparison of two essentially different systems. In site remediation LCA, the functional unit should relate to the production of an amount of treated soil. The majority of site remediation LCA studies consider, as the functional unit, the treatment of an amount of soil or groundwater toward a site-related impact matrix, usually a given regulatory criterion.

The system boundaries for LCA are set based on the characteristics of the site, focusing on the environmental burden related to major technological elements, which will be delineated in detail after the subsequent description of the site. System boundaries in the LCA must be specified in several dimensions: boundaries between the technological system and nature, delimitations of the geographical area and time horizon considered, boundaries between production and production of capital goods, and boundaries between the life cycle of the product studied and related life cycles of other products. Three methods for defining the contents of the analyzed system in this respect are described: process tree, technological whole system, and socioeconomic whole system. The methods are described in the application's multioutput processes and cascade recycling, and examples are discussed. It is concluded that system boundaries must be relevant in relation to the purpose of an LCA, that processes outside the process tree in many cases have more influence on the result than details within the process tree, and that the different methods need to be further compared in practice and evaluated with respect to relevance, feasibility, and uncertainty. The system boundaries determine the unit processes to be included in the LCA study. Defining system boundaries is partly based on a subjective choice, made during the scope phase when the boundaries are initially set. A typical study showing a system boundary is illustrated in Figure 18.3.

The goal and scope of the LCA will determine which categories should be used. Depending on the goal and scope, some inconsistency may be acceptable. If any inconsistency is

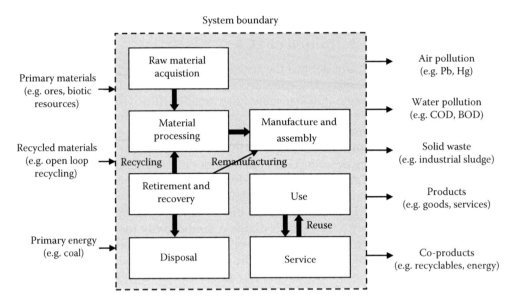

FIGURE 18.3
A typical LCA study showing system boundary.

detected, document the role it played in the overall consistency evaluation. The environmental impacts of a potential product substitution—that is, the choice of one product instead of another (or the choice of a specific product instead of refraining from this product) determine the object of study of an LCA. The functional unit describes those properties of the product, which must be present for the studied product substitution to take place. LCAs are used to compare the impacts of alternative methods of achieving similar outcomes. Understanding the objective allows focus on the pertinent portion of the life cycle:

1. Compare the impact of recycled effluent
2. Compare the impact of fielding recycling of effluent
3. Identify improvements in the life cycle (process, suppliers, and/or customers)
4. Provide credibility for the tagline: the science of sustainability

18.4 Life Cycle Inventory Analysis

Life cycle inventory (LCI) analysis is defined as a phase of LCA involving the compilation and quantification of inputs and outputs for any specific products of a system including the raw material and industrial process at each step that describe the condition for each process and, ultimately, the product obtained (Rebitzer et al. 2004). But, simultaneously the waste as by-product is also generated as a pollutant. Hence, the LCI is the data collection portion of LCA. LCI is the straightforward accounting of everything involved in the "system" of interest. It consists of detailed tracking of all the flows in and out of the product system, including raw resources or materials and energy, by type, water, and emission to air, water, and land by specific substances. An objective, data-based process of quantifying energy and raw material requirements, air emissions, waterborne effluents, solid wastes, and other environmental releases incurred throughout the life cycle of a product, process, or activity.

An inventory may be conducted to aid in decision making by enabling companies or organizations to (a) develop a baseline for a system's overall resource requirement for benchmarking efforts; (b) identify components of the process that are good targets for resource-reduction efforts; (c) aid in the development of new products or processes that will reduce resource requirements or emissions; (d) compare alternative materials, products, processes, or activities within the organization; and (e) compare internal inventory information to that of other manufacturers. A simplified procedure for inventory analysis is shown in Figure 18.4.

Thus, LCI analysis involves creating an inventory of flows from and to nature for a product system. Inventory flows include inputs of water, energy, and raw materials, as well as releases to air, land, and water. To develop the inventory, a flow model of the technical system is constructed using data on inputs and outputs. The flow model is typically illustrated with a flow chart that includes the activities to be assessed in the relevant supply chain and gives a clear picture of the technical system boundaries. The input and output data needed for the construction of the model are collected for all activities within the system boundary, including from the supply chain (referred to as inputs from the technosphere). The data must be related to the functional unit defined in the goal and scope definition. Data can be presented in tables and some interpretations can also be made at this stage. However, the next step of LCI is to develop an LCI checklist that converts most of the decision areas in the performance of an inventory. The checklist is actually the data of all points taken into consideration for LCA. A checklist can be prepared to guide data

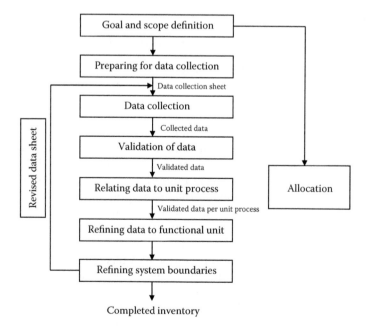

FIGURE 18.4
Simplified procedures for inventory analysis.

collection and validation and to enable construction of a database to store collected data electronically. The following points may be addressed on the inventory checklist sheet:

1. Purpose of the inventory
2. System boundaries
3. Geographical scope
4. Types of data used
5. Data collection procedures
6. Data quality measures
7. Impact assessment condition
8. Phytoremediation process
9. Computational model construction
10. Presentation of results

The result of the inventory is an LCI, which provides information about all inputs and outputs in the form of elementary flow to the environment from all the unit processes involved in the study. Inventory flows can number in the hundreds depending on the system boundary. For product LCAs at either the generic (i.e., representative industry averages) or brand-specific level, those data are typically collected through survey questionnaires. At an industry level, care has to be taken to ensure that questionnaires are completed by a representative sample of producers, leaning toward neither the best nor the worst, and fully representing any regional differences due to energy use, material sourcing, or other factors.

18.5 Methods Used for Phytoremediation of Soil Pollutants

There are several methods of remediation of soil contaminated with industrial and environmental pollutants. Three common approaches are (a) soil excavation, (b) pollutants treated at the natural site, and (c) soil left in the ground and contained is prevented further contamination and risk to environment and humans. Very frequently these treatment methods are also divided into *in situ* and *ex situ* remediation. With *in situ* remediation, soil is left in the ground and treated there; with *ex situ* remediation, soil is excavated and treated elsewhere. In addition, there is another method where no pollutants are removed or treated. Instead, the contaminated area is isolated to prevent migration of contaminants and the possibility of exposure to human and environmental risk. Here pollutants are arrested by the plant microbe activities by growing the potential growth covering on the contaminated soil. This is known as phytocapping. Similarly, other methods are also emerging for reduction of pollutants from the aquifer of contaminated soil, i.e., phytocapping—putting bioventing and putting down zero-valent iron. Phytocapping involves placing a layer of soil material and growing dense vegetation on top of a landfill. In this system, soil acts mainly as the storage to hold and store water, and releases it to the plants when they are in need. Bioventing is a process of stimulating the natural *in situ* biodegradation of contaminants in soil by providing air or oxygen to existing soil microorganisms. Bioventing uses low air flow rates to provide only enough oxygen to sustain microbial activity in the vadose zone. Putting down zero-valent iron involves a permissible reacting barrier with reactive materials.

Presently, phytoremediation is the direct use of living green plants for *in situ*, or in place, removal, degradation, or containment of contaminants in soils, sludge, sediments, surface water, and groundwater (Chandra and Kumar 2017a). Phytoremediation, the use of plants and their associated microorganisms for treatment of contaminated soil and water, is a novel, cost-effective, solar energy-driven cleanup. It is a steadily emerging, eco-friendly green technology with potential for the effective and inexpensive cleanup of a broad range of a wide variety of organic and inorganic compounds from contaminated soil, sludge, sediment, and water (Chandra and Kumar 2015). The phytoremediation is mediated by soil chemistry, plant physiology, and environmental factors (i.e., pH, temperature, light, etc.), as shown in Figure 18.5.

Phytoremediation can be defined as the "efficient use of plants to remove, detoxify or immobilize environmental contaminants in a growth matrix (soil, water, or sediment) through the natural biological, chemical or physical activities and processes of the plants." Plants are unique organisms equipped with remarkable metabolic and absorption capabilities, as well as transport systems that can take up nutrients or contaminants selectively from soil, sludge, sediment, or water. Phytoremediation involves growing plants in a contaminated matrix or facilitating immobilization (binding/containment) or degradation (detoxification) of pollutants. The plant can be subsequently harvested, processed, and disposed. Phytoremediation can be achieved through different methods, such as phytoextraction, rhizofiltration, phytostabilization, phytovolatilization, and phytotransformation/phytodegradation. The ultimate goals of any remediation approach must be to remove the contaminants from the soil and to restore its capacity to function according to its potential. For inorganic pollutants, mechanisms that can be involved are phytostabilization, rhizofiltration, phytoaccumulation, and phytovolatilization. For organic pollutants, this involves phytostabilization, rhizodegradation, rhizofiltration, phytodegradation, and phytovolatilization, as shown in Figure 18.6. Those mechanisms related to the organic contaminant property are not able to be absorbed into the plant tissue.

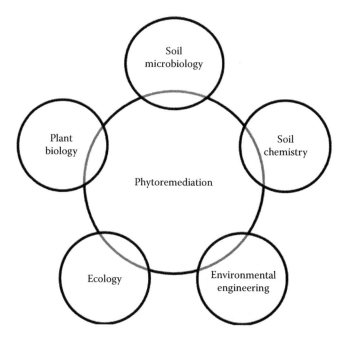

FIGURE 18.5
Schematic showing interdisciplinary nature of phytoremediation.

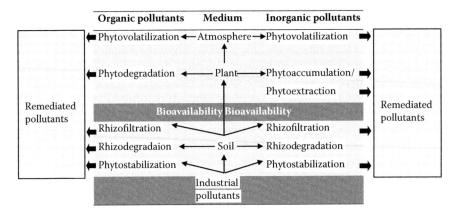

FIGURE 18.6
Different strategies of phytoremediation technology during remediation of industrial pollutants.

TABLE 18.5

Detail of Various Pollutants' Reduction in Various Processes of Phytoremediation

Various Strategies of Phytoremediation	Media	Category of Various Pollutants for Reduction
Phytoextraction	Soil, sludge, sediment	Metals: Ag, Cd, Co, Cr, Cu, Zn, Ni, Mn, Fe, Pb; radionuclides: Sr, Cs, Pb, U
Phytodegradation	Soil, sludge, sediment, groundwater, surface water	BTEX, pentachlorophenol, polychlorinated biphenyl, short-chained aliphatic compounds, chlorinated solvents (tetrachloromethane and dichloromethane)
Phytostabilization	Soil, sludge, sediment	As, Cd, Cr, Cu, Pb, Zn
Phytovolatilization	Soil, sludge, sediment, groundwater	Chlorinated solvent and some inorganic compounds: Hg, Se, As
Rhizofiltration	Groundwater, surface water	Heavy metals, organic chemicals, radionuclides
Rhizodegradation	Soil, sludge, sediment, groundwater, surface water	Organic compounds (TPH, PAH, pesticides, chlorinated solvent, PCBs)

Phytoremediation can be used to clean up a site with shallow, low to moderate levels of contamination. It can be used to clean up metals, pesticides, solvents, explosives, crude oil, polyaromatic hydrocarbons, and landfill leachate. It can also be used for river basin management through the hydraulic control of contaminants. The details for reduction of various pollutants through different strategies of phytoremediation are listed in Table 18.5. Thus, phytoremediation is interdisciplinary in nature and requires background knowledge of ecology, soil chemistry, soil microbiology, and plant biology as well as environmental engineering. In view of the current trends of integration of scientific knowledge worldwide, it is assumed that the many challenging questions also created about commercial application of phytoremediation will be answered in the future.

18.5.1 Types of Phytoremediation

Phytoremediation technology can be divided into two groups based on the physical location of the remedial action.

18.5.1.1 In Situ *Phytoremediation*

In situ phytoremediation involves placement of live plants in contaminated soil, or sediment, or in soil or sediment that is in contact with contaminated groundwater for the purpose of remediation. In this approach, the contaminated material is not removed prior to phytoremediation. *In situ* techniques are favored over *ex situ* techniques due to their low cost and reduced impact on the ecosystem. If the phytomechanism consists of only uptake and accumulation as opposed to transformation of a contaminant, the plants may be harvested and removed from the site after remediation for disposal or recovery of the contaminants. A requirement of the *in situ* approach is that the contaminant must be physically accessible to the roots. This approach generally is the least expensive phytoremediation strategy. Several recent research activities have focused on the application of *in situ* phytoremediation as a sustainable reclamation strategy for bringing soil polluted by organic and inorganic contaminants into productive use (Chandra and Kumar 2017a, Chandra et al. 2017) (Figure 18.7).

18.5.1.2 Ex Situ *Phytoremediation*

For sites where the contaminants are not accessible to plants, such as contaminants in deep aquifers, an alternative method of applying *ex situ* phytoremediation is possible. In this approach, the contaminated media are removed from the actual site using mechanical means and then transferred to a temporary treatment area where they can be exposed to plants selected for optimal phytoremediation. After treatment, the cleaned soil or water

(a) (b) (c)

(a1) (b1) (c1)

FIGURE 18.7
Plant showing *in situ* phytoremediation of industrial waste-contaminated site at initial (a, a1), middle, (b, b1), and final (c, c1) stages of phytoremediation.

can be returned to the original location and the plant may be harvested for disposal if necessary. This approach generally is more expensive than the more passive *in situ* phytoremediation. Several phytoextraction studies have been conducted on pot experiment and laboratory hydroponic solutions to evaluate the phytoremediation potential of plants (Chandra and Yadav 2010, Chandra and Kumar 2017a). Chandra and Yadav (2010) conducted a pot culture experiment to evaluate the heavy-metal (Cu, Pb, Ni, Fe, Mn, and Zn) accumulation pattern of *Typha angustifolia* grown on Cu-, Pb-, Ni-, Fe-, Mn-, and Zn-rich aqueous solutions of phenols and melanoidins (Figure 18.8). They concluded that *T. angustifolia* could be a potential phytoremediator for heavy metals from metal-, melanoidin-, and phenol-containing industrial wastewater at optimized conditions.

Different techniques employed in bioremediation can be divided into two categories (i) *ex situ* and (ii) *in situ* bioremediation (Table 18.6). *Ex situ* bioremediation can occur in two ways: slurry phase phytoremediation or solid phase phytoremediation. Slurry phase phytoremediation is a process where the contaminated soil is mixed with water and other reagents in a large tank known as a bioreactor. It is mixed in order to keep the microorganisms in contact with the toxins present in the soil. Then oxygen and nutrients are added into the mixing so that the microorganisms have an ideal environment to break down the contaminants. Once the process has been completed, the water is separated from the soil and the soil is tested and replaced in the environment. This particular process is comparatively fast compared to other bioremediation techniques.

Solid phase phytoremediation may be performed either by microorganisms (i.e., bacteria, fungi, cynobacteria, and algae) or plant mater (i.e., rhizofiltration or phytoremediation). Solid phase phytoremediation is a process that treats the contaminated soil in an aboveground treatment center. Conditions inside the treatment areas are controlled to ensure

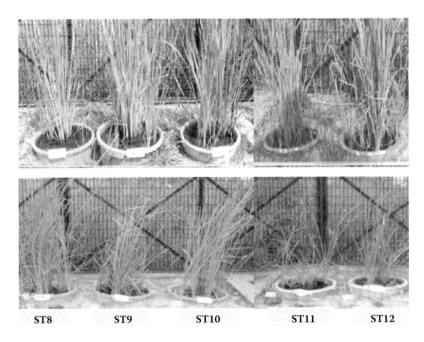

FIGURE 18.8
Pot culture experiment showing phytoremediation potential of *Typha angustifolia* grown on heavy metal–rich aqueous solution of phenol and melanoidins. (Chandra, R., and Yadav, S. 2010. *Ecological Engineering* 36 (10): 1277–1284.)

TABLE 18.6

Limitation and Benefits of *In Situ* and *Ex Situ* Remediation Technology

Technology	Examples	Benefits	Limitations	Factors to Consider
In situ	*In situ* bioremediation; biosparging; bioventing; bioaugumentation	Most cost efficient, noninvasive, relatively passive, natural attenuation, treat soil and water	Environmental constraints; extended treatment time; monitoring difficulties	Biodegradative abilities of indigenous microorganisms; presence of metals and other inorganic environmental parameters; biodegradability of pollutants; chemical solubility, geographical factor, distribution of pollutants
Ex situ	Land forming; composting; biopiles	Cost efficient; low cost; can be done on site	Space requirements; extended treatment time need to control abiotic loss; mass transfer problem; bioavailability limitation	Biodegradative abilities of indigenous microorganisms; presence of metals and other inorganic environmental parameters; biodegradability of pollutants; chemical solubility, geographical factor, distribution of pollutants
Bioreactor	Slurry reactor; aqueous reactor	Rapid degradation; kinetic, optimized environmental parameters; enhanced mass transfer; effective use of inoculants and surfactants	Soil requires excavation; relatively high cost capital investment; relatively high operating cost	Biodegradative abilities of indigenous microorganisms; presence of metals and other inorganic environmental parameters; biodegradability of pollutants; chemical solubility, geographical factor, distribution of pollutants; bioaugumentation; toxicity of amendments; toxic concentrations of contaminants

that optimum treatment can take place. This type of treatment is easy to maintain, but it requires a lot of space, and the process of decontamination will take longer than it would by slurry phase bioremediation. Solid phase soil treatments include land farming, soil biopiles, and composting.

18.5.2 Evaluation of Metal Accumulation in Plants during Phytoremediation

18.5.2.1 Phytoextraction/Phytoaccumulation

In this approach, plants absorb, concentrate, and precipitate toxic metals from contaminated soils, sediment, or sludge into their harvestable parts (shoots and leaves). This can be calculated through an atomic absorption spectrophotometer. To determine metal concentrations, the native plants were washed thoroughly with distilled water to remove sludge particles from the roots, followed by rinsing with a 10 mmol L^{-1} solution of $CaCl_2$. Subsequently, the plant roots, shoots, and leaves were separated and chopped into small pieces and the resulting biomass was oven-dried at 70°C for 7 days until constant weight was achieved. The dried plant parts were ashed in a muffle furnace at 460°C for 6 h. Further, the weighed ash from these samples was digested in 2% HNO_3 and filtered through a 0.45 μm glass fiber filter. One gram of dried and sieved sediments was digested with 10 mL of HNO_3. If brown fumes appeared, 5 mL of HNO_3 was added and digestion continued until no brown fumes were generated. Subsequently, the sample was cooled, 2 mL water and 3 mL of H_2O_2 were added, and digestion continued followed by EPA method 3050-B (EPA 1996). The concentrations of Cr, Zn, Mn, Ni, Cu, Fe, Cd, and Pb were measured using an AAS (ZEEnit 700, Analytic Jena, Germany) (Chandra and Kumar 2017a). The detailed analysis of heavy-metal accumulation in various plant roots from complex organic waste containing stabilized distillery sludge is shown in Table 18.7.

Phytoextraction by accumulation of metals in harvestable plant parts only becomes effective with the presence of plants that either hyperaccumulate metals or have a large annual yield. A plant potential for heavy-metal phytoextraction may be evaluated with a bioconcentration factor, a biological accumulation coefficient, and a translocation factor.

18.5.2.1.1 Bioconcentration Factor (BCF)

The BCF provides an index of the ability of the plant to accumulate metal in the root with respect to metal concentration in the soil. The BCF is calculated as follows (Yoon et al. 2006, Li et al. 2007):

$$BCF = \frac{\text{Concentration of metal in plant root}}{\text{Concentration of metal in contaminated soil}}$$

BCF is a useful parameter to evaluate the potential of the plants in accumulating metals and this value is calculated on a dry weight basis. On the basis of the degree of heavy-metal accumulation in the tissues of the plant and the BCF, it is possible to determine plants belonging to hyperaccumulators, accumulators (BCF > 1), and excluders (BCF < 1).

18.5.2.1.2 Translocation Factor (TF)

The TF is important to evaluating the metabolic activities and health of individual plants growing on any polluted site. The TF (mobilization ratio) of metal is calculated by the ratio of metal concentration in plants' aerial parts (shoots) and metal concentration in plants' roots growing on sludge as reported in earlier studies (Deng et al. 2004, Yoon et al. 2006). It is calculated as follows:

$$TF = \frac{\text{Concentration of metal in plant shoot}}{\text{Concentration of metal in plant root}}$$

An effective process of translocation of heavy metals from the roots of tree shoots occurs at TF > 1. According to the different capacities for metal uptake, a plant with a high BCF (BCF > 1) and TF (TF < 1) has the potential for phytostabilization (Yoon et al. 2006). The BCF and TF of various heavy metals in potential native plants grown on stabilized distillery sludge is presented in Table 18.8.

TABLE 18.7

Metal Content in Roots of Different Native Plants Grown on Stabilized Distillery Sludge after Phytoextraction

Plant Code	Heavy Metals							
	Cr	Zn	Mn	Ni	Cu	Fe	Cd	Pb
A	1.095 ± 0.001e	18.600 ± 0.002i	5.010 ± 0.01e	1.200 ± 0.01a	3.855 ± 0.003i	31.320 ± 0.003g	0.165 ± 1.32g	1.500 ± 0.003g
B	BDL	8.700 ± 0.005d	2.835 ± 0.012c	0.735 ± 0.004a	2.685 ± 0.002f	30.990 ± 0.003f	0.150 ± 0.001f	0.480 ± 0.000c
C	BDL	8.550 ± 0.001c	6.675 ± 0.004h	0.165 ± 0.003a	1.531 ± 0.002c	21.705 ± 0.004d	0.120 ± 0.001e	BDL
D	0.645 ± 0.004d	16.800 ± 0.003h	6.181 ± 0.002g	0.435 ± 0.001a	2.715 ± 0.005f	37.381 ± 0.001j	0.210 ± 0.001i	0.690 ± 0.002d
E	BDL	15.90 ± 0.02f	10.185 ± 0.003k	0.765 ± 0.006a	2.280 ± 0.002d	61.815 ± 0.004n	0.210 ± 0.002i	0.510 ± 0.002
F	BDL	13.800 ± 0.005	6.660 ± 0.004h	0.555 ± 0.001a	2.595 ± 0.006e	41.385 ± 0.01k	0.225 ± 0.005j	1.365 ± 0.006f
G	BDL	27.150 ± 0.008e	5.655 ± 0.008f	0.270 ± 0.003a	3.220 ± 0.004h	31.785 ± 0.02h	0.180 ± 0.004h	0.495 ± 0.01c
H	0.060 ± 0.005a	16.050 ± 0.002g	3.765 ± 0.004d	1.185 ± 0.004a	2.265 ± 0.002d	20.640 ± 0.001c	0.105 ± 0.001d	0.960 ± 0.00e
I	BDL	15.900 ± 0.001f	1.785 ± 0.001b	1.095 ± 0.002a	0.810 ± 0.001a	5.175 ± 0.003a	BDL	1.080 ± 0.001e
J	BDL	4.351 ± 0.001a	1.575 ± 0.000a	0.495 ± 0.003a	0.871 ± 0.002b	5.655 ± 0.003b	BDL	0.315 ± 0.001b
K	0.300 ± 0.00b	7.651 ± 0.001b	5.025 ± 0.001e	0.960 ± 0.001a	2.895 ± 0.003g	22.410 ± 0.001e	BDL	0.600 ± 0.00cd
L	0.510 ± 0.001c	33 ± 0.008m	7.680 ± 0.001j	1.590 ± 0.001a	4.280 ± 0.001k	36.140 ± 0.03i	0.060 ± 0.005c	0.975 ± 0.010e
M	1.250 ± 0.025g	21.750 ± 0.023i	7.290 ± 0.017i	1.960 ± 0.019a	4.070 ± 0.007j	52.860 ± 0.017l	BDL	1.935 ± 0.026i
N	0.740 ± 0.005f	30.750 ± 0.019l	12.090 ± 0.015l	2.010 ± 0.05a	5.72 ± 0.055l	57.5 ± 0.132m	0.030 ± 0.001b	1.730 ± 0.024h
O	2.730 ± 0.019h	64.500 ± 0.009n	20.270 ± 0.078m	2.550 ± 0.081a	13.980 ± 0.010m	167.850 ± 0.238o	0.315 ± 0.002k	4.590 ± 0.153j

Source: Chandra, R., and Kumar, V. 2017a. *Environmental Science and Pollution Research* 24:2605–2619.

Notes: A: *Dhatura stramonium*; B: *Achyranthes sp*.; C: *Kalanchoe pinnata*; D: *Trichosanthes dioica*; E: *Parthenium hysterophorous*; F: *Cannabis sativa*; G: *Amaranthus spinosus* L.; H: *Croton bonplandianum*; I: *Solanum nigrum*; J: *Ricinus communis*; K: *Sacchrum munja*; L: *Basella alba*; M: *Setaria viridis*; N: *Chenopodium album*; O: *Blumea lacera*. All values are mean ($n = 3$) ± SD and presented in milligrams per kilogram. Mean values followed by different letters in the same column are significantly different (one-way analysis of variance; Tukey's test, $p \leq 0.05$); BDL: below detection limit.

TABLE 18.8

Accumulation and Translocation of Heavy Metals in Growing Native Plants

Plant Code	Bioconcentration Factor (BCF)								Translocation Factor (TF)							
	Cr	Zn	Mn	Ni	Cu	Fe	Cd	Pb	Cr	Zn	Mn	Ni	Cu	Fe	Cd	Pb
A	0.050	0.088	0.039	0.089	0.052	0.013	0.114	0.091	0.397	0.564	1.826[a]	1.025[a]	0.692	1.159[a]	0.818	0.87
B	–	0.041	0.022	0.054	0.036	0.012	0.104	0.029	–	0.15	1.232[a]	1.285[a]	1.160[a]	1.089[a]	0.1	1.531[a]
C	–	0.040	0.052	0.012	0.020	0.009	0.083	–	–	1.070	0.092	–	4.163[a]	0.109	0.125	–
D	0.029	0.079	0.048	0.032	0.036	0.015	0.145	0.042	1.651[a]	0.086	0.541	0.896	1.104[a]	0.459	1.142	1.195[a]
E	–	0.075	0.080	0.056	0.030	0.025	0.145	0.031	–	0.103	–	–	–	–	0.285	–
F	–	0.065	0.052	0.041	0.035	0.017	0.156	0.083	–	0.119	–	–	–	–	0.266	0.197
G	–	0.129	0.044	0.020	0.043	0.013	0.125	0.030	–	0.535	0.482	2.388	0.740	0.712	0.25	3.212[a]
H	0.002	0.076	0.029	1.185	0.030	0.008	0.072	0.058	3	1.757[a]	3.561[a]	0.594	2.807[a]	2.881[a]	2.571	1.906[a]
I	–	0.075	0.014	0.081	0.011	0.002	–	0.066	–	0.5	1.512[a]	1.027[a]	3.259[a]	2.182[a]	–	1.180[a]
J	–	0.020	0.012	0.036	0.011	0.002	–	0.019	–	0.896	0.847	1.456[a]	1.206[a]	0.846	–	1.190[a]
K	0.013	0.036	0.039	0.071	0.039	0.009	–	0.036	0.7	2.490[a]	1.370[a]	1.031[a]	1.259[a]	1.091[a]	–	1.125[a]
L	0.023	0.157	0.060	0.118	0.058	0.015	0.041	0.059	0.764	0.477	0.542	0.805	0.497	0.565	0.25	1
M	0.057	0.103	0.057	0.145	0.055	0.021	–	0.118	0.672	1.448*	1.119*	0.913	1.211*	0.940	–	0.511
N	0.033	0.146	0.095	0.149	0.077	0.023	0.020	0.105	2.229*	0.804	0.709	1.024*	0.874	1.090*	–	1.225*
O	0.125	0.306	0.160	0.189	0.189	0.069	0.218	0.281	0.051	0.469	0.424	0.443	0.328	0.261	–	0.193

Source: Chandra, R., and Kumar, V. 2017a. *Environmental Science and Pollution Research* 24:2605–2619.

Notes: A: *Dhatura stramonium;* B: *Achyranthes* sp.; C: *Kalanchoe pinnata;* D: *Trichosanthes dioica;* E: *Parthenium hysterophorous;* F: *Cannabis sativa;* G: *Amaranthus spinosus* L.; H: *Croton bonplandianum;* I: *Solanum nigrum;* J: *Ricinus communis;* K: *Sacchrum munja;* L: *Basella alba;* M: *Setaria viridis;* N: *Chenopodium album;* O: *Blumea lacera.* All values are mean ($n = 3$) ± SD and presented in milligrams per kilogram. Mean values followed by different letters in the same column are significantly different (one-way analysis of variance; Tukey's test, $p \leq 0.05$); BDL: below detection limit.

[a] Values > 1.

18.5.2.1.3 Metal Extraction Ratio (MER)

The metal extraction ratio is used to determine the ability of trees to immobilize heavy metals in their organisms as well as the suitability of a particular plant species in the phytoextraction process (Mertens et al. 2005). The metal extraction ratio is calculated as follows:

$$MER = C_{plant} \times M_{plant} / C_{soil} \times M_{rooted\ zone}) \times 100$$

where

C_{plant} is metal concentration in the aboveground part of the tree (mg kg^{-1})
C_{soil} is metal concentration in the soil (rhizosphere zone) (mg kg^{-1})
M_{plant} is mass of the aboveground part of the tree (g)
$M_{rooted\ zone}$ is mass of the soil (rhizosphere zone) (g)

On the basis of the degree of heavy-metal accumulation in the tissues of the trees and the BAF, it is possible to determine the plant belonging to hyperaccumulators, accumulators (BAF > 1), and excluders (BAF < 1) (Ma et al. 2001, Rezvani and Zaefarian 2011).

However, metal concentration in plants varies with plant species, and plant uptake of heavy metal from soil occurs either passively, with the mass flow of water into the roots, or through active transport across the plasma membrane of root epidermal cells. Also, the presence of organic substances and other co-pollutants influences the bioaccumulation of heavy metals. Moreover, a large number of factors control metal accumulation and its bioavailability associated with soil and climate conditions and agronomic management. Metal solubility in soil is predominantly controlled by pH, amount of metal cation exchange capacity, organic carbon content, and oxidation state of the system (Ghosh and Singh 2005).

In some cases, a breakdown product derived from the rhizodegradation and/or phytodegradation of the parent contaminant along the transpiration pathway may be phytovolatilized. The rate measurement of phytovolatilization is calculated by the following equation given by Limmer and Buken (2016):

$$\text{Phytovolatilization (\%)} = \frac{\text{Mass phytovolatilization}}{\text{Mass phytovolatilized} + \text{mass in harvestable part of plant}}$$

Organic chemical uptake and translocation by plants have been studied intensely since the 1950s. The relative ability of a plant to take up a chemical from soil or groundwater into it roots is described by the root concentration factor (RCF), which is measured as the ratio of the concentration in the root (mg kg^{-1}) to the concentration in the external solution (mg L^{-1}). However, the uptake of organic chemicals into a pass-through root depends on the plant's uptake efficiency, transpiration rate, and concentration of the chemical in soil water (Burken and Schnoor 1997):

$$U = (TSCF)(T)(C)$$

where

U is the rate of organic compound uptake by plant in milligrams per day
TSCF (transpiration stream concentration factor) is the efficiency of uptake of organic compounds
T is the transpiration rate in liters per day
C is the soil water concentration of chemicals in milligrams per liter

TSCF is defined as the ratio of chemical concentration in the xylem pore water to the chemical concentration in the feed solution:

$$\text{TSCF} = \frac{\text{Concentration of xylem sap (mg L}^{-1})}{\text{Root-zone solution concentration (mg L}^{-1})}$$

Higher RCF and TSCF values are an indication of enhanced contaminant uptake by plants and vary directly with the log K_{ow} (octanol–water partition coefficient) of the chemical. Organic compounds with log K_{ow} valued between 1 and 3.5 have been shown to enter plants. On the other hand, organic compounds that are highly polar and water soluble (log K_{ow} < 1.0) are not sufficiently sorbed by the roots, nor are they actively transported through plant membrane due to their higher polarity.

18.5.3 Advantages and Limitations of Phytoremediation

Phytoremediation, like other remediation technologies, has a range of advantages and disadvantages. The most positive aspect of using phytoremediation is that it is (i) more cost effective, (ii) more environmentally friendly, (iii) applicable to a wide range of toxic metals, and (iv) a more aesthetically pleasing method. On the other hand, phytoremediation presents some limitations. It is a lengthy process, so it may take several years or longer to clean up a site, and it is only applicable to surface soils. Advantages and disadvantages of using phytoremediation for remediation of a heavy metal–contaminated area and each mechanism are shown in Table 18.9.

The success of phytoremediation may be seasonal, depending on location. Other climatic factors will also influence its effectiveness. The success of remediation depends on establishing a selected plant community. Introducing new plant species can have widespread ecological ramifications, so it should be studied beforehand and monitored. Additionally, the establishment of the plants may require several seasons of irrigation. It is important to consider extra mobilization of contaminants in the soil and groundwater during this start-up period. If contaminant concentrations are too high, plants may die. Additional research is needed to determine the fate of various compounds in the plant metabolic cycle to ensure that plant droppings and products do not contribute toxic or harmful chemicals to the food chain. Scientists need to establish whether contaminants that collect in the leaves and wood of trees are released when the leaves fall in the autumn.

The toxicity and bioavailability of biodegradation products is not always known. Degradation by-products may be mobilized in groundwater or bioaccumulated in animals. Disposal of harvested plants can be a problem if they contain high levels of heavy metals. The depth of the contaminants limits treatment. The treatment zone is determined by plant root depth. In most cases, it is limited to shallow soils, streams, and groundwater. Pumping the water out of the ground and using it to irrigate plantations of trees may treat contaminated groundwater that is too deep to be reached by plant roots. Where practical, deep tilling to bring heavy metals that may have moved downward in the soil closer to the roots may be necessary. Generally, the use of phytoremediation is limited to sites with lower contaminant concentrations and contamination in shallow soils, streams, and groundwater. However, researchers are finding that the use of trees (rather than smaller plants) allows them to treat deeper contamination because tree roots penetrate more deeply into the ground.

TABLE 18.9

Merits and Constraints of Phytoremediation Processes

S. No.	Advantages	Disadvantages/Limitations
1.	Higher public acceptance	Toxicity and bioavailability of degrading products not known
2.	Generated less secondary wastes	Potential for contaminants to enter food chain through animal consumption
3.	Amendable to a broad range of organic and inorganic pollutants, including many metals with limited alternative options	Restricted to sites with shallow pollutants within rooting zones of remediative plants; ground surface at the site may have to be modified to prevent flooding or erosion
4.	*In situ/ex situ* application possible with effluent/soil substrate respectively; soil can be left at site after pollutants are removed, rather than having to be disposed of or isolated	A long time is often required for remediation; may take up to several years to remediate a contaminated site
5.	*In situ* application decreases amount of soil disturbance compared to conventional methods; can be performed with minimal environmental disturbance; organic pollutants may degrade to carbon dioxide and water, removing environmental toxicity	Restricted to sites with low concentration of pollutants; treatment is generally limited to soils a meter from the surface and groundwater within a few meters of the surface; soil amendments may be required
6.	Reduces the amount of waste to be landfilled (up to 95%); can be further utilized as bio-ore of heavy metals	Harvested plant biomass from phytoextraction may be classified; conditions may restrict the rate of growth of plants that can be utilized
7.	*In situ* application decreased spread of pollutants via air and water; possibly less secondary air and/or water waste is generated than with traditional methods	Climatic conditions are a limiting factor; climatic or hydrologic conditions may restrict the rate of growth of plants that can be utilized
8.	Does not require expensive equipment or highly specialized personnel; cost effective for large volumes of water having low concentration of pollutants and for large areas with low to moderate contamination of surface area	Introduction of non-native species may affect biodiversity
9.	In large-scale applications the potential energy can be utilized to generate thermal energy; plant uptake of contaminated groundwater can prevent off-site migration	Consumption/utilization of contaminated plant biomass is a cause of concern; pollutants may still enter the food chain through animals/insects that eat plant material containing pollutants

18.6 Life Cycle Impact Assessment

Life cycle impact assessment (LCIA) aims at understanding and quantifying the magnitude and significance of the potential environmental impacts of a product or a service throughout its entire life cycle. In the LCIA, the inputs and outputs quantified in the inventory are translated to potential impacts. Grouping and assigning material and energy transfer that exhibit environmental burden and/or intervention to prespecified categories (i.e., characteristics and classification) is the task of life cycle impact assessment (LCA-ISO 14042). These categories represent threatened environmental compartments, states, or relations, such as the consumption of energy resources, global warming, or toxicity to humans (Penington et al. 2004). However, the inventory data are multiplied by a characterization factor (CF) to give indication for the so-called environmental impact categories:

$$IS = \sum_j (Ej \text{ or } Rj) \times CFij$$

where impact category indicator i = indicator value per functional unit of.category j and $CFij$ = the characterization factor for emission j or resource j contributing to impact category i.

The European platform on LCA therefore provides guidance and data to facilitate LCIA. It equally outlines criteria against which models and indicators should be evaluated, covering both scientific aspects and stakeholder acceptability. Several LCIA methodologies exist and can be divided into midpoint or endpoint models. Midpoint impact indicators are located early in the cause–effect chain, whereas endpoint indicators describe damage to the so-called areas of protection, which are human health, natural environment, and natural resources (Finnveden et al. 2009). In addition, the man-made environment may also be included as an endpoint (Udo de Haes et al. 1999). Figure 18.9 illustrates the relation between inventory, midpoint, and endpoint indicators. A midpoint indicator characterizes the elementary flows and other environmental interventions that contribute to the same

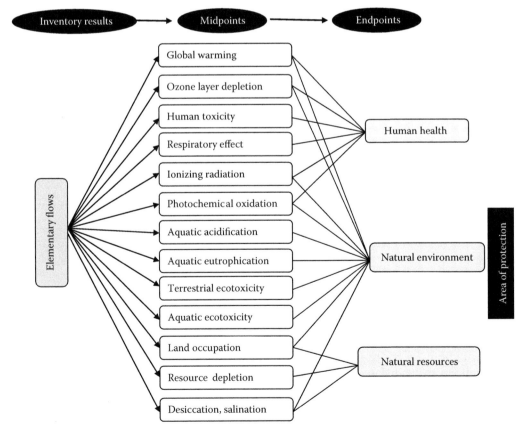

FIGURE 18.9
Illustration of the relation between inventory, midpoint, and endpoint impact assessment, and areas of protection.

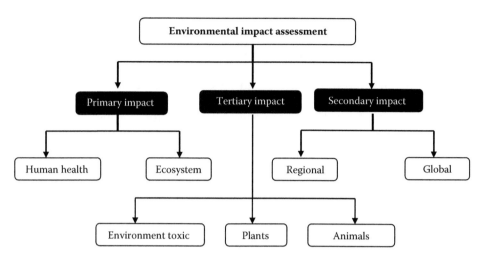

FIGURE 18.10
Types of environmental impact associated with contaminated sites.

impact. The term "midpoint" expresses the fact that this point is located somewhere on an intermediate position between the LCI results and the damage (or endpoint) on the impact pathway. In consequence, a further step may allocate these midpoint categories to one or more damage categories, the latter representing quality changes of the environment. A damage indicator result is the quantified representation of this quality change. In practice, such a result is always a simplified model of a very complex reality, giving only a coarse approximation to the quality status of the item.

Various authors have stated that environmental impact assessments (EIAs) differ fundamentally from product LCAs. The LCA is a specific elaboration of a generic environmental evaluation framework. The EIA is a procedure, rather than a tool, in which LCA certainly may be useful. The EIA is an environmental management tool that focuses on the identification of possible environmental effects and proposes control and mitigation measures for the identified effects. It also predicts whether the effects will have a significant impact in the receiving environment after the mitigation control has been implemented. An EIA successfully assesses both positive and negative effects and takes into consideration the local impact of proposed projects. The remediation of industrial waste-contaminated sites supports the goal of sustainable development but may also have environmental impact at local, regional, and global scales. The environmental impacts related to contaminated soil are usually classified in three categories: primary, secondary, and tertiary impacts (Figure 18.10). In site remediation service systems, primary impacts occur mainly on a local level, while secondary and tertiary impacts can occur at different levels.

18.6.1 Primary Impact

The primary impacts are related to local toxic impacts associated with waste-contaminated sites. The risks for human health and ecosystems are highly dependent on local characteristics such as specific contaminants, site conditions, and level of exposure of receptors. Due to these characteristics, the assessment of primary impacts is usually based on risk assessment. The protection of soil itself is not yet a priority and major concerns are related to groundwater contamination and human exposure via drinking water. This may be due to the lack of specific legislation on this issue.

In addition to the estimation of contaminant emissions, fate and exposure modeling and the assessment of ecotoxic effects are essential issues. A variety of risk assessment tools is used to assess the risk of contaminated soils to human and ecosystem health, which, for humans, can include emissions carcinogenic emissions, climate change, ozone layer depletion, and organic and inorganic substances causing respiratory effects. The concept of disability-adjusted life years (DALYs) was used to quantify human health risk from wastewater treatment plants. DALYs are one of the indicators for measuring aggregated health losses, and they combine years of life lost with years lived with disability standardized by means of severity weights. (Emissions causing health risks from main processes during construction and operation phases have been quantified.)

During the last few years, several impact assessment methods for human health and ecotoxicity have been developed. Pizzol et al. (2011) give an overview of different impact assessment methods of heavy metals on human health, but conclude that LCA practitioners should choose the model that has the highest consensus with regard to their specific problem. A component dealing specifically with human health is often called an "environmental health impact assessment." It is widely held that such impact assessments offer unique opportunities for the protection and promotion of human health. The following components were identified as key elements of an integrated environmental health impact assessment model:

- Project analysis
- Analysis of status quo (including regional analysis, population analysis, and background situation)
- Prediction of impact (including prognosis of future pollution and prognosis of health impact)
- Assessment of impact
- Recommendations
- Communication of results
- Evaluation of the overall procedure

Currently, the coverage of human health aspects in environmental impact assessments still tends to be incomplete, and public health departments often do not participate. The environmental health impact assessment as a tool for health protection and promotion is underutilized. It would be useful to achieve consensus on a comprehensive generic concept. An international initiative to improve the situation seems worth some consideration.

The assessment of the global impact on terrestrial ecosystems within LCIA was limited to the impact on soil organisms only because terrestrial vertebrates require a complex modeling that requires data on toxicity upon terrestrial vertebrates and the multiple exposure pathway that should be considered. In the United States, the USEtox model is recommended for human- and ecotoxicity LCIAs, although its uncertainties are still large and is somewhat limited as it does not consider some metals and is not integrated in any complete LCIA method. For organic contaminants, changes in concentrations over time, such as (bio)degradation and volatilization, are taken into account in risk assessments. For heavy metals, total concentrations in soil are mostly used as an input in risk assessment models. However, heavy metals in contaminated soil are rarely released completely, as only a portion of heavy metals is "bioavailable" or "geoavailable" (which means that the metals can be released and become available for biological uptake).

Assessment of heavy-metal release from soils and sediments over the long term is still controversial. Several methods and procedures have been proposed to estimate the long-term emissions of heavy metals contained in soils, sediments, and waste materials, but there is no consensus on which method performs best. Some hazardous substances produce toxic effects in humans or the environment after a single, episodic release. These toxic effects are referred to as "acute" toxicity. Other hazardous substances produce toxic effects in humans or the environment after prolonged exposure to the substance, which is called "chronic" toxicity. The assessment of ecotoxicity and human toxicity impacts is particularly problematic in site remediation service systems because they are inherently site specific. The site-specific conditions have a dominant influence on the behavior of contaminants in the subsurface, and understanding this behavior is essential for the assessment of ecotoxicity and human toxicity impacts. Depending on the application, the lack of site-specific data could result in misleading conclusions and misdirected decision making.

18.6.2 Secondary Impact

The secondary impacts consider the local, regional, and global impacts arising from extraction, materials use, and end-of-life stages of all consumables, equipment, and energy used for remediation.

LCIA methods include traditional impact categories at the global level (i.e., global warming) and at regional and local levels (such as acidification and ecotoxicity). The regional and global impact assessment processes are a way to identify, predict, and assess the types and scales of potential biodiversity impacts, as well as opportunities to benefit conservation, associated with any business activities or projects. Biodiversity assessment should begin as early as possible, as effective assessment of the biodiversity characteristics of an area—and the potential impacts—may require months or even years to account for seasonal and migration issues.

In addition, early attention to biodiversity issues means that potential impacts can be identified and avoided or mitigated in the earliest stages of planning. Once a project or business activity proceeds, the costs of redesign or resitting will make it more difficult to effectively address biodiversity issues. For example, the projections of precipitation indicate a 3% to 7% overall increase in all India summer monsoon rainfall in the 2030s with respect to the 1970s. However, on a seasonal scale, except for the Himalayan region, all other regions are likely to have lower rainfall in the winter period as well as the presummer period. Spatial patterns of monsoon rainfall indicate a significant decrease in the monsoon rainfall in the future except in some parts of the southern peninsula. There is currently a trend toward making regional impact assessment models on acidification, eutrophication (aquatic and terrestrial), and photochemical oxidation more site dependent, providing pollutant characterization factors for other regions than Western European countries or the United States, or even for regions within these countries.

18.6.3 Tertiary Impact

Tertiary impacts deal with environmental impacts associated with the future use of the site. Lesage et al. (2007a, b) introduced the term, which is related to the postremediation phase and the subsequent use of the site. The site under assessment in their study was an urban brownfield site (i.e., an idle site that is contaminated and cannot be reused for industrial or residential purposes before it is remediated). The tertiary impacts quantified covered the environmental impacts associated with developing and settling on new suburban land. Tertiary impacts are ultimately calculated as the development and occupation impacts of the tracked site less those of the avoided urban or suburban sites. Lesage et al. (2006) studied the environmental impacts associated with the effects of the postrehabilitation fate of the site on regional land use. Like secondary impacts, the environmental consequences of perturbing regional land use patterns will not be restricted to site-specific impact but will have effects regionally and globally.

18.7 Results Interpretation

Life cycle interpretation is a systematic technique to identify, quantify, check, and evaluate information from the results of the LCI and/or the LCIA. The purpose of performing life cycle interpretation is to determine the level of confidence in the final results and communicate them in a fair, complete, and accurate manner. Life cycle interpretation serves as the fourth phase of the LCA process. The international organization for standardization (ISO) has defined the following two objectives of LCA framework interpretation: (i) analyze results, reach conclusions, explain limitations and provide recommendations based on the findings of the preceding phases of the LCA and report the results of the life cycle interpretation in a transparent manner; (ii) provide a readily understandable, complete, and consistent presentation of the results of an LCA study, in accordance with the goal and scope of the study.

While an LCI or LCIA is conducted, assumptions, engineering estimates, and decisions based on personal values and values of stakeholders are made. Each of these decisions must be included and communicated with the final results to clearly and comprehensively explain conclusions drawn from the data. In some cases, it may not be possible to state that one alternative is better than the others because of the uncertainty in the final results. This does not imply that efforts have been wasted. A better understanding of the environmental and health impacts associated with each alternative, the significant impacts of each impact coverage (local, regional, or global), and the relative magnitude of each type of impact in comparison to each of the proposed alternatives included in the study will be gained from the study. This type of information more fully reveals the pros and cons of each alternative. The purpose of conducting an LCA is to better inform an overall decision-making process by providing a particular type of information (often unconsidered): a life cycle perspective of environmental and human health impacts associated with each alternative.

An LCA does not take into account technical performance, cost, or political and social acceptance. Therefore, it is recommended to use an LCA to inform, rather than to make the decision. Results of the interpretation phase should be documented within a comprehensive report documenting the LCA study in a clear and organized manner. This will help to communicate the results of the assessment fairly, completely, and accurately to others interested in the results. The purpose of the report is to present the results, data, methods, assumptions, and limitations in sufficient detail to allow the reader to comprehend the embodied complexities and trade-offs. If the results are to be reported to someone who was not involved in the LCA study (i.e., third parties or stakeholders), this report will serve as a reference document and should be provided to them to help prevent any misrepresentation of the results.

Life cycle interpretation is a systematic technique to identify, quantify, check, and evaluate information from the results of the LCI and/or the LCIA. The results from the inventory analysis and impact assessment are summarized during the interpretation phase. The outcome of the interpretation phase is a set of conclusions and recommendations for the study. According to ISO 14040:2006, the interpretation should include: (i) identification of significant issues based on the results of the LCI and LCIA phases of an LCA; (ii) evaluation of the study considering completeness, sensitivity, and consistency checks; and (iii) conclusions, limitations, and recommendations.

A key purpose of performing life cycle interpretation is to determine the level of confidence in the final results and communicate them in a fair, complete, and accurate manner. Interpreting the results of an LCA is not as simple as "3 is better than 2; therefore alternative A is the best choice!" Interpreting the results of an LCA starts with understanding the accuracy of the results and ensuring they meet the goal of the study. This is accomplished by identifying the data elements that contribute significantly to each impact category, sensitivity analysis of these significant data elements, assessing the completeness and consistency of the study, and drawing conclusions and recommendations based on a clear understanding of how the LCA was conducted and the results were developed. Treated wastewater can be reused as drinking water, in industry (for example, in cooling towers, in artificial recharge of aquifers, in agriculture and in the rehabilitation of natural ecosystems). Organic chemicals usually exist in municipal wastewaters at very low concentrations and ingestion over prolonged periods would be necessary to produce detrimental effects on human health. This is not likely to occur with agricultural/aquaculture use of wastewater, unless cross-connections with potable supplies occur or agricultural workers are not properly instructed, and can normally be ignored.

The principal health hazards associated with the chemical constituents of wastewaters, therefore, arise from the contamination of crops or groundwater. Hillman (1988) has drawn attention to the particular concern attached to the cumulative poisons, principally heavy metals, and carcinogens, mainly organic chemicals. These can be adopted directly for groundwater protection purposes but, in view of the possible accumulation of certain toxic elements in plants (for example, cadmium and selenium), the intake of toxic materials through eating crops irrigated with contaminated wastewater must be carefully assessed. Opportunities to reduce energy, material inputs, or environmental impacts must be evaluated at each stage of the product life cycle. The results are reported in the most informative way possible and the need and opportunities to reduce the impact of the product(s) or service(s) on the environment are systematically evaluated.

18.8 Constraints and Limitations of LCA

LCA methodology can be a very useful tool in this area since it can help in decision-making processes. However, for that same reason, it is important to be aware of constraints and limitations of LCA concerning remediation of contaminated soil. Many LCA studies do not consider in their goal and scope definition the primary and tertiary environmental impacts; this is an important issue because there are interrelated decisions concerning the choice of objectives related to physical state of the site, future of the site, and remediation technology to be used (Morais and Delerue-Matos 2010). Another important issue is the product system definition and the choice of processes to be included in the system. However, this is a common problem when applying LCA methodology and requires a thorough analysis to avoid a rather arbitrary and subjective procedure, especially when the goal of the study is the comparison of alternative products/services.

The lack of inventory data can also be a problem as well as the lack of a relation between inventory data and a specific region, which can be a source of uncertainty. The method and impact categories used also affect the outcome of LCA studies and this choice is very important (Cappuyns 2013). However, there are guidelines to help to make a decision, taking into consideration the goal and scope of the study (Guinée 2002). As mentioned before, many LCA studies do not consider primary impacts, which can be best assessed by risk assessment since it predicts the local risks. For that reason, some authors have applied risk assessment in site remediation LCA studies (Lemming et al. 2012). The assessment of tertiary impacts is also very difficult because it needs the specification of future use of land and this will imply the consideration of different scenarios to be analyzed (Cappuyns 2013).

References

Barakat, M.A. 2011. New trends in removing heavy metals from industrial wastewater. *Arabian Journal of Chemistry* 4:361–377.

Burken, J.G., Schnoor, J.L. 1997. Uptake and metabolism of atrazine by poplar trees. *Environmental Science and Technology* 31:1399–1406.

Cappuyns, V. 2013. LCA based evaluation of site remediation, Opportunities and Limitations. *Green Chemistry Sustainability. Chemistry Today* 31:18–21.

Chandra, R., Bharagava, R.N., Kapley, A., Purohit, H.J. 2011. Bacterial diversity, organic pollutants and their metabolites in two aeration lagoons of common effluent treatment plant (CETP) during the degradation and detoxification of tannery wastewater. *Bioresource Technology* 102:2333–2341.

Chandra, R., Kumar, V. 2015. Mechanism of wetland plant rhizosphere bacteria for bioremediation of pollutants in aquatic ecosystem. In *Advances in Biodegradation and Bioremediation of Industrial Waste*, ed. R. Chandra, 1–29. Boca Raton: CRC Press.

Chandra, R., Kumar, V. 2017a. Phytoextraction of heavy metals by potential native plants and their microscopic observation of root growing on stabilised distillery sludge as a prospective tool for in situ phytoremediation of industrial waste. *Environmental Science and Pollution Research* 24:2605–2619.

Chandra, R., Kumar, V. 2017b. Detection of *Bacillus* and *Stenotrophomonas* species growing in an organic acid and endocrine-disrupting chemical-rich environment of distillery spent wash and its phytotoxicity. *Environmental Monitoring & Assessment* 189(1):1–19.

Chandra, R., Kumar, V. 2017c. Detection of androgenic-mutagenic compounds and potential autochthonous bacterial communities during in situ bioremediation of post methanated distillery sludge. *Frontier in Microbiology* 8:887.

Chandra, R., Saxena, G., Kumar, V. 2015. Phytoremediation of environmental pollutants: An eco-sustainable green technology to environmental management. In *Advances in Biodegradation and Bioremediation of Industrial Waste*, ed. R. Chandra, 1–29. Boca Raton: CRC Press.

Chandra, R., Singh, R. 2012. Decolourisation and detoxification of rayon grade pulp paper mill effluent by mixed bacterial culture isolated from pulp paper mill effluent polluted site. *Biochemical Engineering Journal* 49–58.

Chandra, R., Yadav, S. 2010. Potential of *Typha angustifolia* for phytoremediation of heavy metals from aqueous solution of phenol and melanoidin. *Ecological Engineering* 36(10):1277–1284.

Chandra, R., Yadav, S., Yadav, S. 2017. Phytoextraction potential of heavy metals by native wetland plants growing on chlorolignin containing sludge of pulp and paper industry. *Ecological Engineering* 98:134–145.

Deng, H., Ye, Z.H., Wong, M.H. 2004. Accumulation of lead zinc, copper and cadmium by 12 wetland plant species thriving in metal contaminated site in China. *Environmental Pollution* 132:29–40.

Finnveden, G., Hauschild, M.Z., Ekvall, T., Guinée, J., Heijungs, R., Hellweg, S., Koehler, A., Pennington, D., Suh, S. 2009. Recent developments in life cycle assessment. *Journal of Environmental Management* 91(1):1–21.

Lemming, G., Chambon, J.C., Binning, P.J., Bjerg, P.L. 2012. Is there an environmental benefit from remediation of a contaminated site? Combined assessments of the risk reduction and life cycle impact of remediation. *Journal of Environmental Management* 112:392–403.

Lesage, P., Ekvall, T., Deschenes, L., and Samson, R. (2007a). Environmental assessment of Brownfield rehabilitation using two different life cycle inventory models—Part 1: Methodological approach. *International Journal of Life Cycle Assessment* 12(6):391–398.

Lesage, P., Ekvall, T., Deschenes, L., and Samson, R. 2007b. Environmental assessment of Brownfield rehabilitation using two different life cycle inventory models—Part 2: Case study. *International Journal of Life Cycle Assessment* 12(7):497–513.

Li, M.S., Luo, Y.P., Su, Z.Y. 2007. heavy metal concentration in soil and plant accumulation in a restored management mineland in Guangxi, South China. *Environmental Pollution* 147:168–175.

Limmer, M.A., Burken, J.G. 2016. Phytovolatilization of organic contaminants. *Environmental Science and Technology*. DOI: 10.1021/acs.est.5b04113.

Khasreen, M.M., Banfill, P.F.G., Menzies, G.F. 2009. Life-cycle assessment and the environmental impact of buildings: a review. *Sustainability* 1, 674–701.

Mertens, J., Luyssaert, S., Verheyen, K. 2005. Use and abuse of trace metal concentrations in plants tissue for biomonitoring and phytoextraction. *Environmental Pollution* 138:1–4.

Nie, Z., Gao, F., Gong, X., Wang, Z., Zuo, T. 2011. Recent progress and application of materials life cycle assessment in China. *Progress in Natural Science: Materials International* 21(1):1–11

Onwubuya, K., Cundy, A., Puschenreiter, M., Kumpiene, J., Bone, B., Greaves, J. et al. 2009. Developing decision support tools for the selection of "gentle' remediation approaches. *Science of the Total Environment* 407:6132–6142.

Parkes, O., Lettieri, P., David, I., Bogle, L. 2015. Life cycle assessment of integrated waste management systems for alternative legacy scenarios of the London Olympic Park. *Waste Management* 40:157–166.

Pennington, D.W., Potting, J., Finnveden, G., Lindeijer, E., Jolliet, O., Rydberg, T., Rebitzer, G. 2004. Life cycle assessment part 2: Current impact assessment practice. *Environmental International* 30(5):721–739.

Rebitzer, G., Ekvall, T., Frischknecht, R., Hunkeler, D., Norris, G., Rydberg, T., Schmidt, W.P., Suh, S., Weidema, B.P., Pennington, D.W. 2004. Life cycle assessment part 1: framework, goal and scope definition, inventory analysis, and applications. *Environmental International* 30(5):701–20.

Singh, B.K., Walker, A. 2006. Microbial degradation of organophosphorus compounds. *FEMS Microbiology Review* 30(3):428–471.

USEPA (US Environmental Protection Agency). 2008. Green Remediation: Incorporating Sustainable Environmental Practices into Remediation of Contaminated Sites, Washington, DC: US Environmental Protection Agency. Office of Solid Waste and Emergency Response.

Yoon, J., Cao, X., Zhou, Q., Ma, L.Q. 2006. Accumulation of Pb, Cu, and Zn in native plants growing on a contaminated Florida site. *Science of the Total Environment* 368(2):456–464.

19

Biochemical and Molecular Aspects of Arsenic Tolerance in Plants

Preeti Tripathi, Surabhi Awasthi, Reshu Chauhan, Pradyumna Kumar Singh, Sudhakar Srivastava, and Rudra Deo Tripathi

CONTENTS

19.1 Introduction

Arsenic (As) is a natural contaminant present in various environmental matrices, such as soil, water, and air. Arsenic is a nonessential element and is extremely toxic to all forms of life. The problem of As pollution is receiving increased attention all over the world, especially in Southeast Asia due to its known hazardous effects on plants and humans and owing to the huge scale of the problem.

The permissible limit of As in drinking water is 10 μg L^{-1} (WHO 2001). Elevated levels of As in drinking water extracted from shallow tube wells causes As poisoning in humans (Christen 2001). Plants also depend on As-rich groundwater for irrigation purposes in the affected regions (Christen 2001, Abedin et al. 2002). Irrigation with As-containing groundwater over the years has led to buildup of As levels in soil. The consumption of As-tainted crop products and water by humans and animals has hence become unavoidable.

Arsenic occurs in various forms and one form can be changed to another during transportation in the environment. In nature, As can exist essentially in four oxidation states: (–3), (0), (+3), and (+5). The most common inorganic As (iAs) species in the environment include arsenite [As(III)] and arsenate [As(V)] while the organic species comprise monomethylarsonic acid [MMA: $CH_3AsO(OH)_2$], dimethylarsinic acid [DMA: $(CH_3)_2AsOOH$], trimethylarsine oxide [TMAO: $(CH_3)_3AsO$], and trimethyl arsine [TMA: $(CH_3)_3As$] (Kim et al. 2009, Zhu and Rosen 2009). The International Agency for Research on Cancer (IARC)

471

places iAs as the highest health hazard category (i.e., a group 1 carcinogen), and there is substantial evidence that it increases the risk of cancer of the bladder, lung, skin, and prostate (IARC 2004, Naidu et al. 2006).

Arsenic exposure induces the production of reactive oxygen species (ROS), probably through the interconversion of As(V) and As(III). ROS can damage membranes, proteins, and DNA. They attack membrane lipids, particularly unsaturated fatty acids such as linolenic acid, and produce lipid peroxides, which then cause further damage by initiating radical chain reactions (Mittler 2002). In order to avoid oxidative damage, plants have evolved a complex antioxidant defense system including enzymes and metabolites like phenolics (Dat et al. 2000, Chauhan et al. 2017). In addition to oxidative stress, both inorganic forms of As are highly toxic because As(V) interferes with phosphate metabolism such as phosphorylation and ATP synthesis and As(III) binds to vicinal sulfydryl groups of proteins, affecting their structures and/or catalytic functions (Tripathi et al. 2007, Zhao et al. 2010). Arsenate is readily reduced to As(III) in plants; this is detoxified by complexation with thiol-rich peptides, such as reduced glutathione (GSH) and phytochelatins (PCs) and/or vacuolar sequestration (Zhao et al. 2010). Organic As is also toxic (Zhao et al. 2010); however, its toxicity mechanism in plants remain elusive.

Thus, gradually increasing accumulation of As in crop plants such as rice due to cultivation in anaerobic conditions eventually affects growth profile and yield. Hence, rice cultivation in As-contaminated areas results in low-quality rice grains enriched in As (Dwivedi et al. 2012). There is a critical need to fully elucidate the mechanisms of As assimilation and metabolism to develop the mitigation strategies for reduction in As uptake and accumulation in plants. This would require a detailed understanding of biogeochemical aspects of As bioavailability, factors influencing As uptake and transport, and plant processes involved in metabolism and detoxification of As.

19.2 Arsenic Accumulation and Transport in Plants

19.2.1 Arsenic Accumulation

Three main forms of As in soil are available to plants—namely, arsenate [As(V)], arsenite [As(III)] and methylated As [monomethylarsonic acid (MMA), and dimethylarsinic acid (DMA)]. There is always more than one As species present in plant tissues (Meharg et al. 2002). Transporters play a critical role in As uptake and transport in plants (Zhao et al. 2010). In addition to transporting molecules and minerals essential for plant growth, transporters also transport some nonessential and even harmful components, such as As, owing to ionic mimicry (Schroeder et al. 2013, Awasthi et al. 2017). Plants take up and translocate different As species through distinct transporters and pathways. Element accumulation in plants requires uptake through the roots and then transport through the xylem and the apoplast to sites of usage or storage compartments (Li et al. 2009). There are genotypic differences in different varieties of a plant—for example, rice in As uptake and accumulation (Dwivedi et al. 2010, Wu et al. 2012, Zhao et al. 2013).

The factors influencing As availability in soils also affect As uptake and accumulation in plants. Arao et al. (2009) and Norton et al. (2013) showed that flooding conditions substantially increased straw and grain As content in rice. Hu et al. (2015) demonstrated in a pot experiment that constant and intermittent flooding treatments resulted in 3–16 times greater As concentrations in the soil solution than in aerobic conditions. Wu et al. (2017) reported that oxic conditions also affect the accumulation of As.

19.2.2 Arsenate Uptake

As(V) is the dominant As species in aerobic soils and readily enters into plant roots via phosphate (Pi) transporters because it is structurally analogous to Pi. To date, a number of high-affinity Pi transporters have been identified that are involved in As(V) transport in plants. In *Arabidopsis thaliana*, two high-affinity Pht1 isoforms—AtPht1;1 (Pi transporter 1;1) and AtPht1;4—were initially shown to mediate Pi and As(V) acquisition from soils with both low and high levels of phosphorus (Shin et al. 2004). The semidominant mutant of AtPht1;1, *atpht1;1-3*, exhibits low As(V) uptake rate in the short term, but ultimately accumulates twice the amount of As in comparison to wild type. Furthermore, gene expression analysis revealed that Pi activation genes were suppressed in the *atpht1;1-3* mutant, but As-responsive genes were induced (Catarecha et al. 2007).

Moreover, AtPht1;5 promotes the translocation of both Pi and As(V) from phosphorus sources to sink organs. Loss of AtPht1;5 mitigates As(V) toxicity in plants. Overexpression of AtPht1;7, which is specifically expressed in reproductive tissues of *Arabidopsis*, enhances As(V) accumulation (LeBlanc et al. 2013). AtPht1;9, a predominantly root-expressed Pht1 transporter, mediates inorganic Pi acquisition in the root during Pi starvation. A study by Remy et al. (2012) identified AtPhT1;9 and alsoAtPht1;8 in Pi acquisition and As(V) uptake. Recently, several WRKY transcription factors were found to be involved in As(V) influx. For example, both WRKY6 and WRKY45, which mediate As(V) defense in plants, regulate the expression of AtPht1;1 to modulate As(V) uptake (Castrillo et al. 2013, Wang et al. 2014). In *Oryza sativa*, studies have provided direct evidence for As(V) uptake by roots via high-affinity Pht1 transporters. OsPht1;8, which is expressed in both root and shoot tissue independently of Pi supply, has a high affinity for both Pi and As(V). OsPht1;8 overexpression in rice markedly increases As(V) uptake and translocation (Jia et al. 2011, Wu et al. 2011). The transcription factor OsPHR2 (Pi starvation response 2) regulates the expression of OsPht1;8 (Wu et al. 2011). Moreover, OsPht1;1, a constitutively expressed high-affinity Pht1 transporter located in the plasma membrane, is also involved in As(V) uptake (Sun et al. 2012, Kamiya et al. 2013).

19.2.3 Arsenite Uptake

As(III) is the dominant form of As in anaerobic environments. In plants, members of the nodulin 26-like intrinsic protein (NIP) family of plant aquaporins are involved in As(III) uptake by root (Li et al. 2014, Mukhopadhyay et al. 2014). Bienert et al. (2008) demonstrated that *A. thaliana* NIP5;1 and NIP6;1; *O. sativa* NIP2;1 (predominantly expressed in roots); NIP3;2 and *Lotus japonicus* NIP5;1; and NIP6;1 transport As(III) across cell membranes. Additionally, AtNIP7;1, AtNIP1;1 (highly expressed in roots), AtNIP1;2 (highly expressed in seeds), and AtNIP3;1 (highly expressed in roots) also function in As(III) uptake (Bienert et al. 2008, Isayenkov and Maathuis 2008, Kamiya et al. 2009, Xu et al. 2015). In *O. sativa*, NIP2;1 (also named OsLsi1), a silicon (Si) influx transporter, is involved in a major pathway responsible for As(III) uptake by roots (Ma et al. 2008). However, compared with OsNIP2;1, OsNIP2;2 (OsLsi6), OsNIP1;1, and OsNIP3;1 expression is very weak, which implies that they do not play a significant role in As(III) uptake by rice roots (Ma et al. 2008).

Recently, OsNIP3;3 and HvNIP1;2 were shown to possess As(III) transport activity in yeast cells (Katsuhara et al. 2014). OsNIP3;3 or HvNIP1;2 expression in yeast DACR3 mutants lacking the As(III) efflux transporter ACR3 (arsenical compound resistance 3) increased the sensitivity of the cells to 5 mM As(III), indicating that OsNIP3;3 and HvNIP1;2 are As(III) transporters (Ali et al. 2012). OsNIP3;3 expression was not induced in rice under As(III) treatment conditions. Further investigations are required to determine the function of OsNIP3;3 in rice (Katsuhara et al. 2014).

In addition to NIPs, members of the rice plasma membrane intrinsic protein (PIP) family of aquaporins, including OsPIP2;4, OsPIP2;6, and OsPIP2;7, are also permeable to As(III) and contribute to As(III) tolerance in plants (Mosa et al. 2012). The transcript levels of these genes in the roots and shoots of rice are strongly downregulated in response to As(III) treatment, and all three of these proteins increased As(III) influx when heterologously expressed in *Xenopus laevis* oocytes. Mosa et al. (2012) demonstrated that overexpression of OsPIP2;4, OsPIP2;6, and OsPIP2;7 in *Arabidopsis* enhanced As(III) tolerance and led to both the active influx and efflux of As(III) in plant roots exposed to As(III) for a short duration, although there was no obvious As accumulation in plants in response to long-term As(III) treatment (Mosa et al. 2012).

19.2.4 Uptake of Methylated Arsenic Species

Trace amounts of several methylated As species, mainly MMA and DMA, are present in soils due to use of arsenic pesticide or herbicide and biomethylation of As by microorganisms (Ye et al. 2012, Huang 2014). High concentrations of methylated As have been detected in plants. Li et al. (2009) demonstrated that OsLsi1 (the aquaporin NIP2;1) is critical for the uptake of undissociated methylated As, including MMA and DMA, by rice roots. Moreover, MMA and DMA enter rice roots via the same entry route as glycerol, which is transported into plant cells by aquaporins (Rahman et al. 2011). However, uptake and transport of methylated As species still remains elusive and a lot more work is required to be done.

19.3 Biochemical and Molecular Mechanisms of Arsenic Tolerance in Plants

The toxicity of As is dependent on its speciation, with inorganic arsenicals thought to be more toxic than organic forms (Tamaki and Frankenberger 1992). It must be assumed that As-resistant plants compartmentalize and/or transform As to other less phytotoxic As species in order to withstand high cellular As burdens (Tripathi et al. 2007). Plants adopt various strategies to tolerate As. In some plants, like *Arabidopsis*, As(V) tolerance is achieved by suppression of the phosphate uptake pathway to lower the uptake of As (Abercrombie et al. 2008). In wild grasses, tolerance is achieved by suppression of As(V) uptake through the constitutive suppression of high-affinity phosphate–As(V) plasma-membrane cotransporters. The tolerance gene identified in rice is also a phosphate–As(V) transporter. Tolerance of the grass *Holocus lanatus* and some other plants to As(V) is known to be associated with relatively low uptake of As(V) compared with uptake into nontolerant plants (Abedin et al. 2002, Meharg and Hartley-Whitaker 2002, Zhao et al. 2009). Nevertheless, some plants can tolerate As by effluxing it out from the cell, like in yeast; the As(III) effluxer gene *ACR3* was involved in extrusion of As(III) into the medium. The first step of As(V) metabolism is the reduction of As(V) to As(III) by the aldose reductase (AR) enzyme (Ellis et al. 2006, Duan et al. 2007).

The involvement of AR in As metabolism has also been reported in rice. The rice genome contains two ACR2 (As compound resistant)-like genes, OsACR2.1 and OsACR2.2, and might be involved in regulating As metabolism. Detoxification mechanisms for As(III) include efflux from the roots, sequestration in cell vacuoles, and complexation with thiols such as phytochelatins (PCs), for which As essentially needs to be in the form of As(III) (Figure 19.1). The tolerance of *H. lanatus* is also related to more synthesis of PCs, with higher molar ratio of PC to As to tolerate As (Hartley-Whitaker et al. 2001, 2002). There are relatively few species of plants that are naturally As tolerant. A group of plants including *P. vittata* and other members of the Pteridaceae hyperaccumulate As (Ma et al. 2001, Ellis et al. 2006, Zhao et al. 2010). The growth of these plants is not compromised during times when they are accumulating extremely high levels of As. Furthermore, very little As(III) is effluxed by roots or bound to PCs (Zhao et al. 2003, Tripathi et al. 2012b) in hyperaccumulators. The second step of As detoxification is the complexation of As(III) with GSH or PCs and sequestration into the vacuole. Various species, like *H. lanatus*, *Ceratophyllum*

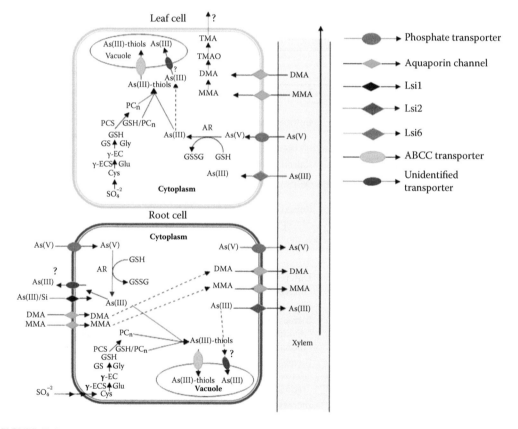

FIGURE 19.1

Arsenic uptake and detoxification mechanism in plants; A simplified schematic diagram of arsenic transport and metabolism in plants. The thickness of the arrow line is indicative of the relative flux. Transporters for As uptake into leaf cells are assumed to be similar to those in roots, but there is little knowledge of their identities. As(III): arsenite; As(V): arsenate; AR: arsenate reductase; Cys: cysteine; DMA: dimethylarsonic acid; Glu: Glutamic acid; Gly: glycine; γ-EC: γ-glutamyl cysteine; γ-*ECS*: γ-glutamyl cysteine synthase; GSH: reduced glutathione; GSSG: oxidized glutathione; GS: glutathione synthase; MMA: monomethylarsonic acid; PC: phytochelatin; PCS: phytochelatin synthase; TMA: trimethyl arsenic acid; TMAO: trimethyl arsine oxide.

demersum, and *Hydrilla verticillata,* are tolerant to As stress by enhanced synthesis of GSH or PCs.

The ability of As(V) to substitute Pi for biological reactions, the tendency of As(III)-based compounds to bind with the –SH group of proteins or enzymes, and the damaging effects of ROS all have direct and important consequences for plant metabolism. The plant response to these factors, predominantly by changes at various antioxidant (biochemical) and genomic (molecular) level 3 is discussed next.

19.3.1 Role of Enzymatic and Nonenzymatic Antioxidants

Several enzymes are involved in ROS defense strategies. Exposure to As upregulates a number of enzymes involved in antioxidant responses (Mylona et al. 1998, Requejo and Tena 2005). A list of As-modulated antioxidants is given in Table 19.1.

Highly reactive superoxide can be converted to less active but longer lasting H_2O_2 through the action of superoxide dismutase (SOD). Based on the metal cofactor used,

TABLE 19.1

Arsenic-Affected Enzymatic and Nonenzymatic Antioxidants in Different Plant Species

Antioxidants Studied	Plant Species	Remarks	Ref.
Proline	*O. sativa, S. oleracea*	Induced during As stress	Mishra and Dubey 2006; Pavlík et al. 2010; Dwivedi et al. 2012
Cysteine	*H. verticillata, C. demersum, B. juncea, O. sativa*	Induced during As stress	Srivastava et al. 2007, 2009; Mishra et al. 2008; Dwivedi et al. 2012
Ascorbic acid	*P. vittata, P. ensiformis*	Raised during As(V) stress	Singh et al. 2006
Nitric oxide	*O. sativa, A. thaliana*	Raised during As stress	Tripathi et al. 2012a, 2015b; Leterrier et al. 2012
Glutathione	*A. thaliana, P. vittata, P. ensiformis, H. verticillata, C. demersum, B. juncea, A. thaliana*	Raised during As(V) stress except in *A. thaliana*	Dhankher et al. 2002; Li et al. 2004; Singh et al. 2006; Srivastava et al. 2007, 2009; Mishra et al. 2008
Phytochelatins	*A. thaliana, H. verticillata, C. demersum*	Different species of PCs were increased during As stress	Dhankher et al. 2002; Li et al. 2004; Srivastava et al. 2007; Mishra et al. 2008
Superoxide dismutase	*P. vittata, H. verticillata, C. demersum*	Increased activity for dismutation of superoxide radicals	Singh et al. 2006; Srivastava et al. 2007; Mishra et al. 2008
Catalase	*P. vittata, H. verticillata, C. demersum*	Increased activity for detoxification of peroxide radicals	Singh et al. 2006; Srivastava et al. 2007; Mishra et al. 2008
Peroxidases— namely, APX, GPX,	*P. vittata, H. verticillata, C. demersum*	Increased activity for detoxification of peroxide radicals	Singh et al. 2006; Srivastava et al. 2007; Mishra et al. 2008
Glutathione reductase	*P. vittata, H. verticillata, C. demersum*	Raised activity for maintaining the balance of GSH	Singh et al. 2006; Srivastava et al. 2007; Mishra et al. 2008
Arsenate reductase	*S. cervisiae, P. vittata, H. verticillata, C. demersum*	Induced activity for conversion of arsenate to arsenite	Duan et al. 2005; Singh et al. 2006; Srivastava et al. 2007; Mishra et al. 2008

SODs are classified into three groups: iron SOD (Fe SOD), manganese SOD (Mn SOD), and copper–zinc SOD (Cu–Zn SOD). Iron SODs are located in the chloroplast, Mn SODs in the mitochondrion and peroxisomes, and Cu–Zn SODs in the chloroplast, cytosol, peroxisomes, and possibly the extracellular space (Alscher et al. 1987, Gratão et al. 2005). SOD activity in plants varies quite widely with As treatment.

H_2O_2 produced in a plant cell (either directly or enzymatically through enzymes such as SOD) can be neutralized by catalase, an enzyme that is often induced by As exposure (Mylona et al. 1998, Srivastava et al., 2005, Geng et al., 2006, Duman et al. 2010). Plants contain a group of nonenzymatic antioxidants that includes glutathione (GSH), phytochelatin (PC), ascorbate, carotenoids, and anthocyanin. These antioxidants generally accumulate during As exposure (Hartley-Whitaker et al. 2001, Bleeker et al. 2003, 2006, Khan et al. 2009, Song et al. 2010). The production of these molecules requires metabolic acclimations, including the diversion of carbon, nitrogen, sulfur, and metabolic energy from normal growth and development.

As nucleophilic scavengers, GSH and ascorbate are fairly unique among the nonenzymatic antioxidants in that they can form a redox cycle. The ROS produced during As treatment typically induces an increase in the oxidation state of the redox active pools of GSH and ascorbate in favor of GSSG dimers and dehydroascorbate over the more reduced GSH and ascorbate (Singh et al. 2006). This shift in redox state arises at two levels (Foyer and Noctor 2011). H_2O_2 can oxidize GSH and ascorbate through the action of specific peroxidases or, in the case of GSH, also through the action GRXs and GSH-S-transferases (GSTs). Like SOD and catalase, GST, GRX and/or peroxidase transcript or protein abundance, or enzymatic activity often increase in response to As exposure (Mylona et al. 1998, Srivastava et al. 2005, Geng et al. 2006, Abercrombie et al. 2008, Ahsan et al. 2008, Norton et al. 2008, Chakrabarty et al. 2009).

The second component of the H_2O_2 neutralizing system is made up of ascorbate peroxidase (APX), monodehydroascorbate reductase (MDHAR), dehydroascorbate reductase (DHAR), and glutathione reductase (GR). Together, these enzymes efficiently recycle oxidized GSH and ascorbate to allow further cycles of H_2O_2 reduction. The reduction of H_2O_2 through the interdependent ascorbate–GSH cycle requires reducing power in the form of NAD(P)H, diverting this energy from other metabolic processes. The enzymes involved in the recycling of oxidized GSH and ascorbate are also often induced in plants upon As exposure (Ahsan et al. 2008, Khan et al. 2009). Thus, the interdependent ascorbate–GSH cycle, when it can be established, has an important role in maintaining ROS balance in plants (Foyer and Noctor 2011), probably even during As exposure.

The oligomerization of GSH to produce PC is also induced during As exposure (Schmöger et al. 2000, Geng et al. 2006, Singh et al. 2006, Khan et al. 2009). However, at higher levels of As, thiol metabolism was disrupted in some overexpressing lines (Wojas et al. 2010). Moreover, the synthesis of PC can deplete cellular GSH pools, decreasing the antioxidant capacity of the cell (De Vos et al. 1992, Sneller et al. 1999, Hartley-Whitaker et al. 2001). These depleted GSH pools can only be rejuvenated by an influx of Glu, Cys, and Gly. Some amino acids—for example, proline, glutamic acid, aspartic acid, and alanine—also increased during As(V) stress in *Spinacia oleracea* (Pavlik et al. 2010). Among stress-responsive amino acids, proline is a much studied molecule and can function as an osmolyte and free radical scavenger and also protects the cell membrane against damage. The level of proline has also been observed to be elevated in *O. sativa* during As(III) stress (Mishra and Dubey 2006).

19.3.2 Molecular Responses in Plants and Biotechnological Approaches

Plants have various strategies to accumulate and cope with excess metals and metalloids. Plants adopt a range of detoxification mechanisms used by plants under heavy-metal stress. Activation of antioxidant enzymes, complexation with organic ligands including phytochelatin-mediated detoxification, and vacuolar sequestration constitute the essential mechanisms (Cobbett and Goldsbrough 2002, Kupper and Kroneck 2005, Verbruggen et al. 2009, Krämer 2010, Tripathi et al. 2015a). *Pteris vittata* is a hyperaccumulator of As and may be used for phytoremediation purposes in As-contaminated soil. It has an efficient mechanism for As uptake and a high proportion of As is translocated to shoot for vacuolar sequestration (Ma et al. 2001). *P vittat* has duplication of *ACR3* and *ACR3;1* genes, which is unique to fern and absent in flowering plants, as a vacuolar arsenite transporter (Indriolo et al. 2010). Transgenic approaches highlighted the role of different transporters and genes related to As accumulation, chelation, and sequestration in plants and their potential for phytoremediation purposes (Dhankher et al. 2002, Duan et al. 2007, Song et al. 2010). Overexpression of arsenate reductase (ArsC) and γ-glutamylcysteine synthetase (γ-*ECS*) genes enhanced As tolerance and accumulation in *Arabidopsis thaliana* (Dhankher et al. 2002). Silencing of AR leads to higher accumulation of arsenic in *Arabidopsis thaliana* (Dhankher et al. 2006). Arsenate reductase (ArsC) and γ-glutamylcysteine synthetase are key enzymes for reduction in the arsenic oxidation state (arsenate to arsenite) and GSH biosynthesis pathway, respectively. In *Arabidopsis thaliana*, overexpression of *GSH1* and *AtPCS1* showed higher accumulation of As and tolerance through As chelation and vacuolar sequestration (Guo et al. 2008, 2012). Moreover, expression analysis indicated that *AtABCC1* and *AtABCC2* are responsible for vacuolar sequestration of the As–thiol complex (Song et al. 2010).

The overexpression of *PvACR3* (arsenite antiporter in *P vittata*) in *A. thaliana* reduced As accumulation in roots by enhancing translocation of arsenic from root to shoot and provided tolerance (Chen et al. 2013). The expression of the metallothionein gene (*AtMT2b*) under 35S cauliflower mosaic virus promoter in *Nicotiana tabacum* enhanced the translocation of As from root to shoot (Grispen et al. 2009). Shri et al. (2014) reported that constitutive expression of *CdPCS1* (*Ceratophyllum demersum* phytochelatin synthase) increased the synthesis of phytochelatins and As content in rice root and shoot. Cd-PCS-1 rice transgenic plants accumulated low As in grains due to more As immobilization through PC complexation in roots of plants (Shri et al. 2014). The studies suggested that complexation of heavy metal with thiol and subsequent vacuolar sequestration are the main strategy for higher accumulation and tolerance in plants and may be used for development of efficient hyperaccumulator plants for phytoremediation and low grain As crops through restricting As in the leaves.

Glutaredoxins are an oxidoreductase that plays an important role in various physiological processes and stresses, such as development, phytohormone pathway, response to pathogen, and cold stress (Meyer et al. 2012, Hu et al. 2015). Sundaram et al. (2009) reported the role of *PvGrx5* (*P. vittata*) in reduction of As content in leaves of *A. thaliana*. Further, recent study showed that overexpression of rice glutaredoxin *OsGrx_C7* and *OsGrx_C2.1* in *A. thaliana* reduced the As accumulation and enhanced tolerance against As stress by maintaining the GSH pool and modulating *NIPs* expression (Verma et al. 2016). The methylation of iAs into organic volatilized form in plants may be another approach to remove As contaminants from soil. The methylation of As converts it into mono-, di-, and trimethylated forms and trimethylarsine, the volatile form of As. But the efficiency of this process in plants is very low.

Genetically engineered plants showed the efficient volatilization of As from aerial parts, thus giving a way to develop transgenic plants for phytoremediation purposes. The expression of arsenic methyltransferase (*arsM*) genes from *Rhodopseudomonas palustris* bacteria in rice highly increased the As methylation and volatilization (Meng et al. 2011). Further, Verma et al. (2016) cloned the arsenic methyltransferase (*WaarsM*) gene from fungus *Westerdykella aurantiaca* and expressed in *A. thaliana*. The expression of *arsM* increased As volatilization, resulting in lowering of As content in shoot and providing tolerance against As stress. HSPs have an important role in abiotic stress such as heat, cold, drought, and heavy-metal stress. Abiotic stress causes misfolding of proteins and inhibits the molecular function of proteins. HSPs are key players in maintaining correct folding of proteins and rescuing molecular functions of proteins during abiotic stress. In this context, Song et al. (2012) highlighted the role of *AtHsp90.3* in As tolerance in *A. thaliana*. The overexpression of alfalfa mitochondrial small heat-shock protein (*Hsp23*) in tall fescue (*Festuca arundinacea*) conferred tolerance against As stress by lowering the ROS level (Lee et al. 2012).

Heavy-metal stress enhances ROS level in plants, which damaged cell membrane, proteins, and cellular structure and caused oxidative stress. The antioxidant enzymes such as SOD, APX, and catalase protect the cell from oxidative stress by reducing ROS level in plants. Lee et al. (2007) established that overexpression of both the Cu–Zn SOD and APX genes in tall fescue chloroplasts enhanced tolerance against As-induced oxidative stress. The understanding of in-depth molecular and genetic mechanisms may be helpful in identifying new adaptive traits in the As-contaminated environment, as well as the novel genes and molecules allowing transgenic approaches to developing efficient, tolerant, and safer plants.

19.4 Conclusion and Future Prospects

In conclusion, arsenic contamination of the environmental food chain is a global concern that is not limited by geological boundaries. In addition to groundwater, rice grain has been identified as a major source of human exposure to As. Genetic responses of plants to As uptake and metabolism may be useful in developing agronomic cultivars suitable for cultivation in As-enriched soils. Further, the development of low-As cultivars with controlled As translocation to seeds would pave the way for reduced dietary exposure to As. A range of mitigation methods, from agronomic measures and plant breeding to genetic modification, may be employed to reduce As contamination of food crops.

Acknowledgments

Dr. Preeti Tripathi is thankful to SERB, New Delhi, for the award of DST Young Scientist. Reshu Chauhan is thankful to the Department of Science and Technology (DST) New Delhi, for the award of DST-INSPIRE Junior/Senior Research Fellowship. Dr. Rudra Deo Tripathi is thankful to the Council of Science and Industrial Research, New Delhi, for the fellowship of Emeritus Scientist.

References

Abedin, M.J., Cotter-Howells, J., and Meharg, A.A. 2002. Arsenic uptake and accumulation in rice (*Oryza sativa* L.) irrigated with contaminated water. *Plant and Soil* 240:311–319.

Abercrombie, J.M., Halfhill, M.D., and Ranjan, P. et al. 2008. Transcriptional responses of *Arabidopsis thaliana* plants to As(V) stress. *BMC Plant Biology* 8:87.

Ahsan, N., Lee, D.G., and Alam, I. et al. 2008. Comparative proteomic study of arsenic-induced differentially expressed proteins in rice roots reveals glutathione plays a central role during As stress. *Proteomics* 8:3561–3576.

Ali, W., Isner, J.C., Isayenkov, S.V., Liu, W.J., Zhao, F.J., and Maathuis, F.J.M. 2012. Heterologous expression of the yeast arsenite efflux system ACR3 improves Arabidopsis thaliana tolerance to arsenic stress. *New Phytologist* 194:716–723.

Alscher, R., Bower, J.L., and Zipfel, W. 1987. The basis for different sensitivities of photosynthesis to SO_2 in two cultivars of pea. *Journal of Experimental Botany* 38:99–108.

Arao, T., Kawasaki, A., Baba, K., Mori, S., and Matsumoto, S. 2009. Effects of water management on cadmium and arsenic accumulation and methylarsinic acid concentrations in Japanese rice. *Environmental Science and Technology* 43:9361–9367.

Awasthi, S., Chauhan, R., Srivastava, S., and Tripathi, R.D. 2017. The journey of arsenic from soil to grain in rice. *Frontiers in Plant Science* 8:1007.

Bienert, G.P., Thorsen, M., and Schüssler, M.D. et al. 2008. A subgroup of plant aquaporins facilitates the bidirectional diffusion of $As(OH)_3$ and $Sb(OH)_3$ across membranes. *BMC Biology* 6:26.

Bleeker, P.M., Hakvoort, H.W., Bliek, M., Souer, E., and Schat, H. 2006. Enhanced arsenate reduction by a CDC25 like tyrosine phosphatase explains increased phytochelatin accumulation in arsenate tolerant Holcus lanatus. *The Plant Journal* 45:917–929.

Bleeker, P.M., Schat, H., Vooijs, R., Verkleij, J.A., and Ernst, W.H. 2003. Mechanisms of arsenate tolerance in Cytisus striatus. *New Phytologist* 157:33–38.

Castrillo, G., Sánchez-Bermejo, E., and de Lorenzo, L. et al. 2013. WRKY6 transcription factor restricts arsenate uptake and transposon activation in Arabidopsis. *Plant Cell* 25:2944–2957.

Catarecha, P., Segura, M.D., and Franco-Zorrilla, J.M. et al. 2007. A mutant of the Arabidopsis phosphate transporter PHT1;1 displays enhanced arsenic accumulation. *Plant Cell* 19:1123–1133.

Chauhan, R., Awasthi, S., and Tripathi, P. et al. 2017. Selenite modulates the level of phenolics and nutrient element to alleviate the toxicity of arsenite in rice (*Oryza sativa* L.). *Ecotoxicology and Environmental Safety* 138:47–55.

Chen, W., Chi, Y., Taylor, N.L., Lambers, H. and Finnegan, P.M. 2010. Disruption of ptLPD1 or ptLPD2, genes that encode isoforms of the plastidial lipoamide dehydrogenase, confers arsenate hypersensitivity in Arabidopsis. *Plant Physiology* 153:1385–1397.

Chen, Y., Xu, W., and Shen, H. et al. 2013. Engineering arsenic tolerance and hyperaccumulation in plants for phytoremediation by a PvACR3 transgenic approach. *Environmental Science and Technology* 47:9355–9362.

Christen, K. 2001. The arsenic threat worsens. *Environmental Science and Technology* 35:286–291.

Cobbett, C., and Goldsbrough, P. 2002. Phytochelatins and metallothioneins: Roles in heavy metal detoxification and homeostasis. *Annual Review of Plant Biology* 53:159–182.

Dat, J., Vandenabeele, S., Vranova, E., Van Montagu, M., Inzé, D., and Van Breusegem, F. 2000. Dual action of the active oxygen species during plant stress responses. *Cell and Molecular Life Science* 57:779–795.

De Vos, C.R., Vonk, M.J., Vooijs, R., and Schat, H. 1992. Glutathione depletion due to copper-induced phytochelatin synthesis causes oxidative stress in Silene cucubalus. *Plant Physiology* 98:853–858.

Dhankher, O.P., Li, Y., and Rosen, B.P. et al. 2002. Engineering tolerance and hyperaccumulation of arsenic in plants by combining arsenate reductase and γ-glutamylcysteine synthetase expression. *Nature Biotechnology* 20:1140–1145.

Dhankher, O.P., Rosen, B.P., McKinney, E.C., and Meagher, R.B. 2006. Hyperaccumulation of arsenic in the shoots of Arabidopsis silenced for arsenate reductase (ACR2). *Proceeding of the National Academy of Science* 103:5413–5418.

Duan, G.L., Zhou, Y., Tong, Y.P., Mukhopadhyay, R., Rosen, B.P., and Zhu, Y.G. 2007. A CDC25 homologue from rice functions as an arsenate reductase. *New Phytologist* 174:311–321.

Duan, G.L., Zhu, Y.G., Tong, Y.P., Cai, C., and Kneer, R. 2005. Characterization of arsenate reductase in the extract of roots and fronds of Chinese brake fern, an arsenic hyperaccumulator. *Plant Physiology* 138:461–469.

Duman, F., Ozturk, F., and Aydin, Z. 2010. Biological responses of duckweed (Lemna minor L.) exposed to the inorganic arsenic species As (III) and As (V): Effects of concentration and duration of exposure. *Ecotoxicology* 19:983–993.

Dwivedi, S., Mishra, A., and Tripathi, P. et al. 2012. Arsenic affects essential and non-essential amino acids differentially in rice grains: Inadequacy of amino acids in rice based diet. *Environmental International* 46:16–22.

Dwivedi, S., Tripathi, R.D., and Tripathi, P. et al. 2010. Arsenate exposure affects amino acids, mineral nutrient status and antioxidants in rice (Oryza sativa L.) genotypes. *Environmental Science and Technology* 44:9542–9549.

Ellis, D.R., Gumaelius, L., Indriolo, E., Pickering, I.J., Banks, J.A., and Salt, D.E. 2006. A novel arsenate reductase from the arsenic hyperaccumulating fern *Pteris vittata*. *Plant Physiology* 141:1544–1554.

Foyer, C.H., and Noctor, G. 2011. Ascorbate and glutathione: The heart of the redox hub. *Plant Physiology* 155:2–18.

Geng, C.N., Zhu, Y.G., Hu, Y., Williams, P., and Meharg, A.A. 2006. Arsenate causes differential acute toxicity to two P-deprived genotypes of rice seedlings (*Oryza sativa* L.). *Plant and Soil* 279:297–306.

Gratão, P.L., Polle, A., Lea, P.J., and Azevedo, R.A. 2005. Making the life of heavy metal-stressed plants a little easier. *Functional Plant Biology* 32:481–494.

Grispen, V.M., Irtelli, B., and Hakvoort, H.W. et al. 2009. Expression of the Arabidopsis metallothionein 2b enhances arsenite sensitivity and root to shoot translocation in tobacco. *Environmental and Experimental Botany* 66:69–73.

Guo, B., Jin, Y., Wussler, C., Blancaflor, E.B., Motes, C.M., and Versaw, W.K. 2008. Functional analysis of the Arabidopsis PHT4 family of intracellular phosphate transporters. *New Phytologist* 177:889–898.

Guo, J., Xu, W., and Ma, M. 2012. The assembly of metals chelation by thiols and vacuolar compartmentalization conferred increased tolerance to and accumulation of cadmium and arsenic in transgenic Arabidopsis thaliana. *Journal of Hazardous Materials* 199:309–313.

Hartley-Whitaker, J., Ainsworth, G., Vooijs, R., Ten Bookum, W., Schat, H., and Meharg, A.A. 2001. Phytochelatins are involved in differential arsenate tolerance in *Holocus lanatus*. *Plant Physiology* 126:299–306.

Hartley Whitaker, J., Woods, C., and Meharg, A.A. 2002. Is differential phytochelatin production related to decreased arsenate influx in arsenate tolerant *Holocus lanatus*? *New Phytologist* 155:219–225.

Hu, D.G., Ma, Q.J., Sun, C.H., Sun, M.H., You, C.X., and Hao, Y.J. 2016. Overexpression of MdSOS2L1, a CIPK protein kinase, increases the antioxidant metabolites to enhance salt tolerance in apple and tomato. *Physiologia plantarum* 156:201–214.

Hu, P., Ouyang, Y., Wu, L., Shen, L., Luo, Y., and Christie, P. 2015. Effects of water management on arsenic and cadmium speciation and accumulation in an upland rice cultivar. *Journal of Environmental Science* 27:225–231.

Huang, J.H. 2014. Impact of microorganisms on arsenic biogeochemistry: A review. *Water Air Soil Pollution* 225:1848.

IARC Working Group on the Evaluation of Carcinogenic Risks to Humans, World Health Organization and International Agency for Research on Cancer, 2004. *Some drinking-water disinfectants and contaminants, including arsenic* (84). IARC.

Indriolo, E., Na, G., Ellis, D., Salt, D.E., and Banks, J.A. 2010. A vacuolar arsenite transporter necessary for arsenic tolerance in the arsenic hyperaccumulating fern *Pteris vittata* is missing in flowering plants. *Plant Cell* 22:2045–2057.

Isayenkov, S.V. and Maathuis, F.J.M. 2008. The Arabidopsis thaliana aquaglyceroporin AtNIP7;1 is a pathway for arsenite uptake. *FEBS Letters* 582:1625–1628.

Jia, H.F., Ren, H.Y., and Gu, M. et al. 2011. The PHOSPHATE TRANSPORTER gene OsPht1;8 is involved in phosphate homeostasis in rice. *Plant Physiology* 156:1164–1175.

Kamiya, T., Islam, M.R., Duan, G.L., Uraguchi, S., and Fujiwara, T. 2013. Phosphate deficiency signaling pathway is a target of arsenate and phosphate transporter OsPT1 is involved in As accumulation in shoots of rice. *Soil Science and Plant Nutrition* 59:580–590.

Kamiya, T., Tanaka, M., Mitani, N., Ma, J.F., Maeshima, M. and Fujiwara, T. 2009. NIP1;1, an aquaporin homolog, determines the arsenite sensitivity of Arabidopsis thaliana. *Journal of Biological Chemistry* 284:2114–2120.

Katsuhara, M., Sasano, S., Horie, T., Matsumoto, T., Rhee, J., and Shibasaka, M. 2014. Functional and molecular characteristics of rice and barley NIP aquaporins transporting water, hydrogen peroxide and arsenite. *Plant Biotechnology* 31:213–219.

Khan, I., Ahmad, A., and Iqbal, M. 2009. Modulation of antioxidant defence system for arsenic detoxification in Indian mustard. *Ecotoxicology and Environmental Safety* 2:626-634.

Kim, K., Moon, J.T., Kim, S.H., and Ko, K.S. 2009. Importance of surface geologic condition in regulating As concentration of groundwater in the alluvial plain. *Chemosphere* 77:478–484.

Krämer, U. 2010. Metal hyperaccumulation in plants. *Annual Review of Plant Biology* 61:517–534.

Kupper, H., and Kroneck, P.M. 2005. Heavy metal uptake by plants and cyanobacteria. *Metal Ions in Biological System* 44:97.

LeBlanc, M.S., McKinney, E.C., Meagher, R.B., and Smith, A.P. 2013. Hijacking membrane transporters for arsenic phytoextraction. *Journal of Biotechnology* 163:1–9.

Lee, K.W., Cha, J.Y., Kim, K.H., Kim, Y.G., Lee, B.H., and Lee, S.H. 2012. Overexpression of alfalfa mitochondrial HSP23 in prokaryotic and eukaryotic model systems confers enhanced tolerance to salinity and arsenic stress. *Biotechnology Letters* 34:167–174.

Lee, Y.P., Kim, S.H., Bang, J.W., Lee, H.S., Kwak, S.S., and Kwon, S.Y. 2007. Enhanced tolerance to oxidative stress in transgenic tobacco plants expressing three antioxidant enzymes in chloroplasts. *Plant Cell Reports* 26:591–598.

Leitenmaier, B., and Küpper, H. 2013. Compartmentation and complexation of metals in hyperaccumulator plants. *Frontier in Plant Science* 4:374.

Leterrier, M., Airaki, M., Palma, J.M., Chaki, M., Barroso, J.B., and Corpas, F.J. 2012. Arsenic triggers the nitric oxide (NO) and S-nitrosoglutathione (GSNO) metabolism in Arabidopsis. *Environmental Pollution* 166:136–143.

Li, G.W., Santoni, V., and Maurel, C. 2014. Plant aquaporins: Roles in plant physiology. *Biochimica et Biophsica Acta* 1840:1574–1582.

Li, R.Y., Ago, Y., and Liu, W.J. et al. 2009. The rice aquaporin Lsi1 mediates uptake of methylated arsenic species. *Plant Physiology* 150:2071–2080.

Li, Y., Dhankher, O.P., and Carreira, L. et al. 2004. Overexpression of phytochelatin synthase in Arabidopsis leads to enhanced arsenic tolerance and cadmium hypersensitivity. *Plant and Cell Physiology* 45:1787–1797.

Ma, J.F., Yamaji, N., and Mitani, N. et al. 2008. Transporters of arsenite in rice and their role in arsenic accumulation in rice grain. *Proceeding of National Academy of Science USA*. 105:9931–9935.

Ma, L.Q., Komar, K.M., Tu, C., Zhang, W., Cai, Y., and Kennelley, E.D. 2001. A fern that hyper accumulates arsenic. *Nature* 409:579–579.

Meharg, A.A., Hartley-Whitaker, J. 2002. Arsenic uptake and metabolism in arsenic resistant and nonresistant plant species. *New Phytologist* 154:29–43.

Meng, X.Y., Qin, J., and Wang, L.H. et al. 2011. Arsenic biotransformation and volatilization in transgenic rice. *New Phytologist* 191:49–56.

Meyer, Y., Belin, C., Delorme-Hinoux, V., Reichheld, J.P., and Riondet, C. 2012. Thioredoxin and glutaredoxin systems in plants: Molecular mechanisms, crosstalks, and functional significance. *Antioxidant & Redox Signaling* 17:1124–1160.

Mishra, S., and Dubey, R.S. 2006. Inhibition of ribonuclease and protease activities in arsenic exposed rice seedlings: Role of proline as enzyme protectant. *Journal of Plant Physiology* 163:927–936.

Mishra, S., Srivastava, S., Tripathi, R.D., and Trivedi, P.K. 2008. Thiol metabolism and antioxidant systems complement each other during arsenate detoxification in *Ceratophyllum demersum* L. *Aquatic Toxicology* 86:205–215.

Mittler, R. 2002. Oxidative stress, antioxidants and stress tolerance. *Trends in Plant Science* 7:405–410.

Mosa, K.A., Kumar, K., and Chhikara, S. et al. 2012. Members of rice plasma membrane intrinsic proteins subfamily are involved in arsenite permeability and tolerance in plants. *Transgenic Research* 21:1265–1277.

Mukhopadhyay, R., Bhattacharjee, H., and Rosen, B.P. 2014. Aquaglyceroporins: Generalized metalloid channels. *Biochimica et Biophysica Acta (BBA)* 1840:1583–1591.

Mylona, P.V., Polidoros, A.N. and Scandalios, J.G. 1998. Modulation of antioxidant responses by arsenic in maize. *Free Radical Biology & Medicine* 25(4):576–585.

Naidu, R., Smith, E., Owens, G., and Bhattacharya, P. 2006. *Managing Arsenic in the Environment: From Soil to Human Health*. CSIRO Publishing.

Norton, G.J., Adomako, E.E., Deacon, C.M., Carey, A.M., Price, A.H. and Meharg, A.A. 2013. Effect of organic matter amendment, arsenic amendment and water management regime on rice grain arsenic species. *Environmental Polluttion* 177:38–47.

Norton, G.J., Lou-Hing, D.E., Meharg, A.A., and Price, A.H. 2008. Rice–arsenate interactions in hydroponics: Whole genome transcriptional analysis. *Journal of Experimental Botany* 59:2267–2276.

Pavlík, M., Pavlíková, D., and Staszková, L. et al. 2010. The effect of arsenic contamination on amino acids metabolism in Spinacia oleracea L. *Ecotoxicology and Environmental Safety* 73:1309–1313.

Raab, A., Williams, P.N., Meharg, A., and Feldmann, J. 2007. Uptake and translocation of inorganic and methylated arsenic species by plants. *Environmental Chemistry* 4:197–203.

Rahman, M.A., Kadohashi, K., Maki, T., and Hasegawa, H. 2011. Transport of DMAA and MMAA into rice (*Oryza sativa* L.) roots. *Environmental and Experimental Botany* 72:41–46.

Remy, E., Cabrito, T.R., Batista, R.A., Teixeira, M.C., Sa´-Correia, I., and Duque, P. 2012. The Pht1;9 and Pht1;8 transporters mediate inorganic phosphate acquisition by the Arabidopsis thaliana root during phosphorus starvation. *New Phytologist* 195:356–371.

Requejo, R., and Tena, M. 2005. Proteome analysis of maize roots reveals that oxidative stress is a main contributing factor to plant arsenic toxicity. *Phytochemistry* 66:1519–1528.

Schmöger, M.E., Oven, M., and Grill, E. 2000. Detoxification of arsenic by phytochelatins in plants. *Plant Physiology* 122:793–802.

Schroeder, J.I., Delhaize, E., Frommer, W.B. et al. 2013. Using membrane transporters to improve crops for sustainable food production. *Nature* 497:60–66.

Shin, H., Shin, H.-S., Dewbre, G.R. and Harrison, M.J. 2004. Phosphate transport in Arabidopsis: Pht1;1 and Pht1;4 play a major role in phosphate acquisition from both low- and high-phosphate environments. *Plant Journal* 39:629–642.

Shri, M., Dave, R., and Diwedi, S. et al. 2014. Heterologous expression of *Ceratophyllum demersum* phytochelatin synthase, CdPCS1, in rice leads to lower arsenic accumulation in grain. *Scientific Reports* 4:5784.

Singh, N., Ma, L.Q., Srivastava, M., and Rathinasabapathi, B. 2006. Metabolic adaptations to arsenic-induced oxidative stress in *Pterisvittata* L and *Pterisensiformis* L. *Plant Science* 170:274–282.

Sneller, F.E.C., Van Heerwaarden, L.M., and Kraaijeveld-Smit, F.J.L. et al. 1999. Toxicity of arsenate in *Silene vulgaris*, accumulation and degradation of arsenate-induced phytochelatins. *New Phytologist* 144:223–232.

Song, H.M., Wang, H.Z., and Xu, X.B. 2012. Overexpression of AtHsp90.3 in Arabidopsis thaliana impairs plant tolerance to heavy metal stress. *Biologia Plantarum* 56:197–199.

Song, W.Y., Park, J., and Mendoza-Cózatl, D.G. et al. 2010. Arsenic tolerance in Arabidopsis is mediated by two ABCC-type phytochelatin transporters. *Proceeding of National Academy of Science* 107:21187–21192.

Srivastava, M., Ma, L.Q., Singh, N., and Singh, S. 2005. Antioxidant responses of hyper-accumulator and sensitive fern species to arsenic. *Journal of Experimental Botany* 56:1335–1342.

Srivastava, S., Mishra, S., Tripathi, R.D., Dwivedi, S., Trivedi, P.K., and Tandon, P.K. 2007. Phytochelatins and antioxidant systems respond differentially during arsenite and arsenate stress in *Hydrilla verticillata* (Lf) Royle. *Environmental Science and Technology* 41:2930–2936.

Srivastava, S., Srivastava, A.K., Suprasanna, P., and D'souza, S.F. 2009. Comparative biochemical and transcriptional profiling of two contrasting varieties of Brassica juncea L. in response to arsenic exposure reveals mechanisms of stress perception and tolerance. *Journal of Experimental Botany* 60:3419–3431.

Stone, R. 2008. Food safety: Arsenic and paddy rice: A neglected cancer risk? *Science* 32:184–185.

Sun, S.B., Gu, M., and Cao, Y. et al. 2012. A constitutive expressed phosphate transporter, OsPht1;1, modulates phosphate uptake and translocation in phosphate-replete rice. *Plant Physiology* 159:1571–1581.

Sundaram, S., Wu, S., Ma, L.Q., and Rathinasabapathi, B. 2009. Expression of a Pteris vittata glutaredoxin PvGRX5 in transgenic Arabidopsis thaliana increases plant arsenic tolerance and decreases arsenic accumulation in the leaves. *Plant, Cell & Environment* 32:851–858.

Tamaki, S., Frankenberger, W.T. 1992. Environmental biochemistry of arsenic. *Reviews of Environmental Contamination and Toxicology* 124:79–110.

Tripathi, P., Singh, P.K., and Mishra, S. 2015a. Recent advances in the expression and regulation of plant metallothioneins for metal homeostasis and tolerance. *Environmental Waste Management* 551–564.

Tripathi, P., Singh, R.P., Sharma, Y.K., and Tripathi, R.D. 2015b. Arsenite stress variably stimulates pro-oxidant enzymes, anatomical deformities, photosynthetic pigment reduction, and antioxidants in arsenic-tolerant and sensitive rice seedlings. *Environmental Toxicology and Chemistry* 34:1562–1571.

Tripathi, P., Mishra, A., Dwivedi, S. et al. 2012a. Differential response of oxidative stress and thiol metabolism in contrasting rice genotypes for arsenic tolerance. *Ecotoxicology and Environmental Safety* 79:189–198.

Tripathi, R.D., Tripathi, P., Dwivedi, S. et al. 2012b. Arsenomics: Omics of arsenic metabolism in plants. *Frontiers in Physiology* 3.

Tripathi, R.D., Srivastava, S., and Mishra, S. et al. 2007. Arsenic hazards: Strategies for tolerance and remediation by plants. *Trends in Biotechnology* 25:158–165.

Verbruggen, N., Hermans, C., and Schat, H. 2009. Molecular mechanisms of metal hyperaccumulation in plants. *New Phytologist* 181:759–776.

Verma, P.K., Verma, S., and Pande, V. et al. 2016. Overexpression of rice glutaredoxin OsGrx_C7 and OsGrx_C2. 1 Reduces intracellular arsenic accumulation and increases tolerance in Arabidopsis thaliana. *Frontier in Plant Science* 7.

Wang, H., Xu, Q., and Kong, Y.H. et al. 2014. Arabidopsis WRKY45 transcription factor activates PHOSPHATE TRANSPORTER1;1 expression in response to phosphate starvation. *Plant Physiology* 164:2020–2029.

Washington, D.C. 2001. *Arsenic in Drinking Water*. NRC (National Research Council) Update: National Academy Press.

WHO, 2001. Arsenic and arsenic compounds. Environmental Health Criteria, vol. 224. World Health Organization, Geneva.

Wojas, S., Clemens, S., Skłodowska, A., and Antosiewicz, D.M. 2010. Arsenic response of AtPCS1- and CePCS-expressing plants—Effects of external As (V) concentration on As-accumulation pattern and NPT metabolism. *Journal of Plant Physiology* 167:169–175.

Wu, C., Huang, L., and Xue, S.G. et al. 2017. Oxic and anoxic conditions affect arsenic (As) accumulation and arsenite transporter expression in rice. *Chemosphere* 168:969–975.

Wu, C., Ye, Z., and Li, H. et al. 2012. Do radial oxygen loss and external aeration affect iron plaque formation and arsenic accumulation and speciation in rice? *Journal of Experimental Botany* 63:2961–2970.

Wu, Z.C., Ren, H.Y., McGrath, S.P., Wu, P., and Zhao, F.J. 2011. Investigating the contribution of the phosphate transport pathway to arsenic accumulation in rice. *Plant Physiology* 157:498–508.

Xu, W., Dai, W., and Yan, H. et al. 2015. Arabidopsis NIP3;1 plays an important role in arsenic uptake and root-to-shoot translocation under arsenite stress conditions. *Molecular Plant* 8:722–733.

Ye, J., Rensing, C., Rosen, B.P., and Zhu, Y.G. 2012. Arsenic biomethylation by photosynthetic organisms. *Trends in Plant Science* 17:155–162.

Zhao, F.J., McGrath, S.P., and Meharg, A.A. 2010. Arsenic as a food chain contaminant: Mechanisms of plant uptake and metabolism and mitigation strategies. *Annual Review of Plant Biology* 61:535–559.

Zhao, F.J., Ma, J.F., Meharg, A.A., and McGrath, S.P. 2009. Arsenic uptake and metabolism in plants. *New Phytologist* 181:777–794.

Zhao, F.J., Wang, J.R., Barker, J.H.A., Schat, H., Bleeker, P.M., and McGrath, S.P. 2003. The role of phytochelatins in arsenic tolerance in the hyper accumulator *Pteris vittata*. *New Phytologist* 159:403–410.

Zhao, F.J., Zhu, Y.G., and Meharg, A.A. 2013. Methylated arsenic species in rice: Geographical variation, origin, and uptake mechanisms. *Environmental Science and Technology* 47:3957–3966.

Zhu, Y.G., and Rosen, B.P. 2009. Perspectives for genetic engineering for the phytoremediation of arsenic-contaminated environments: From imagination to reality? *Current Opinion in Biotechnology* 20:220–224.

Index

Page numbers followed by f and t indicate figures and tables, respectively.